Manoj Kumar Patel, Triloki Nath, Ram Kishor Pandey, and Diwakar Shukla (Eds.)

Advances in Mathematical and Computational Sciences

De Gruyter Proceedings in Mathematics

Advances in Mathematical and Computational Sciences

Proceedings of The ICRTMPCS International Conference 2023

Edited by
Manoj Kumar Patel, Triloki Nath, Ram Kishor Pandey, and Diwakar Shukla

DE GRUYTER

Mathematics Subject Classification 2020
Primary: 11B39, 05A20, 13A15, 16D10, 62A86; Secondary: 16D90, 20C99, 94A60

Editors

Dr. Manoj Kumar Patel
Department of Mathematics (S&H)
National Institute of Technology Nagaland
Dimapur-797103
Nagaland India
mkpitb@gmail.com

Dr. Ram Kishor Pandey
Department of Mathematics and Statistics
Dr. Harisingh Gour Vishwavidyalaya
Sagar-470003
Madhya Pradesh India
pandeywavelet@gmail.com

Dr. Triloki Nath
Department of Mathematics and Statistics
Deen Dayal Upadhyay Gorakhpur University
Gorakhpur-273009
Uttar Pradesh India
tnverma07@gmail.com

Dr. Diwakar Shukla
Department of Mathematics and Statistics
Dr. Harisingh Gour Vishwavidyalaya
Sagar-470003
Madhya Pradesh India
diwakarshukla@rediffmail.com

ISBN 978-3-11-130437-3
e-ISBN (PDF) 978-3-11-131363-4
e-ISBN (EPUB) 978-3-11-131387-0
ISSN 2942-4801

Library of Congress Control Number: 2024941398

Bibliographic information published by the Deutsche Nationalbibliothek
The Deutsche Nationalbibliothek lists this publication in the Deutsche Nationalbibliografie; detailed bibliographic data are available on the Internet at http://dnb.dnb.de.

Typesetting: VTeX UAB, Lithuania

www.degruyter.com

Dedicated to Prof. B. M. Pandeya, Late Prof. S. R. Singh (1960–2017), and Late Prof. O. P. Singh (1951–2016)

About the book

The purpose of this volume is to contribute to the development in mathematical and computational sciences. This research volume contains invited refereed high-quality original research papers with significant contributions in the various branches of mathematics, statistics, and computer science, preferably by the invited speakers, paper presenters, and other participants of our international conference ICRTMPCS, IInd International Conference held during March 17–19, 2023 in Sagar, India. The chapters focus on modern trends and techniques in various branches of pure and applied mathematics, statistics, computer science and highlight their applications in coding theory, cryptography, graph theory, fuzzy theory, variance analysis, data analysis, sample theory, stochastic process, inferences, computer networking, etc. As mathematics, statistics, and computer science continuously experience tremendous growth and diversification, the papers included in this volume display many current trends and techniques in related branches and highlight their applications. They develop some inter relations between them and discuss some mathematical models that deals with our daily life problems. The book serves both the specialist who is looking for the latest results and the novice looking for the appropriate references to access the study and understanding the results presented here.

https://doi.org/10.1515/9783111313634-201

Preface

An International Conference on "Recent Trends in Mathematical, Physical and Computational Sciences (ICRTMPCS-2023)" organized by the Department of Mathematics and Statistics, Dr. Harisingh Gour Vishwavidyalaya, Sagar-470003, India, during March 17-19, 2023. A symposium on "Algebra and Number Theory" is also organized in this conference to celebrate National Science Day and International Day of Mathematics (π-Day) in the collaboration with National Institute of Technology Nagaland, Dimapur, India on March 18, 2023. The purpose of the conference was to present recent progress and new trends in mathematical, physical, and computational sciences.

This is the second international conference on "Recent Trends in Mathematical, Physical and Computational Sciences (ICRTMPCS-2023)" organized by the Department of Mathematics and Statistics, Dr. Harisingh Gour Vishwavidyalaya, Sagar-470003, India, whereas the first international conference on "Recent Trends in Mathematical, Physical & Computational Sciences (ICRTMPCS-2019)" was held during December 12–14, 2019 under the aegis of Silver Jubilee Celebration of International Academy of Physical Sciences, India.

The conference proceedings of the 2nd ICRTMPCS entitled on ''Advances in Mathematical and Computational Sciences" dedicated to our beloved Ph.D. advisors Prof. B. M. Pandeya, Late Prof. S. R. Singh (1960–2017) and Late Prof. O. P. Singh (1951–2016). Their contributions to the research field of their expertise have been invaluable, and their legacies continue to inspire the academic community.

This three-day international conference is partially funded by Dr. Harisingh Gour Vishwavidyalaya Sagar, where around 93 participants were registered. 6 keynote speakers had delivered their talks, among them Polya awardee US mathematician Prof. Bogdan Suceava and Prof. A. R. Alimov from Lomonosov Moscow State University, who also addressed the participants. Almost all the registered participants have presented their papers in various areas. The majority of the registered participants presented their research papers, covering a broad spectrum of topics, including number theory, fluid dynamics, graph theory, complex analysis, numerical analysis, and other interdisciplinary areas of physical and computational sciences.

The special attraction of the conference was the symposium on algebra and number theory wherein, 6 invited talks were delivered by world-renowned algebraists and number theory experts, namely: Prof. Alberto Facchini, University of Padova, Italy, Prof. Ayman Badawi, American University of Sharjah, UAE, Prof. M Ashraf, AMU Aligarh, Prof. Anupam Saikia, IIT Guwahati, Prof. Shanta Laishram, ISI New Delhi and Prof. Pramod Kumar Kewat, IIT-ISM Dhanbad.

This volume comprises the proceedings of the conference, primarily consisting of research papers that reflect the authors' contributions to the conference. The book includes 37 articles that provide an up-to-date account of current research in algebra, number theory, fluid dynamics, graph theory, complex analysis, numerical analysis, and other interdisciplinary related areas of physical and computational sciences.

https://doi.org/10.1515/9783111313634-202

This proceeding book serves both the specialist looking for the latest results and the novice looking for the appropriate references to access the study and understand the results presented here. We greatly appreciate the efforts of the contributors and referees and thank the editorial staff of this proceedings for their patience and assistance.

Guest Editors:

1. Dr. Manoj Kumar Patel
   Associate Professor and Symposia Coordinator
   Department of Mathematics (S&H)
   National Institute of Technology Nagaland, India
   Email ID: mkpitb@gmail.com

2. Dr. Triloki Nath
   Associate Professor and Conference Convener
   Department of Mathematics and Statistics
   Deen Dayal Upadhyay Gorakhpur University, Gorakhpur (U.P.), India
   Email ID: tnverma07@gmail.com

3. Dr. Ram Kishor Pandey
   Assistant Professor and Conference Convener
   Department of Mathematics and Statistics
   Dr. Harisingh Gour Vishwavidyalaya, Sagar (M.P.), India
   Email ID: pandeywavelet@gmail.com

4. Dr. Diwakar Shukla
   Professor and Conference Director
   Department of Mathematics and Statistics
   Dr. Harisingh Gour Vishwavidyalaya, Sagar (M.P.), India
   Email ID: diwakarshukla@rediffmail.com

Contents

About the book —— VII

Preface —— IX

About the Editors —— XV

Anil Kumar Pandey, Laxmipriya Parida, and Ashok Kumar Sahoo
Logarithmic third Hankel determinant for a subclass of starlike functions —— 1

Monika Arora, Mahesh Kumar Sharma, and Kamalpreet Kaur
Effect of gyrotactic microorganisms in rotating magnetic nanofluids in porous media —— 17

R. Krishnaprabha
Some applications of graph spectrum in post quantum cryptography —— 31

Vijay K. Shukla and Prashant K. Mishra
Finite-time stability analysis of neutral type delay Cohen–Grossberg neural network with discontinuous activation function —— 47

S. N. Rai, B. S. Bhadauria, and Anurag Srivastava
Influence of temperature modulation over heat/mass transport of Rivlin–Ericksen viscoelastic nanoliquid in Hele-Shaw cell —— 57

Shabir Ahmad Malik
On an operator preserving polynomial inequality of Paul Turán —— 75

Piu Kundu
Effect of interaction between two finite faults on ground deformation —— 85

Melvin Varghese and G. Sheeja
On the number of bounded regions generated by a class of paths in the complex plane —— 111

Jitender Singh and Renu Bajaj
Binomial and de Moivre revisited through differential equations —— 125

Akshay S. Rane and Kshitij Patil
Modified projection method for second kind Fredholm integral equations with trigonometric polynomials —— 131

Shilpee and B. S. Bhadauria
Effect of heat and mass transfer in binary nanofluids (for silver nanoparticles) saturated in porous medium under the influence of LTNE model and different gravity modulation — 145

Dhriti Raj Talukdar and H. Imo Mani Singha
A comparative analysis of standard Knapsack and Legendre–Knapsack cryptosystem — 169

Tania Bose, Renu Bala, and Madhu Aneja
Study of creep in composite functionally graded disc with non-linear thickness — 179

M. K. Gupta and Abha Sahu
On the $h(hv)$-torsion tensor and the v-curvature tensor of Matsumoto change with an h-vector — 189

Anup Kumar Behera and Priyanka Agrawal
Software reliability growth model considering power-law testing effort using optimization on hybrid PSO-GWO algorithm — 199

Flora Jeba S, Anirban Roy, and Manjil P. Saikia
On near-perfect numbers with five prime factors — 207

Neeti Ghiya and S Manjula
Eulerian integrals associated with the product of multivariable H-function and generalized M-series — 221

R. Bala, T. Bose, and M. Aneja
Upper bounds for the complex growth rates in the Darcy–Brinkman convection in a binary viscoelastic fluid in a saturated porous layer — 233

Kamal Wadhwa, Sanjay Choudhary, and Arvind Kumar Gupta
Some result of fixed-point theorems in complete M-fuzzy metric space — 243

Aronthung S Odyuo, P. Mercy, and Manoj Kumar Patel
FR-labeling on hybrid partition shell graphs — 261

R. K. Singh, R. G. Ghumde, and M. K. Patel
On essential M-Artinian modules — 269

Priti Singh and Sharvan Kumar
Properties of reductions of multiplication modules — 275

S Priyadharsini
Computational analysis of fractional vibrational equations — 285

Preety Kalra and Nisha Malhotra
Analysis of modified fractional order Lotka–Volterra model — 307

Gautham B. Shenoy, A. Ragavaendra, S. B. Chandrakala, Badekara Sooryanarayana, and Muhammad Kamran Siddiqui
Topological indices of two chemical compounds used for drugs in the treatment of H1N1 — 335

M. Kumari, R. Mohanta, K. Prasad, and H. Mahato
On the Gaussian higher-order Mersenne numbers — 355

Sakshi Gupta and Shelly Garg
Petri net: modeling, properties and analysis techniques — 367

Ashish Negi and Ompal Singh
An inventory system with time-dependent linear demand for deteriorating items under uncertainty with a completely backlogged — 379

Deepak Kumar, Sukhchain Singh, and Bharat Bhushan
On $*$-reverse derivations of involutive rings — 389

C. Reena and M. Karthika
More functions via soft pre*-generalized closed sets — 397

Ekta Jain, Dhanpal Singh, and S. D. Ram
Modeling of piston problem for generalized Chaplygin gas — 409

J. K. Kushwaha and R. Vishwakarma
A new estimation of the degree of approximation of functions belonging to Lipschitz class by Borel–Euler summability method of Fourier series — 421

Pankaj Kumar, Raminder Pal Singh, and Dipesh
Modeling the effect of discrete delay in SIR epidemic model with logistic growth — 429

Pintu Bhattacharya
Simple harmonic motion in view of quantum calculus —— 455

Ishfaq Nazir
On Bernstein and Turán-type inequalities for polynomials —— 461

C. Jeslyn Sathya and Carmel Richard
A weak and strong form of $Ng*b$-continuous function in nano topological space —— 473

R. Singh, Suditi, M. K. Patel, N. Bhardwaj, and A. K. Singh
Securing networks with graph theory: an approach to cyber security —— 483

About the Editors

Dr. Manoj Kumar Patel is currently working as Associate Professor of Mathematics at National Institute of Technology Nagaland. He completed his Ph.D. degree in Algebra from one of the premier institution of India IIT-BHU, Varanasi and join Gurukula Kangri University Haridwar as Assistant Professor. He was recipient of Gold Medal in M.Sc. and awarded NBHM Post-Doctoral Fellowship. He had published more than 40 research articles in the journal of international repute and successfully guided 3 Ph.D. students in Algebra. Dr. Patel works as Managing editor of Palestine Journal of Mathematics, Editor of Moroccan Journal of Algebra, Geometry and Applications (MJAGA), Guest Editor of Special Issue in Palestine Journal of Mathematics and volume editor of Proceeding on Advances in Pure and Applied Algebra in De Gruyter Publisher also Algebra and its Application in Springer Nature. He was also appointed as Regional Coordinator of Mathematical Olympiad Program for Nagaland, Centre Head for NBHM Master & Doctoral Scholarship Program for Nagaland and Coordinator of INYAS-North East Local Chapter. He is also a life member of the Indian Mathematical Society, the Indian Science Congress Association, the Ramanujan Mathematical Society, India and many more. More information can be found on his google webpage: https://sites.google.com/view/dr-manoj-kumar-patel/about

Dr. Triloki Nath is currently an Associate Professor of Mathematics at Deen Dayal Upadhyay Gorakhpur University in Gorakhpur, Uttar Pradesh, India. He earned his Ph.D. in Applied Mathematics from the prestigious Banaras Hindu University in Varanasi, Uttar Pradesh, India. Following his Ph.D., he began his academic career as an Assistant Professor at the National Institute of Technology Mizoram and later moved to Dr. Harisingh Gour Vishwavidyalaya in Sagar, Madhya Pradesh, India. Dr. Nath is a distinguished scholar who has been awarded the prestigious CSIR-JRF (NET) and SRF fellowships to support his Ph.D. studies. Additionally, he qualified for the GATE examination, achieving an impressive all-India rank of 4 with a 99.86 percentile score. His research contributions include 10 articles published in internationally reputed journals, and he has successfully guided one Ph.D. student in the field of optimization theory. Dr. Nath has also played a significant role in the academic community by organizing two international conferences as a convener. His primary research interests lie in optimization theory, nonsmooth analysis, geometric functional analysis, and elementary number theory.

Dr. Ram Kishor Pandey is currently serving as an Assistant Professor of Mathematics at Dr. Harisingh Gour Vishwavidyalaya, Sagar, Madhya Pradesh. He completed his Ph.D. in Applied Mathematics from the prestigious IIT-BHU, Varanasi. He was the recipient of CSIR (NET) JRF and SRF fellowships during his research work. Following his Ph.D., he started his academic career as an Assistant Professor at Govt. College, Kondagain in Chhattisgarh. Dr. Pandey was honored with the Chancellor Gold Medal in his M.Sc. for achieving the highest marks. He has made significant contributions to his field, having published 20 research articles in internationally reputed journals and successfully guided one Ph.D. student in Numerical Analysis. Additionally, he has played a key role in the academic community by organizing two international conferences as a convener. His primary research interests include the numerical integration of differential equations of integer and fractional order, geometric numerical integration, and the numerical solution of integro-differential equations.

https://doi.org/10.1515/9783111313634-203

Prof. (Dr.) Diwakar Shukla is currently a senior Professor in the Department of Mathematics and Statistics. He was Head of Department of Computer Science for 7 years and Dean, School of Mathematical and Physical Sciences for 3 years at Dr. HariSingh Gour University, Sagar.
He has more than 34 years of experience in teaching UG and PG classes. He has obtained M.Sc. (Statistics), Ph.D. (Statistics) degrees from Banaras Hindu University and completed the M.Tech. degree in Computer Science from Devi Ahilya University, Indore.
He also served the Devi Ahilya University, Indore as an Assistant Professor for nearly 9 years and thereafter joined the Sagar University as an Associate Professor. He also worked as a professor at Lucknow Universiry, Lucknow, UP for one year.
During his Ph.D. work in BHU, he was recipient of Junior and Senior Research Fellowship from CSIR obtained through all India tests. Till date, he has published more than 155 research papers in journals, 13 papers in book chapters, 17 papers in conference proceedings. In total, he has published more than 180 research publications. He participated in more than 40 seminars/conferences. He has guided 23 Ph.D. theses in mathematics, statistics, and computer science. Currently, two Ph.D. scholars are enrolled under his guidance.
He was the recipient of MPCOST Young Scientist Award, ISAS Young Scientist Award and UGC Career Award. He was a UGC visiting fellow to Amravati University. He has got 7th place in the list of AD Scientific Ranking among 22 selected scientists of the Sagar university in year 2022.
He authored 8 books and is a member of 11 learned bodies in statistics, mathematics, and computer science.
He was a member of UGC Autonomous College Inspection Team, who inspected two colleges at Vijayawada and Trichurapalli.
As administrative experience, he was Incharge Central Library, Incharge IT Cell for seven years. He was the Coordinator Admission Cell, Coordinator G20 Connect Program, Head of Department of Vedic Studies, Chairman Board of Studies in computer Science, Chairman Departmental Research Committee, Chairman Samarth Portal Committee, Chairman Screening Committee for teaching and non-teaching posts, Nodal Officer university Wi-Fi service, Nodal Officer UGC Student Grievance Portal, Nodal officer UGC Central University Portal (CU-Portal).
He was member Academic Council, Member Executive, member Selection Committee for non-teaching post, member Selection Committee for teaching post (Bilaspur) also. He was member Convocation Core Team, member Gour Jayanti Celebration team, member NAD Cell committee, member Revaluation Committee, member CBCS program implementation committee, Secretary Library Committee, member University IT-equipment Purchase Committee.
His areas of research interest are sampling theory, graph theory, stochastic modeling, big data, data mining, computer networks, and computer operating systems.

Anil Kumar Pandey, Laxmipriya Parida, and Ashok Kumar Sahoo

Logarithmic third Hankel determinant for a subclass of starlike functions

Abstract: By using the subordination, a subclass, namely S_{esin}, of univalent functions is constructed. Super ordination part in the subordination condition, involving trigonometric and exponential functions, has its image completely captured in the right-half plane. It is shown that the class is non-empty and validity subordination is shown graphically by using Mathematica software. The bounds of several logarithmic coefficients and the bounds of the third Hankel determinant are calculated.

Keywords: Analytic functions, univalent functions, Carathéodory functions, Hankel determinant, convex functions, logarithmic coefficients, subordination

MSC 2010: Primary 30C45, Secondary 30C80

1 Introduction

Inclusion relation of image of two analytic functions is well established analytically using concept of subordination [15]. Incarnation of the subordination theory, gives powerful tool to realize different properties related to zeros, singularities, and boundary behavior of analytic functions. In subordination theory, certain mathematical transformation is used to get one analytic function from other. The obtained function is called subordinate function of the other function, which is called superordinate function. Different areas in which Subordination theory is used significantly are mathematical physics, which includes the study of differential equations, potential theory, and special functions. Subordination theory is used to model various physical, chemical, and biological phenomena. Application of subordination theory in various field of applied mathematics is well understood from [17, 16, 18]. Using sub-ordination, Srvastava [43] et al. designed a subclass $\mathcal{S}^*(\mathcal{L}, q)$ of univalent function much akin and general to the class of star-like functions using q-th derivative operator. In this work, third-order Hankel determinant $H_3(1)$ us determined for the functions belonging to this class. Shi et al. [40] constructed a subclass of univalent function, which is of bounded turning type, using

Acknowledgement: The authors thank the anonymous referee for helpful comments.

Anil Kumar Pandey, Department of Mathematics, VSSUT, Sambalpur, Odisha, India
Laxmipriya Parida, Basic Science and Humanities, School of Engineering and Technology, GIET University, Gunupur, Rayagada-765022, India
Ashok Kumar Sahoo, Department of Mathematics, Central University of Odisha, Sunabeda, School Of Basic Science And Information Science, Block-1, Koraput-763004, Odisha, India, e-mail: ashokuumt@gmail.com

https://doi.org/10.1515/9783111313634-001

super-ordination function $1 + \sinh^{-1}(z)$ and investigate the sharp estimate of Fekete–Szegö inequality, Zalcman inequality on the logarithmic coefficients and the Hankel determinant. Kowalczyk and Lecko [24] found sharp bounds of the second Hankel determinant of logarithmic coefficients of starlike and convex functions of order α. Wahid et al. [47] obtained the upper bounds of the Toeplitz determinants for the class of functions with bounded turning. Sunthrayuth et al. [44] designed a subclass of bounded turning type taking subordinate to $1 + \sin z$ and investigated bounds of the Zalcman inequality.

We denote by $\mathcal{A}$ the class of analytic (holomorphic) functions f, defined in the region

$$\mathbf{U} = \{z : z \in \mathbb{C} \text{ and } |z| < 1\}$$

and normalized by the conditions

$$f(0) = 0 \quad \text{and} \quad f'(0) = 1.$$

So that each $f \in \mathcal{A}$ has the following Taylor's series expression

$$f(z) = z + \sum_{n=2}^{\infty} a_n z^n \quad z \in \mathbf{U}. \tag{1.1}$$

Furthermore, we will denote by $\mathcal{S}$ the subclass of $\mathcal{A}$ of functions that are univalent in $\mathbf{U}$. This class was introduced by Koebe [22] and has become the heart of the advanced research in this field. Many researchers have shown interest in this concept, but within a short period Bieberbach [7] proposed the famous conjecture. This conjecture states that if $f(z) \in \mathcal{S}$ and has the series form (1.1), then $|a_n| \le n \; \forall n \ge 2$ which was finally settled by L. de-Branges [9]. For two functions $f, g \in \mathcal{A}$, we say that the function f is subordinate to the function g (written as $f \prec g$) if there exists an analytic function ω with the property that

$$|\omega(z)| \le |z| \quad \text{and} \quad \omega(0) = 0$$

such that

$$f(z) = g(\omega(z)), \quad z \in \mathbf{U}.$$

Additionally, if $g \in \mathcal{S}$, then the above condition can be stated as

$$f \prec g \iff f(0) = g(0) \quad \text{and} \quad f(\mathbf{U}) \subset g(\mathbf{U}).$$

Ma and Minda [28] presented a unified version of the class $\mathcal{S}^*$ in terms of subordination as follows:

$$\mathcal{S}^*(\phi) = \left\{ f \in \mathcal{S} : \frac{zf'(z)}{f(z)} \prec \phi(z),\ z \in \mathbf{U} \right\},$$

where $\phi(z)$ is an analytic univalent function with positive real part in $\mathbf{U}$, $\phi(\mathbf{U})$ is symmetric about real axis and star liked with respect to $\phi(0) = 1$ and $\phi'(0) > 0$.

They also obtained some interesting results on the distortions, growth, and the theorem of covering for this family. The most basic of these families are the families of star-like functions obtained by taking $\phi(z) = \frac{1+z}{1-z}$.

Taking various functions φ some recently studied relevant subclasses are: $\mathcal{S}_{\mathcal{L}}^* = \mathcal{S}^*(\sqrt{1+z})$ [42], $\mathcal{S}_{car}^* = \mathcal{S}^*(1+\frac{4}{3}z+\frac{2}{3}z^2)$ [39], $\mathcal{S}_{\rho}^* = \mathcal{S}^*(1+\operatorname{Sinh}^{-1}z)$ [25], $\mathcal{S}_{\exp}^* = \mathcal{S}^*(e^z)$ [41], $\mathcal{S}_{\cos}^* = \mathcal{S}^*(\cos z)$ [6], $\mathcal{S}_{\cosh}^* = \mathcal{S}^*(\cosh z)$ [3], $\mathcal{S}_{\tanh}^* = \mathcal{S}^*(1+\tanh z)$ [46], $\mathcal{S}^*(\frac{2}{1+e^{-z}})$ Goel [13].

Finding bounds for the coefficients in the series expansion of a given function in a class is the most important problem in geometric function theory, as it impacts the geometric features, for instance, the bounds for second coefficients yields information about growth and distortion features.

The Hankel determinant $\mathbf{H_{q,n}(f)}$ $(n, q \in \mathbb{N})$ where $\mathbb{N} = \{1, 2, 3, \dots\}$ for the function $f \in \mathcal{S}$ was introduced by Pommerenke ([35, 36]) defined by:

$$H_{q,n}(f) = \begin{vmatrix} a_n & a_{n+1} & \cdots & a_{n+q-1} \\ a_{n+1} & a_{n+2} & \cdots & a_{n+q} \\ \vdots & \vdots & \cdots & \vdots \\ a_{n+q-1} & a_{n+q} & \cdots & a_{n+2q-2} \end{vmatrix}.$$

For different values of q and n, Hankel determinants of various order are obtained. For example, when $n = 1$ and $q = 2$, we have

$$H_{2,1}(f) = \begin{vmatrix} a_1 & a_2 \\ a_2 & a_3 \end{vmatrix} = a_1a_3 - a_2^2 = a_3 - a_2^2 \quad \text{where } a_1 = 1.$$

Finding upper bound of functional $|a_3 - \mu a_2^2|$, $\mu \in [0, 1]$ is a well-known Fekete–Szego problem studied by Fekete–Szego (see [11]). For more results on Fekete–Szego functional one can see ([19–21]). Furthermore, if we take $n = q = 2$, we get the second Hankel determinant

$$H_{2,2}(f) = \begin{vmatrix} a_2 & a_3 \\ a_3 & a_4 \end{vmatrix} = a_2a_4 - a_3^2.$$

In recent years, many researchers have investigated the upper bound of $|H_{2,2}(f)|$ for different sub-classes of analytic functions. For interesting results on $|H_{2,2}(f)|$ one can see ([30, 14, 31]). Our main aim in this paper is to study third Hankel determinant whose entries are logarithmic coefficients for the class $\mathcal{S}_{esin}$, where

$$\mathcal{S}_{esin} = \left\{ f \in \mathcal{S} : \frac{zf'(z)}{f(z)} \prec e^{\sin z} \right\} \quad (z \in \mathbf{U}). \tag{1.2}$$

Logarithmic coefficients have recently drawn the attention of many researchers. For example, Millin's conjecture highly depends upon logarithmic coefficients, see ([29,

10]). The logarithmic coefficients of $f \in \mathcal{S}$, defined by $\gamma_n = \gamma_n(f)$, are defined by the following series expansion

$$\log\left(\frac{f(z)}{z}\right) = 2\sum_{n=1}^{\infty} \gamma_n z^n.$$

Ali et al. [1] investigated the logarithmic coefficients of some close-to-convex functions, while the third logarithmic coefficients was studied by Cho et al. [8]. Moreover logarithmic coefficients of univalent functions can be found in [12]. For more recent works on logarithmic coefficients, one can see Kowalckzyk and Lecko [23].

For a function f given by (1.1), logarithmic coefficients are as follows

$$\gamma_1 = \frac{1}{2}a_2, \tag{1.3}$$

$$\gamma_2 = \frac{1}{2}\left(a_3 - \frac{1}{2}a_2^2\right), \tag{1.4}$$

$$\gamma_3 = \frac{1}{2}\left(a_4 - a_2a_3 + \frac{1}{3}a_2^3\right), \tag{1.5}$$

$$\gamma_4 = \frac{1}{2}\left(a_5 - a_2a_4 + a_2^2a_3 - \frac{1}{2}a_3^2 - \frac{1}{4}a_2^4\right), \tag{1.6}$$

$$\gamma_5 = \frac{1}{2}\left(a_6 - a_2a_5 - a_3a_4 + a_2a_3^2 + a_2^2a_4 - a_2^3a_3 + \frac{1}{5}a_2^5\right). \tag{1.7}$$

Based on all the ideas described above, we propose the study of Hankel determinant, whose entries are the logarithmic coefficients of $f \in \mathcal{S}$, that is:

$$H_{q,n}(f) = \begin{vmatrix} \gamma_n & \gamma_{n+1} & \cdots & \gamma_{n+q-1} \\ \gamma_{n+1} & \gamma_{n+2} & \cdots & \gamma_{n+q} \\ \vdots & \vdots & \cdots & \vdots \\ \gamma_{n+q-1} & \gamma_{n+q} & \cdots & \gamma_{n+2q-2} \end{vmatrix}.$$

The main aim of this paper is to find upper bounds for $H_{3,1}(f)$ for the class of star-like functions associated with exponential sine function.

Kumar and Vasudev [26] considered three families of close to convex functions $\{f \in \mathcal{A} : G(z) \prec \frac{1+z}{1-z}\}$, where $G(z) = (1-z)f'(z), (1-z^2)f'(z), (1-z+z^2)f'(z)$ and found sharp upper bound for $\gamma_1, \gamma_2, \gamma_3$. Alshehry et al. [4] designed a subclass related to convex function and determined sharp upper bound to second Hankel determinant of logarithmic coefficients. Ali et al. [1] investigated the logarithmic coefficients of some close-to-convex functions, while the third logarithmic coefficients in some subclasses of close-to-convex functions were studied by Cho et al. [32]. Moreover, logarithmic coefficients of univalent functions can be found in [12]. Very recently, Kowalczykand Lecko [23] have studied the Hankel matrices whose entries are logarithmic coefficients of univalent functions and have given sharp bounds for the second Hankel determinant of logarithmic coefficients of convex and starlike functions. For some other related works, see [33, 2, 26, 45, 38].

Using definition of subordination, we have defined the following subclass of univalent functions $\mathcal{S}_{\text{esin}}$, where

$$\mathcal{S}_{\text{esin}} = \left\{ f \in \mathcal{S} : \frac{zf'(z)}{f(z)} \prec e^{\sin z} \right\} \quad (z \in \mathbf{U}). \tag{1.8}$$

In the next subsection, we shall show the subclass defined above is non-empty by designing examples, which are member to the above subclass.

1.1 Examples

Let $t_0(z) = e^{\sin z}$, which is univalent in $\mathbf{U}$ and at origin it agrees with $t_n(z) = e^{\frac{z}{n}}$, $n \in \mathbb{N}$, also $t_n(\mathbf{U}) \subset t_0(\mathbf{U})$ so $\mathcal{R}_{\text{esin}}$ is non-empty.

To justify our claim, we have found graphics of super and subordinate functions (for $n = 2$) and merged it with the help of Mathematica™ software.

Using following code

```
f[z_] := Exp[Sin [z]];
pic1 = ParametricPlot[{Re[f[x + I*y]], Im[f[x + I*y]]}, {x, -1,
    1}, {y, -Sqrt[1 - x^2], +Sqrt[1 - x^2]}, ImageSize -> 300];
Row[{pic1}]
```

we get image (1.1) of $\mathbf{U}$ under $e^{\sin z}$ as below.

Next, we use the code

```
f[z_] := Exp[z/2];
pic1 = ParametricPlot[{Re[f[x + I*y]], Im[f[x + I*y]]}, {x, -1,
    1}, {y, -Sqrt[1 - x^2], +Sqrt[1 - x^2]}, ImageSize -> 300];
Row[{pic1}]\\ and the code Show[{p1, p2}]
                gives confirmation of Subordination.
```

to get image (1.2) of $|z| < 1$ of $t_2(z) = e^{z/2}$, also merging these two figure (1.3), subordination is justified, which is shown below.

In fact, if there exists analytic function $t(z)$ with the property that $t(z) \prec t_0(z)$ $z \in \mathbf{U}$, such that $f(z) = \int_0^z t(s)ds$, then $f \in \mathcal{R}_{\text{esin}}$.

The main aim of this paper is to find upper bounds for $\mathcal{H}_{3,1}(f)$ for the class of bounded turning functions associated with the exponential sine functions.

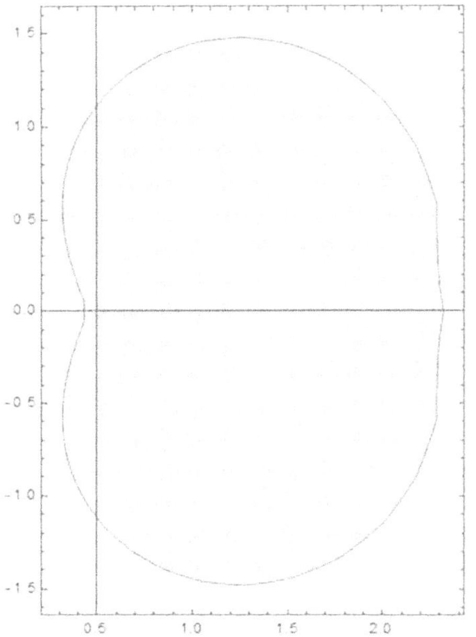

Figure 1.1: The graphics of ($e^{\sin z}$ for $|z| < 1$).

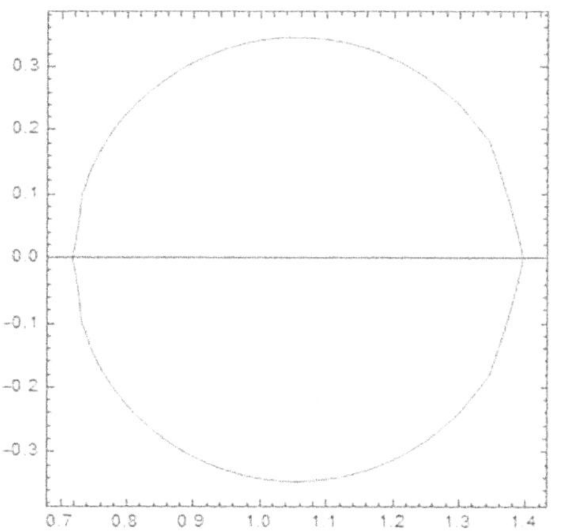

Figure 1.2: The graphics of ($e^{z/2}$ for $|z| < 1$).

2 A set of lemmas

We denote by $\mathcal{P}$ the class of analytic functions p, which are normalized by

$$p(0) = 1 \quad \text{with } \operatorname{Re}(p(z)) > 0 \quad z \in \mathbf{U}$$

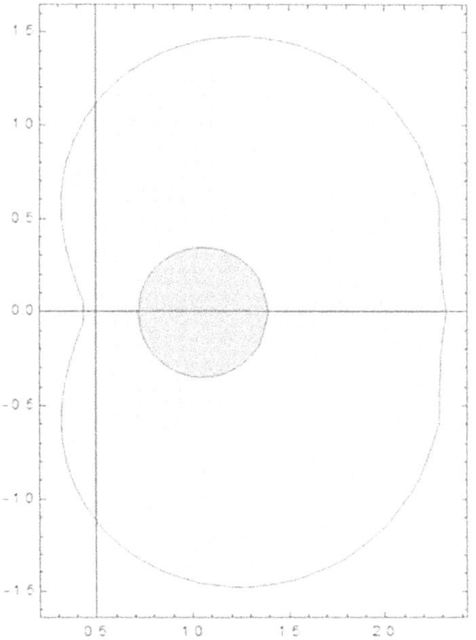

Figure 1.3: The graphics of subordination.

and have the following form:

$$p(z) = 1 + \sum_{n=1}^{\infty} c_n z^n \quad (z \in \mathbf{U}). \tag{1.9}$$

To our main results, we need the following lemmas.

Lemma 1 ([27]). *Let $p \in \mathcal{P}$. Then, there exists x, δ with $|x| \le 1$, $|\delta| \le 1$ such that*

$$2c_2 = c_1^2 + x(4 - c_1^2), \tag{1.10}$$

$$4c_3 = c_1^3 + 2(4 - c_1^2)c_1 x - c_1(4 - c_1^2)x^2 + 2(4 - c_1^2)(1 - |x|^2)\delta. \tag{1.11}$$

Lemma 2. *If $p \in \mathcal{P}$, then the following inequalities hold*

$$|c_k| \le 2 \quad \text{for } k \ge 1 \tag{1.12}$$

$$|c_{n+k} - \mu c_n c_k| < 2 \quad \text{for } 0 \le \mu \le 1 \tag{1.13}$$

$$|c_m c_n - c_k c_l| \le 4 \quad \text{for } m + n = k + l \tag{1.14}$$

$$|c_{n+2k} - \mu c_n c_k^2| \le 2(1 + 2\mu), \quad \text{for } \mu \in \mathbb{R} \tag{1.15}$$

$$\left| c_2 - \frac{c_1^2}{2} \right| \le 2 - \frac{|c_1|^2}{2}, \tag{1.16}$$

and for complex number η, we have

$$|c_2 - \eta c_1^2| < 2\ \max\{1, |2\eta - 1|\}. \tag{1.17}$$

For the inequalities (1.13), (1.14), (1.15), *and* (1.16) *see* [37], *and* (1.17) *is given in* [34].

Lemma 3 ([5, Lemma 2.2]). *If $p \in \mathcal{P}$, then*

$$|Ic_1^3 - Xc_1c_2 + Vc_3| \le 2|I| + 2|X - 2I| + 2|I - X + V|, \tag{1.18}$$

where I, X, V are real numbers.

3 Main results

Theorem 1. *If $f \in \mathcal{S}^*_{\text{esin}}$ and it has the form given in (1.1), then*

$$|\gamma_1| \le \frac{1}{2}, \tag{1.19}$$

$$|\gamma_2| \le \frac{1}{4}, \tag{1.20}$$

$$|\gamma_3| \le \frac{1}{6}, \tag{1.21}$$

$$|\gamma_4| \le \frac{35}{128}, \tag{1.22}$$

$$|\gamma_5| \le \frac{7}{20}. \tag{1.23}$$

Proof. If $f \in \mathcal{S}^*_{\text{esin}}$, then by definition of subordination, there exists a Schwartz function $\omega(z)$ with the property that

$$\omega(o) = 0 \quad \text{and} \quad \omega(z) < 1$$

such that

$$\frac{zf'(z)}{f(z)} = e^{\sin\omega(z)}. \tag{1.24}$$

Define the function

$$p(z) = \frac{1 + \omega(z)}{1 - \omega(z)} = 1 + b_1z + b_2z^2 + b_3z^3 + \cdots.$$

It is clear that $p(z) \in \mathcal{P}$. This implies that

$$\begin{aligned}\omega(z) &= \frac{p(z) - 1}{p(z) + 1}\\ &= \frac{b_1z + b_2z^2 + b_3z^3 + \cdots}{2 + b_1z + b_2z^2 + b_3z^3 + \cdots}\end{aligned}$$

$$= \frac{1}{2}b_1z + \left(\frac{1}{2}b_2 - \frac{1}{4}b_1^2\right)z^2 + \left(\frac{1}{8}b_1^3 - \frac{1}{2}b_1b_2 + \frac{1}{2}b_3\right)z^3 + \left(\frac{1}{2}b_4 - \frac{1}{4}b_2^2 + \frac{3}{8}b_1^2b_2 - \frac{1}{2}b_1b_3 - \frac{1}{16}b_1^4\right)z^4 + \left(\frac{1}{2}b_5 - \frac{1}{2}b_1b_4 - \frac{1}{2}b_2b_3 + \frac{3}{8}b_1^2b_3 + \frac{3}{8}b_1b_2^2 - \frac{1}{4}b_1^3b_2 + \frac{1}{32}b_1^5\right)z^5 + \cdots \tag{1.25}$$

From this we get

$$\sin(\omega(z)) = \frac{1}{2}b_1z + \left(\frac{1}{2}b_2 - \frac{1}{4}b_1^2\right)z^2 + \left(\frac{5}{48}b_1^3 - \frac{1}{2}b_1b_2 + \frac{1}{2}b_3\right)z^3 + \left(\frac{1}{2}b_4 - \frac{1}{2}b_1b_3 + \frac{5}{16}b_1^2b_2 - \frac{1}{4}b_2^2 - \frac{1}{32}b_1^4\right)z^4 + \left(\frac{1}{2}b_5 - \frac{1}{2}b_1b_4 - \frac{1}{2}b_2b_3 + \frac{5}{16}b_1^2b_3 + \frac{5}{16}b_1b_2^2 - \frac{1}{8}b_1^3b_2 + \frac{1}{3840}b_1^5\right)z^5 + \cdots \tag{1.26}$$

Also, we have

$$\frac{zf'(z)}{f(z)} = 1 + a_2z + (2a_3 - a_2^2)z^2 + (a_2^3 - 3a_2a_3 + 3a_4)z^3 + (4a_5 - a_2^4 + 4a_2^2a_3 - 4a_2a_4 - 2a_3^2)z^4 + (5a_6 - 5a_2a_5 - 5a_3a_4 + 5a_2^2a_4 + 5a_2a_3^2 - 5a_2^3a_3 + a_2^5)z^5 + \cdots \tag{1.27}$$

and

$$\begin{aligned} e^{\sin\omega(z)} &= 1 + \frac{1}{1!}\left[\frac{1}{2}b_1z + \left(\frac{1}{2}b_2 - \frac{1}{4}b_1^2\right)z^2 + \left(\frac{1}{2}b_3 - \frac{1}{2}b_1b_2 + \frac{5}{48}b_1^3\right)z^3 + \left(\frac{1}{2}b_4 - \frac{1}{4}b_2^2 - \frac{1}{2}b_1b_3 + \frac{5}{16}b_1^2b_2 - \frac{1}{32}b_1^4\right)z^4 \right.\\ &\quad \left. + \left(\frac{1}{2}b_5 - \frac{1}{2}b_1b_4 - \frac{1}{2}b_2b_3 + \frac{5}{16}b_1^2b_3 + \frac{5}{16}b_1b_2^2 - \frac{1}{8}b_1^3b_2 + \frac{1}{3840}b_1^5\right)z^5 + \cdots\right] + \cdots \\ &= 1 + \frac{1}{2}b_1z + \left(\frac{1}{2}b_2 - \frac{1}{8}b_1^2\right)z^2 + \left(\frac{1}{2}b_3 - \frac{1}{4}b_1b_2\right)z^3 + \left(\frac{3}{128}b_1^4 - \frac{1}{8}b_2^2 - \frac{1}{4}b_1b_3 + \frac{1}{2}b_4\right)z^4 \\ &\quad + \left(-\frac{17}{960}b_1^5 + \frac{3}{32}b_1^3b_2 - \frac{1}{4}b_2b_3 - \frac{1}{4}b_1b_4 + \frac{1}{2}b_5\right)z^5 \cdots \end{aligned} \tag{1.28}$$

From (1.24), (1.27) and (1.28), we have

$$a_2 = \frac{1}{2}b_1, \tag{1.29}$$

$$a_3 = \frac{1}{4}b_2 + \frac{1}{16}b_1^2, \tag{1.30}$$

$$a_4 = \frac{1}{6}b_3 + \frac{1}{24}b_1b_2 - \frac{1}{96}b_1^3, \tag{1.31}$$

$$a_5 = \frac{1}{384}b_1^4 - \frac{5}{192}b_1^2b_2 + \frac{1}{48}b_1b_3 + \frac{1}{8}b_4, \tag{1.32}$$

$$a_6 = -\frac{13}{19200}b_1^5 + \frac{7}{640}b_1^3b_2 - \frac{1}{48}b_1^2b_3 + \frac{1}{80}b_1b_4 - \frac{1}{48}b_1b_2^2 - \frac{1}{120}b_2b_3 + \frac{1}{10}b_5. \tag{1.33}$$

Now from (1.3) to (1.7) and from (1.29) to (1.33), we have

$$\gamma_1 = \frac{1}{4}b_1, \tag{1.34}$$

$$\gamma_2 = \frac{1}{8}b_2 - \frac{1}{32}b_1^2, \tag{1.35}$$

$$\gamma_3 = \frac{1}{12}b_3 - \frac{1}{24}b_1b_2, \tag{1.36}$$

$$\gamma_4 = \frac{3}{1024}b_1^4 - \frac{1}{32}b_1b_3 - \frac{1}{64}b_2^2 + \frac{1}{16}b_4, \tag{1.37}$$

$$\gamma_5 = -\frac{17}{9600}b_1^5 + \frac{3}{320}b_1^3b_2 - \frac{1}{40}b_1b_4 - \frac{1}{40}b_2b_3 + \frac{1}{20}b_5. \tag{1.38}$$

Applying triangle inequality and (1.11) to (1.34), we obtain

$$|\gamma_1| \leq \frac{1}{2}.$$

By rearranging (1.35), we have

$$\gamma_2 = \frac{1}{8}\left(b_2 - \frac{1}{4}b_1^2\right).$$

By applying triangle inequality and (1.12), we get

$$|\gamma_2| \leq \frac{1}{8} \times 2 = \frac{1}{4},$$

so we have

$$|\gamma_2| \leq \frac{1}{4}.$$

Rearranging (1.36), we have

$$\gamma_3 = \frac{1}{12}\left(b_3 - \frac{1}{2}b_1b_2\right).$$

Applying triangle inequality, we get

$$|\gamma_3| \leq \frac{1}{12}\left|b_3 - \frac{1}{2}b_1b_2\right|.$$

After applying (1.12), we obtain

$$|\gamma_3| \le \frac{1}{6}.$$

By rearranging (1.37), we have

$$\gamma_4 = -\frac{1}{32}\left(b_3 - \frac{3}{32}b_1^3\right) + \frac{1}{16}\left(b_4 - \frac{1}{4}b_2^2\right).$$

Applying triangle inequality, (1.12), and (1.14), we have

$$|\gamma_4| \le \frac{35}{128}.$$

By rearranging (1.38), we have

$$\gamma_5 = -\frac{17}{9600}b_1^5 + \frac{3}{320}b_1^3 b_2 - \frac{1}{40}b_1 b_4 - \frac{1}{40}b_2 b_3 + \frac{1}{20}b_5.$$

Applying triangle inequality, we get

$$|\gamma_5| = \frac{3}{320}|b_1|^3\left|b_2 - \frac{17}{90}b_1^2\right| + \frac{1}{20}\left|b_5 - \frac{1}{2}b_1 b_4\right| + \frac{1}{40}|b_2||b_3|.$$

Now applying (1.11) and (1.12), we have

$$|\gamma_5| \le \frac{7}{20}.$$

□

Theorem 2. *If $f \in \mathcal{S}^*_{\text{esin}}$, then*

$$|\gamma_1\gamma_3 - \gamma_2^2| \le \frac{1}{16}.$$

Proof. From (1.34), (1.35), and (1.36), we get

$$\gamma_1\gamma_3 - \gamma_2^2 = -\frac{1}{1024}b_1^4 - \frac{1}{384}b_1^2 b_2 + \frac{1}{48}b_1 b_3 - \frac{1}{64}b_2^2.$$

This implies

$$16(\gamma_1\gamma_3 - \gamma_2^2) = -\frac{1}{64}b_1^4 - \frac{1}{24}b_1^2 b_2 + \frac{1}{3}b_1 b_3 - \frac{1}{4}b_2^2,$$

using Lemma (1) to express b_2 and b_3 in terms of $b_1 = b \in [0, 2]$, we get

$$16(\gamma_1\gamma_3-\gamma_2^2) = -\frac{1}{64}b^4 + \frac{1}{48}b^2x(4-b^2) - \frac{1}{12}b^2x^2(4-b^2) - \frac{1}{16}x^2(4-b^2)^2 + \frac{1}{16}(4-b^2)(1-|x|^2)\sigma b.$$

After applying triangle inequality and taking $|x| = y \le 1$ and using $|\sigma| \le 1$, we get

$$16|\gamma_1\gamma_3 - \gamma_2^2| \le \frac{1}{64}b^4 + \frac{1}{48}b^2y(4-b^2) + \frac{1}{12}b^2y^2(4-b^2) + \frac{1}{16}y^2(4-b_2^2)^2 + \frac{1}{16}(4-b^2)(1-y^2)b = G(b,y) \quad \text{(say)}.$$

Now differentiating partially with respect to y, we get

$$\frac{\partial}{\partial y}G(b,y) = \frac{1}{48}(4-b^2)(b^2+(2b^2-6b+24)y) \ge 0.$$

This implies $G(b,y)$ is an increasing function in y for fixed b in [0,2].

So the maximum value of $G(b,y)$ is attained at $y = 1$. From this, we have

$$16|\gamma_1\gamma_3 - \gamma_2^2| \le G(b,1) = -\frac{5}{192}b^4 - \frac{1}{12}b^2 + 1.$$

Again, differentiating partially with respect to b, we get

$$\frac{\partial}{\partial y}G(b,1) = -\frac{5}{48}b^3 - \frac{1}{6}b \le 0.$$

Which is a decreasing function in $b \in [0,2]$. So the maximum value of $G(b,1)$ is attained at $b = 0$. Therefore

$$\max G(b,1) \le 1.$$

This implies that

$$16|\gamma_1\gamma_3 - \gamma_2^2| \le 1.$$

With this, we conclude that

$$|\gamma_1\gamma_3 - \gamma_2^2| \le \frac{1}{16}.$$

□

Theorem 3. *If $f \in \mathcal{S}^*_{\text{esin}}$, then*

$$|\gamma_1\gamma_4 - \gamma_2\gamma_3| \le \frac{1}{8}.$$

Proof. From (1.34) to (1.37), we get

$$\gamma_1\gamma_4 - \gamma_2\gamma_3 = \frac{3}{4096}b_1^5 - \frac{1}{192}b_1^2b_3 + \frac{1}{768}b_1b_2^2 + \frac{1}{64}b_1b_4 - \frac{1}{96}b_2b_3 - \frac{1}{768}b_1^3b_2.$$

After some simple calculation, we have

$$32 \times (\gamma_1\gamma_4 - \gamma_2\gamma_3) = \frac{3}{128}b_1^5 - \frac{1}{6}b_1^2b_3 + \frac{1}{24}b_1b_2^2 + \frac{1}{2}b_1b_4 - \frac{1}{3}b - 2b_3 - \frac{1}{24}b_1^3b_2.$$

By rearranging, we get

$$32 \times (\gamma_1\gamma_4 - \gamma_2\gamma_3) = -\frac{1}{24}b_1^3\left(b_2 - \frac{9}{16}b_1^2\right) + \frac{1}{2}b_1\left(b_4 - \frac{1}{3}b_1b_3\right) - \frac{1}{3}b_2\left(b_3 - \frac{1}{8}b_1b_2\right).$$

Applying triangle inequality, we obtain

$$32 \times |\gamma_1\gamma_4 - \gamma_2\gamma_3| \leq \frac{1}{24}|b_1|^3\left|b_2 - \frac{9}{16}b_1^2\right| + \frac{1}{2}|b_1|\left|b_4 - \frac{1}{3}b_1b_3\right| + \frac{1}{3}|b_2|\left|b_3 - \frac{1}{8}b_1b_2\right|.$$

Now applying (1.11) and (1.12), we obtain

$$|\gamma_1\gamma_4 - \gamma_2\gamma_3| \leq \frac{1}{8}.$$ □

Theorem 4. *If* $f \in \mathcal{S}^*_{\text{esin}}$, *then*

$$|\gamma_2\gamma_4 - \gamma_3^2| \leq \frac{85}{768}.$$

Proof. From (1.35), (1.36), and (1.37), we get

$$\gamma_2\gamma_4 - \gamma_3^2 = \frac{1}{16}\left\{-\frac{3}{2048}b_1^6 + \frac{3}{512}b_1^4b_2 + \frac{7}{144}b_1b_2b_3 - \frac{1}{32}b_2^3 + \frac{1}{8}b_2b_4 + \frac{1}{64}b_1^3b_3 - \frac{23}{1152}b_1^2b_2^2 - \frac{1}{32}b_1^4b_4 - \frac{1}{9}b_3^2\right\},$$

which by rearranging becomes

$$16 \times (\gamma_2\gamma_4 - \gamma_3^2) = \frac{3}{512}b_1^4\left(b_2 - \frac{1}{4}b_1^2\right) + \frac{7}{144}b_1b_2\left(b_3 - \frac{23}{56}b_1b_2\right) + \frac{1}{8}b_2\left(b_4 - \frac{1}{4}b_2^2\right) - \frac{1}{32}b_1^2\left(b_4 - \frac{1}{2}b_1b_3\right) - \frac{1}{9}b_3^2.$$

Now, by applying triangle inequality, we get

$$16 \times |\gamma_2\gamma_4 - \gamma_3^2| \leq \frac{3}{512}|b_1|^4\left|b_2 - \frac{1}{4}b_1^2\right| + \frac{7}{144}|b_1||b_2|\left|b_3 - \frac{23}{56}b_1b_2\right| + \frac{1}{8}|b_2|\left|b_4 - \frac{1}{4}b_2^2\right| + \frac{1}{32}|b_1^2|\left|b_4 - \frac{1}{2}b_1b_3\right| + \frac{1}{9}|b_3|^2,$$

which by applying (1.11) and (1.12) gives

$$|\gamma_2\gamma_4 - \gamma_3^2| \leq \frac{85}{768}.$$ □

Theorem 5. *If $f \in \mathcal{S}^*_{\text{esin}}$, then we have*

$$|H_{3,1}(f)| \leq 0.07.$$

Proof. Since

$$H_{3,1}(f) = \begin{vmatrix} \gamma_1 & \gamma_2 & \gamma_3 \\ \gamma_2 & \gamma_3 & \gamma_4 \\ \gamma_3 & \gamma_4 & \gamma_5 \end{vmatrix} = \gamma_3(\gamma_2\gamma_4 - \gamma_3^2) - \gamma_4(\gamma_1\gamma_4 - \gamma_2\gamma_3) + \gamma_5(\gamma_1\gamma_3 - \gamma_2^2),$$

applying triangle inequality, we get

$$|H_{3,1}(f)| \leq |\gamma_3||\gamma_2\gamma_4 - \gamma_3^2| + |\gamma_4||\gamma_1\gamma_4 - \gamma_2\gamma_3| + |\gamma_5||\gamma_1\gamma_3 - \gamma_2^2|.$$

Now, by applying Theorems (2), (3), (4), (1.21), (1.22), and (1.23), we get

$$|\mathcal{H}_{3,1}(f)| \leq \frac{3433}{46080} \approx 0.07.$$ □

4 Conclusion

The new subclass $\mathcal{S}_{\text{esin}}$ of the class $\mathcal{A}$ that we introduced using the concept of subordination is quite a typical subclass of the class of functions with bounded turning. The non-emptiness of the class is justified giving a sequence of examples with their geometrical shape. Corresponding to each $f \in \mathcal{S}_{\text{esin}}$, we have defined a function $\log(\frac{f(z)}{z})$ whose coefficients are denoted by γ_i, $i = 1, 2, 3\cdots$, which are expressed over coefficients of $f \in \mathcal{S}_{\text{esin}}$, a_i, $i = 1, 2, 3\cdots$. We have obtained various results on bounds of coefficients mentioned in results of Theorem 1 to 5.

Further research work can be taking in the direction of finding sharp bound of Toeplitz determinants and bounds of the Zalcman inequality of the above class.

Compliance with Ethical Standards:
The authors declare that there is no conflict of interest.

This article does not contain any studies with human participants or animals performed by any of the authors.

Bibliography

[1] M. Ali and A. Vasudevarao, On logarithmic coefficients of some close-to-convex functions, *Proc. Am. Math. Soc.* **146** (2018), 1131–1142.

[2] M. Ali, A. Vasudevarao and D. Thomas, On the third logarithmic coefficients of close-to-convex functions, in: *Current research in mathematical and computer sciences II*, A. Lecko, ed., Olsztyn, Poland, 2018, pp. 271–278.

[3] A. Alotaibi, M. Arif, M. Alghamdi and S. Hussain, Starlikness associated with cosine hyperbolic function, *Mathematics* **8**(7) (2020).

[4] R. S. A. S. Alshehry and A. Bariq, The second Hankel determinant of logarithmic coefficients for starlike and convex functions involving four-leaf-shaped domain, *J. Funct. Spaces*, **2022** (2022), Article ID 2621811.

[5] M. Arif, M. Raza, H. Tang, S. Hussain and H. Khan, Hankel determinant of order three for familiar subsets of analytic functions related with sine function, *Open Math.* **17**(1) (2019) 1615–1630.

[6] K. Bano and M. Raza, Starlike functions associated with cosine function, *Bull. Iran. Math. Soc.* **47** (2020), 1513–1532.

[7] L. Bieberbach, Über dié koeffizienten derjenigen potenzreihen, welche eine schlichte abbildung des einheitskreises vermitteln, *Sitzungsberichte Preuss. Akad. Der Wiss* **138** (1916), 940–955.

[8] N. Cho, B. Kowalczyk, O. Kwon, A. Lecko and Y. Sim, On the third logarithmic coefficient in some subclasses of close-to-convex functions, *Rev. R. Acad. Cienc. Exactas Fís. Nat. (Esp.)* **114**(52) (2020).

[9] L. De Branges, A proof of the Bieberbach conjecture, *Acta Math.* **154**(2) (1985), 137–152.

[10] P. Duren, *Univalent functions*, Springer, New York, NY, USA, 1983.

[11] M. Fekete and G. Szego, Eine bemerkung uber ungerade schlichte funktionen, *J. Lond. Math. Soc.* **8**(2) (1933), 85–89.

[12] D. Girela, Logarithmic coefficients of univalent functions, *Ann. Acad. Sci. Fenn., Math.* **25** (2000), 337–350.

[13] P. Goel and S. S. Kumar, Certain class of starlike functions associated with modified sigmoid function, *Bull. Malays. Math. Sci. Soc.* **43** (2020), 957–991.

[14] W. Hayman, On the second Hankel determinant of mean univalent functions, *Proc. Lond. Math. Soc.* (1968), 77–94.

[15] R. W. Ibrahim, On the subordination and super-ordination concepts with applications, *J. Number Theory* **14**(5) (2017), 3037–3046.

[16] R. W. Ibrahim and D. Baleanu, Analytic solution of the Langevin differential equations dominated by a multibrot fractal set, *Fractal Fract.* **88**(3) (2021), 520–524.

[17] R. W. Ibrahim and D. Maslina, Subordination and superordination for univalent solutions for fractional differential equations, *J. Math. Anal. Appl.* **345**(2) (2008), 871–879.

[18] R. W. Ibrahim, M. E. Rafida and J. O. Suzan, Generalized Briot-Bouquet differential equation based on new differential operator with complex connections, *Axioms* **9**(2) (2020), 157–188.

[19] F. Keogh and E. Merkes, A coefficient inequality for certain classes of analytic functions, *Proc. Am. Math. Soc.* **20** (1969), 8–12.

[20] W. Keopf, On the Fekete-Szegö problem for close-to-convex functions, *Proc. Am. Math. Soc.* **101** (1987), 89–95.

[21] M. Khan, B. Ahmad, G. Moorthy, R. Chinram and W. Mashwani, Applications of modified sigmoid functions to a class of starlike functions, *J. Funct. Spaces* (2020), 8.

[22] P. Koebe, Über die uniformisierung der algebraischen kurven. I, *Math. Ann.* **67** (1909).

[23] B. Kowalczyk and A. Lecko, Second Hankel determinant of logarithmic coefficients of convex and starlike functions, *Bull. Aust. Math. Soc.* **105**(3) (2021), 458–467.

[24] B. Kowalczyk and A. Lecko, Second Hankel determinant of logarithmic coefficients of convex and starlike functions of order alpha, *Bull. Malays. Math. Soc.* **45** (2021), 727–740.

[25] S. Kumar and K. Arora, Starlike functions associated with a petal shaped domain, *Bull. Korean Math. Soc.* **59**(4) (2022), 993–1010.

[26] U. Kumar and A. Vasudevarao, Logarithmic coefficients for certain subclasses of close-to-convex functions, *Monatshefte Math.* **187** (2018), 543–563.

[27] R. Libera and E. Zlotkiewicz, Early coefficients of the inverse of a regular convex function, *Proc. Am. Math. Soc.* **85** (1982), 225–230.
[28] W. Ma and D. Minda, A unified treatment of some special classes of univalent functions, in: *Proceeding of Conference on Complex Analytic*, International Press, New York, 1994, pp. 157–169.
[29] I. Milin, *Univalent functions and orthonormal systems*, Translations of Mathematical Monographs, Vol. 49, Nauka, Moscow, 1977.
[30] J. Noonan and D. Thomas, On the second Hankel determinant of a really mean p-valent functions, *Trans. Am. Math. Soc.* **223** (1976), 337–346.
[31] H. Orhan, N. Magesh and J. Yamini, Bounds for the second Hankel determinant of certain bi-univalent functions, *Turk. J. Math.* **40**(3) (2016), 679–687.
[32] L. Parida, and A. K Sahoo, Coefficients Bounds on a Certain Class of Multivalent Analytic Functions, *J. Math. Comput. Sci.*, **8** (2022).
[33] L. Parida, A. K. Sahoo and T. B. Bulboca, Hankel Determinants for a New Subclasses of Analytic Functions Involving a Linear Operator, *Kragujev. J. Math.*, **46** (2022), 605–616.
[34] J. Patel and A. K Sahoo, On a Subclass of Analytic Functions Related to a Hyperbola, *Int. J. Math. Sci.* **2014** (2014).
[35] C. Pommerenke, On the coefficients and Hankel determinants of univalent functions, *J. Lond. Math. Soc.* **41**(1) (1966), 111–122.
[36] C. Pommerenke, On the Hankel determinants of univalent functions, *Mathematika* **14**(1) (1967), 108–112.
[37] C. Pommerenke, *Univalent functions*, 1975.
[38] A. K Sahoo and J. Patel, Inclusion and neighborhood properties of certain subclasses of p-valent analytic functions of complex order involving a linear operator, *Bull. Korean Math. Soc.* **51** (2014), 1625–1647.
[39] K. Sharma, N. Jain and V. Ravichandran, Starlike functions associated with a cardioid, *Afr. Math.* **27** (2016), 923–939.
[40] L. Shi, A. Muhammad, A. Rafiq, A. Muhammad and J. Iqbal, Sharp bounds of Hankel determinant on logarithmic coefficients for functions of bounded turning associated with petal-shaped domain, *Mathematics* **10**(11) (2022).
[41] L. Shi, H. Srivastava, M. Arif, S. Hussain and H. Khan, An investigation of the third Hankel determinant problem for certain subfamilies of univalent functions involving the exponential function, *Symmetry* (2019).
[42] J. Sokół and J. Stankiewicz, Radius of convexity of some subclasses of strongly starlike functions, *Bull. Sci. Math.* **167** (2021).
[43] H. Srivastava, B. Khan, N. Khan, M. Tahir, S. Ahmad and N. Khan, Upper bound of the third Hankel determinant for a subclass of q-starlike functions associated with the q-exponential function, *Bull. Sci. Math.* **167** (2021).
[44] P. Sunthrayuth, N. Iqbal, M. Naeem, Y. Jawarneh and S. K. Samura, The sharp upper bounds of the Hankel determinant on logarithmic coefficients for certain analytic functions connected with eight-shaped domains, *J. Funct. Spaces* **2022** (2022).
[45] D. K. Thomas, On logarithmic coefficients of close to convex functions, *Proc. Am. Math. Soc.* **144** (2016) 1681–1687.
[46] K. Ullah, S. Zainab, M. Arif, M. Darus and M. Shutaywi, Radius problems for starlike functions associated with the tan hyperbolic function, *J. Funct. Spaces* (2021).
[47] A. Wahid, D. Mohamad, M. Kamarozzaman and A. A. Shahminan, Toeplitz determinants for the class of functions with bounded turning, *Eur. J. Pure Appl.* **15** (2022) 1937–1947.

Monika Arora, Mahesh Kumar Sharma, and Kamalpreet Kaur

Effect of gyrotactic microorganisms in rotating magnetic nanofluids in porous media

Abstract: This paper deals with the linear stability analysis to study the onset of bioconvection process under the effect of rotation in porous media. Here, we consider water-based magnetic nanofluid (W_{mnf}) containing gyrotactic microorganisms in the presence of gravity. The inertial effects of microorganisms are redundant while considering their locomotion, as microorganisms considered here are motile with small size. The model used integrates the effects of two important slip mechanisms, namely Brownian motion and thermophoresis, as suggested by J. Buongiorno (J. Buongiorno, "Convective transport in nanofluids", *Journal of Heat Transfer*, 128(3), 240–250, 2006) along with magnetophoresis. Numerical method is employed using MATLAB's EIG function to solve the resulting eigen value problem. For the analysis, critical Rayleigh number is determined in terms of the governing parameters through neutral stability curves (NSCs). Taylor number (T_A) is used to see the effect of rotation and it is observed that T_A stabilizes the system.

Keywords: Rotation, bioconvection, magnetic nanofluids, microorganisms, porous media

MSC 2010: 76E06, 76R07, 76E15

1 Introduction

Bioconvection is a process of spontaneous formation of pattern in the suspension of upwardly swimming microorganisms with density greater than water, for instance soil bacterium *Bacillus subtilis* and heavy-bottom algae *Chlamydomonas nivalis* [25]. The phenomenon of bioconvection takes place due to the unstable density stratification that occurs when the upwardly swimming microorganisms, which are denser than the fluid, accumulate at the top layer. The clustering of these microorganisms makes the top layer denser than the lower surrounding area and instability is induced, which ultimately leads to several pattern formation [2, 5–7, 11, 13, 14]. Once the microorganisms stop swimming, the process of bioconvection ends. While swimming in the upward direction, the

Monika Arora, Department of Mathematics, School of Chemical Engineering and Physical Sciences, Lovely Professional University, Block 55, Agriculture School, Phagwara-144411, Punjab, India, e-mail: monikaarora1879@gmail.com
Mahesh Kumar Sharma, Department of Mathematics, School of Basic and Applied Sciences, Maharaja Agrasen University, D Block, Baddi-174103, H.P., India, e-mail: mksharma.maths@gmail.com
Kamalpreet Kaur, Lambton College at Queen's College of Business, Technology and Public Safety, 1457, London Road, Mississauga-L4Z 3E9, Ontario, Canada, e-mail: kaurkamalpreet189@gmail.com

https://doi.org/10.1515/9783111313634-002

microorganisms tend to respond to certain stimuli. These responses are referred to as taxes. The paper relies on gyrotaxis. Bioconvection can also occur in the suspension involving nanoparticles with low concentration. The low concentration of nanoparticles is considered to avoid the drastic enhancement in the viscosity of base fluid [21]. Kuznetsov and Avramenko [21] researched the influence of small particles on the stability of the system containing motile gyrotactic microorganisms. The conclusion was–bioconvection speeds up the mixing process as well as assists in sedimentation of small particles. Further, onset of convection is advanced with the increase in the volumetric fraction of small particles.

Kuznetsov and Geng [22] numerically investigated the influence of bioconvection on the mixing of small particles found that bioconvection process gets decelerated by the small particles of ideal size, while those with inconsequential "settling velocity" do not significantly effect bioconvection. Moreover, extremely large particles with negligible diffusivity and extremely heavy ones with high settling velocity have no effect on bioconvection. Kuznetsov [16] studied oscillatory and non-oscillatory bio-thermal convection in a suspension with nanoparticles and gyrotactic microorganisms and found that bottom-heavy distribution of nanoparticle tends to stabilize the system, but top-heavy distribution has opposite effect. Also, upswimming microorganisms have destabilizing effect. Kuznetsov [17] studied bioconvection in porous media saturated with nanofluid and concluded that Bioconvection Péclet number (Q) determines the effect of microorganisms on the stability of the suspension. It was found that microorganisms have stabilizing effect for $Q > 3.0$, while effect is opposite for $Q < 3.0$ using one-term Galerkin approximation analysis. Sanjalee and Sharma [27] considered a layer of nanofluid heated from below containing gyrotactic microorganisms and studied the effect of externally applied magnetic field and heat on the bioconvection. It was found that magnetic field tends to stabilize the system, while microorganisms speeds up the onset of convection.

Kaur et al. [10] considered a W_{mnf} infinite layer with microorganisms and investigated the effect of several important parameters on the onset of convection for (R_r) boundary. It was concluded that Langevin parameter (a_L) and bioconvection Lewis number (Lb) stabilize the system, whereas Q, nanoparticle concentration Rayleigh number (Rn), bioconvection Rayleigh number (Rb) and gyrotactic number (G) have opposite effect. Kessler [12] reported that development of convective instability in algal suspensions is prevented by porous medium which suggests that a critical value of permeability of porous medium exists. If permeability is less than critical value, bioconvection does not occur and system is stable, whereas if permeability exceeds critical value, bioconvection may take place. Kuznetsov and Avramenko [19] concluded that increase in cell diffusivity and fluid viscosity tends to increase critical permeability, while it decreases with the increase in density difference, average swimming velocity of cells, number density of the cells and volume of the cell. Kuznetsov [18] investigated a nanofluid containing oxytactic motile microorganisms and reported that stabilizing nanoparticle distribution and destabilizing vertical temperature variation favor oscillatory instability. Moreover, suspension is less stable in the presence of oxytactic microorganisms and they favor non-oscillatory instability by destroying oscillatory instability. Hill and Pedley [8] observed that the suspension would not attain significantly high concentration for the occurance of bioconvection if the porosity is small. This fact is favored by the stability analysis in [20], which states that the phenomenon of bioconvection takes place solely in porous media with high value of porosity.

The aim of the present work is to study the onset of bioconvection in magnetic nanofluid (*MNF*) under the assumption that system is rotated through angular velocity **Ω** along vertical axis. The effect of rotation is exhibited graphically through the Taylor number T_A along with the effect of other significant non-dimensional parameters on the onset of convection. The work is presented as follows: in Section 2, physical model of the problem is given. Section 3 is devoted to mathematical formulation and the system is made dimensionless here. In Section 4, we find steady state solution. Section 5 demonstrates the linear stability problem. Results and discussion part is in Section 6 and conclusions are presented in Section 7.

2 Physical model of the flow

An infinite horizontal layer of incompressible W_{mnf} containing gyrotactic microorganisms is considered. The layer is heated from below in the presence of uniform vertical magnetic field under the effect of rotation about the vertical axis. The system rotates at the angular velocity **Ω** (see Fig. 2.1).

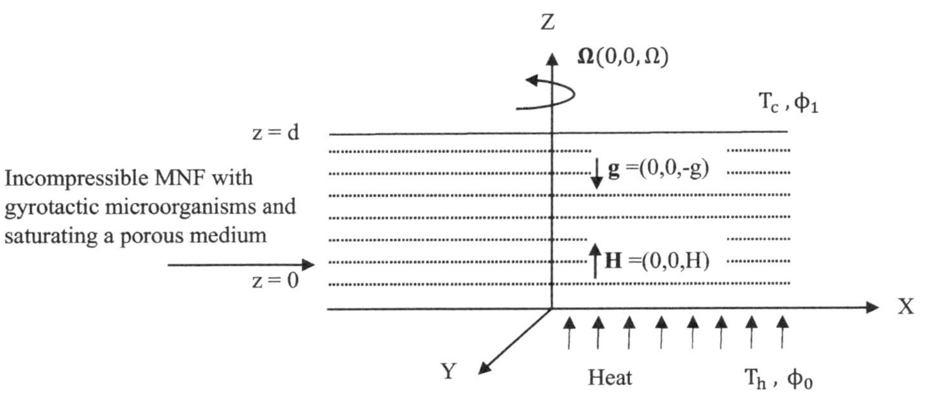

Figure 2.1: Geometric structure.

3 Mathematical formulation of the problem

Following [10, 15, 23], the governing equations for the present problem, known as equation of continuity, equation of momentum, equation for nanoparticles, equation of thermal energy, Maxwell's equations, magnetization equation and cell conservation equation for microorganisms are respectively:

$$\nabla \cdot \mathbf{u} = \mathbf{0}, \tag{2.1}$$

$$\frac{\rho_f}{\epsilon}\frac{\partial \boldsymbol{u}}{\partial t} = -\nabla p - \frac{\mu}{K}\boldsymbol{u} + \mu_0(\boldsymbol{M} \cdot \nabla)\boldsymbol{H} - \rho g \boldsymbol{k} + \frac{2\rho_f}{\epsilon}(\boldsymbol{u} \times \boldsymbol{\Omega}), \tag{2.2}$$

where K is the medium permeability, ϵ is the porosity, $\boldsymbol{u}$ is the filter velocity, t is the time, p is the pressure, μ is the viscosity, $\boldsymbol{M}$ is the magnetization and ρ is the overall density of the *MNF*, given by the following expression

$$\rho = \{\phi\rho_p + (1-\phi)\rho_f[1-\alpha(T-T_c)]\},$$

here ρ_p is the magnetic nanoparticle density, ρ_f is the base fluid density, α is the thermal volumetric expansion coefficient, ϕ is the nanoparticle volume fraction and T_c is the reference temperature.

$$\frac{\partial \phi}{\partial t} + \frac{1}{\epsilon}\boldsymbol{u} \cdot \nabla\phi = \nabla \cdot \left(D_B \nabla\phi + D_T\frac{\nabla T}{T_c} - D_H \nabla H\right), \tag{2.3}$$

where D_B, D_T and D_H are the Brownian diffusion coefficient, thermophoretic diffusion coefficient and magnetophoretic diffusion coefficient, respectively.

$$(\rho c)_m\left(\frac{\partial T}{\partial t}\right) + (\rho c)_f(\boldsymbol{u} \cdot \nabla T) = \nabla \cdot (k_1 \nabla T) + \epsilon\rho_p c_p\left(D_B \nabla T \cdot \nabla\phi + D_T\frac{\nabla T \cdot \nabla T}{T_c} - D_H \nabla T \cdot \nabla H\right), \tag{2.4}$$

here T, c_f are the nanofluid temperature and nanofluid specific heat, respectively.

$$\nabla \cdot B = 0, \quad \nabla \times H = 0, \quad B = \mu_0(M + H), \tag{2.5}$$

$$M_{\text{eq}} = \frac{\boldsymbol{H}}{H}[M_0 + \chi(H - H_0) - K_m(T - T_h) + K_p(\phi - \phi_0)], \tag{2.6}$$

here H_0 is the uniform magnetic field of the magnetic nanofluid layer when placed in an external magnetic field $\boldsymbol{H} = H_0^{\text{ext}}\boldsymbol{k}$, χ is tangent magnetic susceptibility, $M_0 = \chi_2 H_0$, χ_2 is chord magnetic susceptibility, $K_m = \frac{\chi H_0}{T_h}$ and $K_p = \frac{\chi H_0}{\phi_0}$. Here K_m represents a function of temperature and magnetic field, and K_p represents a function of particle concentration and magnetic field.

$$\epsilon\frac{\partial n}{\partial t} = -\nabla \cdot \boldsymbol{j}, \tag{2.7}$$

here

$$\boldsymbol{j} = (n\boldsymbol{u} + nw_m\hat{\boldsymbol{p}} - D_m\nabla n)$$

is the total flux of microorganisms due to the three physical mechanisms of microorganisms, viz. fluid convection, self-propelled motion (swimming) and diffusion. Here $\hat{\boldsymbol{p}}$ is the unit vector pointing in the average swimming direction of the microorganism, w_m is the constant swimming speed of the microorganism relative to the *MNF* and D_m is the diffusivity of microorganisms, which we also assume to be isotropic and constant.

Boundary Conditions (Bc_s):

$$\left.\begin{aligned} &w = 0, \quad T = T_h, \quad \phi = \phi_0, \quad \boldsymbol{j}.\hat{\boldsymbol{k}} = 0 \quad \text{for} \quad z = 0, \\ &w = 0, \quad T = T_c, \quad \phi = \phi_1, \quad \boldsymbol{j}.\hat{\boldsymbol{k}} = 0 \quad \text{for} \quad z = d. \end{aligned}\right\} \tag{2.8}$$

Equations (2.1)–(2.7) in dimensionless form are:

$$\nabla\cdot\boldsymbol{u} = 0, \tag{2.9}$$

$$\begin{aligned} \frac{1}{V_a}\frac{\partial \boldsymbol{u}}{\partial t} = &-\nabla p - \boldsymbol{u} + \lambda_1(\boldsymbol{M}\cdot\nabla)\boldsymbol{H} - \Big(\mathrm{Rn}\,\phi - \mathrm{Ra}\,T + \mathrm{Ra}_N\,N_\phi T\phi \\ &+ \frac{\mathrm{Rb}}{\mathrm{Lb}\,\vartheta}n - \rho_1\phi + \rho_2\Big)\boldsymbol{k} + T_A^{1/2}(\boldsymbol{u}\times\boldsymbol{k}), \end{aligned} \tag{2.10}$$

$$\frac{\partial\phi}{\partial t} + \frac{\mathrm{Pr}}{D_aV_a}\boldsymbol{u}\cdot\nabla\phi = \frac{1}{\mathrm{Le}}\nabla^2\phi + \frac{N_a}{\mathrm{Le}}\nabla^2T - \frac{N_a'}{\mathrm{Le}}\nabla^2H, \tag{2.11}$$

$$A\frac{\partial T}{\partial t} + \boldsymbol{u}\nabla T = \nabla^2T + \frac{N_b}{\mathrm{Le}}(\nabla T\cdot\nabla\phi) + \frac{N_aN_b}{\mathrm{Le}}(\nabla T\cdot\nabla T) - \frac{N_a'N_b}{\mathrm{Le}}(\nabla T\cdot\nabla H), \tag{2.12}$$

$$\chi_2\nabla\cdot\boldsymbol{M} + \nabla\cdot\boldsymbol{H} = 0, \tag{2.13}$$

$$\boldsymbol{M} = \frac{\boldsymbol{H}}{H}\frac{(1+\chi)}{\chi^2}\left(\frac{\chi}{1+\chi}H - \frac{M_{p_1}}{M_{p_3}}T + \frac{M_{p_1}'}{M_{p_3}'}\phi + \frac{\chi_2-\chi}{1+\chi}\right), \tag{2.14}$$

$$\frac{D_aV_a}{\mathrm{Pr}}\frac{\partial n}{\partial t} = -\nabla\cdot\left(n\boldsymbol{u} + \frac{Q}{\mathrm{Lb}}n\boldsymbol{p} - \frac{1}{\mathrm{Lb}}\nabla n\right), \tag{2.15}$$

where

$$\lambda_1 = \frac{\mu_0M_0H_0K}{\kappa\mu}, \quad \rho_1 = \frac{Kd\rho_f}{\kappa\mu}(\phi_0-\phi_1)\alpha T_c g, \quad A = \frac{(\rho c)_m}{(\rho c)_f},$$

$$N_\phi = \frac{\phi_0-\phi_1}{1-\phi_0}, \quad \rho_2 = \frac{Kd\rho_f}{\kappa\mu}(1+\alpha T_c)g.$$

The Bc_s (2.8) become,

$$\left.\begin{array}{llllll} w=0, & T=\dfrac{T_h}{T_h-T_c}, & \phi=\dfrac{\phi_0}{\phi_0-\phi_1}, & Q_n=\dfrac{dn}{dz} & \text{for} & z=0, \\ w=0, & T=\dfrac{T_c}{T_h-T_c}, & \phi=\dfrac{\phi_1}{\phi_0-\phi_1}, & Q_n=\dfrac{dn}{dz} & \text{for} & z=1. \end{array}\right\} \tag{2.16}$$

Non-dimensional parameters:

$$\mathrm{Ra}=\frac{\rho_f gadK(T_h-T_c)}{\mu\kappa}\quad\text{(Thermal Rayleigh number)},\qquad \mathrm{Pr}=\frac{\mu}{\rho_f\kappa}\quad\text{(Prandtl number)},$$

$$D_a=\frac{K}{d^2}\quad\text{(Darcy number)},\qquad V_a=\frac{\epsilon\,\mathrm{Pr}}{D_a}\quad\text{(Vadasz number)},$$

$$\mathrm{Le}=\frac{\kappa}{D_B}\quad\text{(Lewis number)},\qquad M_{p_1}=\frac{\mu_0\chi^2H_0^2(T_h-T_c)}{\rho_f gad(1+\chi)T_h^2}\quad\text{(Magnetic parameter)},$$

$$M'_{p_1}=\frac{\mu_0\chi^2H_0^2(\phi_0-\phi_1)}{\rho_f gad(1+\chi)\phi_0^2}\quad\text{(Magnetic parameter)},$$

$$M_{p_3}=\frac{\mu_0\chi H_0^2}{\rho_f gadT_h}\quad\text{(Magnetic parameter)},\qquad M'_{p_3}=\frac{\mu_0\chi H_0^2}{\rho_f gad\phi_0}\quad\text{(Magnetic parameter)},$$

$$N_b=\frac{\epsilon(\rho C)_p(\rho_0-\rho_1)}{(\rho C)_f}\quad\text{(Modified particle density increment)},$$

$$\mathrm{Rn}=\frac{(\rho_p-\rho_f)(\phi_0-\phi_1)gdK}{\mu\kappa}\quad\text{(Concentration Ra)},$$

$$N_a=\frac{D_T(T_h-T_c)}{D_BT_c(\phi_0-\phi_1)}\quad\text{(Modified diffusivity ratio)},$$

$$N'_a=\frac{D_HH_0}{D_B(\phi_0-\phi_1)}\quad\text{(Modified diffusivity ratio)},$$

$$\mathrm{Rb}=\frac{\Delta\rho_m g\vartheta dK}{\mu D_m}\quad\text{(Bioconvection Ra)},\qquad \mathrm{Lb}=\frac{\kappa}{D_m}\quad\text{(Bioconvection Le)},$$

$$Q=\frac{w_md}{D_m}\quad\text{(Bioconvection Péclet number)},\qquad T_A=\frac{4\Omega^2K^2}{\epsilon^2\nu^2}\quad\text{(Taylor number)}.$$

Here ϑ is a non-dimensional number defined in (4), $\nu=\frac{\mu}{\rho_f}$ is the kinematic viscosity and $\mathrm{Ra}_N=(1-\phi_0)\,\mathrm{Ra}$.

4 The steady state

Here

$$\boldsymbol{u}_b=\boldsymbol{0},\quad \hat{\boldsymbol{p}}=\hat{\boldsymbol{k}}$$

and p_b, T_b, ϕ_b, $\boldsymbol{M}_b$, $\boldsymbol{H}_b$, and N_b all are functions of z only.

Therefore, the Equations (2.10)–(2.15) reduce to

$$-\frac{dp_b}{dz}+\lambda_1 M_b\frac{dH_b}{dz}-\text{Rn}\,\phi_b+\text{Ra}\,T_b-\text{Ra}_N\,N_\phi T_b\phi_b-\frac{\text{Rb}}{\text{Lb}\,\vartheta}n+\rho_1\phi_b+\rho_2=0, \tag{2.17}$$

$$\frac{d^2\phi_b}{dz^2}+N_a\frac{d^2T_b}{dz^2}-N_a'\frac{d^2H_b}{dz^2}=0, \tag{2.18}$$

$$\frac{d^2T_b}{dz^2}+\left\{\frac{N_b}{\text{Le}}\frac{d\phi_b}{dz}+\frac{N_aN_b}{\text{Le}}\frac{dT_b}{dz}-\frac{N_a'N_b}{\text{Le}}\frac{dH_b}{dz}\right\}\frac{dT_b}{dz}=0, \tag{2.19}$$

$$\chi_2\frac{dM_b}{dz}+\frac{dH_b}{dz}=0, \tag{2.20}$$

$$M_b=\frac{1+\chi}{\chi_2}\left\{\frac{\chi}{1+\chi}H_b-\frac{M_{p_1}}{M_{p_3}}T_b+\frac{M_{p_1}'}{M_{p_3}'}\phi_b+\frac{\chi_2-\chi}{1+\chi}\right\}, \tag{2.21}$$

$$\frac{d^2N_b}{dz^2}-Q\frac{dN_b}{dz}=0. \tag{2.22}$$

Using the Bc_s (2.16), and following [1, 3, 24], we solve the equations (2.17)–(2.22), to obtain the solution of steady state. The following results are obtained:

$$\begin{gathered}\boldsymbol{u}_b=\mathbf{0},\quad p=p_b(z),\quad T_b=\frac{T_h}{(T_h-T_c)}-z,\\ \phi_b=\frac{\phi_0}{(\phi_0-\phi_1)}-z,\quad H_b=1-\frac{M_{p_1}}{M_{p_3}}z+\frac{M_{p_1}'}{M_{p_3}'}z,\\ M_b=1+\frac{1}{\chi_2}\left(\frac{M_{p_1}}{M_{p_3}}\right)z-\frac{1}{\chi_2}\left(\frac{M_{p_1}'}{M_{p_3}'}\right)z,\\ N_b=\vartheta e^{Qz},\quad \vartheta=\frac{\bar{n}Q}{e^Q-1},\quad \bar{n}=\int_0^1 N_b(z)dz.\end{gathered} \tag{2.23}$$

5 Linear analysis

Considering small perturbations to $\boldsymbol{u}, p, T, n, \boldsymbol{M}, \phi, \boldsymbol{H}, \hat{\boldsymbol{p}}$ and substituting the perturbed variables in the equations (2.10)–(2.15) and then linearizing them gives the following set of equations.

$$\begin{aligned}\frac{1}{V_a}\frac{\partial\nabla^2 w}{\partial t}=&-\nabla^2 w-\left\{\text{Ra}\,\frac{M_{p_3}M_{p_1}'}{M_{p_3}'}-Ng-\text{Ra}_N(1+N_\phi z)\right\}\nabla_1^2\theta-\left\{\text{Ra}\,\frac{M_{p_3}M_{p_1}'}{M_{p_3}'}-\text{Ra}_s\,M_{p_1}'\right.\\ &\left.+\text{Rn}+\text{Ra}_N\,N_\phi(1-z)\right\}\nabla_1^2\phi-\{\text{Ra}\,M_{p_3}-\text{Ra}_s\,M_{p_3}'\}\frac{\partial\nabla_1^2\psi}{\partial z}-\frac{\text{Rb}}{\text{Lb}\,\vartheta}\nabla_1^2 n-T_A^{1/2}\frac{\partial\xi}{\partial z},\end{aligned} \tag{2.24}$$

$$\frac{1}{V_a}\frac{\partial\xi}{\partial t}=-\xi+T_A^{1/2}\frac{\partial w}{\partial z}, \tag{2.25}$$

$$\frac{\partial\phi}{\partial t} = \frac{1}{\epsilon}w + \frac{1}{\text{Le}}\nabla^2\theta - \frac{N'_a}{\text{Le}}\frac{\nabla^2\psi}{\partial z}, \tag{2.26}$$

$$A\frac{\partial\theta}{\partial t} = w + \nabla^2\theta - \left\{\frac{N_b}{\text{Le}} + \frac{2N_aN_b}{\text{Le}} - \frac{N_bN'_aM_{p_1}}{\text{Le}\,M_{p_3}} + \frac{N_bN'_aM'_{p_1}}{\text{Le}\,M'_{p_3}}\right\}\frac{\partial\theta}{\partial z} - \frac{N_b}{\text{Le}}\frac{\partial\phi}{\partial z} + \frac{N_bN'_a}{\text{Le}}\frac{\partial^2\psi}{\partial z^2}, \tag{2.27}$$

$$\frac{\partial^2\psi}{\partial z^2} = \frac{M_{p_1}}{M_{p_3}}\frac{\partial\theta}{\partial z} - \frac{M'_{p_1}}{M'_{p_3}}\frac{\partial\phi}{\partial z} - \frac{(1+\chi_2)}{(1+\chi)}\nabla_1^2\psi, \tag{2.28}$$

$$\frac{D_aV_a}{\text{Pr}}\frac{\partial n}{\partial t} = -\frac{Q}{\text{Lb}}\frac{\partial n}{\partial z} + \frac{1}{\text{Lb}}\nabla^2 n - wve^{Qz}Q + ve^{Qz}Q\left\{G(1-\alpha_0)\nabla_1^2 w + G(1+\alpha_0)\frac{\partial^2 w}{\partial z^2}\right\}, \tag{2.29}$$

where $\nabla_1^2 = \frac{\partial^2}{\partial x^2} + \frac{\partial^2}{\partial y^2}$, $Ng = M_{p_1}$ Ra and $\text{Ra}_s = \frac{\rho_f gadK(\phi_0-\phi_1)}{\mu\kappa}$.

We obtain equation (2.24) by applying $\nabla X\nabla$ of linearized momentum equation and then taking its vertical component. Equation (2.25) represents the effect of rotation which is obtained by considering k^{th} component of curl of linearized momentum equation.

Further, we assume

$$\{w,\theta,\phi,\psi,n\} = \{w(z),\theta(z),\phi(z),\psi(z),n(z)\}\exp\{\sigma t + i(k_xx + k_yy)\}, \tag{2.30}$$

where k_x, k_y are the wave numbers in x-direction and y-direction, respectively, and $k = \sqrt{k_x^2 + k_y^2}$.

By the substitution of (2.30) into the equations (2.24)–(2.29), we get

$$\begin{aligned}&\frac{\sigma}{V_a}(4D^2 - k^2)w(z)\\ &\quad= -(4D^2 - k^2)w(z) - \left\{Ng - \frac{\text{Ra}\,M_{p_3}M'_{p_1}}{M'_{p_3}} + \text{Ra}_N\left\{1 + N_\phi\right.\right.\\ &\quad\times\left.\left.\left(\frac{z+1}{2}\right)\right\}\right\}k^2\theta(z) + \left\{\frac{\text{Ra}\,M_{p_3}M'_{p_1}}{M'_{p_3}} - \text{Ra}_s\,M'_{p_1} + \text{Rn} + \text{Ra}_N\,N_\phi\frac{(1-z)}{2}\right\}k^2\phi(z)\\ &\quad+ \{\text{Ra}\,M_{p_3} - \text{Ra}_s\,M'_{p_3}\}2k^2D\psi(z) + \frac{\text{Rb}}{\text{Lb}\,\vartheta}k^2n(z) - 2T_A^{1/2}D\xi(z),\end{aligned} \tag{2.31}$$

$$\frac{\sigma}{V_a}\xi(z) = -\xi(z) + 2T_A^{1/2}Dw(z), \tag{2.32}$$

$$\sigma\phi(z) = \frac{\text{Pr}}{D_aV_a}w(z) + \frac{1}{\text{Le}}(4D^2 - k^2)\phi(z) + \frac{N_a}{\text{Le}}(4D^2 - k^2)\theta(z) - \frac{2N'_a}{\text{Le}}(4D^2 - k^2)D\psi(z), \tag{2.33}$$

$$\begin{aligned}A\sigma\theta(z) &= w(z) + (4D^2 - k^2)\theta(z) + \left\{-\frac{N_b}{\text{Le}} - \frac{2N_aN_b}{\text{Le}} + \frac{N_bN'_aM_{p_1}}{\text{Le}\,M_{p_3}} - \frac{N_bN'_aM'_{p_1}}{\text{Le}\,M'_{p_3}}\right\}2D\theta(z)\\ &\quad- \frac{2N_b}{\text{Le}}D\phi(z) + \frac{4N_bN'_a}{\text{Le}}D^2\psi(z),\end{aligned} \tag{2.34}$$

$$\left\{4D^2 - \frac{k^2(1+\chi_2)}{(1+\chi)}\psi(z)\right\} - \frac{2M_{p_1}}{M_{p_3}}D\theta(z) + \frac{2M'_{p_1}}{M'_{p_3}}D\phi(z) = 0, \tag{2.35}$$

$$\frac{D_a V_a}{\Pr}\sigma n(z) = -\frac{2Q}{\mathrm{Lb}}Dn(z) + \frac{1}{\mathrm{Lb}}(4D^2 - k^2)n(z) + \vartheta e^{Q(z+1)/2}Q\{4G(1+\alpha_0)D^2 w(z) - G(1-\alpha_0)k^2 w(z) - w(z)\}. \tag{2.36}$$

We employ Chebyshev pseudospectral method [4] to find the solution of the eigen value problem generated by the set of equations (2.31)–(2.36).

6 Results and discussion

The important flow governing parameters such as Taylor number, which characterizes the effect of rotation, Lewis number, which is defined as the ratio of thermal diffusivity and mass diffusivity and is used to characterize fluid flows, where there is simultaneous heat and mass transfer, and Rayleigh number, which is the parameter to characterize heat transfer by natural convection, are discussed here. Dimension of nanoparticles and thickness of the layer are considered as 10 nm and 1 mm, respectively, and [9, 26] are the sources of physical quantities.

In Figures 2.2–2.4, NSCs have been plotted for T_A, $\Delta\phi$, Rb, Lb, Q and G. To study the stability of the system, we check the behavior of critical Rayleigh number Ra_c by changing the values of these parameters. The system gets stabilized if on increasing the value of any one of these parameters, the NSC shifts upwards. If the NSC shifts downwards with increasing value of any one of these parameters, then system becomes unstable. In other words, if Ra_c value goes up with increasing the value of any of these parameters,

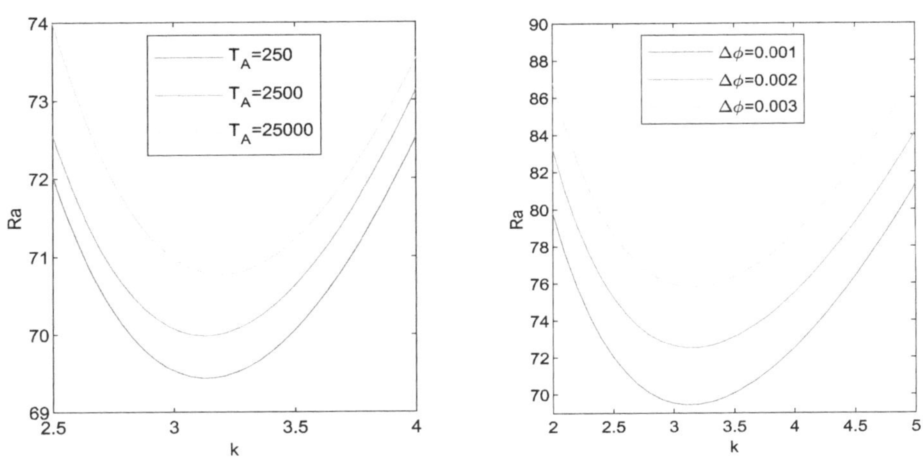

Figure 2.2: NSCs for the different values of T_A and ϕ.

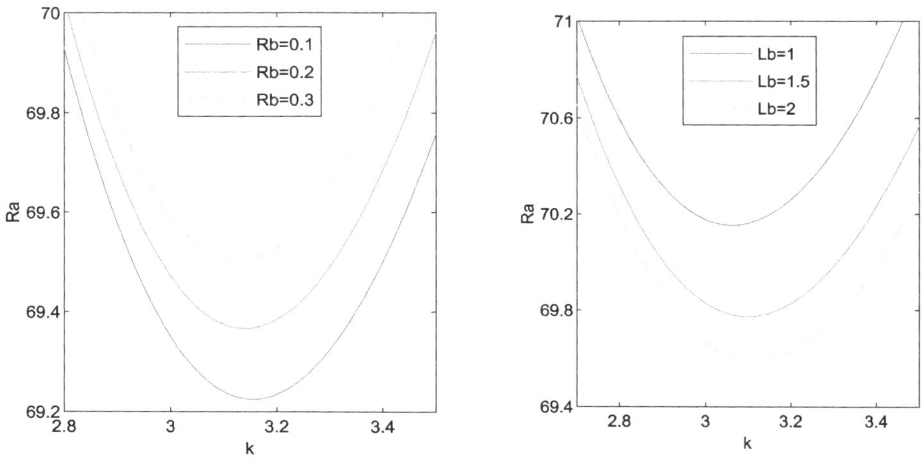

Figure 2.3: NSCs for the different values of Rb and Lb.

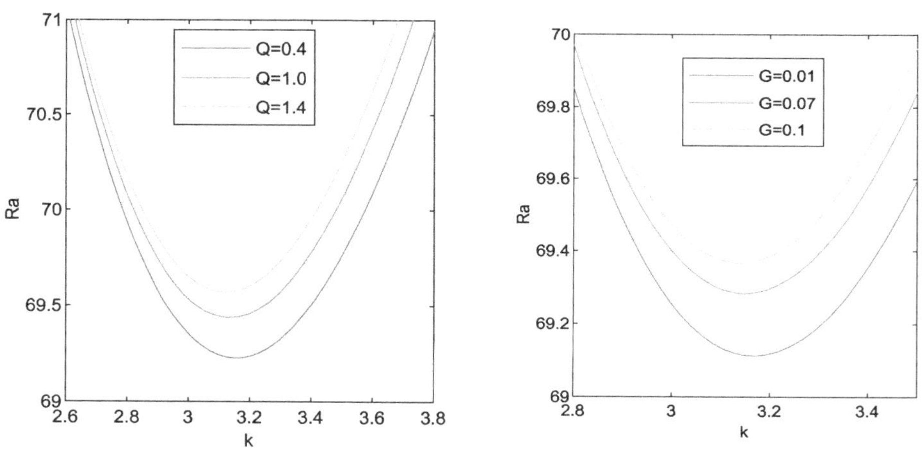

Figure 2.4: NSCs for the different values of Q and G.

then the convection gets delayed. On the other hand, if Ra_c decreases with increasing the value of any of these parameters, then that parameter destabilizes the system. In view of this argument, we see from Figures 2.2–2.4, T_A, $\Delta\phi$, Rb, Q and G delay the onset of convection, while Lb hastens the convection process.

As in Figure 2.2, value of Ra_c increases with $\Delta\phi$, therefore system gets stabilized. Temperature gradient produces top-heavy, unstable configuration. With an increase in particle concentration near the bottom of MNF layer, density distribution (of particles) reduces with height. This stabilizes the system and leads to high value of Ra_c. Hence, increase in $\Delta\phi$ stabilizes the system. Effect of rotation is also shown in Table 2.1 for three

Table 2.1: Effect of rotation (T_A) for three different values of Rb.

Rb	T_A	k_c	Ra_c
0.1	400	3.15	69.504
	600	3.14	69.663
	800	3.13	69.746
	1000	3.13	69.799
	2000	3.13	69.925
	3000	3.13	69.99
	4000	3.13	70.039
	5000	3.14	70.082
	10000	3.15	70.268
	100000	3.38	72.933
	1000000	4.39	86.854
0.2	400	3.14	69.592
	600	3.13	69.721
	800	3.13	69.79
	1000	3.14	69.835
	2000	3.13	69.943
	3000	3.13	70.002
	4000	3.13	70.048
	5000	3.13	70.089
	10000	3.15	70.272
	100000	3.38	72.933
	1000000	4.40	86.855
0.3	400	3.12	69.678
	600	3.13	69.78
	800	3.13	69.834
	1000	3.12	69.869
	2000	3.13	69.961
	3000	3.13	70.014
	4000	3.13	70.057
	5000	3.14	70.097
	10000	3.16	70.276
	100000	3.38	72.934
	1000000	4.40	86.855

different values of Rb. Stabilizing effect of T_A is found on the system. Ra_c increases more rapidly when values of T_A are larger, that is system is becoming more stable with higher rotation rates. There is no qualitative change in the propensity of T_A on changing the value of Rb. However negligible quantitative change in the value of Ra_c could be observed for the particular values of T_A for three distinct values of Rb.

7 Conclusion

The study of the onset of convection in a thin horizontal layer of a W_{mnf} heated from below, saturating a porous media and containing gyrotactic microorganisms, under the effect of rotation using linear stability analysis has been performed. Analysis has been done including the effects of three important slip mechanisms viz. Brownian motion, thermophoresis and magnetophoresis. The effect of magnetic nanoparticles, gyrotactic microorganisms and rotation in the gravity environment have been studied through neutral stability curves.

The following conclusions are drawn:
(i) The parameters T_A, $\Delta\phi$, Rb, Q and G stabilize the system.
(ii) Lb destabilizes the system.

Bibliography

[1] M. Arora, R. Singh and M. K. Panda, *J. Eng. Math.* (2016). https://doi.org/10.1007/s10665-016-9855-9.
[2] M. A. Bees and N. A. Hill, *J. Exp. Biol.* (1997). https://doi.org/10.1242/jeb.200.10.1515.
[3] J. Buongiorno, *J. Heat Transf.* (2006). https://doi.org/10.1115/1.2150834.
[4] C. Canuto, M. Y. Hussaini, A. Quateroni and T. A. Zang, *Spectral Methods in Fluid Dynamics*, Springer, Berlin, Heidelberg, 1998. https://doi.org/10.1007/978-3-642-84108-8.
[5] S. Ghorai and N. A. Hill, *J. Fluid Mech.* (1999). https://doi.org/10.1017/S0022112099006473.
[6] S. Ghorai and N. A. Hill, *Phys. Fluids* (2000). https://doi.org/10.1063/1.870249.
[7] A. Harashima, M. Watanabe and I. Fujishiro, *Phys. Fluids* (1988). https://doi.org/10.1063/1.866812.
[8] N. A. Hill and T. J. Pedley, *Fluid Dyn. Res.* (2005). https://doi.org/10.1016/j.fluiddyn.2005.03.002.
[9] P. N. Kaloni and J. X. Lou, *Phys. Rev. E* (2004). https://doi.org/10.1103/PhysRevE.70.026313.
[10] K. Kaur, M. Danesh and M. Arora, *J. Phys. Conf. Ser.* **1849** (2021), Article ID: 012008. https://doi.org/10.1088/1742-6596/1849/1/012008.
[11] J. O. Kessler, *Contemp. Phys.* (1985). https://doi.org/10.1080/00107518508210745.
[12] J. O. Kessler, *Prog. Phycol. Res.* (1986), https://scholar.google.co.uk/citations?view_op=view_citation&hl=en&user=5dqWEZoAAAAJ&citation_for_view=5dqWEZoAAAAJ:W7OEmFMy1HYC.
[13] J. O. Kessler, in: *Proceedings of the Workshop "Traffic and Granular Flow '97"*, M. Schreckenburg and D. E. Wolf, eds, Springer, New York, 1997, https://cir.nii.ac.jp/crid/1574231874898972288.
[14] J. O. Kessler, G. D. Burnett and K. E. Remick, in: *Nonlinear Science at the Dawn of the 21st Century*, P. L. Christiansen, M. P. Sorensen and A. C. Scott, eds, Springer, New York, 2000. https://doi.org/10.1007/3-540-46629-0_21.
[15] A. V. Kuznetsov, *Int. Commun. Heat Mass Transf.* (2010). https://doi.org/10.1016/j.icheatmasstransfer.2010.08.015.
[16] A. V. Kuznetsov, *Eur. J. Mech. B, Fluids* (2011). https://doi.org/10.1016/j.euromechflu.2010.10.007.
[17] A. V. Kuznetsov, *J. Porous Media* (2012). https://doi.org/10.1615/JPorMedia.v15.i1.20.
[18] A. V. Kuznetsov, *J. Porous Media* (2012). https://doi.org/10.1615/JPorMedia.v15.i3.30.
[19] A. V. Kuznetsov and A. A. Avramenko, *Transp. Porous Media* (2003). https://doi.org/10.1023/A:1023582001592.
[20] A. V. Kuznetsov and A. A. Avramenko, *Int. Commun. Heat Mass Transf.* (2003). https://doi.org/10.1016/S0735-1933(03)00097-6.
[21] A. V. Kuznetsov and A. A. Avramenko, *Int. Commun. Heat Mass Transf.* (2004). https://doi.org/10.1016/S0735-1933(03)00196-9.

[22] A. V. Kuznetsov and P. Geng, *Int. J. Numer. Methods Heat Fluid Flow* (2005). https://doi.org/10.1108/09615530510590597.

[23] A. Mahajan, M. K. Sharma and Sunil, *J. Nanofluids* (2014). https://doi.org/10.1166/jon.2014.1120.

[24] D. A. Nield and A. V. Kuznetsov, *Eur. J. Mech. B, Fluids* (2010). https://doi.org/10.1016/j.euromechflu.2010.02.003.

[25] T. J. Pedley and J. O. Kessler, *Annu. Rev. Fluid Mech.* (1992). https://doi.org/10.1146/annurev.fl.24.010192.001525.

[26] R. E. Rosensweig, *Ferrohydrodynamics*, Courier Dover Publications, 1997, https://books.google.com/books/about/Ferrohydrodynamics.html?id=uSa5nJGXYicC.

[27] Sanjalee and R. Sharma, *J. Nanofluids* (2022). https://doi.org/10.1166/jon.2022.1894.

R. Krishnaprabha

Some applications of graph spectrum in post quantum cryptography

Abstract: Having access to an elliptic curve with a specified number of points over a specific finite field is a prerequisite for many contemporary cryptography systems. The isogenies, which are surjective morphisms, play a significant role as specific mappings between these curves. Isogeny volcanoes are graphs with elliptic curves as their vertices and l-isogenies as their edges. Two identical elliptic curves over F_q share the same trace t and cardinality. David Kohel examines the relationships between isogenies of degree l and curves in $\mathrm{Ell}_t(F_q)$, the collection of curves defined over F_q with trace t, in his thesis. More precisely, he describes the structure of the graph of l-isogenies defined on $\mathrm{Ell}_t(F_q)$. There is a relation between isogeny graphs and orders in O_K and uses modular polynomials to find the conductor of $\mathrm{End}(E)$. The connected components of this graph are isogeny volcanoes and it is possible to travel through these structures using modular polynomials, even without knowing the cardinality of the curve. By using this information computation of the l-adic valuation of the trace t, for $l|g$ possible and hence obtain some information on the cardinality of the curve. Isogeny graphs of supersingular elliptic curves plays a major role in cryptography. Isogeny based cryptography, studies cryptosystems whose security is based on the difficulty of finding a path in isogeny graphs of supersingular elliptic curves. This study investigates the use of graph adjacency matrices in maintaining the security of isogeny based key exchange protocols.

Keywords: Elliptic curve, isogeny graph, expander graph, supersingular isogeny graph, adjacency matrix, spectrum, SIDH

1 Introduction

In the 1980s, cryptographers believed that if computers could benefit from quantum physics' features, they would be able to carry out complex computations considerably quicker than traditional binary computers.

Shor's method can operate on superposition and entanglement, and it is effective at finishing certain difficult computations in an hour; nevertheless, a computer would need several years to do the same calculation. Pre-quantum cryptography converts read-

Acknowledgement: I acknowledge Dr. Praveen I, Assistant Professor (Sr. Grade), Amrita Vishwa Vidyapeetham Ettimadai Coimbatore India, for his valuable suggestions.

R. Krishnaprabha, Amrita Vishwa Vidyapeetham, Mathematics, 534 Ettimadai, 641112 Coimbatore, Tamilnadu, India, e-mail: krishnaprabha.kpr@gmail.com

https://doi.org/10.1515/9783111313634-003

able input into secret code using a particular kind of encryption known as an algorithm. Pre-quantum cryptography has a problem in making encryption cyphers simple to comprehend yet challenging to decipher. Cryptologists all across the globe started to consider what a post-quantum cryptography system might like when mathematician Peter Shor demonstrated in the 1990s that a theoretical quantum computer could quickly crack the method used for public key encryption.

Information in quantum bits is processed by quantum computers using the principles of quantum physics (qubits). A quantum computer may process variables exponentially quicker than a conventional, binary computer, since each qubit can be a mixture of 0s and 1s. The majority of cryptography techniques are based on mathematical issues that, while challenging for classical computers, are easily resolved by quantum algorithms. We must create cryptographic primitives for common activities like encryption and authentication that withstand quantum assaults in order to be ready for the development of quantum computers.

The security of the discrete logarithm issue on elliptic curves, which is used in elliptic curve cryptography, is predicated on the difficulty of that problem. Nonetheless, this hardness assumption is false in the case of quantum computers. As a result, standard elliptic curve cryptography cannot be used as a base to create cryptosystems that are immune to quantum attacks. As a result, alternative elliptic curve cryptosystems based on hardness assumptions other than discrete logarithms have been proposed for use in settings where quantum resistance is desired. One early proposal by Stolbunov, based on isogenies between ordinary elliptic curves, was subsequently shown by Childs to offer only subexponential difficulty against quantum computers. Following these developments, De Feo proposed a new collection of quantum-resistant public-key cryptographic protocols for entity authentication, key exchange, and public-key cryptography, based on the difficulty of computing isogenies between supersingular elliptic curves.

De Feo, Jao, and Plut suggested cryptographic primitives [5] for building quantum-resistant public-key encryption, key exchange, and entity authentication protocols that make use of isogenies across supersingular elliptic curves. While geometric cyphers are used in quantum cryptography to convert readable data into uncrackable secret code, the technique depends on the physical characteristics of atoms. As quantum physics is still a very new field of study and prototype quantum computers are exceedingly expensive to construct and run, this presents a significant hurdle for post-quantum cryptography.

Shor's method was successfully implemented in 2016 by MIT and University of Innsbruck researchers, who also developed a tiny quantum computer that was capable of discovering the factors of the number 15. The world's most popular encryption techniques might be cracked by a quantum computer as soon as researchers showed that Shor's quantum algorithm could be utilised to return the right factors with a confidence level that surpassed 99 %. In the same year, the National Institute of Standards and Technology (NIST) started accepting proposals for algorithms that would take the role of key encapsulation mechanisms (KEMs), digital signatures, and public key encryption.

The Elliptic Curve Digital Signature Algorithm (ECDSA), Elliptic Curve Diffie–Hellman Key Exchange (ECDH), and Digital Signature Algorithm (DSA) cryptosystems, as well as the Rivest–Shamir–Adleman (RSA) algorithm were developed as alternatives to integer factorization and the discrete logarithmic problems they used. The post-quantum algorithm used in Google's post-quantum cryptography studies is coupled with a traditional elliptic curve method. The theory is that the adoption of an elliptic curve method still offers some level of security even if quantum cryptography turns out to be breakable.

Two hash functions by Charles–Goren–Lauter [2] were proposed in 2006 and were predicated on how difficult it is to locate pathways in Ramanujan graphs. One is based on Supersingular Isogeny Graphs, and the other is based on Lubotzky–Phillips–Sarnak (LPS) graphs. Based on LPS graphs, Petit–Lauter–Quisquater discovered strategies for defeating the hash function in 2008. The security of some post-quantum cryptosystems still depends on the difficulty of discovering isogenies between supersingular elliptic curves in the situation of supersingular isogeny graphs. A cryptographic system based on the Supersingular Isogeny Diffie–Hellman problem was presented by De Feo–Jao–Plut in 2011 [5]. The security of the SIDH proposal depends on the difficulty of the Supersingular Isogeny Graphs path-finding issue. Most cryptographic methods, particularly Ramanujan and Caley, utilise pathways in unique graphs. Graphs having a huge girth are used in post-quantum cryptography.

There are certain encryption techniques where the key is a graph. The process of encryption involves drawing a route on that graph. The plain text is composed of a series of vertices along the key graph's route. The encrypted text is made up of a series of edges connecting those vertices. Cryptography uses the second-largest eigen value more frequently.

By demonstrating a reduction between the Supersingular Isogeny Diffie Hellmann (SIDH) problem and the path-finding issue in SSIG, we demonstrate in Part 1 of the study how the security of the Key Exchange protocol depends on the difficulty of the problem.

In the second section of the study, we investigate the number-theoretic properties of the LPS and Pizer, Caley, and Ramanujan graphs.

Lattice schemes appear to be the most promising, possibly because it is very challenging to calculate the shortest vector of a large lattice when the shortest vector is quantum and can exist in more than one dimension. Quantum-resistant algorithms can be built using lattice, code-based, and multivariate schemes. Within ten years, according to cryptographers, quantum supremacy will be realized, rendering RSA and other asymmetrical algorithms useless for protecting sensitive data.

NIST is working harder to develop a post-quantum encryption standard. Quantum key distribution and post-quantum cryptography are two distinct concepts. A secret cryptographic key must be exchanged between two distant parties using QKD in a way that makes key interception obvious. Companies may benefit from photons' no-change and no-cloning properties by using quantum key distribution to send data instead of bits. As a result, a secret key sent between two parties using this method cannot be surreptitiously duplicated or intercepted. With this approach, the photon carrying the key changes state and automatically fails, warning the two parties that their connection is not secure. This happens if a third party listening in on their interactions tries to learn about the key being established.

Using QKD, businesses and government organisations may exchange sensitive, vital data in an unbreakable, totally secure manner. The security of communication is guaranteed by quantum events in quantum cryptography. We are working to create cryptographic algorithms, particularly public-key algorithms, in post-quantum cryptography that are resistant to attack by a quantum computer. Post-quantum cryptography focuses on modernizing current mathematically based standards and algorithms in order to get ready for the age of quantum computing.

2 Preliminaries

In a graph $G = (V, E)$, two points $x_i, x_j \in V$ are adjacent or neighbors if $\{x_i, x_j\} \in E$. If all the vertices of G are pairwise adjacent, then we say G is complete [7]. A complete graph with n vertices is denoted as K^n. The degree $d(v)$ of a vertex v is the number of vertices in G that are adjacent to v.

Adjacency matrix associated with a graph G is a square matrix A_G.

The entries $a_{i,j}$ are given by

$$a_{i,j} = \begin{cases} 1 & \text{if } \{x_i, x_j\} \in E \\ 0 & \text{otherwise.} \end{cases}$$

The adjacency matrix of a graph is to calculate the number of walks of different length connecting two vertices in the graph. A vertex $v \in V$ is incident with an edge $\{x_i, x_j\} \in E$ if $v = v_i$ or $v = v_j$. A walk on a graph is an alternating series of vertices and edges, beginning and ending with a vertex, in which each edge is incident with the vertex immediately preceding it and the vertex immediately following it. A walk between two vertices u and v is called a u-v walk. The length of a walk is the number of edges it has. In a walk, we count repeated edges as many times as they appear. The eigen value of a graph is the eigen value of the corresponding adjacency matrix. Since the diagonal of the adjacency matrix is all zeros, the sum of all eigenvalues of a graph is always 0. An expander graph [6] is a graph in which every subset of the vertices has many neighbors. An undirected graph $G = (V, E)$ is c-expanding if for every $S \subseteq V$, $|S| \leq \frac{|V|}{(2c)}$, and $|\Gamma(S)| \geq (c-1)|S|$, where $\Gamma(S)$ is the set of neighbors of S (not including S). The most effective graphs for preserving communication secrecy are the Ramanujan and Caley graphs. Since it is difficult to locate pathways in these networks, these graphs are crucial in cryptography. For supersingular elliptic curves, the path finding issue and the endomorphism ring calculation problem are identical. The mixing qualities of Ramanujan graphs make them ideal expander graphs. The goal is to build a collision-resistant, efficiently calculable hash function. The mixing features of these graphs are very helpful for building verifiable collision-resistant hash functions from expander graphs. According to the quick mixing property, the output of hash functions closely resembles the uniform distribution and, as a result, is essentially identical to random bits. Expander graphs' ability to swiftly approximate the uniform distribution using random walks is their most valuable feature for cryptography. Several methods for calculating collision-resistant hash functions employ the input as walking directions around a graph, and the output of the hash function is the last vertex. [2] Any expander graph may use these constructions, however, locating cycles in such a graph is a challenging challenge. We can maintain a short path length in a Ramanujan graph. The difficulty of reversing a walk on a graph is largely used in the construction of cryptographic hash functions.

Path-finding problem

For a fixed starting and ending vertices representing the start and end points of a walk on the graph of a fixed length, find a path between them. A hash function can be defined by using the input to the function as directions for walking around the graph: the output is the label for the ending vertex of the walk. Finding collisions for the hash function is equivalent to finding cycles in the graph, and finding pre-images is equivalent to path-finding in the graph. Backtracking is not allowed in these walks. Charles–Goren–Lauter in 2006 proposed two hash functions based on the hardness of finding paths in Lubotzky–Phillips–Sarnak (LPS) graphs and the next one is based on Supersingular Isogeny Graphs (Pizer) [2]. In 2008, Petit–Lauter–Quisquater breaks the hash function based on LPS graphs. As a result, Supersingular Isogeny graphs have lately attracted the attention of scholars. We transmit the images of the elements in the torsion basis through the isogeny in quantum-resistant public-key cryptosystems based on the difficulty of finding isogenies between supersingular elliptic curves in order to construct a shared commutative square, even though the endomorphism ring is not commutative in this situation. Charles, Goren, and Lauter introduced supersingular isogeny graphs to cryptography in 2006 [2]. They are optimal expander graphs, that is Ramanujan graphs, in these graphs walks of length approximately equal to the logarithm of the graph size. Isogeny graphs for supersingular elliptic curves were first considered by Mestre. Pizer showed that these graphs have the Ramanujan property. The path-finding problem in supersingular isogeny graphs is the basis for the CGL cryptographic hash function.

Elliptic curve

Let F_q be a finite field, where $q = p^a$, p a large prime [11]. Short Weierstrass form of an elliptic curve E over F_q is the set of points $E(F_q) = \{(x,y), x, y \in F_q : y^2 = x^3+ax+b\} \cup \{0_E\}$,

$a, b \in F_q$, $0_E = 0 = (0 : 1 : 0)$ the point at infinity. The j-invariant of an elliptic curve given by the Weierstrass equation $y^2 = x^3 + ax + b$ is given by the formula

$$j(E) = 1728 \frac{4a^3}{4a^3 + 27b^2}.$$

For the elliptic curve E_0: $y^2 = x^3 + x$ the j-invariant is $j(E_0) = 1728$. Isomorphic curves have the same j-invariant; over an algebraically closed field, two curves with the same j-invariant are isomorphic. Let $E : y^2 = x^3 + ax + b$ be an elliptic curve. Let $P_1 = (x_1, y_1)$ and $P_2 = (x_2, y_2)$ be two points on E different from the point at infinity, then we define the binary operation $\oplus$ on E as $P \oplus 0 = 0 \oplus P = P$ for any point $P \in E$. If $x_1 = x_2$ and $y_1 = -y_2$, then $P_1 \oplus P_2 = 0$; Otherwise set

$$\lambda = \begin{cases} \frac{y_2 - y_1}{x_2 - x_1}, & \text{if } P \neq Q \\ \frac{3x_1^2 + a}{2y_1}, & \text{if } P = Q \end{cases}$$

the point $(P_1 \oplus P_2) = (x_3, y_3)$ is defined by

$$\begin{aligned} x_3 &= \lambda^2 - x_1 - x_2, \\ y_3 &= -\lambda x_3 - y_1 + \lambda x_1. \end{aligned}$$

Under the binary operation $\oplus$, the set of points on the elliptic curve E forms an abelian group. Over finite fields, they form finite Abelian groups. Let E/F_q be an elliptic curve, and suppose that $\#E(Fq) = n$ is relatively prime to q. Then $E(F_q)$ is either cyclic or isomorphic to the direct product of two cyclic groups. If F is a number field, then $E(F)$ is finitely generated. The m torsion group $E[m]$ is the set of all points $P \in E(\overline{F})$ such that $mP = 0$, where $\overline{F}$ is the algebraic closure of F. If F is a field whose characteristic is p, then E is super singular if $E[p] \cong O$ and it is ordinary if $E[p] \cong \frac{Z}{pZ}$. An elliptic curve E/F_p is supersingular if $\#E(F_p) = p + 1$, otherwise it is ordinary.

Theorem. *Let E/F_q be an elliptic curve, with $q = p^r$ for some prime p. Then $E(F_q)$ is cyclic, if one of the following conditions hold:*

1. *$\#E(F_q) = p^k$ for some integer k.*
2. *E/F_q is supersingular and $q \geq 5$ is prime with $q \equiv 1 \bmod 4$.*

An *isogeny* between two elliptic curves E and E' is a surjective morphism [3]. Isogeny may not be injective. If $E = E'$, we call an isogeny an endomorphism. The collection of all endomorphisms on E along with the zero map is called the endomorphism ring, denoted by $\mathrm{End}(E)$. An example for an endomorphism on the elliptic curve E is the multiplication-by-m map, i. e., $[m] : E \to E$ defined by $P \to [m]P$, $P \in E(\overline{F})$.

Definition. For elliptic curves E_1 and E_2 over K, an isogeny is a homomorphism from $E_1(\overline{K})$ to $E_2(\overline{K})$ that is given by rational functions: $\phi : E_1(\overline{K}) \to E_2(\overline{K})$ such that $\phi : (x,y) \to (\frac{f_1(x,y)}{g_1(x,y)}, \frac{f_2(x,y)}{g_2(x,y)})$, where f_1, f_2, g_1, g_2 are polynomials in variables x, y with coefficients in K.

Degree of an isogeny

Let $\phi : E_1 \to E_2$ be an isogeny, then the degree of ϕ is the maximum of the degrees of the polynomials $f_1(x,y)$ and $f_2(x,y)$. If the degree of an isogeny is l, then it is known as an l-isogeny. Dual isogeny of the l isogeny $\phi : E_1 \to E_2$ is an l isogeny $\hat{\phi} : E_2 \to E_1$ such that $\phi \circ \hat{\phi} = \hat{\phi} \circ \phi = [l]$, where $[l]$ is the multiplication by l map. Elliptic curves related by an isogeny of degree n are said to be n-isogenous. Two elements j_1, j_2 of a field k are n-isogenous over k if there are n-isogenous elliptic curves E_1; E_2 over k with $j(E_1) = j_1$ and $j(E_2) = j_2$. "A *sparse graph* is a graph in which there are less edges overall than there are at its maximum. A sparsely populated, highly linked graph is known as an Expander graph." Imagine a simple graph G with 7 vertices and 2 edges originating from each vertex. There are 14 edges in this graph. If this graph is complete every vertex connected to every other vertex, so 7! edges. This is an example of a sparse graph, since $7! > 14$. A supersingular isogeny network is a graph whose edges are special mappings (isogenies) between such (classes of) curves and whose nodes are supersingular curves (isomorphism classes of) curves. Isogeny graphs are undirected, since each isogeny has a dual isogeny. It is hard to find a path of a given length between two random nodes in the isogeny graph. This hardness is the basis for isogeny-based cryptosystems [4].

3 Supersingular isogeny graph

To define a supersingular isogeny graph, fix a finite field K of characteristic p, a supersingular elliptic curve E over K, and a prime $l \neq p$. Then the corresponding isogeny graph is constructed in such a way that, the vertices are the $\overline{K}$-isomorphism classes of elliptic curves that are $\overline{K}$-isogenous to E. Each vertex is labeled with the j-invariant of the curve. The edges of the graph correspond to the l-isogenies between the elliptic curves [12]. As the vertices are isomorphism classes of elliptic curves, isogenies that differ by composition with an automorphism of the image are identified as edges of the graph. Consider F_{p^2} and supersingular elliptic curves. In the supersingular isogeny graph, vertices are all isogenous elliptic curves over F_{p^2}. Edges are isogenies of a fixed prime degree l. Suppose $l = 2$. Every finite subgroup of $E_1(\overline{k})$ is the kernel of a separable isogeny over k that is uniquely determined (up to isomorphism) and this isogeny can be computed using Vélu's algorithm. Also, every n-isogeny $\phi : E_1 \longrightarrow E_2$ has a unique dual isogeny $\hat{\phi} : E_2 \longrightarrow E_1$ that satisfies $\phi \circ \hat{\phi} = \hat{\phi} \circ \phi = [n]$ where $[n]$ is the multiplication by n map that sends $P \in E_1$ to $nP \in E_1$. The kernel of the multiplication by n map is the n-torsion subgroup. If E_0, E_1 are $\overline{K}$-isogenous elliptic curves, $\phi : E_0 \longrightarrow E_1$ is an l-isogeny and

$\Phi \in \mathrm{Aut}(E_1)$ is an automorphism, then ϕ and $\Phi \circ \phi$ are identified and correspond to the same edge of the graph. If $p \equiv 1 \bmod 12$, we can uniquely identify an isogeny with its dual to make it an undirected graph. It is a multigraph in the sense that there can be multiple edges if no extra conditions are imposed on p. Let l be a positive integer. Every separable l-isogeny $\phi : E_0 \longrightarrow E_1$ is found out by using its kernel up to composition with an automorphism of the elliptic curve E_1. Hence the edge corresponding to this ϕ in the isogeny graph is uniquely determined by $\ker(\phi)$ and vice versa. The kernel is a subgroup of the l-torsion $E_0[l]$, and the latter is isomorphic to $Z/lZ \times Z/lZ$ if l is coprime to the characteristic of the field. Choose a prime l and a finite field F_q such that the characteristic not equal to l, the number of l-isogenies ϕ: $E_0 \longrightarrow E_1$ that correspond to different edges of the graph is equal to the number of subgroups of $Z/lZ \times Z/lZ$ of order l, and is equal to $l+1$. E is l-isogenous to $l+1$ elliptic curves. Some of these l-isogenous curves may be isomorphic. In the isogeny graph since the vertices are represented by isomorphism classes of elliptic curves, E has degree $l+1$ and may have $l+1$ neighbors or fewer. The edges between vertices are calculated by using Velu's formula from the kernel of the isogeny. Let P, Q are generators of $E_0[l]$, then the subgroups of order l are generated by Q and $P+iQ$ for $i = 0, 1, 2 \ldots (l-1)$. Ramanujan graphs are optimal expanders, where Expander graphs are such graphs where small sets of vertices have many neighbors [10]. For constructing Ramanujan graph, choose two distinct primes p and l for an $(l+1)$-regular graph its size depends on the prime p. Let B be a quaternion algebra defined over $\mathbb{Q}$ and ramified at p and at ∞, also let $\mathbb{O}$ be an order of the quaternion algebra. Let $\mathbb{A}$ denote the adeles of $\mathbb{Q}$ and $\mathbb{A}_f$ denote the finite adeles. For a finite prime p, let $\mathbb{Q}_p$ denote the field of p-adic numbers and $\mathbb{Z}_p$ its ring of integers. Let $Q_\infty = R$. The adele ring of $\mathbb{Q}$ is $\mathbb{A} = \prod'_p \mathbb{Q}_p = \{(a_p) \in \prod_p \mathbb{Q}_p : a_p \in \mathbb{Z}_p$ for all but a finite number of $p < \infty\}$. The finite adeles of $\mathbb{Q}$, $\mathbb{A}_f = \prod'_{p<\infty} \mathbb{Q}_p = \{(a_p) \in \prod_{p<\infty} \mathbb{Q}_p : a_p \in \mathbb{Z}_p$ for all, but a finite number of $p\}$. If the expander graph G is an $(l+1)$-regular graph, then the adjacency matrix of G has $l+1$ as an eigen value. All eigenvalues λ satisfy $-(l+1) \leq \lambda \leq (l+1)$, G is bipartite if and only if $-(l+1)$ is an eigen value of the adjacency matrix of G. The absolute value of the second largest eigenvalue has important properties. The smaller the value, the more expander G is. An $(l+1)$-regular graph G is Ramanujan if the second largest eigenvalue is less than or equal to $2\sqrt{l}$. Three important properties of these graphs are

1. The graph is connected for any $l \neq p$.
2. A supersingular isogeny graph has roughly $\frac{p}{12}$ vertices.
3. Supersingular isogeny graphs are optimal expander graphs, in particular they are Ramanujan [9, 10].

Because it is challenging to locate pathways, expander graphs, particularly Ramanujan graphs, are tremendously helpful in cryptography. No known subexponential algorithms, whether on a conventional or quantum computer, have been found to tackle this issue. As a result, post-quantum cryptography systems based on the difficulty of challenges on supersingular isogeny graphs are being considered. In SIDH, Alice (resp. Bob) executes a random walk without backtracking in l_A (resp. l_B) isogeny graph according to her (resp. his) secret key, where l_A and l_B are distinct small prime numbers. These random walks represent isogenies $E \longrightarrow E_A$ of degree e_A and $E \longrightarrow E_B$ of degree e_B respectively. The security of SIDH relies on the hardness of the path-finding problem in this graph. Therefore, studying the structure of isogeny graphs in SIDH is important. Here, two main problems will arise:

1. Is there a shorter path to E_A than the secret path? That is, is there exist an isogeny $\phi : E \longrightarrow E_A$ whose degree is a power of l_A and less than $l_A^{e_A}$.
2. Is there another path to E_A of the same degree as the secret path?

A short path is easier to find than the path corresponding to a secret key. Suppose that there is a shorter path, and that we could find a shorter path $\phi : E \longrightarrow E_A$. Let $\phi_A : E \longrightarrow E_A$ be the isogeny corresponding to Alice's secret key, and let $\phi : E \longrightarrow E_A$ a shorter path, and l_A^f the degree of ϕ. Then $\phi_A \circ \hat{\phi}$ is an endomorphism on E_A of degree l^{e_A+f} [6]. The security of SIDH is now reduced to find this endomorphism [8]. Since we know the structure of End(E), we can find endomorphism ring End(E_A) of E_A. Hence, $\phi_A \circ \hat{\phi}$ can be calculated by solving Diophantine equation. This weakens the security of SIDH. In the case of supersingular elliptic curve modulo p with isogeny edges, the vertices of the graph are supersingular elliptic curve defined over F_P or F_p2, and the vertices are labeled with the j-invariants of the curve that is by $j(E) = 1728\frac{4a^3}{4a^3+27b^2}$ in the case of the curve $y^2 = x^3 + ax + b$. The edges are degree l isogenies between elliptic curves. These graphs are $l + 1$ regular, Ramanujan and undirected if $p \equiv 1 \bmod 12$. The degree of a separable isogeny is the size of its kernel [13]. Velu in his paper suggested one Formula for finding isogeny between elliptic curves. Supersingular isogeny Diffie–Hellman, is a key-exchange protocol based on isogenies of supersingular elliptic curves. It is modeled as a random walk in the graphs $G_{l_A}(p)$ and $G_{l_B}(p)$ for distinct small primes l_A, l_B. SIDH protocol is for a prime p of the form $p = l_A^{e_A} l_B^{e_B} - 1$, for some small primes l_A, l_B and positive integers e_A, e_B such that $l_A^{e_A} \approx l_B^{e_B} \approx 2^{2\lambda}$ (λ: a security parameter). Let E be a supersingular curve and $\{P_A, Q_A\}$ (resp. $\{P_B, Q_B\}$) a generator of $E[l_A^{e_A}]$ (resp. $E[l_B^{e_B}]$). On the public parameters p, e_A, e_B, l_A, l_B and (E, P_A, Q_A, P_B, Q_B), Alice and Bob execute key exchange in the following procedure (we describe Alice's side only since Bob's execution is similar without index):

(1) Alice chooses $n_A, m_A \in \mathbb{Z}/l_A^{e_A}\mathbb{Z}$ as a secret key
(2) Alice computes the l_A^e isogeny $\phi_A : E \longrightarrow E_A$ such that its kernel is the subgroup of $E[l_A^{e_A}]$ generated by $n_A P_A + m_A Q_A$. Alice publishes $(E_A, \phi_A(P_B), \phi_A(Q_B))$.
(3) By using Bob's public key, Alice computes the isogeny $\phi_{BA} : E_B \longrightarrow E_{BA}$ such that its kernel is the subgroup of $E_A[l_B]$ generated by $n_A \phi_B(P_A), m_A \phi_B(P_B)$
(4) Alice and Bob share the key, the j-invariant of $E_{BA} \approx E_{AB}$.

The secret information $n_A, m_A \in \mathbb{Z}/l_A^{e_A}\mathbb{Z}$ corresponds to the path of random walk from the starting curve E to the public information E_A The security of SIDH depends on the hardness of the path-finding problem in supersingular isogeny graphs. Each endomorphism ring has a basis. To generate endomorphism rings, we can use loops in the isogeny graph. Path finding problem in the supersingular isogeny graph reduces to a computation of the maximal order corresponding to the endomorphism ring, also it helps us to understand how the generators of that order acts on the torsion subgroups of the curve. Thus, if we have given a path in the supersingular isogeny graph, we can compute the corresponding isogeny. Also from a given isogeny we can compute the corresponding path in the graph.

Expander graph

A type of graph, known as an expander graph, possesses the seemingly incompatible qualities of being both sparse and well-connected. These graphs end up having a wide range of uses in computer science, including the creation of network topologies that are efficient and the creation of graph-based cryptographic hash functions [1]. An expander graph is a graph in which every subset S of vertices is connected to many vertices in the complementary set $\overline{S}$ of vertices. Expander graphs are sparse graphs with a variety of advantageous characteristics, including a small diameter, high connectedness, and a large chromatic number.

4 Graph spectrum

Let $G = (V, E)$ be an undirected graph on n vertices. Then the adjacency matrix A_G is an $n \times n$ matrix with elements defined by: $a(i,j) = \{(v_i, v_j) \in E\}$ Since G is undirected, edges in E are unordered pairs, which means $a(i,j) = a(j,i)$ for all $1 \leq i \leq n$ and for all $1 \leq j \leq n$. As a result, A_G is a symmetric matrix, which means that the eigenvalues of A_G are all real. That is, all values $\lambda_m(G)$ that satisfy the relation $A_G v_m = \lambda_m(G) v_m$, where $v_m \in R^v$ is called the eigenvector of A_G corresponding to the eigenvalue λ_m, are real. For a symmetric graph, $v_m \neq 0$, and the eigenvectors form an orthonormal basis for R^v. The set of eigenvalues of A_G is called the spectrum of the graph [13]. Let $G = (V, E)$ be a k-regular undirected multigraph, and let A_G be its corresponding adjacency matrix. Let $\lambda_1 \geq \lambda_2 \geq \cdots \geq \lambda_n$ be the real eigenvalues of A_G. Then 1. $\lambda_1 = k$ and $\lambda_n \geq -k$. 2. $\lambda_2 = k$ if and only if G is disconnected. 3. $\lambda_n = -k$ if and only if at least one of the connected components of G is bipartite the edge expansion ratio of a graph is related closely to the second largest eigenvalue, λ_2, of A_G. If $G = (V, E)$ is an undirected graph with eigenvalues $\lambda_1 \geq \lambda_2 \geq \cdots \geq \lambda_n$, then $\Delta(G) \equiv \lambda_1 - \lambda_2$ is the spectral gap of G. This gives the relationship between $h(G)$ and λ_2. Cheeger's inequality: The edge expansion ratio $h(G)$ for a k-regular graph is related to the spectral gap $\Delta(G)$ by: $\frac{\Delta(G)}{2} \leq h(G) \leq \sqrt{2k\Delta(G)}$. For a k-regular graph $\lambda_1 = k$, $\frac{k-\lambda_2}{2} \leq h(G) \leq \sqrt{2k(k-\lambda_2)}$. The Cheeger constant for graphs is actually analogous to a theory in Riemannian geometry concerning Riemannian manifolds. As such, Cheeger's inequality for graphs can be considered a discrete version of Cheeger's inequality for Riemannian manifolds. A k-regular finite graph G is called a Ramanujan graph, if $\lambda(G) \leq 2\sqrt{(k-1)}$, $\lambda(G) = \max_{|\lambda_i|<k} |\lambda_i|$ [7].

5 Expander hash

An Expander hash is a hash function where the input to the hash is used as directions for walking around an expander graph without backtracking, and the output of the hash function is the ending vertex of the walk. For a fixed hash function, the walk starts at a fixed vertex in the given expander graph, and a family of hash functions can be defined by allowing this starting vertex to vary. To execute a walk on a k-regular undirected expander graph, the input to the hash function must be broken into chunks of size c such that $2^c = k-1$ [2]. Starting at the first vertex, each step of the walk chooses an edge emanating from that vertex to follow to get to the next vertex, where the choice of the edge to follow is determined by the next c bits of the input. Since we do not allow backtracking, there are only $k-1$ choices for the next edge at each step. Directed k-regular graphs, on the other hand, cannot have backtracking in them unless the graph has multiple edges, so the input of the hash function must be broken into chunks of size c such that $2^c = k$, as there are k choices for the next edge at each step. Let l be a small prime and let p be a large prime such that $p \equiv 1 (\mathrm{mod}\, 12)$. The vertices of the Pizer graph $\prod_{l,p}$ are the members of the set V of all supersingular elliptic curves over the finite field F_{p^2} up to isomorphism. This [9] set has $\lfloor p/12 \rfloor + \varepsilon$ elements, where $\varepsilon \in \{0, 1, 2\}$ depending on the congruence class of p modulo 12. For the case where $p \equiv 1(\mathrm{mod}\, 12)$, $\varepsilon = 0$. The elements of V can be labeled by their j-invariants, where we write $E(j_i)$ for an elliptic curve with j-invariant j_i. There is an edge from j_1 to j_2 if and only if there is an l-isogeny from $E(j_1)$ to $E(j_2)$. The Pizer graph $\prod_{l,p}$ is a Ramanujan $(l+1)$-regular graph, and it is undirected without multiple edges if $p \equiv 1(\mathrm{mod}\, 12)$. Let E be an elliptic curve defined over a finite field F_q. An isogeny of E to E' is a surjective morphism between them. An isogeny [8] of E to itself is called an endomorphism of E. The set of endomorphisms of E defined over F_q together with the zero map is called the endomorphism ring of E, and is denoted by End(E). If the endomorphism ring of E is non-commutative, E is called a supersingular elliptic curve. Otherwise we call E ordinary. Every supersingular elliptic curve over a field of characteristic p has a model that is defined over F_{p^2} because the j-invariant of such a curve is in F_{p^2}. Let E, E_0 be two supersingular elliptic curves defined over F_{p^2}. For each prime $l \neq p$, E and E_0 are connected by a chain of isogenies of degree l. Supersingular isogeny graph, whose nodes correspond to (isomorphism classes of) supersingular curves and whose edges are special maps (isogenies) between such (classes of) curves. Isogeny graphs are undirected ($p \equiv 1(\mathrm{mod}\, 12)$) since each isogeny has a dual isogeny. It is hard to find a path of a given length between two random nodes in the isogeny graph. This hardness is the basis for isogeny-based cryptosystems. "*The Deuring correspondence* [6] states that there is bijective correspondence, given by associating a supersingular j-invariant to a maximal order in $B_{p,\infty}$ isomorphic to $\mathrm{End}(E(j))$." We compute an isogeny from a supersingular elliptic curve, then utilise the target curve as public data and the isogeny as secret data to build public-key cryptosystems. Isogeny-based cryptography is the name for public-key encryption in this fashion. Let l be a prime number, then the supersingular l-isogeny graph [8] is such that the vertices are isomorphism classes of supersingular elliptic curves represented by their j-invariant, and two vertices are connected by an edge if and only if they are l-isogenous. It is a connected $(l + 1)$-regular Ramanujan expander graph. If $k = F_q$ is a finite field, then G_k, is a finite graph that is the disjoint union of ordinary connected components and supersingular connected components. The ordinary components were studied in Kohel's PhD thesis. Due to their regular structure, these components later became known as *isogeny volcanoes* [1]. In general, the supersingular components may bear no similarity at all to the volcanoes of the ordinary case. Following Pizer, one instead studies the unique supersingular component of G_k, where $k = F_q$, which turns out to be a finite $(q + 1)$-regular Ramanujan graph and forms the basis for the SIDH protocol. By using generalized Riemann hypothesis, we can prove that "The path-finding problem in l-isogeny graphs and the endomorphism ring problem for supersingular elliptic curves" are equivalent [6]. Isogeny graphs are optimal expander graphs. This implies that random walks quickly reach the uniform distribution. Starting from an elliptic curve E, one can compute a chain of random l-isogenies until the endpoint E' is uniformly distributed. Then, given only E and E', it seems hard to recover a path connecting them in supersingular l-isogeny graph. The l-isogeny path problem is [13], given a prime p, and two supersingular elliptic curves E and E' over F_{p^2}, find a path from E to E' in the l-isogeny graph. An endomorphism is an isogeny ϕ from E to itself. The set of all endomorphisms of E, $\mathrm{End}(E)$, is a ring, where the addition is pointwise and multiplication is given by composition. Loops in l-isogeny graphs provide endomorphisms, hence there is some connection between path-finding problems and computing endomorphism rings. Since the curves considered are supersingular, the endomorphism rings are always generated by four elements (as a lattices), and they are isomorphic to subrings of a quaternion algebra $B_{p,\infty}$, called maximal orders. An isomorphism is an isogeny $\phi : E_1 \longrightarrow E_2$ of degree 1. We say that E_1 and E_2 are isomorphic over k (an extension of k) if there is an isomorphism between them that is defined over k. The j-invariant of E is $j(E) = \frac{256.27.A^3}{4A^3+27B^2}$. $j(E_1) = j(E_2)$ if and only if E_1 and E_2 are isomorphic over the algebraic closure of k. The expander graphs are used to produce hash functions that are collision-resistant. The input to the hash function is used as directions for walking around a graph (without backtracking), and the output of the hash function is the ending vertex of the walk. For a fixed hash function, the walk starts at a fixed vertex in the given graph. A family of hash functions can be defined by allowing the starting vertex to vary. We execute a walk on a k-regular expander graph by converting the input for the hash function to a base-$(k - 1)$ number, whose digits then dictate which edge to take at each step. Starting at the first vertex, each step of the walk chooses an edge emanating from that vertex to follow to get to the next vertex. At each step of the walk, the choice of the edge to follow is determined by the next digit of the (converted) input. We do not allow backtracking in the walk, so only $k - 1$ choices for the next edge are allowed at each step. Let p and l be two distinct prime numbers. Define the graph $G(p, l)$ to have vertex set, V, the set of isomorphism classes of supersingular elliptic curves over the finite field F_{p^2}. We label vertices with their j-invariants, which can be computed directly from the curve equation and are a priori elements of F_{p^2}. The number of vertices of $G(p, l)$ is $\lfloor p/12 \rfloor + \varepsilon$, where $\varepsilon \in \{0, 1, 2\}$ depends on the congruence class of p modulo 12. We will impose $p \equiv 1 (\mathrm{mod}\, 12)$, in which case $\varepsilon = 0$. Since there are roughly $p/12$ distinct j-invariants, we will choose a linear congruential function to map j-invariants from F_{p^2} to F_p for the output of the hash function. Thus the output of the hash function will be just $\log_2(p)$ bits. We propose to use a graph of cryptographic size $p \approx 2^{256}$. Here are the steps to compute the output of the hash function when using supersingular elliptic curves and 2-isogenies (i. e., $l = 2$). Since there are 3 edges emanating from each vertex, and no backtracking is allowed in a walk, there are two choices of which edge to follow next from each vertex, and this can be determined by 1 bit as follows. Start at a vertex E_1. Subgroups of E_1 of order 2 are each given by a single two-torsion point on the elliptic curve $E_1 : y^2 = f(x)$. The 3 non-trivial 2-torsion points are $P_i = (x_i, 0)$, where the cubic $f(x)$ factors as $(x - x_1)(x - x_2)(x - x_3)$ over an extension field. As an example, when computing the isogeny ϕ, which corresponds to taking the quotient by $\langle p_1 \rangle$, both of the other 2-torsion points are mapped to the same 2-torsion point $\phi(P_2) = \phi(P_3)$ on the isogenous elliptic curve E_2. In turn, the isogeny which corresponds to taking the quotient of E_2 by the subgroup generated by $\phi(P_2)$ is the dual isogeny $\hat{\phi}$ and leads back to E_1. So to choose the next step from E_2, it suffices to choose between the two other 2-torsion subgroups different from $\langle \phi(p_2) \rangle$. An efficient way to determine the 2 new 2-torsion points on E_2 is to keep $\tilde{x}$ the x-coordinate of $\phi(P_2)$, and to factor $(x - \widetilde{x_2})$ out of the new cubic $f_2(x)$, leaving a quadratic to be factored. The roots of the quadratic can be ordered according to some convention, and one bit suffices to choose between them for the next step in the walk. So if the input bit length is n, then the hash function takes a walk of length n steps. The supersingular isogeny graph's path finding issue is analogous to the supersingular endomorphism ring calculation problem. Using walks on expander graphs, we can create hash functions. The security of the Elliptic curve discrete logarithm issue and the Expander graph's expansion features are strongly linked. There are several unresolved issues in this area, including the complexity of the path-finding issue in Supersingular Isogeny (Pizer) graphs and the difficulty of the group word problems in Cayley-type Ramanujan graphs may or may not be related. The technique employed for Cayley graphs can be used to examine the hardness of the path-finding issue in Pizer's graphs if it is able to connect the two problems theoretically or schematically. It is necessary to look at much more versatile renditions of explicit Ramanujan graph constructions as a part of these methodologies. The family of $(2p + 1)$-regular graphs, where p is an Eichler prime based on the quaternion algebra and an explicit construction of the Eichler order with class number 1, is currently being created. Now, using analogous justifications in graphs of the LPS type, we examine the Ramanujan-ness of these graphs. Additionally, it is unclear whether a Pizer's graph with a sufficient number of vertices is a Cayley graph or not, even though it is difficult to predict that Pizer's graph can be represented as a Cayley graph over a group with respect to a suitable generating set (in fact, all graphs with a small number of vertices, suggested as examples in Pizer 1998 are not Cayley graphs). Because creating a path on a graph is a key component of encryption, there is a close relationship between graph theory and cryptography. The key graph's path is made up of a number of vertices, which make up the plain text. The edges joining those vertices make up the encrypted text.

6 Conclusion

While studying quantum algorithms using quantum computers, eigenvalues are created. The eigenvector of an extreme eigenvalue of an exponentially large matrix may be found using all quantum algorithms, which were initially introduced as a novel method of designing quantum algorithms. Nevertheless, gaps between eigenvalues have an impact on how long the procedure takes to construct this eigenvector. So, learning about methods for bounding eigenvalues is a necessary step if we are to advance in creating quantum algorithms. So, these kinds of research projects are crucia for maintaining the security of various post-quantum public key cryptosystems in the post-quantum age.

Bibliography

[1] S. Arpin, M. Chen, K. E. Lauter, R. Scheidler, K. E. Stange and H. T. N. Tran, Orienteering with one endomorphism, Cryptology ePrint Archive, Paper 2022/098, 2022, https://eprint.iacr.org/2022/098.

[2] D. X. Charles, K. E. Lauter and E. Z. Goren, Cryptographic hash functions from expander graphs, *J. Cryptol.* **22**(1) (2009), 93–113.

[3] C. Costello, B-sidh: supersingular isogeny Diffie-Hellman using twisted torsion, in: *Advances in Cryptology–ASIACRYPT 2020: 26th International Conference on the Theory and Application of Cryptology and Information Security, Daejeon, South Korea, December 7–11, 2020, Proceedings, Part II 26*, Springer, 2020, pp. 440–463.

[4] S. D. Galbraith et al., Supersingular curves in cryptography, in: *Asiacrypt*, Vol. 2248, Springer, 2001, pp. 495–513.

[5] D. Jao and L. De Feo, Towards quantum-resistant cryptosystems from supersingular elliptic curve isogenies, in: *Post-Quantum Cryptography: 4th International Workshop, PQCrypto 2011, Taipei, Taiwan, November 29–December 2, 2011. Proceedings 4*, Springer, 2011, pp. 19–34.

[6] A. Maricq, Applications of expander graphs in cryptography, pg. 1-75, 2014, https://www.whitman.edu/documents/academics/majors/mathematics/2014/maricqaj.pdf.

[7] M. R. Murty, Ramanujan graphs: An introduction, *Indian J. Discrete Math.* **6**(2) (2020), 91–127.

[8] C. Petit and K. Lauter, Hard and easy problems for supersingular isogeny graphs, 2017, Cryptology ePrint Archive, Paper 2017/962, https://eprint.iacr.org/2017/962.

[9] A. K. Pizer, Ramanujan graphs and Hecke operators, *Bull. Am. Math. Soc.* **23**(1) (1990), 127–137.

[10] J. Siegel, Expander graphs, 2014, https://math.uchicago.edu/~may/REU2014/REUPapers/Siegel.pdf.

[11] J. H. Silverman, *The Arithmetic of Elliptic Curves*, Vol. 106, Springer, 2009.

[12] B. Wesolowski, The supersingular isogeny path and endomorphism ring problems are equivalent, in: *2021 IEEE 62nd Annual Symposium on Foundations of Computer Science (FOCS)*, IEEE, 2022, pp. 1100–1111.

[13] G. Xiao, L. Luo and Y. Deng, Constructing cycles in isogeny graphs of supersingular elliptic curves, *J. Math. Cryptol.* **15**(1) (2021), 454–464.

Vijay K. Shukla and Prashant K. Mishra

Finite-time stability analysis of neutral type delay Cohen–Grossberg neural network with discontinuous activation function

Abstract: In this article, we will discuss finite-time stability analysis of neutral type Cohen–Grossberg neural network with mixed delay. To investigate the stability analysis of neutral type network, we consider the discontinuous activation function. The theory of differential inclusion, set-valued map and inequality technique are used to tackle the network. A modified Lyapunov–Krasovskii functional has been designed to deal with mixed time-delays. Sufficient conditions are derived to examine stability of proposed neural network. Further, the settling-time also estimated. Finally, numerical example is discussed to validate the established results.

Keywords: Finite-time, stability analysis, Cohen–Grossberg neural network, mixed delay, discontinuous activation function

MSC 2010: 11A25, 11A67

1 Introduction

The neural network generally interconnected biological nervous systems and works like machines to connect the human brain with biological neurons have ability to learn. The neural network produce an output pattern of sorting one group of neurons to another when presented with an input [15]. Neural networks are like an excellent fuzzy logic systems that perform the same type of information. The importance of neural network is that traditional forms of the back propagation occur during processing and violating the locality of restriction [13]. Electronic data processing have more importance in clinical diagnosis from last several decades. The main reasons are their use is inconvenient and time-consuming.

Further, artificial neural network has pattern recognition and is successfully used in complex environments to apply in the clinical setting [3]. A CNN is identified as a neural network without transmission delays but with symmetric interconnections. In addition, Hopfield neural network is a particular case of CNN [20]. Firstly, Hopfield and Tank [7]

Acknowledgement: The authors thank the anonymous referee for helpful comments.

Vijay K. Shukla, Department of Mathematics, Shiv Harsh Kisan P.G. College, Basti-272001, India
Prashant K. Mishra, Department of Mathematics, P.C. Vigyan Mahavidyalaya, Jai Prakash University, Chhapra-841301, India

https://doi.org/10.1515/9783111313634-004

work on neural networks and developed a new method to obtain solutions for parallel processing. This method is generalized to obtain feasible solutions of unsolved problems. In 1983, Cohen and Grossberg [4] discussed a new neural networks and this model attracted the researchers due to applications in parallel computing. From last several decades, different types of neural networks for example, CNN and Hopfield neural network, have been successfully applied in science and mathematics. Among these, CNN has tremendous popularity because it includes several neural networks in its particular form [17].

The stability analysis of chaotic systems and neural network have been a hot research area in current time. However, Lyapunov theory has been used by several researchers [8, 11, 21] to study stability, but it is unable to reflect the transient state and showed bad effect on the system. Further, study of asymptotic stability of time-dependent systems is more challenging. The concept of global asymptotic stability is applied for those networks which possess only one equilibrium and all trajectories converges to equilibrium point. Further, the global stability used for those networks that have more than one equilibrium points [19]. Neutral-type systems are often considered to explain the rate of current state according to the past states.

From last few years, the stability analysis of neutral-type neural network is a hot topic among researchers [14, 16, 18, 22]. In some problems, it is necessary to design a neural circuit that possess a unique equilibrium point, also known as mono stability. Forti and Nistri [5] discussed some neural network models with discontinuous activation functions. The activation functions are generally used in high-gain limit and applied to negligible the connection of neural network energy function [6, 10]. In addition, Arvind and Balasubramaniam [1] analyzed stability of Cohen–Grossberg neural networks with delays and derived some sufficient conditions with the help of LMIs. Further, synchronization of neural networks in finite-time has been studied by Li et al. [12] by adaptive method. Meanwhile, Cohen–Grossberg neural networks with time delays discontinuous activation function have been studied by several researchers [2, 9]. In this paper, we have studied the stability analysis of Cohen–Grossberg neural network of neutral type with discontinuous activation function and mixed delay in finite-time. The aforementioned discussion inspired to authors for this research. To the best of the authors' knowledge, this work has not been yet published.

2 Model description and preliminaries

In this section, some basic results and system description are provided, which are standard and being used in our study. Now, neural network with discontinuous activation function and mixed delay is defined as

$$\dot{\varsigma}_i(t)+\sum_{j=1}^{n}e_{ij}(t)\dot{\varsigma}_i(t-\pi_j(t))=-p_i(\varsigma_i(t))(q_i(t,\varsigma_i(t))-\sum_{j=1}^{n}r_{ij}g_j(\varsigma_j(t))$$
$$-\sum_{j=1}^{n}s_{ij}g_j(\varsigma_j(t-\beth_j))+\sum_{j=1}^{n}A_{ij}(\varsigma_i(t))\int_{t-\varepsilon_j(t)}^{t}g_j(\varsigma_j(\theta))dt+J_i,\tag{4.1}$$

where $\varsigma_i(t)\in R$ is the state of ith neuron, $g_j(\cdot)$ is the activation function. The time-delays $\pi_j(t)$, $\beth_j(t)$ and ε_j represent the leakage delay, time varying delay and distributive delay respectively. The time delays $\beth_j(t)$ and ε_j satisfy $0\le\beth_i(t)\le\beth$, $\dot{\beth}_i(t)\le\ell_{\beth}$, $0\le\varepsilon_i(t)\le\varepsilon$, $\dot{\varepsilon}_i(t)\le\ell_{\varepsilon}$, where $\beth$, $\ell_{\beth}$, ε and ℓ_{ε} are positive constants and J_i is the external input. The solution of the neutral type Cohen–Grossberg neural network has been considered in Filippov's sense. Throughout the paper R represents a set of real numbers, R^n stands for a real vector space of dimension n and R^{nm} stands for collection of $n\times m$ matrices. $c\bar{o}[\widehat{\Omega}_1,\breve{\Omega}_1]$ denotes the closure of the convex hull generated by $\widehat{\Omega}_1$ and $\breve{\Omega}_1$, where, $\widehat{\Omega}_1,\breve{\Omega}_2\in R$. The E^T represents the transpose matrix of E and $\|\|_1$ is the representation of standard 1-norm. Let us consider $G_C=(G[-C,0],R^n)$ as a Banach space of continuous functions $\lambda:[-C,0]\to R^n$ with norm $\|\lambda\|=\sup_{\theta\in[-C,0]}\|\lambda(\theta)\|$ and $C=\max\{\beth,\varepsilon,\pi\}$. Now, consider the Lebesgue measurable space $([0,\varpi],L)$ and an n-dimensional real Euclidean space $(R^n,\|(\cdot)\|)$, $(n>1)$ with induced Euclidean norm $\|\|$. Assume $Z\subseteq R^n$, initially, we introduce the following notations

$$\Gamma_0(Z)=\{G\subset Z:G\text{ is non}-\text{empty}\},\quad\Gamma(Z)=\Gamma_0(Z)\bigcup\{\Psi\},$$
$$\Gamma_{f(r)}(Z)=\{G\subset Z:G\text{ is a non}-\text{empty closed convex set}\},$$
$$\Gamma_{\lambda(r)}(Z)=\{E\subset Y:E\text{ is a non}-\text{empty compact convex set}\}.$$

For convenience, we denote $2^Z=\Gamma_0(Z)$. For a given $G\subset\Gamma_f(Z)$, $x_3\in Z$, the distance function can be defined as $\text{dist}(x_3,G)=\inf\{\|x_3-r\|:r\in G\}$. The Hausdorff metric on $\Gamma_f(Z)$ is defined as

$$S_M(G,H)=\max\{B(G,H),\ B(H,G)\},$$

where $B(G,H)=\sup\{\text{dist}(x_1,H):x_1\in G\}$, and $B(F,E)=\sup\{\text{dist}(x_2,H):x_2\in H\}$.

Suppose $Z\subseteq R^n$, if there exist a non-empty set $O(x_3)\subset R^n$ for each $x_3\in Z$, then a map $x_3\mapsto O(x_3)$ is said to be a set valued map $Z\hookrightarrow R^n$. An upper semi-continuous (USC) map is a set valued map O with non-empty values at $x_{30}\in Z$, if $B(O(x_3),O(x_{30}))\to 0$ as $x_3\to x_{30}$. $O(x_3)$ is said to have a closed (compact, convex) image if for each $x_3\in Z$, $O(x_3)$ is closed (compact, convex). We say $O:[0,\varpi]\to\Gamma_f(R^n)$, a set-valued map is measurable, if for all $x_3\in R^n$, a positive real-valued function $t\mapsto O(x_3(t))$ is measurable, i. e., $\text{Graph}(O)=\{(t,v)\in[0,\varpi]\times R^n,v\in O(t)\}\in L\times F(R^n)$ is measurable, where $L[0,\varpi]$ stands for Lebesgue π – field, $F(R^n)$ represents a borel π – field of R^n.

The following basic assumptions have been taken for the sake of convenience:

1. The continuous function $g_i : R \to R$ for $i = 1, 2, \ldots, n$, i. e., except the points $\{\mu_k^i\}$ with $g_i^+(\mu_k^i)$ and $g_i^-(\mu_k^i)$. Additionally, g_i has finite number of discontinuities in any interval of R.
2. There exist a non-negative constant E_i s. t.

$$\sup_{C_i \in \overline{co}[g_i(\varsigma_i)]} |C_i| \leq E_i|\varsigma_i| + F_i \quad \forall\, \varsigma_i \in R, \quad i = 1, 2, \ldots, n,$$

where $\overline{co}[g_i(\varsigma_i)] = [\min\{g_i^-(\varsigma_i), g_i^+(\varsigma_i)\}, \max\{g_i^-(\varsigma_i), g_i^+(\varsigma_i)\}]$.

3. There exit positive constant s_i and $\bar{s}_i$ such that $0 < s_i \leq p_i(\varsigma) \leq \bar{s}_i$, $\forall\, \varsigma \in R$.
4. For each ς, $q_i(t, \varsigma)$ is continuous, $q(t, 0) = 0$ and there exist two positive constants $\underline{\pi_i}$ and $\overline{\pi_i}$ s. t. $0 < \underline{\pi_i} \leq \frac{q_i(t,\varsigma) - q_i(t,\xi)}{\varsigma - \xi} = \frac{|q_i(t,\varsigma) - q_i(t,\xi)|}{|\varsigma - \xi|} \leq \overline{\pi_i}$, $\varsigma, \xi \in R, \varsigma \neq \xi, i = 1, 2, \ldots, n$, there exist a continuous function $\Delta_i(t) > 0$ such that $|q_i(t, \varsigma)| \leq \Delta_i(t)|\varsigma|$, $\forall\, \varsigma \in R$.
5. The time-delays $\mathfrak{I}_{ij}(t)$ are continuous differentiable function and satisfy $\dot{\mathfrak{I}}_{ij}(t) \neq 1$, $i, j = 1, 2, \ldots, n$.
6. For each $i = 1, 2, \ldots, n$ there exist non-negative constant E_i, F_i such that here $Q[g_j](w) \triangleq \bigcap_{\delta > 0} \bigcap_{\rho(N)=0} \overline{co}[g_j(t, F(\rho, \varepsilon) \backslash N\}$, $\overline{co}(t)$ is the closure of the convex hull of set P, $\rho(N)$ is the Lebesgue measure of set N and $F(\rho, \varepsilon)$ is the open ball with center at $\rho \in R$ and the radius $\phi \in R$.

The measurability of function g is not enough to guarantee the existence of solutions for the Cauchy problem $\dot{x}_1(t) = f(x_1(t))$, $\forall\, t \geq 0$, $x_1(0) = x_{10}$. This can be fixed by substituting the function f with its Filippov regularization O i. e. the differential inclusion $\dot{x}_1(t) \in O(x_1(t))$, is always solvable. The set-valued maps are said to be Filippov-representable if they can be obtained from Filippov regularization of a single-valued measurable function. A single-valued map O is identical to f if the function f is continuous. For the purpose of getting solutions to discontinuous differential equations, the Krasovskii and Filippov regularization approaches are introduced.

Definition 1 ([18]). The origin of system (4.1) is finite-time stable if the origin of system (4.1) is stable and finite-time convergent, i. e., for $\psi(\theta) \in \Pi$, there exist $0 \leq T(\psi) < +\infty$, such that every solution of system (4.1) satisfies $\varsigma(t, \psi) = 0$, $\forall\, t > T(\psi)$. The functional $T_0(\psi) = \inf\{T(\psi) \geq 0 : \varsigma(t, \psi) = 0, \forall t \geq T(\psi)\}$ is known as settling-time.

Lemma 1 ([18]). *Let $W : [0, +\infty) \times \Pi \to R^+$ is continuous function and v, h are such that $W(t, 0) = 0$, $v(\|\varsigma\|) \leq W(t, \varsigma)$, $t \in [0, +\infty)$ and $H^+W(t, \varsigma) \leq -h(W(t, \varsigma))$ with $\int_0^\delta dx_3/h(x_3) < +\infty$, $\forall\, \delta > 0$, $\varsigma \in \Pi$. Then system (4.1) is finite-time stable with a settling-time satisfying the inequality $T_0(\psi) \leq \int_0^{W(0,\psi)} dx_3/h(x_3)$.*

Lemma 2 ([16]). *Suppose $W(\varsigma)$ is G-regular and $\varsigma(t)$ is absolutely continuous on $[0, +\infty)$. Then $W(\varsigma(t))$ and $\varsigma(t)$ are differentiable for almost everywhere $t \in [0, +\infty)$ and $\frac{dW(\varsigma(t))}{dt} =$*

$\xi^T \dot{\varsigma}(t)$, $\forall\, \xi \in \partial W(\varsigma(t))$, where $\partial W(\varsigma) = \bar{co}[\lim_{k\to\infty} \nabla W(\varsigma_k) : \varsigma_k \to \infty, \varsigma_k \notin N, \varsigma_k \notin \Omega_2]$. Here $\Omega_2 \subset R^n \times R$ is the set of points, and $N \subset R^n \times R$ is an arbitrary set with measure zero.

3 Finite time stability analysis

In this section, Lyapunov–Krasovskii functional is design for finite-time stability analysis. Now define feedback controller as

$$w_i(t) = \operatorname{sgn}(\varsigma_i(t))\left[-\phi_i - \pi_i|\varsigma_i(t)| - B_i|\varsigma_i(t - \mathfrak{I}_{ij})| - \chi_i \int_{t=\varepsilon_{ij}(t)}^{t} |\varsigma_i(\theta)|d\theta\right],$$

where $i = 1, 2, \ldots, n$, ϕ_i, π_i, χ_i are unknown gain coefficients.

Theorem 1. *Suppose that Assumptions* (1–6) *holds and*

$$\Pi_i = \lim_{t\to\infty} \inf\left\{-\phi_i + \overline{p_i}|I_i(t)| + \overline{p_i}\sum_{j=1}^{n} F_J(|r_{ij}(t)|) + |s_{ij}(t)| + |A_{ij}(t)||\varepsilon_{ij}(t)|\right\} \geq 0,$$

$$\lim_{t\to\pm\infty} \sup\left\{-\pi_i + \bar{p}_i\Delta_i(t) + \sum_{j=1}^{n} [\overline{p}_i\ E_i|r_{ij}(t)| + \frac{|s_{ji}(\psi_{ji}^{-1}(t))|}{1 - \dot{\mathfrak{I}}_{ij}(\psi_{ji}^{-1}(t))} + |\varepsilon_{ij}(t)|\right\} \leq 0,$$

$$\lim_{t\to\infty} \sup\left\{-\chi_i + \sum_{j=1}^{n}[-|(s_{ij}(t)| + \overline{p_i}E_j(|s_{ij}(t)|)]\right\} \leq 0,$$

$$\lim_{t\to\infty} \sup\left\{\chi_i + \sum_{j=1}^{n}[\overline{p_i}E_i(|A_{ij}(t)|) + \varepsilon_{ji}(t)]\right\} \leq 0.$$

Then, the system (4.1) *is finite time stable and the setting time satisfies $T_\Delta = \frac{W(0)}{\Delta}$, $\Delta = \min_{i\in I}\{\Delta_i\}$, where*

$$W(0) = \sum_{i-1}^{n}|\varsigma_i(0)| + \sum_{i-1}^{n}\sum_{j-1}^{n}\int_{-\mathfrak{I}_{ij}(0)}^{0} \frac{|s_{ij}(\psi_{ij}^{-1}(\theta))|}{1 - \dot{\mathfrak{I}}_{ij}(\psi_{ij}^{-1}(\theta))}|\varsigma_j(\theta)|d\theta + \sum_{i=1}^{n}\sum_{j=1}^{n}\int_{-\varepsilon_{ij}(0)}^{0}\int_{\theta}^{0}|\varsigma_j(w)|dw\, d\theta. \quad (4.2)$$

Proof. Since system (4.1) is discontinuous, we use the theory of differential inclusion and set-valued map and obtain

$$\begin{aligned}\dot{\varsigma}_i(t) - \sum_{i=1}^{n} e_{ij}(t)\dot{\varsigma}_i(t - \pi_i(t)) &= -p_i(\varsigma_i(t))(q_i(t, \varsigma_i(t)) - \sum_{j=1}^{n} r_{ij}(t)Q[g_j](\varsigma_j(t)) \\ &\quad + \sum_{j=1}^{n} A_{ij}(\varsigma_i(t)) \int_{t-\varepsilon_{ij}(t)}^{t} Q[g_j](\varsigma_j(t))\, d\theta + I_i(t) + w_i(t). \quad (4.3)\end{aligned}$$

Applying measure selection Theorem [2], then there exist a measurable function $C = (C_1, C_2, \ldots, C_n)^T : [-\varsigma, \infty] \to R^n$ where $C_j(t) \in Q[g_j](\varsigma_j(t))$ such that

$$\dot{\varsigma}_i(t) = -p_i(\varsigma_i(t))\left\{q_i(t, \varsigma_i(t)) - \sum_{j=1}^{n} p_j(t)C_j(t) - \sum_{j=1}^{n} s_{ij}(t)C_j(t - \daleth_{ij}(t))\right.$$
$$\left. - \sum_{j=1}^{n} A_{ij}(t) \int\limits_{t-\varepsilon_{ij}(t)}^{t} C_i d\theta + \sum_{j=1}^{n} \dot{e}_{ij}(t)\dot{\varsigma}_i(t - \pi_i(t)) - I_i(t)\right\} + \overline{w_i}(t) \tag{4.4}$$

$$\dot{\varsigma}_i(t) + \sum_{j=1}^{n} e_{ij}(t)\dot{\varsigma}_i(t - \pi_i(t)) = -p_i(\varsigma_i(t))\left(q_i(t, \varsigma_i - (t)) \sum_{j=1}^{n} r_{ij}C_j(t) - \sum_{j=1}^{n} s_{ij}(t)C_j(t - \daleth_{ij}(t))\right.$$
$$\left. - \sum_{j=1}^{n} A_{ij}(t) \int\limits_{t-\varepsilon_{ij}(t)}^{t} C_j(\theta)d\theta - I_i(t)\right) + \overline{w_i}(t). \tag{4.5}$$

where $\overline{w_i}(t) \in Q[w_i(t)]$, and

$$Q[w_i(t)] = Q[\operatorname{sgn} \varsigma_i(t)]\left(-\phi_i - \pi_i|\varsigma_i(t)| - B_i|\varsigma_i(t - \daleth_{ij}(t))| - \chi_i \int\limits_{t-\varepsilon_{ij}(t)}^{t} |\varsigma_i(\theta)|d\theta \right)$$

and

$$Q[\operatorname{sgn}(\varsigma_i(t))] = \begin{cases} 1, & \varsigma_i(t) > 0 \\ [-1, 1], & \varsigma_i(t) = 0 \\ -1, & \varsigma_i(t) < 0. \end{cases} \tag{4.6}$$

The Lyapunov–Krasovskii functional is expressed as

$$W(t) = \sum_{i=1}^{n}|\varsigma_i(t)| + \sum_{i=1}^{n}\sum_{j=1}^{n} \int\limits_{t-\daleth_{ij}(t)}^{t} \frac{|s_{ij}(\psi_{ij}^{-1}(\theta))|}{1 - \dot{\daleth}_{ij}(\psi_{ij}^{-1}(\theta))}|\varsigma_j(\theta)|d\theta$$
$$+ \sum_{i=1}^{n}\sum_{j=1}^{n} \int\limits_{-\theta_{ij}(t)}^{0} \int\limits_{t+\theta}^{t} |\varsigma_j(w)|dwd\theta + \sum_{i=1}^{n}\sum_{j=1}^{n} e_{ij}(t)|\varsigma_j(t - \pi)|. \tag{4.7}$$

Here ψ_{ij}^{-1} is the inverse function of $\psi_{ij}(t) = t - \daleth_{ij}(t)$. Obviously, $W(t)$ is a G-regular function. Now calculating the time derivative of $W(t)$ along the trajectories of (4.3) as

$$\dot{W}(t) = \sum_{i=1}^{n} \operatorname{sgn}(\varsigma_i(t))\left\{-p_i(\varsigma_i(t))\, [q_i(t, \varsigma_i(t)) - \sum_{j=1}^{n} r_{ij}(t)C_i(t))\right.$$

$$-\sum_{j=1}^{n} s_{ij}(t)C_{ij}(t-\beth_{ij}(t)) - \sum_{j=1}^{n} A_{ij}(t)\int_{t-\varepsilon_{ij}(t)}^{t} C_j(\theta)d\theta\Bigg\}$$

$$-\sum_{i=1}^{n}\sum_{j=1}^{n}|s_{ij}(t)||\varsigma_j(t-\beth_{ij}(t))| + \sum_{i=1}^{n}\sum_{j=1}^{n}\dot{\varepsilon}_{ij}(t)\int_{t-\varepsilon_{ij}(t)}^{t}|\varsigma_j(w)|dw$$

$$+\sum_{i=1}^{n}\sum_{j=1}^{n}\int_{-\varepsilon_{ij}(t)}^{0}|\varsigma_j(t)| - |\varsigma_j(t+\phi)|d\theta, \tag{4.8}$$

$$\dot{W}(t) \le \sum_{i=1}^{n}\left\{\overline{p_i}\Delta_i(t) + \sum_{j=1}^{n}\overline{p_i}E_i|r_{ji}(t)| + \frac{|s_{ij}(\psi_{ji}^{-1}(t))|}{1-\dot{\beth}_{ij}(\psi_{ji}^{-1}(t))} + |\varepsilon_{ij}(t)|\right\}|\varsigma_i(t)|$$
$$+\sum_{i=1}^{n}\left\{-B_i + \sum_{j=1}^{n}[-|s_{ij}(t)| + \overline{p_i}E_j(|s_{ij}(t)|)]\right\}|\varsigma_i(t-\beth_{ij}(t))|$$
$$+\sum_{i=1}^{n}\left\{-\phi_i + \overline{p_i}|I_i(t)| + \overline{p_i}\sum_{j=1}^{n}[F_j(|r_{ij}(t)|) + |s_{ij}(t)| + |A_{ij}(t)||\varepsilon_{ij}(t)|]\right\}$$
$$+\sum_{i=1}^{n}\left\{-\chi_i + \sum_{j=1}^{n}\overline{p_i}E_i(|A_{ij}(t)|) + \dot{\varepsilon}_{ji}(t)\right\}\int_{t-\varepsilon_{ji}(t)}^{t}|\varsigma_i(\theta)|d\theta. \tag{4.9}$$

Since $\overline{\varepsilon_{ij}}$ is continuously differentiable for $i,j = 1,2,\dots,n$, we can have $\dot{W}(t) \le -\Delta$, where $\Delta = \min_{i\in I}\{\Delta_i\}$ for $i,j = 1,2,\dots,n$ and $\Delta_i = \phi_i - \overline{p_i}|I_i(t)| - \overline{p_i}\sum_{j=1}^{n}[F_j(|G_{ij}(t)|) + |s_{ij}(t)| + |A_{ij}(t)||\varepsilon_{ij}(t)|]$ there exists a constant $T_0 = W(0)/\Delta$ such that $W(t) = 0\ \forall\ t \ge T_0$. Therefore, system (4.1) is finite-time stable under the designed controller. □

4 Numerical result and discussion

This section authenticates the efficiency of the theoretic results through numerical simulation. Consider a discontinuous Cohen–Grossberg neural network of neutral type with mixed delay

$$\dot{\varsigma}_i(t) + \sum_{j=1}^{n} e_{ij}(t)\dot{\varsigma}_i(t-\pi_i(t)) = -p_i(t,\varsigma_i(t))\Bigg\{q_i(t,\varsigma_i(t)) - \sum_{j=1}^{n} r_{ij}g_j(\varsigma_j(t))$$
$$-\sum_{j=1}^{n} s_{ij}g_j(\varsigma_j(t-\beth_j)) + \sum_{j=1}^{n} A_{ij}(\varsigma_i(t))\int_{t-\varepsilon(t)}^{t} g_j(\varsigma_j(\theta))\,dt\Bigg\}, \tag{4.10}$$

where $i,j = 1,2$ and $p_1(t,\varsigma) = p_2(t,\varsigma) = 2 + \frac{1}{2(1+\varsigma^2)}$, $\beth_{ij} = 0.5\sin t$, $\varepsilon_{ij} = 0.4\cos t$, $\forall\ i,j$.

$$q_1(t,\varsigma) = (1.5 + 0.3\sin t)\varsigma, q_2(t,\varsigma) = (1.5 + 0.5\text{cast})\varsigma,$$
$$\Delta_1(t) = 1.5 + 0.4\sin t, \Delta_2(t) = 1 + 0.9\cos t.$$
$$(r_{ij}(t))_{2\times 2} = \begin{pmatrix} 0.7 + 0.2\cos t & 0.9 + 0.6\sin t \\ -2 + 0.5\sin t & -1.8 + 0.3\cos t \end{pmatrix},$$
$$(s_{ij}(t))_{2\times 2} = \begin{pmatrix} -0.4 + 0.2\cos t & 0.7 + 0.3\sin t \\ -1.6 + 0.3\sin t & 0.3 + 0.1\cos t \end{pmatrix},$$
$$(A_{ij}(t))_{2\times 2} = \begin{pmatrix} -2.4 + 0.6\sin t & -1.4 + 0.7\cos t \\ -1.2 + 0.8\cos t & -1 + 0.3\sin t \end{pmatrix}.$$

The numeric values of the parameters for (4.1) are taken as follows

$$\overline{p_1} = \overline{p_2} = \frac{5}{2}, \quad \beth = 0.5, \quad \varepsilon = 0.4, \quad E_1 = E_2 = 0.5, \quad F_1 = F_2 = 0.1,$$
$$\Pi_1 = 0.79, \quad \Pi_2 = 1.59, \quad \pi_1 = 9.43, \quad \pi_2 = 9.23, \quad \phi_1 = 9, \quad \phi_2 = 5.2,$$
$$B_1 = 0.95, \quad B_2 = 0.8, \quad \chi_1 = 2.5, \quad \chi_2 = 2.95,$$
$$g_1(\varsigma) = g_2(\varsigma) = \begin{cases} 0.5\tanh\varsigma - 0.5, & \varsigma \geq 1 \\ 0.1\varsigma + 0.2\cos\varsigma, & \varsigma < 1. \end{cases}$$

It can be observed that the activation functions $g_j(\varsigma_j)$ are non-increasing and discontinuous at $\varsigma_j = 1$ and $c\bar{o}[g_j(1)] = [g_j^+(1), g_j^-(1)] = [-0.12, 0.20]$, $j = 1, 2$. Thus, the assumptions 1, 2 and Theorem 1 are satisfied. It indicates that the origin of (4.1) is stable in finite-time.

5 Conclusion

This article gave a beautiful result about stability analysis of discontinuous Cohen–Grossberg neural network with mixed delay of neutral-type. The Filippov regularization technique has been utilized to deal the discontinuity of the addressed system. The settling-time has been estimated with the help of theory of differential inclusion, inequality technique and non-smooth Lyapunov–Krasovskii functional. Finally, to authenticate the accuracy of the obtained results numerical example is discussed.

Bibliography

[1] R. V. Aravind and P. Balasubramaniam, Stability criteria for memristor-based delayed fractional-order Cohen-Grossberg neural networks with uncertainties, *J. Comput. Appl. Math.* **420** (2023), 114764. https://doi.org/10.1016/j.cam.2022.114764, ISSN: 0377-0427.

[2] J. P. Aubin and A. Cellina, A Differential inclusions: set-valued maps and viability theory, *Acta Appl. Math.* **6**(2) (1986), 215–217. https://doi.org/10.1007/978-3-642-69512-4, ISSN: 0072-7830.

[3] W. G. Baxt, Use of an artificial neural network for the diagnosis of myocardial infarction, *Ann. Intern. Med.* **115**(11) (1991), 843–848. https://doi.org/10.7326/0003-4819-115-11-843, ISSN: 1539-3704.

[4] M. A. Cohen and S. Grossberg, Absolute stability of global pattern formation and parallel memory storage by competitive neural networks, *IEEE Trans. Syst. Man Cybern.* **SMC-13**(5) (1983), 815–826. https://doi.org/10.1109/TSMC.1983.6313075, ISSN: 2168-2232.

[5] M. Forti and P. Nistri, Global convergence of neural networks with discontinuous neuron activations, *IEEE Trans. Circuits Syst. I* **50** (2003), 1421–1435, https://doi.org/10.1109/TCSI.2003.818614, ISSN: 1549-8328.

[6] J. J. Hopfield and D. W. Tank, Computing with neural circuits: A model, *Science* **233** (1986), 625–633, ISSN: 0036-8075.

[7] J. J. Hopfield and D. W. Tank, "Neural" computation of decisions in optimization problems, *Biol. Cybern.* **90** (2014), 186–194. https://doi.org/10.1007/BF00339943, ISSN:1432-0770.

[8] J. Huang, C. Li, T. Huang and X. He, Finite-time lag synchronization of delayed neural networks, *Neurocomputing* **139** (2014), 145–149, https://doi.org/10.1016/j.neucom.2014.02.050, ISSN: 0925-2312.

[9] F. Kong and Q. Zhu, Anti-periodic dynamical behaviors of discontinuous neutral-type Cohen-Grossberg neural networks with mixed time delays, *Comput. Intell.* **36**(2) (2020), 698–719. https://doi.org/10.1111/coin.12262, ISSN: 1875-6883.

[10] F. Kong, Q. Zhu and R. Sakthivel, Finite-time stabilization of discontinuous fuzzy inertial Cohen-Grossberg neural networks with mixed time-varying delays, *Nonlinear Anal.: Modelling and Control* **26**(5) (2021), 759–780, https://doi.org/10.15388/namc.2021.26.23935, ISSN: 1392-5113.

[11] D. Li and J. Cao, Finite-time synchronization of coupled networks with one single time-varying delay coupling, *Neurocomputing* **166** (2015), 265–270. https://doi.org/10.1016/j.neucom.2015.04.013, ISSN: 0925-2312.

[12] X. Li, W. Zhang, J. A. Fang and H. Li, Finite-time synchronization of memristive neural networks with discontinuous activation functions and mixed time-varying delays, *Neurocomputing* **340** (2019), 99–109, https://doi.org/10.1016/j.neucom.2019.02.051, ISSN: 0925-2312.

[13] C. T. Lin and C. S. G. Lee, Neural-network-based fuzzy logic control and decision system, *IEEE Trans. Comput.* **40**(12) (1991), 1320–1336, https://doi.org/10.1109/12.106218, ISSN: 0018-9340.

[14] G. Nagamani, T. Radhika and P. Balasubramaniam, A delay decomposition approach for robust dissipativity and passivity analysis of neutral-type neural networks with leakage time-varying delay, *Complexity* **21**(5) (2016), 248–264, https://doi.org/10.1002/cplx.21652, ISSN: 1076-2787.

[15] P. Picton, *Introduction to Neural Networks*, Palgrave, London, 1994.

[16] W. Wang, L. Li, H. Peng, W. Wang, J. Kurths, J. Xiao and Y. Yang, Anti-synchronization of coupled memristive neutral-type neural networks with mixed time-varying delays via randomly occurring control, *Nonlinear Dyn.* **83**(4) (2016), 2143–2155. https://doi.org/10.1007/s11071-015-2471-9, ISSN: 1562-8353.

[17] Z. Wang, Y. Liu, M. Li and X. Liu, Stability analysis for stochastic Cohen-Grossberg neural networks with mixed time delays, *IEEE Trans. Neural Netw.* **17**(3) (2006), 814–820. https://doi.org/10.1109/TNN.2006.872355. ISSN: 2162-2388.

[18] J. Xia, J. H. Park and H. Zeng, Improved delay-dependent robust stability analysis for neutral-type uncertain neural networks with Markovian jumping parameters and time-varying delays, *Neurocomputing* **149** (2015), 1198–1205, https://doi.org/10.1016/j.neucom.2014.09.008, ISSN: 0925-2312.

[19] H. Ye, N. Micheal and K. Wang, Robust stability of nonlinear time-delay systems with applications to neural networks, *IEEE Trans. Circuits Syst. I, Fundam. Theory Appl.* **43**(7) (1996), 532–543, ISSN: 1057-7122.

[20] H. Ye, A. N. Michel and K. Wang, Qualitative analysis of Cohen-Grossberg neural networks with multiple delays, *Phys. Rev. E* **851**(3) (1995), 2611. https://doi.org/10.1103/PhysRevE.51.2611, ISSN: 2470-0045.

[21] M. Zheng, L. Li, H. Peng, J. Xiao, Y. Yang and H. Zhao, Finite-time stability analysis for neutral-type neural networks with hybrid time-varying delays without using Lyapunov method, *Neurocomputing* **238** (2017), 67–75. https://doi.org/10.1016/j.neucom.2017.01.037, ISSN: 0925-2312.

[22] W. Zhou, Q. Zhu, P. Shi, H. Su, J. Fang and L. Zhou, Adaptive synchronization for neutral-type neural networks with stochastic perturbation and Markovian switching parameters, *IEEE Trans. Cybern.* **44**(12) (2014), 2848–2860, https://doi.org/10.1109/TCYB.2014.2317236, ISSN: 2168-2267.

S. N. Rai, B. S. Bhadauria, and Anurag Srivastava

Influence of temperature modulation over heat/mass transport of Rivlin–Ericksen viscoelastic nanoliquid in Hele-Shaw cell

Abstract: This manuscript includes weakly-nonlinear analysis to estimate the heat and mass transportation in the Rivlin–Ericksen nanoliquid within a Hele-Shaw (HS) cell under the effect of thermal modulation. The stability analysis is done by means Nusselt numbers based on the cubic Ginzburg–Landau (GBL) equation. We have considered three phases of thermal modulation namely symmetric temperature modulation, asymmetric temperature modulation, and lower-boundary temperature modulation. It is found that Rivlin–Ericksen viscoelastic parameter has stabilizing impact on the system in all three phases of thermal modulation, while maximum heat/mass transport can be observed in the case of out-phase modulation. Moreover, both heat as well as mass transport show a decrement for an increment in the HS-number.

Keywords: Rivlin–Ericksen nanoliquid, thermal modulation, non-linear instability, Hele-Shaw cell

MSC 2010: 11A25, 11A67

1 Introduction

In order to understand and visualize the hydrodynamical circulation penetration into earth, a laboratory apparatus was set up. The mathematical representation of this laboratory apparatus was first given by Hele-Shaw [21] in 1898. He studied a slow two-dimensional laminar flow in a uniform porous medium in a thin slot squeezed between equidistant walls. This entire system is known as Hele-Shaw (HS) cell. Wooding [39] investigated thermal instability in a HS-cell using a viscous fluid having variable density. Hartline and Lister [20] studied thermal convection process in a HS-cell and described

Acknowledgement: Author S. N. Rai thanks the University Grants Commission (UGC) Bahadur Shah Zafar Marg, New Delhi-110002 India, for financial assistance as Senior Research fellowship.

S. N. Rai, Department of Applied Science and Humanities (Mathematics), B. P. Mandal College of Engineering, B(I)-104, Professor Quarter, Madhepura, Bihar-852128, India, e-mail: shobhnathrai802@gmail.com
B. S. Bhadauria, Department of Mathematics, Babasaheb Bhimrao Ambedkar University, 107, Old Academic Building, BBAU, Lucknow-226025, U.P., India, e-mail: mathsbsb@yahoo.com
Anurag Srivastava, Department of Mathematics, Chandigarh University, 313, DD1, Mohali, Punjab-140413, India, e-mail: anurag.091@gmail.com

https://doi.org/10.1515/9783111313634-005

that for marginal stability the system of equations are identical to that describing flow through a porous media. Aniss et al. [2] derived the Ginzburg–Landau equation from the hydrodynamic equations for an infinite HS-cell. Further work in this area can be found in the references [7, 15, 32, 38]. Yadav [40] investigated the combined effect of magnetic field and through-flow over thermal instability of a nanoliquid in a HS-cell. Bhadauria and Kumar [11] recently studied linear as well as non-linear stability analysis on thermal instability in a HS-cell saturated by a nanoliquid under the influence of through-flow and gravity modulation using Brinkman model.

Ample of research works are reported within the last one decade due to the higher thermal conductivity of nanoliquids, which are the combination of base-liquid with suspended materials having diameter below 50 nm. Since nanoparticles possess greater surface area, they not only improve heat transport capacity, but also enhance the consistency of the suspensions. Abrasion-related properties can also be improved by nanoliquids as compared to the usual solid and liquid mixtures. The name of “nanofluid” is introduced by Choi [19] and Buongiorno [16] described the two-phase mathematical model for nanofluids. Using this model, many authors [14, 35] studied the issue of temperature instability in nanoliquids. Nowadays both Newtonian and non-Newtonian fluids are used as basefluids for the preparation of nanofluids. Keeping eye on the aim of our paper, we mainly focus on non-Newtonian fluids, especially Rivlin–Ericksen (R–E) fluid. A non-Newtonian fluid, viscoelastic fluid have interesting characteristics to generate both viscous as well as elastic properties at the same time. Due to these properties, they can be visualized [30, 34] in different types of liquids, colloids, polymers, organic alloys, polymer alloys, biological fluids, etc. Mogilevskii [24] recently investigated time-independent flow of non-Newtonian fluid using sinusoidal micro topography over the plane.

R–E fluid, one type of viscoelastic fluid, is theoretically introduced by Rivlin–Ericksen [27]. Srivastava [33] examined the unsteady flow of R–E fluid in the presence of pressure gradient (time-dependent). Rana and Kumar [25] examined the temperature instability of a R–E elastico-viscous rotating under variable gravity in a porous medium. Rana and Thakur [26], investigated the effect of suspended particles on thermal convection in R–E fluid using Brinkman model. Chand and Rana [17] studied the thermal instability of R–E nanoliquid soaked porous medium. Chand et al. [18] extended the previous work and presented a revised model. Rana et al. [1] further studied the more realistic model of temperature instability of a R–E nanoliquid soaked by a Darcy–Brinkman porous medium. Saini and Sharma [31] did the similar study with non-Darcy model with the effect of vertical through-flow. Awasthi et al. [3] recently published an article discussing instability of R–E fluid and viscous fluid interface.

Temperature modulation can be defined as a sinusoidal perturbation (time-dependent) to the wall temperatures. Venezian [37] was the first person to study Rayleigh–Bénard convection with the effect of temperature modulation. Rosenblat and Herbert [29] did the similar study for the temperature modulation of low frequency. Roppo et al. [28] further extended the idea of Venezian and presented a non-linear analysis for the same. Bhadauria and Bhatia [6] presented four different profiles for temperature modulation. Bhadauria et al. [8] further extended the similar study and presented a weakly non-linear analysis. Umavathi [36] analytically investigated the impact of temperature modulation over convective instability of a nanoliquid saturated porous medium. Bhadauria and Kiran [9] investigated the effect of temperature modulation over double-diffusive magnetoconvection using Newtonian liquid. Bhadauria and Kiran [10] performed the similar study with viscoelastic liquid stratum porous medium. Kiran and Narasimhulu [23] studied the impact of thermal-modulation and internal heating on chaotic-convection. Further work in the area of thermal-modulation can be the references [5, 12, 13, 22]. Bansal and Suthar [4] published an article discussing the effect of temperature modulation using a local thermal non-equilibrium approach between fluid and porous medium.

Motivated by the studies that used thermal modulation, HS-cell, and Rivlin–Ericksen viscoelastic nanofluid, in this paper, we concentrate on studying the heat and mass transport in R–E viscoelastic nanofluid in a HS-cell under the effect of temperature modulation. To the best of our information, no similar investigation is done in the literature discussing these important aspects simultaneously.

2 Problem formulation

Let us consider a R–E viscoelastic nanoliquid stratum of depth d vertically restricted by two side by side free-free isothermal boundaries confined at $z = 0$ and $z = d$. We consider such a system, which is extended in x-direction at the same time restricted in the y-direction, it is supposed to be restricted by upright (vertical) impermeable confines (phase sides) at $y = 0$ and $y = b(\ll d)$. The presentation is given in the Fig. 5.1.

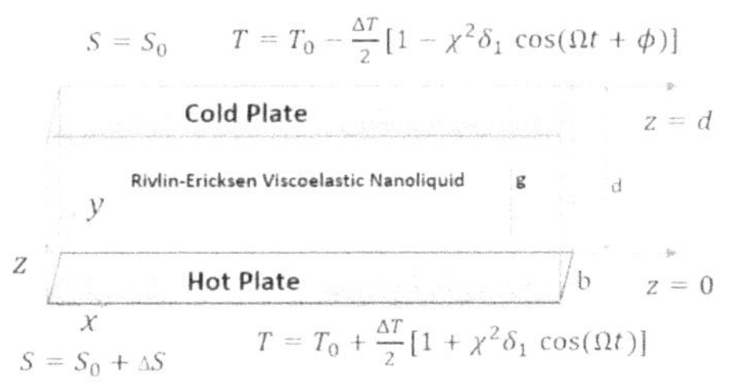

Figure 5.1: Physical presentation.

2.1 Mathematical equations

The mathematical equations are taken from Bhadauria and Kumar [11] and Rana et al. [1]:

$$\nabla \cdot \mathbf{V} = 0, \tag{5.1}$$

$$\rho_0\left[\frac{\partial}{\partial t}+(\mathbf{V}\cdot\nabla)\right]\mathbf{V}=-\nabla p-\frac{1}{K}\left(\mu+\tilde{\mu}\frac{\partial}{\partial t}\right)\mathbf{V}+\mu\nabla^2\mathbf{V}+\rho\mathbf{g}, \tag{5.2}$$

$$\left[\frac{\partial}{\partial t}+(\mathbf{V}\cdot\nabla)\right]T=\alpha\nabla^2T+\frac{(\rho c)_p}{(\rho c)_f}\left[D_B\nabla S\cdot\nabla T+\left(\frac{D_T}{T_0}\right)\nabla T\cdot\nabla T\right], \tag{5.3}$$

$$\left[\frac{\partial}{\partial t}+(\mathbf{V}\cdot\nabla)\right]S=D_B\nabla^2S+\left(\frac{D_T}{T_0}\right)\nabla^2T, \tag{5.4}$$

$$\rho=S\rho_p+\rho_0(1-S)\{1-\gamma(T-T_0)\}. \tag{5.5}$$

2.2 Boundary conditions

The boundary conditions (BC's) are considered from Umavathi [36]:

$$\left.\begin{array}{ll}\mathbf{V}=\mathbf{0},\quad T=T_0+\frac{\Delta T}{2}[1+\chi^2\delta_1\cos(\Omega t)],\quad S=S_0+\Delta S, & \text{at } z=0,\\ \mathbf{V}=\mathbf{0},\quad T=T_0-\frac{\Delta T}{2}[1-\chi^2\delta_1\cos(\Omega t+\phi)],\quad S=S_0, & \text{at } z=d.\end{array}\right\} \tag{5.6}$$

Where, t is time, $\mathbf{V}$ is the velocity of the Rivlin–Ericksen viscoelastic nanoliquid, S is volume fraction of the nanoparticles, $K=b^2/12$ is permeability of the nanoliquid-flow, D_B is coefficient of Brownian diffusion, p is hydrodynamic pressure, d is distance to the dimensional layer, D_T is coefficient of thermophoresis diffusion, $(\rho c)_f$ is heat capacity of liquid, $(\rho c)_p$ heat capacity of nanoparticle, γ is factor of proportionality, $\tilde{\mu}$ is kinematic visco-elasticity of nanoliquid, μ is viscosity of nanoliquid, ρ_0 is density of baseliquid, ρ is density of the nanoliquid, ρ_p is density of nanoparticle, T is temperature of nanoliquid, $\mathbf{g}$ is gravitational acceleration, δ_1 is amplitude of modulation, Ω is modulating-frequency, ϕ is angle of phase, $\alpha=\frac{k}{(\rho c)_f}$ is thermal diffusivity of nanoliquid, k is effective thermal conductivity of nanoliquid and $\nabla^2\equiv\frac{\partial^2}{\partial x^2}+\frac{\partial^2}{\partial y^2}+\frac{\partial^2}{\partial z^2}$ is operator.

2.3 Basic solution

$$\mathbf{V}=\mathbf{V_b}=\mathbf{0},\quad T=T_b(z,t),\quad S=S_b(z),\quad p=p_b(z,t),\quad \rho=\rho(z,t), \tag{5.7}$$

$$\rho\mathbf{g}+\nabla p_b=0, \tag{5.8}$$

$$\frac{\partial T}{\partial t}=\alpha\frac{\partial^2T_b}{\partial z^2}, \tag{5.9}$$

$$\frac{d^2S_b}{dz^2}=0. \tag{5.10}$$

Using the BC's eq. (5.6) in eqs. (5.9) and (5.10), the solutions of the conduction state are:

$$T_b=T_s(z)+\chi\,T_1(z,t), \tag{5.11}$$

where,

$$T_s(z) = \frac{\Delta T}{2}\left(1 - \frac{2z}{d}\right) + T_0, \tag{5.12}$$

$$T_1(z,t) = \mathrm{Re}\left[\left(a(\lambda)e^{\frac{\lambda z}{d}} + a(-\lambda)e^{\frac{-\lambda z}{d}}\right)e^{-i\Omega t}\right], \tag{5.13}$$

$$\lambda = (1-i)\left(\frac{\Omega d^2}{2\alpha}\right)^{\frac{1}{2}}, \tag{5.14}$$

and

$$a(\lambda) = \frac{\Delta T}{2}\left(\frac{e^{-i\phi} - e^{-\lambda}}{e^{\lambda} - e^{-\lambda}}\right), \tag{5.15}$$

$$S_b(z) = \frac{\Delta S}{2}\left(1 - \frac{z}{d}\right) + S_0. \tag{5.16}$$

2.4 Perturbation solution

Superimposing infinitesimal perturbation on the conduction state, we have:

$$\left.\begin{aligned} \mathbf{V} = \mathbf{V_b} + \chi\mathbf{V}' = \chi(u', v', w'), \quad p = p_b + \chi p', \quad T = T_b + \chi T, \\ S = S_b + \chi S', \quad \rho = \rho_b + \chi\rho', \end{aligned}\right\} \tag{5.17}$$

where, $\mathbf{V}', p', T', S'$ and ρ' shows the perturbed quantities.

Using eq. (5.17) into eqs. (5.1)–(5.5), we obtain:

$$\nabla \cdot \mathbf{V}' = 0, \tag{5.18}$$

$$\rho_0\left[\frac{\partial}{\partial t} + (\mathbf{V}' \cdot \nabla)\right]\mathbf{V}' = -\nabla p - \frac{1}{K}\left(\mu + \tilde{\mu}\frac{\partial}{\partial t}\right)\mathbf{V}' + \mu\nabla^2\mathbf{V}' + \rho'\mathbf{g}, \tag{5.19}$$

$$\left[\frac{\partial}{\partial t} + (\mathbf{V}' \cdot \nabla)\right]T' + w'\frac{\partial T_b}{\partial z} = \alpha\nabla^2 T' + \frac{(\rho c)_p}{(\rho c)_f}\left[D_B\nabla S' \cdot \nabla T' + \left(\frac{D_T}{T_0}\right)\nabla T' \cdot \nabla T'\right], \tag{5.20}$$

$$\left[\frac{\partial}{\partial t} + (\mathbf{V}' \cdot \nabla)\right]S' + w'\frac{\partial S_b}{\partial z} = D_B\nabla^2 S' + \left(\frac{D_T}{T_0}\right)\nabla^2 T', \tag{5.21}$$

$$\rho' = S'(\rho_p - \rho_0 - \rho_0\gamma T_0) + T'(-\rho_0\beta). \tag{5.22}$$

Furthermore, we study only 2-dimensional perturbations in the investigation and therefore the stream-function ψ is familiar by $(u', 0, w') = (\frac{\partial\psi}{\partial z}, 0, -\frac{\partial\psi}{\partial x})$. By removing the density/pressure terms as eqn. (5.19).

Introducing dimension-free variable quantities:

$$\left.\begin{aligned} (x', y', z') = d(x^*, y^*, z^*), \quad t' = \frac{d^2}{\alpha}t^*, \quad V' = \frac{\alpha}{d}V^*, S' = \Delta S.S^*, \\ T' = \Delta T.T^*, \quad \nabla' = \frac{\nabla^*}{d}, \quad \Omega = \frac{\alpha}{d^2}\Omega^*, \quad \psi = \alpha\psi^* \quad \text{where, } \alpha = \frac{k}{(\rho c)_f}. \end{aligned}\right\} \tag{5.23}$$

Finally, omit the asterisk for legibility and let the dimensionless mathematical system of equations be:

$$\frac{\mathrm{Hs}}{\mathrm{Pr}}\frac{\partial}{\partial t}(\nabla^2\psi) - \mathrm{Hs}\,\nabla^4\psi + \left[1 + F_1\frac{\partial}{\partial t}\right]\nabla^2\psi + \mathrm{Ra}\,\frac{\partial T}{\partial x} - \mathrm{Rn}\,\frac{\partial S}{\partial x} = \frac{\mathrm{Hs}}{\mathrm{Pr}}\frac{\partial(\psi,\nabla^2\psi)}{\partial(x,z)}, \tag{5.24}$$

$$\left(\frac{\partial}{\partial t} - \nabla^2\right)T - \frac{\partial\psi}{\partial x}\frac{\partial T_b}{\partial z} = \frac{\partial(\psi,T)}{\partial(x,z)} + \left(\frac{N_B}{\mathrm{Le}}\right)\nabla S\cdot\nabla T + \left(\frac{N_A N_B}{\mathrm{Le}}\right)\nabla T\cdot\nabla T, \tag{5.25}$$

$$\left(\frac{\partial}{\partial t} - \frac{1}{\mathrm{Le}}\nabla^2\right)S - \frac{\partial\psi}{\partial x}\frac{\partial S_b}{\partial z} = \frac{N_A}{\mathrm{Le}}\nabla^2 T + \frac{\partial(\psi,S)}{\partial(x,z)}, \tag{5.26}$$

where,

$$\begin{aligned} &S_b = 1 - z, \quad \frac{\partial T_b}{\partial z} = -1 + \chi^2\delta_1 f(z,t), \\ &f(z,t) = \mathrm{Re}\big[(A(\lambda)e^{\lambda z} + A(-\lambda)e^{-\lambda z})\big]e^{-i\Omega t}, \\ &A(\lambda) = \frac{\lambda}{2}\left[\frac{e^{-i\phi} - e^{-\lambda}}{e^{\lambda} - e^{-\lambda}}\right], \quad \text{and} \quad \lambda = (1-i)\sqrt{\frac{\Omega}{2}}, \end{aligned} \tag{5.27}$$

and

$\mathrm{Ra} = \dfrac{\rho_0 g\gamma K d\Delta T}{\mu a}$ is HS–Rayleigh number,

$\mathrm{Rn} = \dfrac{\rho_0 g\gamma K d\Delta S[(\rho_p - \rho_0 - \gamma\rho_0 T_0)]}{\mu a}$ is nanoparticle-concentration HS–Rayleigh number,

$\mathrm{Pr} = \dfrac{\nu}{a}$ is Prandtl number,

$F_1 = \dfrac{\tilde{\mu}}{\mu}\dfrac{a}{d^2}$ is kinematic viscoelasticity permeablity,

$\mathrm{Hs} = \dfrac{K}{d^2}$ is HS-number,

$\mathrm{Le} = \dfrac{a}{D_B}$ is Lewis number,

$N_A = \dfrac{D_T\Delta T}{D_B T_0\Delta S}$ is modified-diffusivity ratio,

$N_B = \dfrac{(\rho c)_p}{(\rho c)_f}\Delta S$ is modified particle density increment.

The desired system of eqns. (5.24)–(5.26) is solved under the assumption of isothermal and stress free BC's:

$$\psi = \nabla^2\psi = T = S = 0 \quad \text{at } z = 0, 1. \tag{5.28}$$

3 Stability exploration

Here and now, given the variation in time slowed by the measure $\tau = \chi^2 t$, then the eqns. (5.24)–(5.26) get the form:

$$\frac{\chi^2 \,\mathrm{Hs}}{\mathrm{Pr}}\frac{\partial}{\partial \tau}(\nabla^2\psi) - \mathrm{Hs}\,\nabla^4\psi + \left[1 + F_1\chi^2\frac{\partial}{\partial \tau}\right]\nabla^2\psi + \mathrm{Ra}\,\frac{\partial T}{\partial x} - \mathrm{Rn}\,\frac{\partial S}{\partial x} = \frac{\mathrm{Hs}}{\mathrm{Pr}}\frac{\partial(\psi, \nabla^2\psi)}{\partial(x,z)}, \tag{5.29}$$

$$\left(\chi^2\frac{\partial}{\partial \tau} - \nabla^2\right)T - (-1 + \chi^2\delta_1 f(z,\tau))\frac{\partial \psi}{\partial x} = \frac{\partial(\psi, T)}{\partial(x,z)} + \left(\frac{N_B}{\mathrm{Le}}\right)\nabla S\cdot\nabla T + \left(\frac{N_A N_B}{\mathrm{Le}}\right)\nabla T\cdot\nabla T, \tag{5.30}$$

$$\left(\chi^2\frac{\partial}{\partial \tau} - \frac{1}{\mathrm{Le}}\nabla^2\right)S + \frac{\partial \psi}{\partial x} - \frac{N_A}{\mathrm{Le}}\nabla^2 T = \frac{\partial(\psi, S)}{\partial(x,z)}. \tag{5.31}$$

The matrix representation of eqs. (5.29)–(5.31) is given by:

$$\begin{bmatrix} \frac{\chi^2\,\mathrm{Hs}}{\mathrm{Pr}}\frac{\partial\nabla^2\psi}{\partial\tau} - \mathrm{Hs}\,\nabla^4 + [1 + F_1\chi^2\frac{\partial}{\partial\tau}]\nabla^2 & \mathrm{Ra}\,\frac{\partial}{\partial x} & -\mathrm{Rn}\,\frac{\partial}{\partial x} \\ -(-1+\chi^2\delta_1 f(z,\tau))\frac{\partial}{\partial x} & (\chi^2\frac{\partial}{\partial\tau} - \nabla^2) & 0 \\ \frac{\partial}{\partial x} & -\frac{N_A}{\mathrm{Le}}\nabla^2 & (\chi^2\frac{\partial}{\partial\tau} - \frac{1}{\mathrm{Le}}\nabla^2) \end{bmatrix} \begin{bmatrix} \Psi \\ T \\ S \end{bmatrix} \tag{5.32}$$

$$= \begin{bmatrix} \frac{\mathrm{Hs}}{\mathrm{Pr}}\frac{\partial(\psi,\nabla^2\psi)}{\partial(x,z)} \\ \frac{\partial(\psi,T)}{\partial(x,z)} \\ \frac{\partial(\psi,S)}{\partial(x,z)} \end{bmatrix}. \tag{5.33}$$

Now, we present the later asymptotic expansions used for physical variable quantities:

$$\left.\begin{aligned} \mathrm{Ra} &= \mathrm{Ra}_0 + \chi^2\,\mathrm{Ra}_2 + \chi^4\,\mathrm{Ra}_4 + \cdots \\ \psi &= \chi\psi_1 + \chi^2\psi_2 + \chi^3\psi_3 + \cdots \\ T &= \chi T_1 + \chi^2 T_2 + \chi^3 T_3 + \cdots \\ S &= \chi S_1 + \chi^2 S_2 + \chi^3 S_3 + \cdots \end{aligned}\right\}, \tag{5.34}$$

where, Ra_0 is the critical HS–Rayleigh number at which convective heat-transport occurs in absence of modulation, χ is the expansion parameter.

At the first order:

Using eq. (5.34) into the eqs. (5.33) and collecting the linear terms only, we get the following system:

$$\begin{bmatrix} \nabla^2 - \mathrm{Hs}\,\nabla^4 & \mathrm{Ra}_0\,\frac{\partial}{\partial x} & -\mathrm{Rn}\,\frac{\partial}{\partial x} \\ \frac{\partial}{\partial x} & -\nabla^2 & 0 \\ \frac{\partial}{\partial x} & -\frac{N_A}{\mathrm{Le}}\nabla^2 & -\frac{1}{\mathrm{Le}}\nabla^2 \end{bmatrix} \begin{bmatrix} \psi_1 \\ T_1 \\ S_1 \end{bmatrix} = \begin{bmatrix} 0 \\ 0 \\ 0 \end{bmatrix}. \tag{5.35}$$

The solution at the first order will be:

$$\psi_1 = A(\tau)\sin(ax)\sin(\pi z), \tag{5.36}$$

$$T_1 = \frac{-a}{\delta^2} A(\tau)\cos(ax)\sin(\pi z), \tag{5.37}$$

$$S_1 = \frac{a}{\delta^2}(\mathrm{Le} - N_A) A(\tau)\cos(ax)\sin(\pi z), \tag{5.38}$$

where, $\delta^2 = a^2 + \pi^2$ is the total wave number. The expression for critical Rayleigh number is as follows:

$$\mathrm{Ra}_0 = \frac{a^2\,\mathrm{Rn}(\mathrm{Le} - N_A) + \delta^4 + \delta^6\,\mathrm{Hs}}{a^2}. \tag{5.39}$$

At the second order:

$$\begin{bmatrix} \nabla^2 - \mathrm{Hs}\,\nabla^4 & \mathrm{Ra}_0\frac{\partial}{\partial x} & -\mathrm{Rn}\frac{\partial}{\partial x} \\ \frac{\partial}{\partial x} & -\nabla^2 & 0 \\ \frac{\partial}{\partial x} & -\frac{N_A}{\mathrm{Le}}\nabla^2 & -\frac{1}{\mathrm{Le}}\nabla^2 \end{bmatrix} \begin{bmatrix} \psi_2 \\ T_2 \\ S_2 \end{bmatrix} = \begin{bmatrix} R_{21} \\ R_{22} \\ R_{23} \end{bmatrix}, \tag{5.40}$$

where,

$$R_{21} = 0, \tag{5.41}$$

$$R_{22} = -\frac{\pi a^2}{2\delta^2} A^2(\tau)\sin(2\pi z), \tag{5.42}$$

$$R_{23} = \frac{\pi a^2}{2\delta^2}(\mathrm{Le} - N_A) A^2(\tau)\sin(2\pi z). \tag{5.43}$$

Then the solution of 2nd-order system with BC's (5.28) can be written as:

$$\psi_2 = 0, \tag{5.44}$$

$$T_2 = \frac{-a^2}{8\pi\delta^2} A^2(\tau)\sin(2\pi z), \tag{5.45}$$

$$S_2 = \frac{a^2}{8\pi\delta^2}[\mathrm{Le}(N_A - \mathrm{Le}) - N_A] A^2(\tau)\sin(2\pi z). \tag{5.46}$$

The nanoliquid-Nusselt number $\mathrm{Nu}_T(\tau)$ and concentration-Nusselt number $\mathrm{Nu}_S(\tau)$ for stationary mode of convection are:

$$\mathrm{Nu}_T(\tau) = 1 + \left[\frac{\int_0^{\frac{2\pi}{a}} (\frac{\partial T_2}{\partial z}) dx}{\int_0^{\frac{2\pi}{a}} (\frac{\partial T_b}{\partial z}) dx}\right]_{z=0}. \tag{5.47}$$

Applying eq. (5.27) and (5.45) in eq. (5.47), we obtain

$$\mathrm{Nu}_T(\tau) = 1 + \frac{a^2}{4\delta^2}A^2(\tau), \tag{5.48}$$

$$\mathrm{Nu}_S(\tau) = 1 + \left[\frac{\int_0^{\frac{2\pi}{a}}(\frac{\partial S_2}{\partial z})dx}{\int_0^{\frac{2\pi}{a}}(\frac{\partial S_b}{\partial z})dx}\right]_{z=0}. \tag{5.49}$$

Applying eqs. (5.27) and (5.46) in eq. (5.49), we obtain

$$\mathrm{Nu}_S(\tau) = 1 - \frac{a^2(\mathrm{Le}(\mathrm{Le} - N_A) - N_A)}{4\delta^2}A^2(\tau). \tag{5.50}$$

At the third order:

$$\begin{bmatrix} \nabla^2 - \mathrm{Hs}\,\nabla^4 & \mathrm{Ra}_0\frac{\partial}{\partial x} & -\mathrm{Rn}\frac{\partial}{\partial x} \\ \frac{\partial}{\partial x} & -\nabla^2 & 0 \\ \frac{\partial}{\partial x} & -\frac{N_A}{\mathrm{Le}}\nabla^2 & -\frac{1}{\mathrm{Le}}\nabla^2 \end{bmatrix}\begin{bmatrix} \psi_3 \\ T_3 \\ S_3 \end{bmatrix} = \begin{bmatrix} R_{31} \\ R_{32} \\ R_{33} \end{bmatrix}, \tag{5.51}$$

where,

$$R_{31} = -\frac{\mathrm{Hs}}{\mathrm{Pr}}\frac{\partial}{\partial \tau}(\nabla^2\psi_1) - \mathrm{Ra}_2\frac{\partial T_1}{\partial x} - F_1\frac{\partial}{\partial \tau}\nabla^2\psi_1, \tag{5.52}$$

$$R_{32} = \delta_1 f(z,\tau)\frac{\partial \psi_1}{\partial x} - \frac{\partial T_1}{\partial \tau} + \frac{\partial \psi_1}{\partial x}\frac{\partial T_2}{\partial z}, \tag{5.53}$$

$$R_{33} = -\frac{\partial S_1}{\partial \tau} + \frac{\partial \psi_1}{\partial x}\frac{\partial S_2}{\partial z}. \tag{5.54}$$

The solvability-condition is used for the 3rd-order solution to get cubic GBL equation with time-dependent coefficients as follows:

$$P_1A'(\tau) + P_2A(\tau) + P_3A^3(\tau) = 0, \tag{5.55}$$

where,

$$P_1 = \left(\frac{\pi\delta^2(\mathrm{Hs} + F_1\,\mathrm{Pr})}{2a\,\mathrm{Pr}} + \frac{\pi a\,\mathrm{Rn}(\mathrm{Le} - N_A)^2}{2\delta^4} + \frac{\pi a\,\mathrm{Ra}_0}{2\delta^4}\right),$$

$$P_2 = \left(\frac{\pi a\delta_1\,\mathrm{Ra}_0}{\delta^2}\tilde{I} - \frac{\pi a\,\mathrm{Ra}_2}{2\delta^2}\right),$$

$$P_3 = \left(\frac{\pi a^3\,\mathrm{Rn}(\mathrm{Le} - N_A)(\mathrm{Le}^2 - (\mathrm{Le} + 1)N_A)}{16\delta^4} + \frac{\pi a^3\,\mathrm{Ra}_0}{16\delta^4}\right),$$

$$\text{and} \quad \tilde{I} = \int_{z=0}^{1}\frac{\partial f}{\partial z}\sin^2(\pi z)dz.$$

4 Results and discussion

Weakly non-linear exploration is done using GBL approach to study the heat/mass transport in the Rivlin–Ericksen nanofluid under temperature modulation. The analysis is done for Le = 4, $N_A = 5$, Rn = 50, Pr = 5, $F_1 = 0.0$ to 1.5, Hs = 0.01 to 0.9, $\delta = 0.0$ to 0.5, and $\Omega = 3.0$ to 4.0 (Bhadauria and Kumar [11], Rana et al. [1]). Three phases of thermal modulation are considered:

1. Symmetric oscillating temperature field, i. e., in-phase ($\phi = 0$) modulation,
2. Asymmetric oscillating temperature field, i. e., out-phase ($\phi = \pi$) modulation,
3. Lower-boundary temperature modulation ($\phi = -i\infty$).

The Nusselt numbers for transportation of heat and mass (Nu_T and Nu_S) are evaluated as a function of time and all results are presented graphically.

4.1 Heat-transportation

Figure 5.2 demonstrates the heat transfer in the R–E nanofluid in case of symmetric oscillating temperature field ($\phi = 0$). In Fig. 5.2a, it is concluded that on rising the kinematic visco-elastic parameter F_1 the outset of convection delays. It is also notable that convection starts quite earlier in case of Newtonian liquids as compared to non-Newtonian liq-

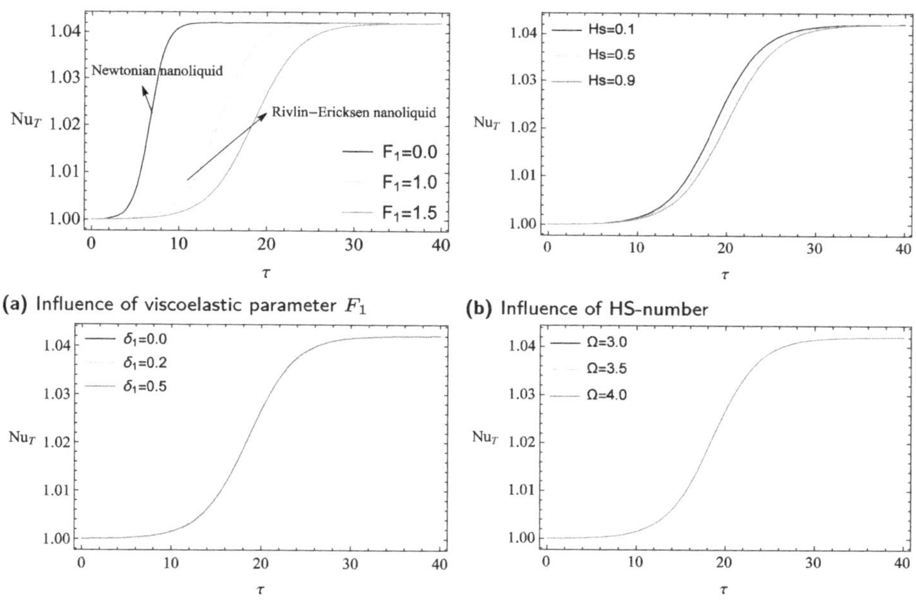

(a) Influence of viscoelastic parameter F_1

(b) Influence of HS-number

(c) Influence of amplitude of thermal modulation

(d) Influence of frequency of thermal modulation

Figure 5.2: Heat transport for in-phase modulation with Hs = 0.01, $\delta_1 = 0.2$, $\Omega = 3$, $F_1 = 1.5$.

uids. The physical reason behind such result is that for higher values of F_1 the nanofluid becomes more viscous and therefore heat transfer by conduction increases. Figure 5.2b describes that on increasing the HS-number, the convection gets delayed. This is because on rising the HS-number, the cell width increases which reduces the heat transport. We do not observe any significant impact of frequency and amplitude of in-phase temperature modulation over the heat transport in the nanofluid (Fig. 5.2c and 5.2d).

In Fig. 5.3, the transport of heat in the system is presented for asymmetric oscillating temperature field ($\phi = \pi$). An increment in heat transport is observed in case of out-phase temperature modulation. Fig. 5.3a shows that Newtonian liquid permits more heat transport as compared to Rivlin–Ericksen non-Newtonian nanoliquid. On increasing the kinematic visco-elastic parameter F_1, heat transport reduces in the system. This is because of the increased viscosity of the nanoliquid for higher values of F_1. Figure 5.3b depicts that on increasing the HS-number, heat transport reduces in the system because of the increased cell width. Similar result was also described by Bhadauria and Kumar [11]. Figure 5.3c describes that in case of $\phi = \pi$, heat transport increases in the nanoliquid with the amplitude of temperature modulation because of the increased temperature gradient. A decrement in heat transport can be observed for increasing values of frequency of temperature modulation (see Fig. 5.3d), because high frequency neutralizes the effect of increased temperature gradient. And the amount of heat transport tends to that of the unmodulated case for very high frequency of temperature modulation.

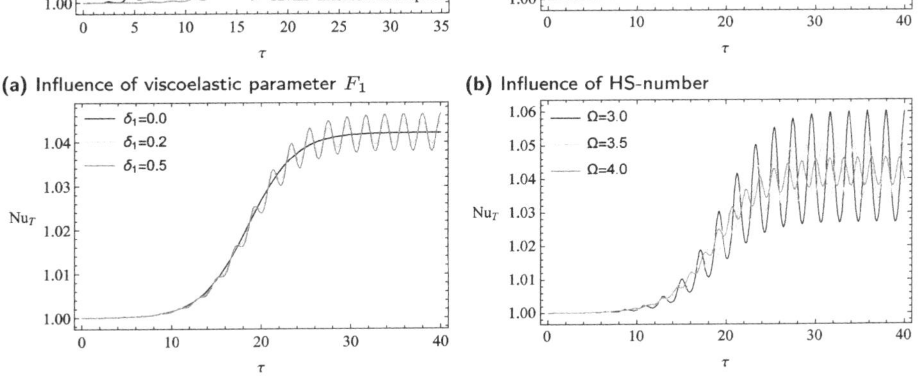

(a) Influence of viscoelastic parameter F_1 **(b)** Influence of HS-number

(c) Influence of amplitude of thermal modulation **(d)** Influence of frequency of thermal modulation

Figure 5.3: Transport of heat for out-phase modulation with Hs = 0.01, $\delta_1 = 0.2$, $\Omega = 3$, $F_1 = 1.5$.

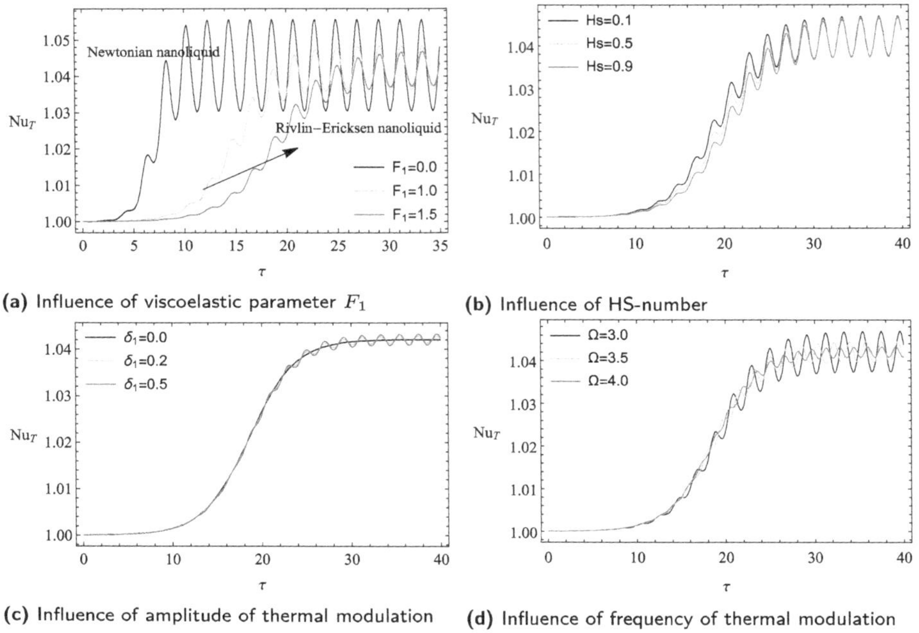

(a) Influence of viscoelastic parameter F_1

(b) Influence of HS-number

(c) Influence of amplitude of thermal modulation

(d) Influence of frequency of thermal modulation

Figure 5.4: Heat transport for lower-boundary temperature modulation with Hs $= 0.01$, $\delta_1 = 0.2$, $\Omega = 3$, $F_1 = 1.5$.

In Fig. 5.4, the case of heat transport in the system is discussed when only lower boundary is modulated ($\phi = -i\infty$). One common thing can be easily noticed that heat transport in this case is larger than in the case of symmetric oscillating temperature field and smaller than in the case of asymmetric oscillating temperature field. Similar result was also reported by Bhadauria et al. [12]. Fig. 5.4a describes one interesting result that although the steady-state heat transport in this case less than the corresponding heat transport in the case of $\phi = \pi$ yet the transport of heat by convection near the onset point is more than the corresponding heat transport in the case of $\phi = \pi$. Figures 5.4b, 5.4c, and 5.4d depict the effect of variation of HS-number, amplitude, and frequency of temperature modulation respectively over the transport of heat in the system with respect to time. Similar results are obtained, as in the previous case of asymmetric oscillating temperature field, except a decrement in overall heat transport.

4.2 Mass transportation

Figure 5.5 represents the transportation of mass in the system for the case of symmetric oscillating temperature field. It is observed that mass transport by convection starts earlier in the case of Newtonian nanoliquid as compared to the Rivlin–Ericksen nano-

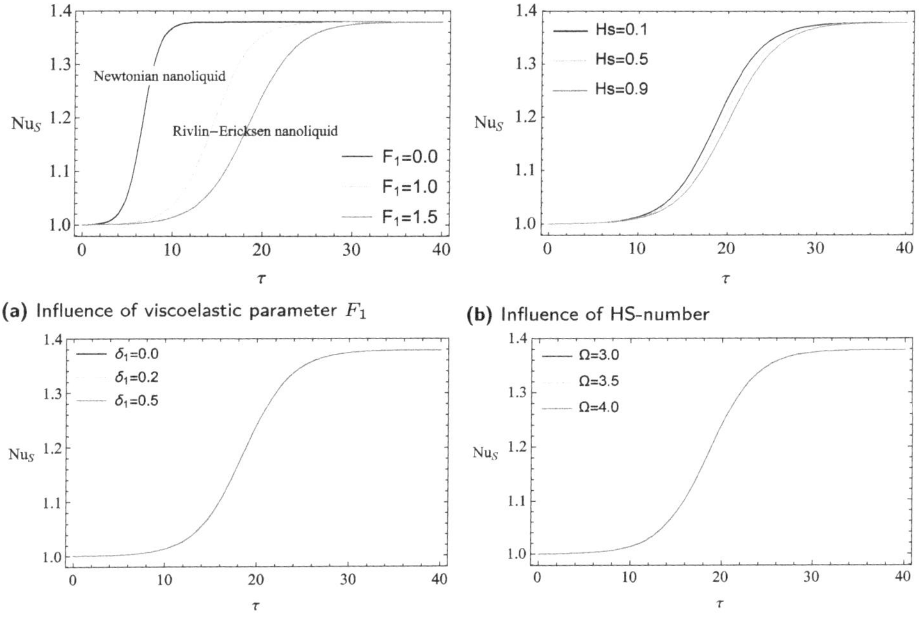

(a) Influence of viscoelastic parameter F_1 **(b)** Influence of HS-number

(c) Influence of amplitude of thermal modulation **(d)** Influence of frequency of thermal modulation

Figure 5.5: Mass transport for in-phase modulation with Hs = 0.01, $\delta_1 = 0.2$, $\Omega = 3$, $F_1 = 1.5$.

liquid. This delay in transportation of mass by convection in Rivlin–Ericksen nanoliquid is because of its viscosity (see Fig. 5.5a). Figure 5.5b explains that at higher values of HS-number, there is a delay in mass transport by convection because of the increased HS-cell width. Figures 5.5c and 5.5d describe that in case of symmetric oscillating temperature field, the amplitude and frequency of temperature modulation has no effect on mass transportation.

Figure 5.6 depicts that in there is a significant increment in the amount of mass transport in case of asymmetric oscillating temperature field. Figure 5.6a explains that on increasing the viscoelastic parameter F_1, the mass transport reduces. This is because for higher values of F_1 the fluid becomes more viscous. Figure 5.6b claims that for larger values of HS-number, i. e., for larger width of HS-cell, the mass transport reduces in the system. This is because the process of convection slows-down for higher width of HS-cell. Figures 5.6c and 5.6d represent the effect of amplitude and frequency of temperature modulation over mass transport when $\phi = \pi$. It is observed that mass transport increases with the amplitude of temperature modulation, while frequency of temperature modulation has an opposing impact.

Figure 5.7 displays the mass transport in the system when the temperature of only lower boundary is modulated. It can be seen that mass transport in this case is larger than the mass transport in case of in-phase modulation and smaller than the mass trans-

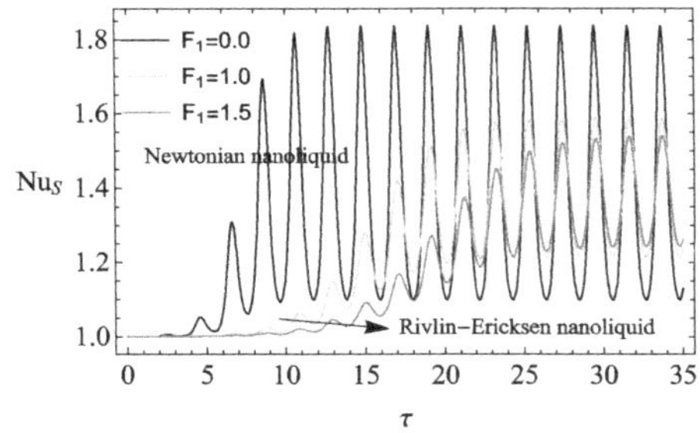

(a) Influence of viscoelastic parameter F_1

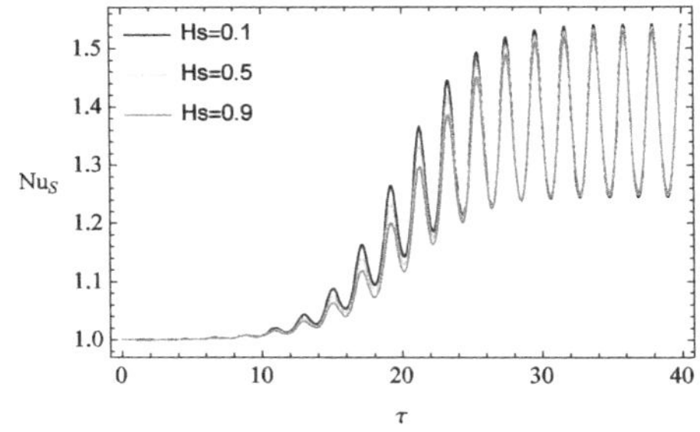

(b) Influence of HS-number

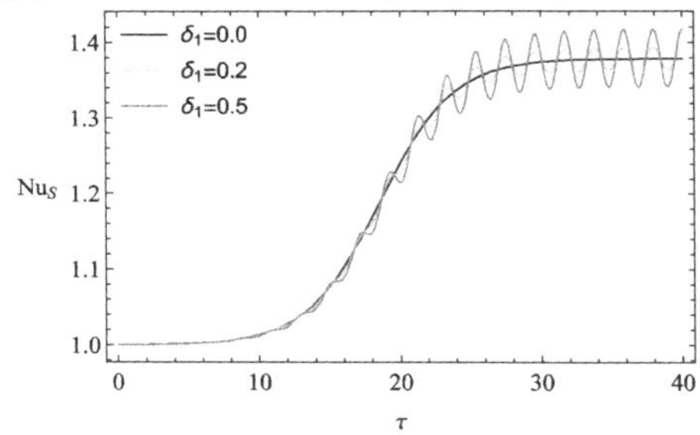

(c) Influence of amplitude of thermal modulation

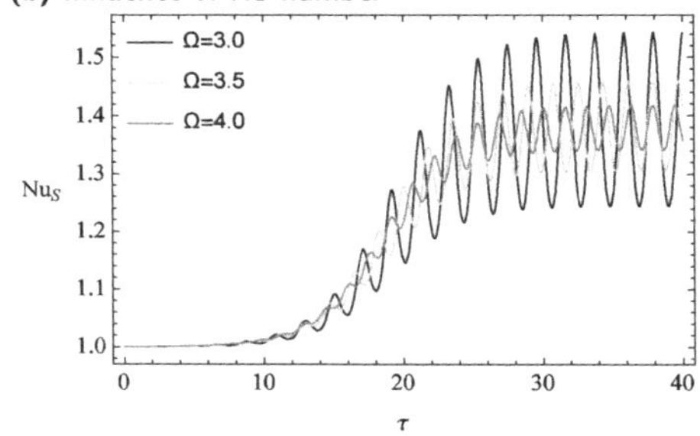

(d) Influence of frequency of thermal modulation

Figure 5.6: Mass transport for out-phase modulation with Hs = 0.01, $\delta_1 = 0.2$, $\Omega = 3$, $F_1 = 1.5$.

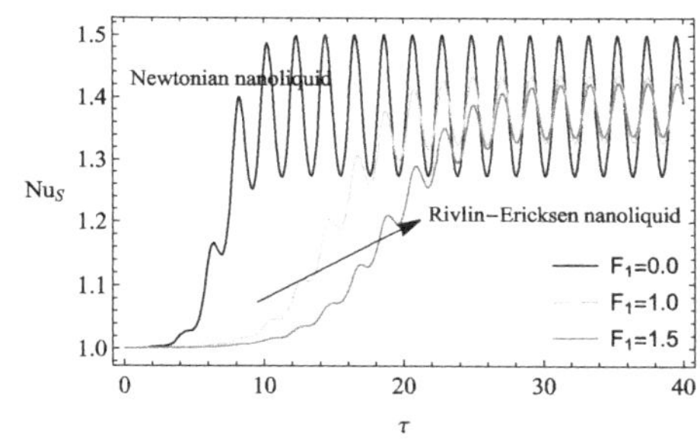

(a) Influence of viscoelastic parameter F_1

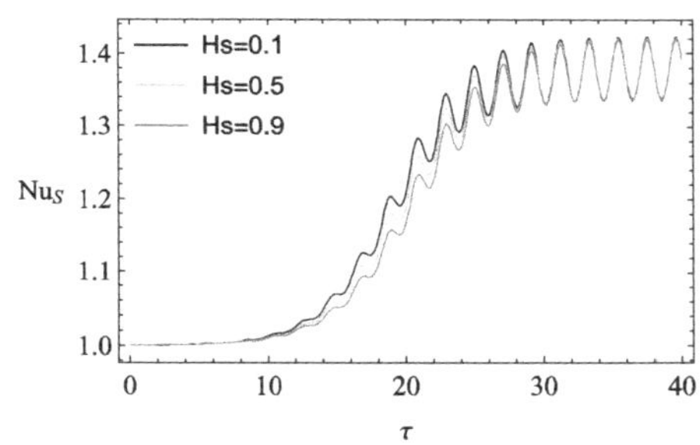

(b) Influence of HS-number

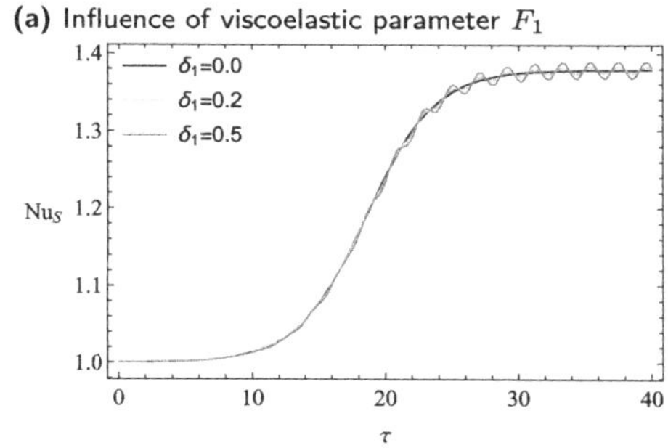

(c) Influence of amplitude of thermal modulation

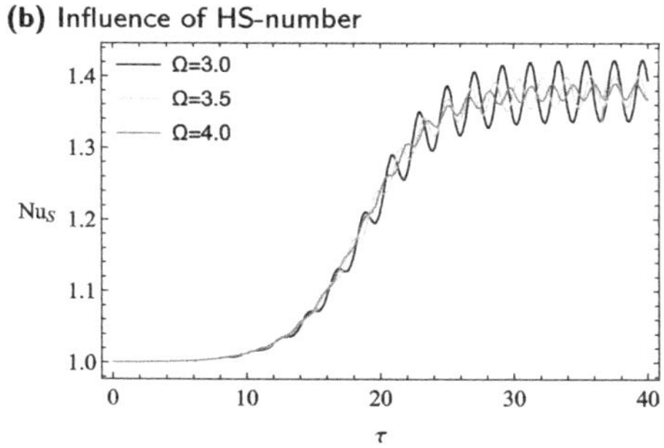

(d) Influence of frequency of thermal modulation

Figure 5.7: Mass transport for lower-boundary temperature modulation with Hs = 0.01, $\delta_1 = 0.2$, $\Omega = 3$, $F_1 = 1.5$.

port in case of out-phase modulation. Furthermore it is noteworthy that mass transport in this case ($\phi = -i\infty$) starts earlier (see Fig. 5.7a) as compared to the previous case when $\phi = \pi$. Figures 5.7b, 5.7c, and 5.7d describe the similar story as discussed in the case of asymmetric oscillating temperature field, except a decrement in mass transportation.

5 Conclusions

- The viscoelastic parameter F_1 has stabilizing effect in all three types of temperature modulation.
- More heat and mass transport is observed in case of Newtonian nanoliquid as compared to Rivlin–Ericksen nanoliquid.
- Heat/mass transport reduces in the system for increasing values of F_1.
- $\mathrm{Nu}_{T_{(\phi=0)}} < \mathrm{Nu}_{T_{(\phi=-i\infty)}} < \mathrm{Nu}_{T_{(\phi=\pi)}}$.
- $\mathrm{Nu}_{S_{(\phi=0)}} < \mathrm{Nu}_{S_{(\phi=-i\infty)}} < \mathrm{Nu}_{S_{(\phi=\pi)}}$.
- Heat/mass transport reduces in the system for increasing values of Hs.
- More convective heat/mass transport is observed near the onset point in case of $\phi = -i\infty$ as compared to the other two cases.
- No change in heat and mass transport is observed in the case of symmetric oscillating temperature field for varying amplitude and frequency of temperature modulation.
- In case of $\phi = \pi$ and $\phi = -i\infty$, heat and mass transport increase with the amplitude of temperature modulation, while reduce with the frequency of temperature modulation.

Bibliography

[1] G. C. Rana, R. Chand and V. Sharma, Thermal instability of a Rivlin–Ericksen nanofluid saturated by a Darcy-Brinkman porous medium: a more realistic model, *Eng. Trans.* **64**(3) (2016), 271–286.

[2] S. Aniss, M. Souhar and J. P. Brancher, Asymptotic study and weakly nonlinear analysis at the onset of Rayleigh-Bénard convection in Hele-Shaw cell, *Phys. Fluids* **7**(5) (1995), 926–934.

[3] M. K. Awasthi and D. Yadav, Instability of Rivlin–Ericksen fluid film with heat and mass transfer, *Int. Commun. Heat Mass Transf.* **135** (2022), 106085.

[4] A. Bansal and O. P. Suthar, A study on the effect of temperature modulation on Darcy-Bénard convection using a local thermal non-equilibrium model, *Phys. Fluids* 34(4) (2022).

[5] B. S. Bhadauria, Throughflow effect on weakly nonlinear oscillatory convection in a viscoelastic fluid saturating porous medium under temperature modulation, *J. Porous Media* **20**(6) (2017), 513–529.

[6] B. S. Bhadauria and P. K. Bhatia, Time-periodic heating of Rayleigh-Bénard convection, *Phys. Scr.* **66**(1) (2002), 59.

[7] B. S. Bhadauria, P. K. Bhatia and L. Debnath, Convection in Hele-Shaw cell with parametric excitation, *Int. J. Non-Linear Mech.* **40**(4) (2005), 475–484.

[8] B. S. Bhadauria, P. K. Bhatia and L. Debnath, Weakly non-linear analysis of Rayleigh–Bénard convection with time periodic heating, *Int. J. Non-Linear Mech.* **44**(1) (2009), 58–65.

[9] B. S. Bhadauria and P. Kiran, Weak nonlinear double diffusive magneto-convection in a Newtonian liquid under temperature modulation, *Int. J. Eng. Math.* **2014** (2014), 01–14.
[10] B. S. Bhadauria and P. Kiran, Weakly nonlinear oscillatory convection in a viscoelastic fluid saturating porous medium under temperature modulation, *Int. J. Heat Mass Transf.* **77** (2014), 843–851.
[11] B. S. Bhadauria and A. Kumar, Throughflow and gravity modulation effect on thermal instability in a Hele-Shaw cell saturated by nanofluid, *J. Porous Media* **24**(6) (2021), 31–51.
[12] B. S. Bhadauria, P. G. Siddheshwar, J. Kumar and O. P. Suthar, Weakly nonlinear stability analysis of temperature/gravity-modulated stationary Rayleigh-Bénard convection in a rotating porous medium, *Transp. Porous Media* **92** (2012), 633–647.
[13] B. S. Bhadauria, P. G. Siddheshwar and O. P. Suthar, Nonlinear thermal instability in a rotating viscous fluid layer under temperature/gravity modulation, *ASME J. Heat Transf.* **134**(10) (2012), 102502.
[14] B. S. Bhadauria and A. Srivastava, Combined effect of internal heating and through-flow in a nanofluid saturated porous medium under local thermal nonequilibrium, *J. Porous Media* **25**(2) (2022), 75–95.
[15] T. Boulal, S. Aniss, M. Belhaq and A. Azouani, Effect of quasi-periodic gravitational modulation on the convective instability in Hele-Shaw cell, *Int. J. Non-Linear Mech.* **43**(9) (2008), 852–857.
[16] J. Buongiorno, Convective transport in nanofluids, *ASME J. Heat Transf.* **128** (2006), 240–250.
[17] R. Chand and G. C. Rana, Thermal Instability of Rivlin–Ericksen elastico-viscous nanofluid saturated by a porous medium, *J. Fluids Eng.* **134**(12) (2012), 121203.
[18] R. Chand, G. C. Rana and K. Singh, Thermal instability in a Rivlin–Ericksen elastico-viscous nanofluid in a porous medium: a revised model, *Int. J. Nanosci. Nanotechnol.* **2**(1) (2015), 1–5.
[19] S. U. S. Choi, Enhancing thermal conductivity of fluids with nanoparticles, in: *Development and Applications of Non-Newtonian flows, ASME FED 231/MD*, D. A. Signier and H. P. Wang, eds, Vol. 66, 1995, pp. 99–105.
[20] B. K. Hartline and C. R. B. Lister, Thermal convection in a Hele-Shaw cell, *J. Fluid Mech.* **79**(2) (1977), 379–389.
[21] H. S. Hele-Shaw, Experiments on the nature of surface resistance of water and streamline motion under certain experimental conditions, *Trans. Inst. Naval Archtects* **40** (1898), 21–46.
[22] P. Kiran and B. S. Bhadauria, Nonlinear throughflow effects on thermally modulated porous medium, *Ain Shams Eng. J.* **7**(1) (2016), 473–482.
[23] P. Kiran and Y. Narasimhulu, Internal heating and thermal modulation effects on chaotic convection in a porous medium, *J. Nanofluids* **7**(3) (2018), 544–555.
[24] E. I. Mogilevskii, Non-Newtonian fluid film flowing down an inclined plane with a periodic topography, *Fluid Dyn.* **56** (2021), 786–798.
[25] G. C. Rana and S. Kumar, Thermal instability of Rivlin–Ericksen elasto-viscous rotating fluid permeating with suspended particles under variable gravity field in porous medium, *Stud. Geotech. Mech.* **32**(4) (2010), 39–54.
[26] G. C. Rana and R. C. Thakur, Effect of suspended particles on thermal convection in Rivlin–Ericksen fluid in a Darcy-Brinkman porous medium, *J. Mech. Eng. Sci.* **2** (2012), 162–171.
[27] R. S. Rivlin and J. L. Ericksen, Stress-deformation relations for isotropic materials, in: *Collected Papers of RS Rivlin: Volume I and II*, 1997, pp. 911–1013.
[28] M. N. Roppo, S. H. Davis and S. Rosenblat, Bénard convection with time-periodic heating, *Phys. Fluids* **27**(4) (1984), 796–803.
[29] S. Rosenblat and D. M. Herbert, Low-frequency modulation of thermal instability, *J. Fluid Mech.* **43**(2) (1970), 385–398.
[30] J. Rotheiser, *Joining of Plastics*, Hanser, 1999.
[31] S. Saini and Y. D. Sharma, The effect of vertical throughflow in Rivlin–Ericksen elastico-viscous nanofluid in a non-Darcy porous medium, *Nanosyst.: Phys. Chem. Math.* **8**(5) (2017), 606–612.
[32] K. Souhar and S. Aniss, Effect of Coriolis force on the thermosolutal convection threshold in a rotating annular Hele-Shaw cell, *Heat Mass Transf.* **48**(1) (2012), 175–182.

[33] L. P. Srivastava, Unsteady flow of Rivlin–Ericksen fluid with uniform distribution of dust particles through channels of different cross sections in the presence of time dependent pressure gradient, *Istanbul Teknil Univer. Bul.* **194** (1971), 19.
[34] M. J. Troughton, *Handbook of Plastics Joining: a Practical Guide*, William Andrew, 2008.
[35] D. Y. Tzou, Thermal instability of nanofluids in natural convection, *Int. J. Heat Mass Transf.* **51** (2008), 2967–2979.
[36] J. C. Umavathi, Effect of thermal modulation on the onset of convection in a porous medium layer saturated by a nanofluid, *Transp. Porous Media* **98** (2013), 59–79.
[37] G. Venezian, Effect of modulation on the onset of thermal convection, *J. Fluid Mech.* **35**(2) (1969), 243–254.
[38] A. Wakif, Z. Boulahia and R. Sehaqui, The effect of the rotation on the onset of convection in a Hele-Shaw cell saturated by a Newtonian nanofluid: A revised model, *Elixir Therm. Eng.* **92** (2016), 38976–38985.
[39] R. A. Wooding, Instability of a viscous liquid of variable density in a vertical Hele-Shaw cell, *J. Fluid Mech.* **7**(4) (1960), 501–515.
[40] D. Yadav, The effect of pulsating throughflow on the onset of magneto convection in a layer of nanofluid confined within a Hele-Shaw cell, *Proc. Inst. Mech. Eng., E J. Process Mech. Eng.* **233**(5) (2019), 1074–1085.

Shabir Ahmad Malik

On an operator preserving polynomial inequality of Paul Turán

Abstract: Let $P(z) = \sum_{\nu=0}^{n} a_\nu z^\nu$ be a polynomial of degree n and $D_\alpha P(z) = nP(z) + (\alpha - z)P'(z)$ is the polar differentiation of $P(z)$ with respect to point α, $\alpha \in \mathbb{C}$. Note that $D_\alpha P(z)$ is a polynomial of degree at most $n-1$. In this paper, we present some results to establish the lower bound estimates for the maximum modulus of the tth polar derivative $D_{\alpha_t} \ldots D_{\alpha_2} D_{\alpha_1}$ of polynomial $P(z)$, where $\alpha_t \in \mathbb{C}, 1 \le t < n$, under the assumption that the underlying polynomial $P(z)$ does not vanish in $|z| > k, k \ge 1$. In addition, we discuss their relative importance with respect to existing results as well.

Keywords: Polynomial, polar derivative, maximum modulus principle, inequalities in the complex plane, zeros

MSC 2010: 30A10, 30A99, 30C10, 30C15, 30E10

1 Introduction

Let $\|f\|$ denote the supremum norm of a function f on $|z| = 1$. Let $\mathcal{P}$ be the linear space of all polynomials $P(z) = \sum_{\nu=0}^{n} a_\nu z^\nu$ over $\mathbb{C}$ of degree n and $P'(z)$ be the derivative of $P(z)$. The Turán's classical inequality [13] asserts that

$$\|P'\| \ge \frac{n}{2}\|P\| \tag{6.1}$$

holds for all polynomials $P \in \mathcal{P}$ having all zeros in $|z| \le 1$. This result is the best possible and the extremal polynomial for (6.1) is $P(z) = z^n + 1$. The relationships between the bounds, their refinements and extensions, and the distribution of zeros of P in a certain region of $\mathbb{C}$ has been studied extensively and has deeply influenced such type of inequalities. One would expect, the development of the lower bound estimates in (6.1) under the condition when there is no any zero of P on $|z| = 1$. This assertion was observed in [1] in which the authors proved the following inequality under the same hypothesis as in (6.1).

$$\|P'\| \ge \frac{n}{2}\left\{\|P\| + \min_{|z|=1}|P(z)|\right\}. \tag{6.2}$$

Inequalities (6.1) and (6.2) are very useful to prove some well-known polynomial inequalities.

Acknowledgement: The author is thankful to the anonymous referee for helpful comments.

Shabir Ahmad Malik, Department of Mathematics, University of Kashmir, 223 Kulbugh, Budgam, Srinagar-190006, India, e-mail: shabir2101@gmail.com

https://doi.org/10.1515/9783111313634-006

For polynomials of a complex variable, we also have the following more general result, due to Govil [5], which is one of the best-known polynomial inequality in this direction after Malik's inequality [11] and will be useful for our results. In fact, he showed that the inequality

$$\|P'\| \geq \frac{n}{1+k^n}\|P\| \tag{6.3}$$

holds for all polynomials $P \in \mathcal{P}$ having all zeros in $|z| \leq k$, where $k \geq 1$. It is easy to see that (6.3) becomes an equality when $P(z) = z^n + k^n$. Again, one would expect that if we exclude the class of polynomials having all zeros on $|z| = k$, then it may be possible to improve the bound in (6.3). In this direction, under the same hypothesis as of (6.3), it was shown by Govil [4] that the following inequality holds

$$\|P'\| \geq \frac{n}{1+k^n}\left\{\|P\| + \min_{|z|=k}|P(z)|\right\}. \tag{6.4}$$

The research on mathematical objects associated with Turán-type inequalities has been active over a period; there are many research papers published in a variety of journals each year and different approaches have been taken for different purposes. The present article is concerned with Turán type inequalities for the polar derivative of a polynomial with restricted multiplicity of zeros. Before moving to our main results, we first introduce the concept of the operator known as *polar derivative*. For a polynomial P of degree n, we define

$$\begin{aligned} D_\alpha P(z) &= -\left[\frac{P(z)}{(z-\alpha)^n}\right]'(z-\alpha)^{n+1} \\ &= nP(z) + (\alpha - z)P'(z), \end{aligned}$$

the polar derivative of $P(z)$ with respect to point α. Note that $D_\alpha P(z)$ is a polynomial of degree at most $n-1$ and it generalizes the ordinary derivative in the sense:

$$\lim_{\alpha\to\infty}\frac{D_\alpha P(z)}{\alpha} := P'(z) \tag{6.5}$$

uniformly with respect to z for $|z| \leq R, R > 0$.

In the polar derivative setting, all the above inequalities have been widely investigated, the research in this field has sprawled in many directions and generates a plethora of publications, see, e. g., [6, 9, 14, 7]. In this paper, our interest is mainly motivated upon the study of various versions of inequalities (6.3) and (6.4), their refinements, strengthening and generalizations in the polar derivative setting by introducing constraints on the zeros of $P \in \mathcal{P}$, restrictions on the folded zeros, the modulus of largest root of $P \in \mathcal{P}$, restrictions on coefficients, etc. In this contexture, the inequality

$$\max_{|z|=1}|D_\alpha P(z)| \geq \frac{n(|\alpha|-k)}{1+k^n}\|P\| \tag{6.6}$$

holds for all polynomials $P \in \mathcal{P}$, which has all its zeros in $|z| \le k$, $k \ge 1$ and for every $\alpha \in \mathbb{C}$ with $|\alpha| \ge k$. This result is ascribed to Aziz and Rather [2].

In 2010, Dewan et al. [3] established an improvement as well as an extension of inequality (6.6) by introducing restrictions on the multiplicity of zero at $z = 0$. In fact, they proved the following result under the assumption of having information about the distribution of zeros of $P \in \mathcal{P}$ in $|z| \le k$, where $k \ge 1$.

Theorem A. *If $P(z) = z^s H(z)$ is a polynomial of degree n, where $0 \le s \le n$ and $H(z)$ is a polynomial of degree $n - s$ having all its zeros in $|z| \le k$, where $k \ge 1$, then for every real or complex number α with $|\alpha| \ge k$*

$$\max_{|z|=1}|D_\alpha P(z)| \ge \frac{n(|\alpha| - k)}{1 + k^{n-s}}\|P\|. \tag{6.7}$$

Another result in this direction ascribed to Govil and Mctume [8] acts as a refinement of (6.6) and states that:

Theorem B. *If $P(z) = \sum_{v=0}^{n} a_v z^v$ is a polynomial of degree n having all its zeros in $|z| \le k$, where $k \ge 1$, then for every real or complex number α with $|\alpha| \ge 1 + k + k^n$*

$$\max_{|z|=1}|D_\alpha P(z)| \ge n\left(\frac{|\alpha| - k}{1 + k^n}\right)\|P\| + n\left(\frac{|\alpha| - (1 + k + k^n)}{1 + k^n}\right)\min_{|z|=k}|P(z)|. \tag{6.8}$$

Now corresponding to a given nth degree polynomial $P(z)$. We can construct a sequence of polar derivatives or so-called higher order derivatives with respect to finitely many poles, as given below

$$\begin{aligned}
D_{\alpha_1}P(z) &= nP(z) + (\alpha_1 - z)P'(z),\\
D_{\alpha_2}D_{\alpha_1}P(z) &= (n-1)D_{\alpha_1}P(z) + (\alpha_2 - z)(D_{\alpha_1}P(z))',\\
\dots \quad \dots \quad &\dots\\
D_{\alpha_t}\dots D_{\alpha_2}D_{\alpha_1}P(z) &= (n-t+1)D_{\alpha_{t-1}}\dots D_{\alpha_1}P(z) + (\alpha_t - z)(D_{\alpha_{t-1}}\dots D_{\alpha_1}P(z))',\\
&\text{for } 2 \le t \le n.
\end{aligned}$$

Like the tth ordinary derivative $P^{(t)}(z)$ of $P(z)$, it is clear that the tth polar derivative $D_{\alpha_t}\dots D_{\alpha_2}D_{\alpha_1}P(z)$ of $P(z)$ is a polynomial of degree at most $n - t$.

2 Main results

In the present paper, we shall obtain some lower bound estimates for the maximum modulus of $D_{\alpha_t}\dots D_{\alpha_2}D_{\alpha_1}P(z)$, which in particular yield the inequalities (6.6), (6.7) and (6.8) as special cases. For the sake of simplicity, we use the following notations:

$$A^k_{\alpha_t} = (|\alpha_1| - k)(|\alpha_2| - k) \dots (|\alpha_t| - k),$$
$$N_t = n(n-1)(n-2) \dots (n-t+1).$$

More precisely, we first prove the following result.

Theorem 1. *If $P(z) = z^s H(z)$ is a polynomial of degree n, where $0 \le s \le n$, and $H(z)$ is a polynomial of degree $n-s$ having all its zeros in $|z| \le k$, where $k \ge 1$, then for every real or complex numbers $\alpha_1, \alpha_2, \dots, \alpha_t$ with $|\alpha_i| \ge k$, $i = 1, 2, \dots, t$,*

$$\max_{|z|=1} |D_{\alpha_t} \dots D_{\alpha_2} D_{\alpha_1} P(z)| \ge \frac{N_t}{K(t,s)} A^k_{\alpha_t} \max_{|z|=1} |P(z)|, \tag{6.9}$$

where $K(t,s) = (1 + k^{n-s})(1 + k^{n-s-1}) \dots (1 + k^{n-s-t+1})$.

Remark 1. For $t = 1$, we easily get Theorem A from Theorem 1. For $t = 1$ and $s = 0$, we get inequality (6.6), similarly for every $t = 2, 3, \dots, n-1$, we get different types of results from Theorem 1. For $t = n$, inequality (6.9) does not hold because $D_{\alpha_n} \dots D_{\alpha_2} D_{\alpha_1} P(z)$ is a constant polynomial.

If we choose $\alpha_1 = \alpha_2 = \dots = \alpha_t = \alpha$, then by dividing both sides of inequality (6.9) by $|\alpha|^t$ and letting $|\alpha| \to \infty$, therefore taking (6.5) into consideration, we obtain the following result.

Corollary 1. *If $P(z) = z^s H(z)$ is a polynomial of degree n, where $0 \le s \le n$ and $H(z)$ is a polynomial of degree $n-s$ having all its zeros in $|z| \le k$, where $k \ge 1$, then*

$$\max_{|z|=1} |P^{(t)}(z)| \ge \frac{N_t}{K(t,s)} \max_{|z|=1} |P(z)|, \tag{6.10}$$

where $K(t,s) = (1 + k^{n-s})(1 + k^{n-s-1}) \dots (1 + k^{n-s-t+1})$.

Remark 2. For $t = 1$ and $s = 0$, we get inequality (6.3) from Corollary 1.

Next, we prove the following generalization of Theorem B.

Theorem 2. *If $P(z) = z^s H(z)$ is a polynomial of degree n, where $0 \le s \le n$, and $H(z)$ is a polynomial of degree $n-s$ having all its zeros in $|z| \le k$, where $k \ge 1$, then for every real or complex number $\alpha_1, \alpha_2, \dots, \alpha_t$ with $|\alpha_i| \ge 1 + k + k^n$, $i = 1, 2, \dots, t$,*

$$\max_{|z|=1} |D_{\alpha_t} \dots D_{\alpha_2} D_{\alpha_1} P(z)| \ge \frac{N_t}{K(t,s)} \left[A^k_{\alpha_t} \|P\| + \{A^k_{\alpha_t} - K(t,s)\} \min_{|z|=k} |P(z)| \right], \tag{6.11}$$

where $K(t,s) = (1 + k^{n-s})(1 + k^{n-s-1}) \dots (1 + k^{n-s-t+1})$.

Remark 3. For $s = 0$, Theorem 2 reduces to a result ascribed to Malik et al. [10]. For $t = 1$ and $s = 0$, Theorem 2 gives us inequality (6.8) of Theorem B as a special case.

If we choose $\alpha_1 = \alpha_2 = \cdots = \alpha_t = \alpha$, then by dividing both sides of inequality (6.11) by $|\alpha|^t$ and letting $|\alpha| \to \infty$, therefore taking (6.5) into consideration, we obtain the following result.

Corollary 2. *If $P(z) = z^s H(z)$ is a polynomial of degree n, where $0 \le s \le n$ and $H(z)$ is a polynomial of degree $n - s$ having all its zeros in $|z| \le k$, where $k \ge 1$, then*

$$\max_{|z|=1} |P^{(t)}(z)| \ge \frac{N_t}{K(t,s)} \Big[\|P\| + \min_{|z|=k} |P(z)| \Big], \tag{6.12}$$

where $K(t,s) = (1 + k^{n-s})(1 + k^{n-s-1}) \ldots (1 + k^{n-s-t+1})$.

Remark 4. Inequality (6.12) is a refinement of inequality (6.10) because the inequality

$$\|P\| + \min_{|z|=k} |P(z)| \ge \|P\|$$

is always true. For $t = 1$ and $s = 0$, inequality (6.12) reduces to inequality (6.4). For $t = 1$, $s = 0$, $k = 1$ and $\min_{|z|=k} |P(z)| = 0$, inequality (6.12) reduces to inequality (6.1), which is classical inequality due to Turán.

3 Lemmas

For the proofs of main results we require the following lemma, which follows by repeated applications of Laguerre's theorem [12].

Lemma 1. *If all the zeros of an nth degree polynomial $P(z)$ lie in a circular region C and if none of the points $\alpha_t, \alpha_{t-1}, \ldots, \alpha_1$ lie in the region C, then each of the polar derivatives*

$$D_{\alpha_t} \ldots D_{\alpha_2} D_{\alpha_1} P(z), \quad t = 1, 2, \ldots, n-1$$

has all of its zeros in region C.

4 Proofs of the theorems

Proof of Theorem 1. Theorem follows trivially if $|\alpha_i| = k$ for at least one i, $1 \le i \le t \le n$. Therefore, from now on we assume that

$$|\alpha_i| > k.$$

We shall prove this result by mathematical induction. Result is true for $t = 1$ by inequality (6.7), i. e., if $|\alpha_1| > k$, then

$$\left|D_{\alpha_1}P(z)\right| \geq \frac{n(|\alpha_1| - k)}{1 + k^{n-s}}\left|P(z)\right|. \tag{6.13}$$

Now for $t = 2$. Except for one value (say α) of α_1, $D_{\alpha_1}P(z)$ is a polynomial of degree $(n - 1)$. Let us take any α_1 ($\alpha_1 \neq \alpha$ if $|\alpha| > k$) with $|\alpha_1| > k$ and fix it up. Thus $D_{\alpha_1}P(z)$ is a polynomial of degree $(n - 1)$ and by Lemma 1 it has all its zeros in $|z| \leq k$. Therefore on applying inequality (6.7) to $D_{\alpha_1}P(z)$ with

$$\alpha = \alpha_2, \quad |\alpha_2| > k$$

we get,

$$\left|D_{\alpha_2}(D_{\alpha_1}P(z))\right| \geq \frac{n-1}{1 + k^{n-s-1}}(|\alpha_2| - k)\left|D_{\alpha_1}P(z)\right|.$$

Using (6.13), we have

$$\left|D_{\alpha_2}D_{\alpha_1}P(z)\right| \geq \frac{n(n-1)}{(1 + k^{n-s})(1 + k^{n-s-1})}(|\alpha_1| - k)(|\alpha_2| - k)\left|P(z)\right|.$$

It follows result is true for $t = 2$. We assume that result is true for

$$t = q$$

i. e., for $|z| = 1$ and $|\alpha_i| > k$, $i = 1, 2, \ldots, q$

$$\left|D_{\alpha_q}\ldots D_{\alpha_2}D_{\alpha_1}P(z)\right| \geq \frac{N_q}{(1 + k^{n-s})(1 + k^{n-s-1})\ldots(1 + k^{n-s-q+1})}A^k_{\alpha_q}\left|P(z)\right|, \tag{6.14}$$

and we will prove that result is true for

$$t = q + 1, (< n).$$

Again, except for one value (say α_1') of α_1, $D_{\alpha_1}P(z)$ will be a polynomial of degree $(n-1)$. Let us take any α_1 ($\alpha_1 \neq \alpha_1'$ if $|\alpha_1'| > k$) with $|\alpha_1| > k$ and fix it up. Thus $D_{\alpha_1}P(z)$ is a polynomial of degree $(n-1)$. Now $D_{\alpha_2}D_{\alpha_1}P(z)$ will be a polynomial of degree $(n-2)$ for every α_2, except for one value α_2', (say), of α_2. Let us take any α_2 ($\alpha_2 \neq \alpha_2'$ if $|\alpha_2'| > k$) with $|\alpha_2| > k$ and fix it up. Therefore, $D_{\alpha_2}D_{\alpha_1}P(z)$ is a polynomial of degree $(n - 2)$. Likewise, one can continue and say that $D_{\alpha_q}\ldots D_{\alpha_2}D_{\alpha_1}P(z)$ will be a polynomial of degree $(n - q)$ for every α_q, except for one value α_q', (say), of α_q. Let us take any α_q ($\alpha_q \neq \alpha_q'$ if $|\alpha_q'| > k$) with $|\alpha_q| > k$ and fix it up. Thus $D_{\alpha_q}\ldots D_{\alpha_2}D_{\alpha_1}P(z)$ is a polynomial of degree $(n - q)$ with fixed α_q ($|\alpha_q| > k$), fixed α_{q-1} ($|\alpha_{q-1}| > k$), $\ldots$, fixed α_1 ($|\alpha_1| > k$) and by Lemma 1 $D_{\alpha_q}\ldots D_{\alpha_2}D_{\alpha_1}P(z)$ has all its zeros in $|z| \leq k$. Therefore on applying inequality (6.7) to $D_{\alpha_q}\ldots D_{\alpha_2}D_{\alpha_1}P(z)$ with

$$\alpha = \alpha_{q+1}, \quad |\alpha_{q+1}| > k$$

we get,

$$|D_{\alpha_{q+1}}(D_{\alpha_q}\dots D_{\alpha_2}D_{\alpha_1}P(z))| \geq \frac{n-q}{1+k^{n-q-s}}(|\alpha_{q+1}|-k)|D_{\alpha_q}\dots D_{\alpha_2}D_{\alpha_1}P(z)|,$$

which on being combined with (6.14) gives for $|z|=1$ and $|\alpha_{q+1}|>k$

$$|D_{\alpha_{q+1}}D_{\alpha_q}\dots D_{\alpha_2}D_{\alpha_1}P(z)| \geq \frac{n(n-1)\dots(n-q)}{(1+k^{n-s})(1+k^{n-s-1})\dots(1+k^{n-s-q})}A^k_{\alpha_{q+1}}|P(z)|. \tag{6.15}$$

Further, α_q ($\alpha_q \neq \alpha'_q$ if $|\alpha'_q|>k$) with $|\alpha_q|>k$ was a fixed element but was chosen arbitrarily. Accordingly, (6.15) is true for every α_{q+1} ($|\alpha_{q+1}|>k$) and every α_q ($\alpha_q \neq \alpha'_q$ if $|\alpha'_q|>k$) with $|\alpha_q|>k$, i. e.,

$$|D_{\alpha_{q+1}}D_{\alpha_q}\dots D_{\alpha_2}D_{\alpha_1}P(z)| \geq \frac{n(n-1)\dots(n-q)}{(1+k^{n-s})(1+k^{n-s-1})\dots(1+k^{n-s-q})}A^k_{\alpha_{q+1}}|P(z)|,$$
$$\text{for } |z|=1, |\alpha_{q+1}|>k \,\&\, |\alpha_q|>k(\alpha_q \neq \alpha'_q \text{ if } |\alpha'_q|>k). \tag{6.16}$$

Now if $|\alpha'_q|>k$, then for a fixed α_{q+1} ($|\alpha_{q+1}|>k$), $D_{\alpha_{q+1}}D_{\alpha_q}\dots D_{\alpha_2}D_{\alpha_1}$ is continuous function of α_q and hence by continuity we can say (6.16) will be true for a fixed α_{q+1} and α'_q and accordingly (6.16) will be true for every α_{q+1} ($|\alpha_{q+1}|>k$) and α'_q. Thus (6.16) is true for every α_{q+1} ($|\alpha_{q+1}|>k$) and every α_q ($|\alpha_q|>k$), that is

$$|D_{\alpha_{q+1}}D_{\alpha_q}\dots D_{\alpha_2}D_{\alpha_1}P(z)| \geq \frac{n(n-1)\dots(n-q)}{(1+k^{n-s})(1+k^{n-s-1})\dots(1+k^{n-s-q})}A^k_{\alpha_{q+1}}|P(z)|,$$
$$\text{for } |z|=1, |\alpha_{q+1}|>k \,\&\, |\alpha_q|>k. \tag{6.17}$$

As argued for α_q and α'_q, we can argue for α_{q-1} and α'_{q-1} and say that

$$|D_{\alpha_{q+1}}D_{\alpha_q}\dots D_{\alpha_2}D_{\alpha_1}P(z)| \geq \frac{n(n-1)\dots(n-q)}{(1+k^{n-s})(1+k^{n-s-1})\dots(1+k^{n-s-q})}A^k_{\alpha_{q+1}}|P(z)|,$$
$$\text{for } |z|=1, |\alpha_{q+1}|>k, |\alpha_q|>k \,\&\, |\alpha_{q-1}|>k.$$

One can continue similarly and obtain

$$|D_{\alpha_{q+1}}D_{\alpha_q}\dots D_{\alpha_2}D_{\alpha_1}P(z)| \geq \frac{n(n-1)\dots(n-q)}{(1+k^{n-s})(1+k^{n-s-1})\dots(1+k^{n-s-q})}A^k_{\alpha_{q+1}}|P(z)|,$$
$$\text{for } |z|=1, |\alpha_{q+1}|>k, |\alpha_q|>k, |\alpha_{q-1}|>k,\dots,|\alpha_2|>k, |\alpha_1|>k.$$

Hence, the result is true for

$$t=q+1.$$

So we conclude with the help of mathematical induction that result is true for $t \geq 1$. This proves Theorem 1 completely. □

Proof of Theorem 2. Let

$$m = \min_{|z|=k} |P(z)|,$$

then $|P(z)| \geq m$ on $|z| = k$. Therefore, for every λ with $|\lambda| < 1$

$$|P(z)| > |\lambda m| \quad \text{on } |z| = k.$$

If $P(z)$ has a zero on $|z| = k$, then $m = 0$ and the result follows from Theorem 1. Therefore, from now on we will assume that $P(z)$ has all its zeros in $|z| < k$, where $k \geq 1$. By Rouche's theorem the polynomial

$$F(z) = P(z) + \lambda m$$

also has all its zeros in $|z| < k$. Thus, on applying Theorem 1 to $F(z)$, we obtain for $|\alpha_1| \geq k$, $|\alpha_2| \geq k, \ldots, |\alpha_t| \geq k$

$$|D_{\alpha_t} \ldots D_{\alpha_2} D_{\alpha_1} F(z)| \geq \frac{N_t}{(1 + k^{n-s})(1 + k^{n-s-1}) \ldots (1 + k^{n-s-t+1})} A^k_{\alpha_t} |F(z)|,$$

i. e.,

$$|D_{\alpha_t} \ldots D_{\alpha_2} D_{\alpha_1} P(z) + m\lambda n(n-1) \ldots (n-t+1)| \geq \frac{N_t}{K(t,s)} A^k_{\alpha_t} |P(z) + \lambda m|, \tag{6.18}$$

where $K(t,s) = (1 + k^{n-s})(1 + k^{n-s-1}) \ldots (1 + k^{n-s-t+1})$.

If we choose the argument of λ such that

$$|P(z) + \lambda m| = |P(z)| + |\lambda| m,$$

then from inequality (6.18), we get

$$|D_{\alpha_t} \ldots D_{\alpha_2} D_{\alpha_1} P(z) + m\lambda n(n-1) \ldots (n-t+1)| \geq \frac{N_t}{K(t,s)} A^k_{\alpha_t} \{|P(z)| + |\lambda| m\},$$

this gives for $|z| = 1$, that

$$|D_{\alpha_t} \ldots D_{\alpha_2} D_{\alpha_1} P(z)| + mN_t |\lambda| \geq \frac{N_t}{K(t,s)} A^k_{\alpha_t} \{|P(z)| + |\lambda| m\}.$$

Equivalently,

$$|D_{\alpha_t} \ldots D_{\alpha_2} D_{\alpha_1} P(z)| \geq \frac{N_t}{K(t,s)} [A^k_{\alpha_t} |P(z)| + |\lambda| \{A^k_{\alpha_t} - K(t,s)\} m]. \tag{6.19}$$

Now letting $|\lambda| \to 1$ in (6.19), we get

$$\max_{|z|=1}|D_{\alpha_t}\dots D_{\alpha_2}D_{\alpha_1}P(z)| \geq \frac{N_t}{K(t,s)}\Big[A^k_{\alpha_t}\|P\| + \{A^k_{\alpha_t} - K(t,s)\}\min_{|z|=k}|P(z)|\Big].$$

This proves Theorem 2 completely. □

Bibliography

[1] A. Aziz and Q. M. Dawood, Inequalities for a polynomial and its derivative, *J. Approx. Theory* **54** (1988), 306–313.

[2] A. Aziz and N. A. Rather, A refinement of a theorem of Paul Turán concerning polynomials, *Math. Inequal. Appl.* **1** (1998), 231–238.

[3] K. K. Dewan, A. Ahuja and S. Hans, Turán-type inequalities for polynomials with restricted zeros, *J. Interdiscip. Math.* **13** (2010), 217–224.

[4] N. K. Govil, Some inequalities for derivatives of polynomials, *J. Approx. Theory* **66** (1991), 29–35.

[5] N. K. Govil, On the derivative of a polynomial, *Proc. Am. Math. Soc.* **41** (1973), 543–546.

[6] N. K. Govil and P. Kumar, On Bernstein-type inequalities for the polar derivative of a polynomial, in: *Progress in Approximation Theory and Applicable Complex Analysis*, Vol. 117, 2017, pp. 41–74.

[7] N. K. Govil and P. Kumar, On sharpening of an inequality of Turán, *Appl. Anal. Discrete Math.* **13** (2019), 711–720.

[8] N. K. Govil and G. N. Mctume, Some generalizations involving the polar derivative for an inequality of Paul Turán, *Acta Math. Hung.* **104**(1–2) (2004), 115–126.

[9] P. Kumar, On Zygmund-type inequalities involving polar derivative of a lacunary type polynomial, *Bull. Math. Soc. Sci. Math. Roum.* **61** (2019), 155–164.

[10] S. A. Malik and B. A. Zargar, On a generalization of an operator preserving Turán-type inequality for complex polynomials, *J. Contemp. Math. Anal.* **58** (2023), 341–346.

[11] M. A. Malik, On the derivative of a polynomial, *J. Lond. Math. Soc.* **2** (1969), 57–60.

[12] M. Marden, *Geometry of Polynomials*, 2nd edn, Mathematical Surveys, Amer. Math. Soc., Providence R. I., 1966.

[13] P. Turán, Über die Ableitung von polynomen, *Compos. Math.* **7** (1939), 89–95.

[14] A. Zireh and M. Bidkham, Inequalities for the polar derivative of a polynomial with restricted zeros, *Kragujev. J. Math.* **40** (2016), 113–124.

Piu Kundu

Effect of interaction between two finite faults on ground deformation

Abstract: In this paper, we consider a model of two interacting faults that are finite in length and aseismic, locked and inclined. The faults are situated in visco-elastic half-space and the medium is Burger's rheology. The movement of the faults helps us to study the stress-strain accumulation in the neighboring fault due to movement across other faults. The solutions of displacement, stress and strain due to interaction between the faults are obtained using the method of integral transform, Green's function technique and the correspondence principle. The effect of the inclination of both the faults, distance between the faults and their slip magnitude on ground deformation have been described graphically.

Keywords: Finite strike-slip fault, interacting faults, locked, integral transform, Burger's rheology

MSC 2010: 11A25, 11A67

1 Introduction

At present, the observation of the earthquake fault model is one of the main topic of research in theoretical seismology. Occurrence of an earthquake is the cause of a sudden movement of tectonic plates, which always move slowly. Due to friction, the tectonic plates get stuck at their edges. When the stress at the edge overcomes the friction, some energy is released in waves form travelling through the earth's crust and cause earthquake. When a fault in a seismically active region moves in aseismic period (period between two consecutive seismic events), stress-strain accumulates or is released in this region. In different parts of the world, there are some major fault systems in consist neighboring faults instead of a single fault. For example, Hayward faults, Garlock and San Jacinto faults are in the neighborhood of SanAndreas fault system in southern California.

"Some theoretical models of the lithosphere-asthenosphere system in seismically active regions during aseismic periods with two interacting faults" have been presented by Mukhopadhyay et al. (1978b [18], 1979c [19]), Ghosh et al. (1992a [6], 1992b [7]), Mukhopadhya and Mukherji (1984 [20], 1986 [21]) and Ghosh and Sen (2011 [8]). Kato, N.

Acknowledgement: The authors thank the anonymous referee for helpful comments.

Piu Kundu, Department of Mathematics, School of Advance Sciences, VIT-AP University, CB-205A, Inavolu, Beside AP Secretariat, 522237 Amravati AP, India, e-mail: piukundu91@gmail.com

https://doi.org/10.1515/9783111313634-007

and Lei, X. (2001 [11]) demonstrated “a numerical simulation of the activities of many parallel strike-slip faults embedded in an elastic layer over a viscoelastic half-space”. Parsons et al. (2003 [22]) studied “the potential interaction of the Hayward and Rodgers Creek faults at the extensional step-over that separates the two faults underneath San Pablo Bay”. Debnath, P. and Sen, S. (2014 [4]) have been described “a model of interaction between two neighboring strike-slip faults in standard linear solids”. Segall and Pollard (1980 [25]) investigated interacting strike-slip fault model. Mavco (1982 [16]) modelled has explained the recordsof creep on the San Andreas fault near Hollister, California. Shao, B. and Guiting H. (2019 [26]) have established “the interactions of fault patterns and stress fields during active faulting in Central North China Block by using numerical simulation”. Attanayake, J. et al. (2019 [1]) have explained about “interacting intraplate fault systems in Australia”. Debnath, S. K (2013 [5]) has described “a model of two interacting creeping vertical finite strike-slip faults in a visco-elastic half-space of the lithosphere”. “A model of interaction between two long inclined strike-slip faults in layered medium” developed by Manna, K. and Sen, S. (2017 [13]). Kundu, P. and Sarkar (Mondal), S. (2020 [12]) have described “the ground deformation by two interacting faults of one infinite and other finite in visco-elastic half-space of Maxwell medium”.

The representation of lithosphere-asthenosphere system by the material of Maxwell or Standard linear solid has been considered by most of the researchers. Hu. Y. et al. (2016 [9]) suggested that “Burger-type visco-elastic material may be a suitable representation of the lithosphere-asthenosphere system in the seismically active region during aseismic period”. Mondal, D. et al. (2019 [17]) developed “the effect of movement of a long, buried, inclined fault in visco-elastic half-space of Burger’s rheology”. Burger’s rheology is described in Fig. 7.1 which is the combination of the Maxwell and Kelvin-Voiget type material. The solutions of displacements, stresses and strains obtained in Burger’s rheology depends on time. For a details study of ground deformation in a viscoelastic medium of Burger’s rheology, a model of finite, buried, inclined, interacting strike-slip faults with sudden movement have been considered. After the sudden movement, the faults become locked, i. e., there is no movement across the faults. The solutions for displacements, stresses and strains are obtained before the fault movement, after the movement of the first fault when the other fault remains locked and after the interaction between both the faults by using Green’s function technique and integral transforms.

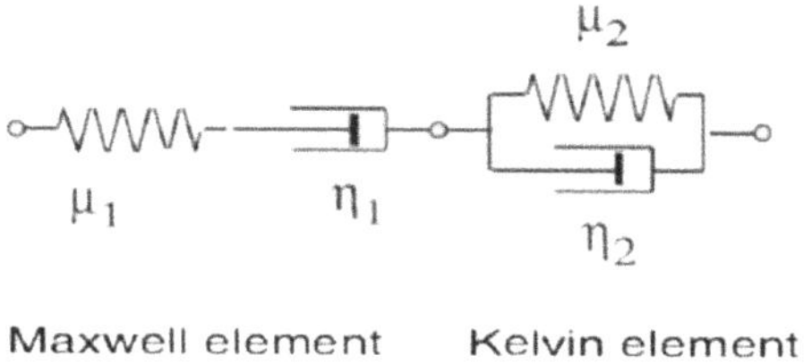

Figure 7.1: Burger’s model.

Numerical computation and graphical illustrations have been carried out for describing displacement, stress and strain components.

The rest of the paper is formed as follows: the model description and the formulation of the problem have been discussed in Section 2. The solutions have been obtained in Section 3. In Section 4, the results have been discussed with some numerical computations. Finally, the conclusion and the future scope of our work are given in Section 5.

2 Description of the model and formulation of the problem

To formulate the present problem, an interacting, buried, inclined and strike-slip fault model with interaction between two finite faults has been depicted through Fig. 7.2. The faults are defined as F_1 and F_2. The material of the visco-elastic half-space where the fault is situated, is Burger's Rheology. The inclinations are θ_1 and θ_2 across F_1 and F_2 respectively. d_1 and d_2 are taken as depth from the free surface of the faults F_1 and F_2 respectively. Width of the faults is assumed as D_1 and D_2 respectively and D be the distance between them. The length of finite fault F_1 is $2L_1$ and F_2 is $2L_2$. The fault F_1 moves after time $t = T_1$ (> 0) and F_2 moves after time $t = T_2$ (> 0), where $T_1 < T_2$. The movement across both the faults is sudden. After the movement at time T_1 and T_2 the faults becomes locked, that is there is no movement across the faults.

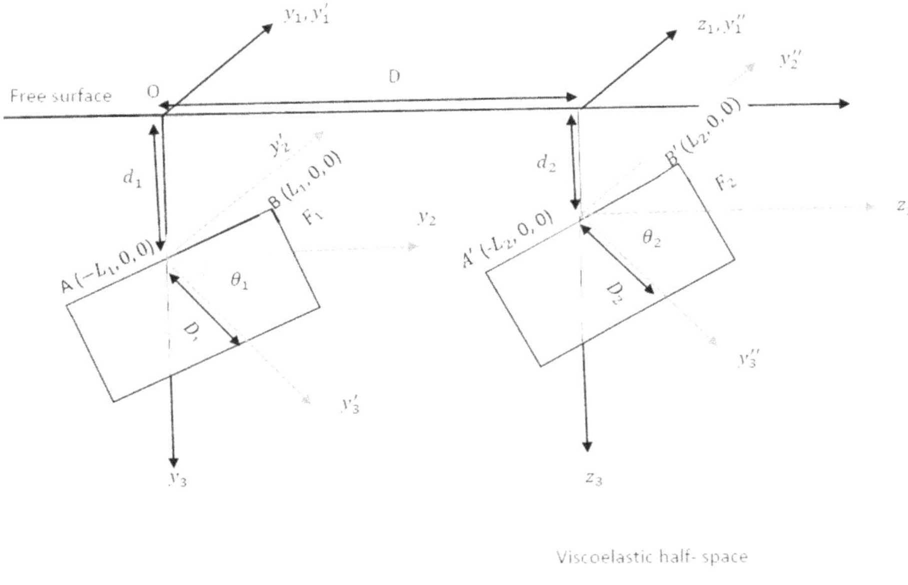

Figure 7.2: Two finite interacting buried strike-slip fault in visco-elastic half-space of Burger medium.

Two sets of rectangular Cartesian co-ordinate system (y_1, y_2, y_3) and (y_1', y_2', y_3') have been chosen as for fault F_1 and two other sets (z_1, z_2, z_3) and (y_1'', y_2'', y_3'') for F_2 as described in Fig. 7.2. y_3 and z_3 axes are pointing into the half-space with $y_3 = 0$ and $z_3 = 0$ as the free surface. y_1 and z_1 axes are vertically above the upper edges of the fault. These systems are associated by the following relations:

$$\left.\begin{aligned} y_1' &= y_1 \\ y_2' &= y_2 \sin\theta_1 - (y_3 - d_1)\cos\theta_1 \\ y_3' &= y_2 \cos\theta_1 + (y_3 - d_1)\sin\theta_1 \end{aligned}\right\} \tag{7.1}$$

and

$$\left.\begin{aligned} y_1'' &= z_1 \\ y_2'' &= z_2 \sin\theta_2 - z_3\cos\theta_2 \\ y_3'' &= z_2 \cos\theta_2 + z_3\sin\theta_2 \end{aligned}\right\}, \tag{7.2}$$

where $z_2 = y_2 - D$ and $z_3 = y_3 - d_2$.

The co-ordinates of the two end points of the upper edge of F_1 and F_2 are A, B and A', B' are respectively given by A $(-L_1, 0, 0)$, B $(L_1, 0, 0)$ and A' $(-L_2, 0, 0)$, B' $(L_2, 0, 0)$ w. r. t. prime co-ordinate system (y_1', y_2', y_3') and (y_1'', y_2'', y_3''). Then the faults F_1 and F_2 are given by

$$\begin{aligned} F_1 &: (-L_1 \le y_1' \le L_1, y_2' = 0, 0 \le y_3' \le D_1), \\ F_2 &: (-L_2 \le y_1'' \le L_2, y_2'' = 0, 0 \le y_3'' \le D_2). \end{aligned}$$

The displacement, stress and strain components are u_i $(i = 1, 2, 3)$, τ_{ij} $(i, j = 1, 2, 3)$ and e_{ij} $(i, j = 1, 2, 3)$.

2.1 Constitutive equations

For visco-elastic medium of Burger's rheology, the constitutive equations have been taken as

$$\tau_{ij} + p_1 \frac{\partial}{\partial t}(\tau_{ij}) + p_2 \frac{\partial^2}{\partial t^2}(\tau_{ij}) = 2q_1 \frac{\partial}{\partial t}(e_{ij}) + 2q_2 \frac{\partial^2}{\partial t^2}(e_{ij}) \quad (i, j = 1, 2, 3), \tag{7.3}$$

where

$$p_1 = \frac{\eta_1}{\mu_1} + \frac{\eta_2}{\mu_2} + \frac{\eta_1}{\mu_2}, \quad p_2 = \frac{\eta_1\eta_2}{\mu_1\mu_2}, \quad q_1 = \eta_1, \quad q_2 = \frac{\eta_1\eta_2}{\mu_2}.$$

Here constants η_1, η_2 denotes the respective effective viscosities and μ_1, μ_2 represents the respective effective rigidities of the material.

2.2 Quasi-static equilibrium equation

The quasi-static equilibrium equations are:

$$\left.\begin{aligned} \frac{\partial}{\partial y_1}(\tau_{11}) + \frac{\partial}{\partial y_2}(\tau_{12}) + \frac{\partial}{\partial y_3}(\tau_{13}) = 0 \\ \frac{\partial}{\partial y_1}(\tau_{21}) + \frac{\partial}{\partial y_2}(\tau_{22}) + \frac{\partial}{\partial y_3}(\tau_{23}) = 0 \\ \frac{\partial}{\partial y_1}(\tau_{31}) + \frac{\partial}{\partial y_2}(\tau_{32}) + \frac{\partial}{\partial y_3}(\tau_{33}) = 0 \end{aligned}\right\} \tag{7.4}$$

2.3 Boundary conditions

Since the faults are finite then boundary conditions near the ends of F_1 at time t equal to zero are taken as

$$\left.\begin{aligned} &\lim_{y_1'\to L_1-} \tau_{11} = \lim_{y_1'\to L_1+} \tau_{11} = \tau_{L_1}(\text{say}) \\ \text{and}\quad &\lim_{y_1'\to(-L_1)-} \tau_{11} = \lim_{y_1'\to(-L_1)+} \tau_{11} = \tau_{-L_1}(\text{say}) \end{aligned}\right\} \tag{7.5}$$

for $y_2' = 0, 0 \le y_3' \le D_1, t \ge 0$.

It is assuming that at the tip of the fault the stresses maintain constant value τ_{L_1} and τ_{-L_1}.

Also

$$\tau_{12} \to \tau_\infty(t) \quad \text{as } |y_2| \to \infty \quad (-L_1 \le y_1 \le L_1, y_3 \ge 0, t \ge 0), \tag{7.6}$$

where τ_{12} represents as the main driving force for strike-slip fault. $\tau_\infty(t)$ is the value of τ_{12} far away from the fault maintained by tectonic forces and independents of y_3. If it changes with time $\tau_\infty(t) = \tau_\infty(0)(1 + kt)$ (say), where $k \ge 0$ is a constant, then

$$\tau_{12} \to \tau_\infty(t) = \tau_\infty(0)(1 + kt), \quad k > 0 \text{ as } |y_2| \to \infty \text{ for } y_3 \ge 0, t \ge 0 \tag{7.7}$$

$\tau_\infty(0)$ = the value of $\tau_\infty(t)$ at $t = 0$.

On $y_3 = 0$ (free surface),

$$\tau_{13} = \tau_{23} = \tau_{22} = \tau_{33} = 0 \quad (-\infty < y_2 < \infty, t \ge 0). \tag{7.8}$$

As $y_3 \to \infty$,

$$\tau_{13} = \tau_{23} = \tau_{22} = \tau_{33} \to 0 \quad (-\infty < y_2 < \infty, t \ge 0). \tag{7.9}$$

2.4 Initial conditions

At time $t = 0$,

$$
\begin{aligned}
u_i = (u_i)_0, \quad (u_i)_t = \frac{\partial}{\partial t}(u_i) = 0 \quad \text{for } i = 1,2,3,\\
\tau_{ij} = (\tau_{ij})_0, \quad \frac{\partial}{\partial t}(\tau_{ij}) = 0 \quad \text{for } i,j = 1,2,3,\\
e_{ij} = (e_{ij})_0, \quad \frac{\partial}{\partial t}(e_{ij}) = 0 \quad \text{for } i,j = 1,2,3.
\end{aligned}
$$

3 Solution

Differentiating 1st, 2nd and 3rd equation of (7.3) w. r. t. y_1, y_2, y_2 respectively, 2nd, 5th, 4th equation of (7.3) w. r. t. y_1, y_2, y_3 respectively and 3rd, 4th, 6th equation of (7.3) w. r. t. y_1, y_2, y_3 respectively and then finally using equation (7.4) we get,

$$\nabla^2 U_1 = 0, \tag{7.10}$$

$$\nabla^2 U_2 = 0, \tag{7.11}$$

$$\nabla^2 U_3 = 0, \tag{7.12}$$

whose Laplace transform w. r. t. t gives

$$\nabla^2 \bar{U}_i = 0, \quad \text{where } \bar{U}_i = \bar{u}_i - \frac{(u_i)_0}{s} \text{ for } i = 1,2,3. \tag{7.13}$$

The displacement components satisfy Laplace's equation and these are the general solution.

The solution for displacements, stresses and strains after the movement due to interaction between F_1 and F_2 are given by

$$
\left.
\begin{aligned}
u_i &= (u_i)_1 + (u_i)_2 + (u_i)_3 \quad (i = 1,2,3)\\
\tau_{ij} &= (\tau_{ij})_1 + (\tau_{ij})_2 + (\tau_{ij})_3, \quad (i,j = 1,2,3)\\
e_{12} &= (e_{12})_1 + (e_{12})_2 + (e_{12})_3\\
e_{13} &= (e_{13})_1 + (e_{13})_2 + (e_{13})_3
\end{aligned}
\right\} \tag{7.14}
$$

where $(u_i)_1$ $(i = 1,2,3)$, $(\tau_{ij})_1$ $(i,j = 1,2,3)$, $(e_{12})_1$, $(e_{13})_1$ are displacement, stress and strain components before the fault movement, $(u_i)_2$ $(i = 1,2,3)$, $(\tau_{ij})_2$ $(i,j = 1,2,3)$, $(e_{12})_2$, $(e_{13})_2$ are displacement, stress and strain components after the fault movement across F_1 fault and $(u_i)_3$ $(i = 1,2,3)$, $(\tau_{ij})_3$ $(i,j = 1,2,3)$, $(e_{12})_3$, $(e_{13})_3$ are displacement, stress and strain components after the interaction between F_1 and F_2. Since the displacement components u_2 and u_3 are continuous after the fault movement across F_2, $(u_2)_3 = 0$ and $(u_3)_3 = 0$ for $(t > T_2)$.

3.1 Displacement, stress and strain components before the fault movement

The displacement, stress, and strain components are continuous throughout the system. All the constitutive equations and boundary conditions (7.3) to (7.9) are valid.

For solving the boundary value problem, it is assumed that

$$\left.\begin{aligned} \bar{u}_1 &= \frac{(u_1)_0}{s} + A_1 y_1 + B_1 y_2 + C_1 y_3 \\ \bar{u}_2 &= \frac{(u_2)_0}{s} + A_2 y_1 + B_2 y_2 + C_2 y_3 \\ \bar{u}_3 &= \frac{(u_3)_0}{s} + A_3 y_1 + B_3 y_2 + C_3 y_3 \end{aligned}\right\} . \tag{7.15}$$

Taking the Laplace Transform on all constitutive equations and boundary conditions and then using (7.15), we get

$$A_1 = \frac{\tau_L}{2s^2(q_1 + q_2 s)}, \quad B_1 = A_2 = \frac{(1 + p_1 + p_2 s^2)}{2s(q_1 + q_2 s)} \left[\tau_\infty(s) - \frac{(p_1 + p_2 s)\tau_\infty(0)}{(1 + p_1 + p_2 s^2)} \right]$$

and $C_2 = B_2 = C_3 = B_3 = 0$.

Therefore (7.15) becomes

$$\left.\begin{aligned} \bar{u}_1 &= \frac{(u_1)_0}{s} + \frac{\tau_L y_1}{2s^2(q_1 + q_2 s)} + \frac{(1 + p_1 + p_2 s^2)}{2s(q_1 + q_2 s)} \left[\tau_\infty(s) - \frac{(p_1 + p_2 s)\tau_\infty(0)}{(1 + p_1 + p_2 s^2)} \right] y_2 \\ \bar{u}_2 &= \frac{(u_2)_0}{s} + \frac{(1 + p_1 + p_2 s^2)}{2s(q_1 + q_2 s)} \left[\tau_\infty(s) - \frac{(p_1 + p_2 s)\tau_\infty(0)}{(1 + p_1 + p_2 s^2)} \right] y_1 \\ \bar{u}_3 &= \frac{(u_3)_0}{s} \end{aligned}\right\} \tag{7.16}$$

By taking inverse Laplace transform, we get

$$\begin{aligned} (u_1)_1 = (u_1)_0 &+ \frac{\tau_L y_1}{2q_1} \left[t + \frac{q_2}{q_1} (e^{-\frac{q_1}{q_2} t} - 1) \right] + \frac{\tau_\infty(0)}{2q_1} \Bigg[t - \frac{q_2}{q_1} (1 - e^{-\frac{q_1}{q_2} t}) \\ &+ k \left\{ \frac{t^2}{2} + \frac{q_2}{q_1} \left(\frac{q_2}{q_1} (1 - e^{-\frac{q_1}{q_2} t}) - t \right) + p_1 \left(t - \frac{q_2}{q_1} (1 - e^{-\frac{q_1}{q_2} t}) \right) + p_2 (1 - e^{-\frac{q_1}{q_2} t}) \right\} \Bigg] y_2, \\ (u_2)_1 = (u_2)_0 &+ \frac{\tau_\infty(0)}{2q_1} \Bigg[t - \frac{q_2}{q_1} (1 - e^{-\frac{q_1}{q_2} t}) \\ &+ k \left\{ \frac{t^2}{2} + \frac{q_2}{q_1} \left(\frac{q_2}{q_1} (1 - e^{-\frac{q_1}{q_2} t}) - t \right) + p_1 \left(t - \frac{q_2}{q_1} (1 - e^{-\frac{q_1}{q_2} t}) \right) + p_2 (1 - e^{-\frac{q_1}{q_2} t}) \right\} \Bigg] y_1, \\ (u_3)_1 &= (u_3)_0. \end{aligned}$$

Then

$$\begin{aligned} (\tau_{11})_1 = &\frac{(\tau_{11})_0}{A} [(p_1 - p_2 r_1) e^{-r_1 t} - (p_1 - p_2 r_2) e^{-r_2 t}] \\ &+ \tau_L [1 - \{(p_1 - p_2 r_1) e^{-r_1 t} - (p_1 - p_2 r_2) e^{-r_2 t}\}], \end{aligned}$$

$$
\begin{aligned}
(\tau_{12})_1 &= \frac{(\tau_{12})_0}{A}[(p_1 - p_2r_1)e^{-r_1t} - (p_1 - p_2r_2)e^{-r_2t}] \\
&\quad + \left[\tau_\infty(t) - \frac{\tau_\infty(0)}{A}\{(p_1 - p_2r_1)e^{-r_1t} - (p_1 - p_2r_2)e^{-r_2t}\}\right], \\
(\tau_{13})_1 &= \frac{(\tau_{13})_0}{A}[(p_1 - p_2r_1)e^{-r_1t} - (p_1 - p_2r_2)e^{-r_2t}], \\
(\tau_{23})_1 &= \frac{(\tau_{23})_0}{A}[(p_1 - p_2r_1)e^{-r_1t} - (p_1 - p_2r_2)e^{-r_2t}], \\
(\tau_{22})_1 &= \frac{(\tau_{22})_0}{A}[(p_1 - p_2r_1)e^{-r_1t} - (p_1 -, p_2r_2)e^{-r_2t}], \\
(\tau_{33})_1 &= \frac{(\tau_{33})_0}{A}[(p_1 - p_2r_1)e^{-r_1t} - (p_1 - p_2r_2)e^{-r_2t}],
\end{aligned}
$$

$$(e_{11})_1 = \frac{1}{2}(e_{11})_0,$$

$$
\begin{aligned}
(e_{12})_1 &= \frac{1}{2}(e_{12})_0 + \frac{\tau_\infty(0)}{4q_1}\left[t - \frac{q_2}{q_1}(1 - e^{-\frac{q_1t}{q_2}})\right. \\
&\quad \left. + k\left\{\frac{t^2}{2} + \frac{q_2}{q_1}\left(\frac{q_2}{q_1}(1 - e^{-\frac{q_1t}{q_2}})\right) + p_1\left(t - \frac{q_2}{q_1}(1 - e^{-\frac{q_1t}{q_2}})\right) + p_2(1 - e^{-\frac{q_1t}{q_2}})\right\}\right]
\end{aligned}
$$

$$(e_{13})_1 = \frac{1}{2}(e_{13})_0,$$

where $r_1 = \frac{(p_1-A)}{2p_2}$, $r_2 = \frac{(p_1+A)}{2p_2}$, $A = (p_1^2 - 4p_2)^{\frac{1}{2}}$ and p_1, p_2 are same as given in equation (7.3).

Now, τ'_{12} is the stress component, which tends to cause strike-slip movement across the fault F_1.

$$
\begin{aligned}
\tau'_{12} &= \tau_{12}\sin\theta_1 - \tau_{13}\cos\theta_1 = \frac{(\tau'_{12})_0}{A}[(p_1 - p_2r_1)e^{-r_1t} - (p_1 - p_2r_2)e^{-r_2t}] \\
&\quad + \left[\tau_\infty(t)\sin\theta_1 - \frac{\tau_\infty(0)\sin\theta_1}{A}\{(p_1 - p_2r_1)e^{-r_1t} - (p_1 - p_2r_2)e^{-r_2t}\}\right],
\end{aligned}
$$

where $(\tau'_{12})_0 = (\tau_{12})_0\sin\theta_1 - (\tau_{13})_0\cos\theta_1$.

It is observed that for the fault F_1 the relevant stress component τ'_{12} increases with time and finally tends to $\tau_\infty\sin\theta_1$. The fault F_1 starts to slip when the magnitude of stress τ'_{12} reaches a threshold value say τ_{c_1} $(< \tau_\infty\sin\theta_1)$ after a time T_1.

3.2 Displacement, stress and strain components after the fault movement

It is assumed that the fault F_1 moves at time $t = T_1$ (> 0), while the other fault F_2 remains locked. The equations (7.3) to (7.9) remain valid for $t \geq T_1$. The condition for characterization of the slip across F_1:

$$[u_1]_{F_1} = U_1(t_1)H(t_1)f_1(y'_1, y'_3) \quad (y'_2 = 0, 0 \leq y'_3 \leq D_1, t_1 \geq 0, t_1 = t - T_1), \tag{7.17}$$

where $[u_1]_{F_1}$ is the relative displacement across F_1 and the discontinuity of displacement u_1 is given by

$$[u_1]_{F_1} = \lim_{(y_2' \to 0^+)} (u_1) - \lim_{(y_2' \to 0^-)} (u_1). \tag{7.18}$$

In equation (7.17), $H(t_1)$ is the Heaviside function, $f_1(y_1', y_3')$ is a continuous function representing the surface of the crack, which gives the spatial dependence of the slip movement along the fault F_1, and since after a sudden movement across F_1, it becomes locked then $U_1(t_1) = \text{constant} = U_1$.

Taking Laplace transform on equation (7.17) gives

$$[\bar{u}_1]_{F_1} = \frac{U_1}{s} f_1(y_1', y_3'),$$

where $\bar{u}_1$ represents Laplace transform of u_1.

Clearly, $[(u_1)]_{F_1} = 0$ for $t_1 \leq 0$, where $t_1 = t - T_1$.

$(u_1)_2$ satisfy the dislocation condition (7.18). Since u_2 and u_3 are continuous after the fault movement across F_1 then $(u_2)_2 = 0$ and $(u_3)_2 = 0$ for $t > T_1$.

We get

$$\left.\begin{aligned}
(u_1)_2 &= \frac{U_1}{2\pi} H(t_1) \phi'(y_1, y_2, y_3, t) \\
(u_2)_2 &= 0 \\
(u_3)_2 &= 0 \\
(\tau_{11})_2 &= \frac{U_1}{\pi} H(t_1) \phi_1' \left[\left(\frac{q_2}{2p_2} + \frac{q_1}{A} + \frac{q_2 p_1}{2Ap_2} \right) e^{-r_1 t_1} + \left(\frac{q_2}{2p_2} - \frac{q_1}{A} - \frac{q_2 p_1}{2Ap_2} \right) e^{-r_2 t_1} \right] \\
(\tau_{12})_2 &= \frac{U_1}{2\pi} H(t_1) \phi_2' \left[\left(\frac{q_2}{2p_2} + \frac{q_1}{A} + \frac{q_2 p_1}{2Ap_2} \right) e^{-r_1 t_1} + \left(\frac{q_2}{2p_2} - \frac{q_1}{A} - \frac{q_2 p_1}{2Ap_2} \right) e^{-r_2 t_1} \right] \\
(\tau_{13})_2 &= \frac{U_1}{2\pi} H(t_1) \phi_3' \left[\left(\frac{q_2}{2p_2} + \frac{q_1}{A} + \frac{q_2 p_1}{2Ap_2} \right) e^{-r_1 t_1} + \left(\frac{q_2}{2p_2} - \frac{q_1}{A} - \frac{q_2 p_1}{2Ap_2} \right) e^{-r_2 t_1} \right] \\
(\tau_{23})_2 &= 0 \\
(\tau_{33})_2 &= 0 \\
(e_{11})_2 &= \frac{U_1}{2\pi} H(t_1) \phi_1' \\
(e_{12})_2 &= \frac{U_1}{4\pi} H(t_1) \phi_2' \\
(e_{13})_2 &= \frac{U_1}{4\pi} H(t_1) \phi_3'
\end{aligned}\right\} \tag{7.19}$$

where the expression of ϕ' is given in Appendix and $\phi_2' = \frac{\partial \phi'}{\partial y_2}$, $\phi_3' = \frac{\partial \phi'}{\partial y_3}$.

Next the movement across F_2 has been occurred after time T_2 ($T_2 > T_1$) when the accumulated stress near it exceeds the critical value τ_{c_2} (say).

The slip condition is characterized by:

$$[u_1]_{F_2} = U_2(t_2)f_2(y''_1, y''_3)H(t_2) \quad (-L \le y''_1 \le L, y''_2 = 0, 0 \le y''_3 \le D_2,\ t_2 \ge 0, t_2 = t - T_2), \tag{7.20}$$

where $H(t_2)$ is the Heaviside function, $f_2(y''_1, y''_3)$ gives the spatial dependence of movement along the fault F_2 and since after a sudden movement across F_2 at $t = T_2$, it also becomes locked then $U_2(t_2) = \text{constant} = U_2$.

$[u_1]_{F_2}$ is the relative displacement across F_2 given by

$$[u_1]_{F_2} = \lim_{(y''_2 \to 0^+)} (u_1) - \lim_{(y''_2 \to 0^-)} (u_1). \tag{7.21}$$

Laplace transform on equation (7.20) gives

$$[\bar{u}_1]_{F_2} = \frac{U_2}{s} f_2(y''_1, y''_3).$$

Since the fault movement occurs across F_2 after time T_2 hence $[(u_1)]_{F_2} = 0$ for $t_2 \le 0$, where $t_2 = t - T_2$.

Since the displacement components u_2 and u_3 are continuous after the fault movement across F_2, $(u_2)_3 = 0$ and $(u_3)_3 = 0$ for $(t > T_2)$.

So, we get

$$\left.\begin{aligned}
(u_1)_3 &= \frac{U_2}{2\pi} H(t_2)\phi(z_1, z_2, z_3, t)\\
(u_2)_3 &= 0\\
(u_3)_3 &= 0\\
(\tau_{11})_3 &= \frac{U_2}{\pi} H(t_2)\phi_1\left[\left(\frac{q_2}{2p_2} + \frac{q_1}{A} + \frac{q_2 p_1}{2Ap_2}\right)e^{-r_1 t_2} + \left(\frac{q_2}{2p_2} - \frac{q_1}{A} - \frac{q_2 p_1}{2Ap_2}\right)e^{-r_2 t_2}\right]\\
(\tau_{12})_3 &= \frac{U_2}{2\pi} H(t_2)\phi_2\left[\left(\frac{q_2}{2p_2} + \frac{q_1}{A} + \frac{q_2 p_1}{2Ap_2}\right)e^{-r_1 t_2} + \left(\frac{q_2}{2p_2} - \frac{q_1}{A} - \frac{q_2 p_1}{2Ap_2}\right)e^{-r_2 t_2}\right]\\
(\tau_{13})_3 &= \frac{U_2}{2\pi} H(t_2)\phi_3\left[\left(\frac{q_2}{2p_2} + \frac{q_1}{A} + \frac{q_2 p_1}{2Ap_2}\right)e^{-r_1 t_2} + \left(\frac{q_2}{2p_2} - \frac{q_1}{A} - \frac{q_2 p_1}{2Ap_2}\right)e^{-r_2 t_2}\right]\\
(\tau_{23})_3 &= 0\\
(\tau_{33})_3 &= 0\\
(e_{11})_3 &= \frac{U_2}{2\pi} H(t_2)\phi_1\\
(e_{12})_3 &= \frac{U_2}{4\pi} H(t_2)\phi_2\\
(e_{13})_3 &= \frac{U_2}{4\pi} H(t_2)\phi_3
\end{aligned}\right\} \tag{7.22}$$

where the expression of ϕ given in Appendix and $\phi_1 = \frac{\partial\phi}{\partial y_1}$, $\phi_2 = \frac{\partial\phi}{\partial y_2}$, $\phi_3 = \frac{\partial\phi}{\partial y_3}$.

Finally, the complete solutions for displacement, stress and strain after the interaction between F_1 and F_2 are obtained (from equation (7.14)) as follows:

$$
\left.
\begin{aligned}
u_1 &= (u_1)_0 + \frac{\tau_L y_1}{2q_1}\left[t + \frac{q_2}{q_1}\left(e^{-\frac{q_1}{q_2}t} - 1\right)\right] + \frac{\tau_\infty(0)}{2q_1}\left[t - \frac{q_2}{q_1}\left(1 - e^{-\frac{q_1}{q_2}t}\right) + k\left\{\frac{t^2}{2} + \frac{q_2}{q_1}\left(\frac{q_2}{q_1}\left(1 - e^{-\frac{q_1}{q_2}t}\right) - t\right)\right.\right. \\
&\quad \left.\left. + p_1\left(t - \frac{q_2}{q_1}\left(1 - e^{-\frac{q_1}{q_2}t}\right)\right) + p_2\left(1 - e^{-\frac{q_1}{q_2}t}\right)\right\}\right] y_2 + \frac{U_1}{2\pi}H(t_1)\phi'(y_1, y_2, y_3, t) + \frac{U_2}{2\pi}H(t_2)\phi(z_1, z_2, z_3, t) \\
u_2 &= (u_2)_0 + \frac{\tau_\infty(0)}{2q_1}\left[t - \frac{q_2}{q_1}\left(1 - e^{-\frac{q_1}{q_2}t}\right) + k\left\{\frac{t^2}{2} + \frac{q_2}{q_1}\left(\frac{q_2}{q_1}\left(1 - e^{-\frac{q_1}{q_2}t}\right) - t\right) + p_1\left(t - \frac{q_2}{q_1}\left(1 - e^{-\frac{q_1}{q_2}t}\right)\right)\right.\right. \\
&\quad \left.\left. + p_2\left(1 - e^{-\frac{q_1}{q_2}t}\right)\right\}\right] y_1 \\
u_3 &= (u_3)_0 \\
\tau_{11} &= \frac{(\tau_{11})_0}{A}\left[(p_1 - p_2 r_1)e^{-r_1 t} - (p_1 - p_2 r_2)e^{-r_2 t}\right] + \left[\tau_L - \frac{\tau_L}{A}\left\{(p_1 - p_2 r_1)e^{-r_1 t} - (p_1 - p_2 r_2)e^{-r_2 t}\right\}\right] \\
&\quad + \frac{U_1}{\pi}H(t_1)\phi_1'\left[\left(\frac{q_2}{2p_2} + \frac{q_1}{A} + \frac{q_2 p_1}{2Ap_2}\right)e^{-r_1 t_1} + \left(\frac{q_2}{2p_2} - \frac{q_1}{A} - \frac{q_2 p_1}{2Ap_2}\right)e^{-r_2 t_1}\right] \\
&\quad + \frac{U_2}{\pi}H(t_2)\phi_1\left[\left(\frac{q_2}{2p_2} + \frac{q_1}{A} + \frac{q_2 p_1}{2Ap_2}\right)e^{-r_1 t_2} + \left(\frac{q_2}{2p_2} - \frac{q_1}{A} - \frac{q_2 p_1}{2Ap_2}\right)e^{-r_2 t_2}\right] \\
\tau_{12} &= \frac{(\tau_{12})_0}{A}\left[(p_1 - p_2 r_1)e^{-r_1 t} - (p_1 - p_2 r_2)e^{-r_2 t}\right] \\
&\quad + \left[\tau_\infty(t) - \frac{\tau_\infty(0)}{A}\left\{(p_1 - p_2 r_1)e^{-r_1 t} - (p_1 - p_2 r_2)e^{-r_2 t}\right\}\right] \\
&\quad + \frac{U_1}{2\pi}H(t_1)\phi_2'\left[\left(\frac{q_2}{2p_2} + \frac{q_1}{A} + \frac{q_2 p_1}{2Ap_2}\right)e^{-r_1 t_1} + \left(\frac{q_2}{2p_2} - \frac{q_1}{A} - \frac{q_2 p_1}{2Ap_2}\right)e^{-r_2 t_1}\right] \\
&\quad + \frac{U_2}{2\pi}H(t_2)\phi_2\left[\left(\frac{q_2}{2p_2} + \frac{q_1}{A} + \frac{q_2 p_1}{2Ap_2}\right)e^{-r_1 t_2} + \left(\frac{q_2}{2p_2} - \frac{q_1}{A} - \frac{q_2 p_1}{2Ap_2}\right)e^{-r_2 t_2}\right] \\
\tau_{13} &= \frac{(\tau_{13})_0}{A}\left[(p_1 - p_2 r_1)e^{-r_1 t} - (p_1 - p_2 r_2)e^{-r_2 t}\right] \\
&\quad + \frac{U_1}{2\pi}H(t_1)\phi_3'\left[\left(\frac{q_2}{2p_2} + \frac{q_1}{A} + \frac{q_2 p_1}{2Ap_2}\right)e^{-r_1 t_1} + \left(\frac{q_2}{2p_2} - \frac{q_1}{A} - \frac{q_2 p_1}{2Ap_2}\right)e^{-r_2 t_1}\right] \\
&\quad + \frac{U_2}{2\pi}H(t_2)\phi_3\left[\left(\frac{q_2}{2p_2} + \frac{q_1}{A} + \frac{q_2 p_1}{2Ap_2}\right)e^{-r_1 t_2} + \left(\frac{q_2}{2p_2} - \frac{q_1}{A} - \frac{q_2 p_1}{2Ap_2}\right)e^{-r_2 t_2}\right] \\
\tau_{23} &= \frac{(\tau_{23})_0}{A}\left[(p_1 - p_2 r_1)e^{-r_1 t} - (p_1 - p_2 r_2)e^{-r_2 t}\right] \\
\tau_{22} &= \frac{(\tau_{22})_0}{A}\left[(p_1 - p_2 r_1)e^{-r_1 t} - (p_1 - p_2 r_2)e^{-r_2 t}\right] \\
\tau_{33} &= \frac{(\tau_{33})_0}{A}\left[(p_1 - p_2 r_1)e^{-r_1 t} - (p_1 - p_2 r_2)e^{-r_2 t}\right] \\
e_{11} &= \frac{1}{2}(e_{11})_0 + \frac{U_1}{2\pi}H(t_1)\phi_1' + \frac{U_2}{2\pi}H(t_2)\phi_1 \\
e_{12} &= \frac{1}{2}(e_{12})_0 + \frac{\tau_\infty(0)}{4q_1}\left[t - \frac{q_2}{q_1}\left(1 - e^{\frac{q_1 t}{q_2}}\right) + k\left\{\frac{t^2}{2} + \frac{q_2}{q_1}\left(\frac{q_2}{q_1}\left(1 - e^{-\frac{q_1 t}{q_2}}\right) - t\right) + p_1\left(t - \frac{q_2}{q_1}\left(1 - e^{-\frac{q_1 t}{q_2}}\right)\right)\right.\right. \\
&\quad \left.\left. + p_2\left(1 - e^{-\frac{q_1 t}{q_2}}\right)\right\}\right] + \frac{U_1}{4\pi}H(t_1)\phi_2' + \frac{U_2}{4\pi}H(t_2)\phi_2 \\
e_{13} &= \frac{1}{2}(e_{13})_0 + \frac{U_1}{4\pi}H(t_1)\phi_3' + \frac{U_2}{4\pi}H(t_2)\phi_3
\end{aligned}
\right\}
\qquad (7.23)
$$

4 Numerical computations

For the numerical calculation and graphical representation of displacement, stress and strain components due to interaction between two finite faults, the following mechanical properties (Table 7.1) are used (Cathles (1975 [2]), Clift. P. et al. (2002 [3]) and Karato (2010 [10])).

Table 7.1: Value of the parameters used in this study.

Parameter	Symbol used	Value taken
Rigidity	μ_1	3.5×10^{10} N/m^2 (Pascal)
Rigidity	μ_2	3×10^{10} N/m^2 (Pascal)
Viscocity	η_1	3.5×10^{19} Pascal-s
Viscocity	η_2	3×10^{19} Pascal-s
Width	D_1, D_2	10 km
Depth from the free surface	d_1, d_2	5 km
Distance between F_1 and F_2	D	8 km
Length of F_1	$2L_1$	20 km
Length of F_2	$2L_2$	40 km
Total time	t	150 year
Time of movement of F_1 fault	T_1	50 year
Time of movement of F_2 fault	T_2	100 year
Initial stress	$(\tau_{12})_0, (\tau_{13})_0$	20×10^5 N/m^2 (Pascal)
Initial strain	$(e_{12})_0, (e_{13})_0$	20×10^5 N/m^2 (Pascal)
Stress at infinity position at time t	$\tau_\infty(t)$,	2×10^7 N/m^2 (Pascal)
Stress at infinity position at time 0	$\tau_\infty(0)$,	20×10^5 N/m^2 (Pascal)
Constant	k	10^{-9}
Inclination	θ_1, θ_2	$\pi/6, \pi/4, \pi/3, \pi/2$
Slip magnitude	U_1, U_2	1.6, 3.7, 5.6 meter

The slip function which represents the surface of the crack of the faults are

$$\text{for } F_1 \text{ fault,} \quad f_1(y_1', y_3') = R\left(1 - \frac{1}{L_1{}^2 y_1'^2}\right)\left(1 - \frac{3y_3'^2}{D_1{}^2} + \frac{2y_3'^3}{D_1{}^3}\right),$$

$$\text{for } F_2 \text{ fault,} \quad f_2(y_1'', y_3'') = R\left(1 - \frac{1}{L_2^2 y_1''^2}\right)\left(1 - \frac{3y_3''^2}{D_2{}^2} + \frac{2y_3''^3}{D_2{}^3}\right),$$

where R is a constant and taken as 3 m, L_1, L_2 are length and D_1, D_2 are width of the faults F_1 and F_2 respectively.

4.1 Result and discussion

4.1.1 Effect of inclination of the faults

The change in surface displacement has been plotted in Figure 7.3(a) in y_2 direction due to sudden movement across F_1 only ($T_1 < t < T_2$), there being no movement across F_2 for different inclinations of F_1 fault. It is found that the surface displacement attains its extreme value near F_1. For $y_2 < 0$, it increases with increasing value of θ_1 and for $y_2 > 0$, it increases with decreasing value of θ_1. For $\theta_1 = 90^0$, surface displacement is anti-symmetrical with respected to $y_2 = 0$. However, for $\theta_1 \neq 90^0$, there is no such anti-symmetry. In Figure 7.3(b), surface displacement has been plotted due to sudden movement across both F_1 and F_2 (for $t > t_2$) for different values of θ_1 when $\theta_2 = 60^0$. Surface displacement attains its maximum value near both the fault F_1 and F_2. This value near F_1 fault is greater than F_2 fault. It increases with increasing value of θ_1 near F_1 fault

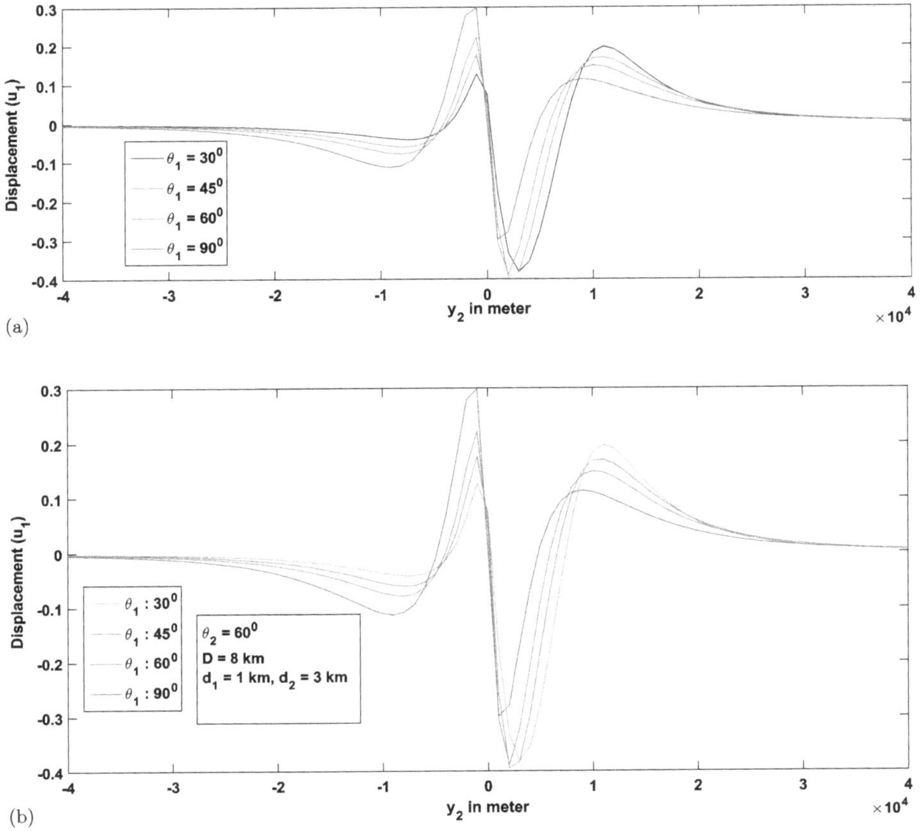

Figure 7.3: Effect of different inclinations of the faults on displacement with y_2 due to fault movement (a) only across F_1 (no movement across F_2) and (b) across both F_1 and F_2.

and decreases with increasing value of θ_1 near F_2 fault. Since a sudden movement occur across F_1 fault then for $y_2 \to 0^-$ the fault displace maximum value and then it stop. Again since at $t = T_2$, a sudden movement occur across F_2 then at $y_2 = D = 8$ km, F_2 fault displace maximum value and then for $y_2 \gg D$, it decrease rapidly towards zero.

The change of shear stress (τ_{12}) against y_2 for different inclination of F_1 fault when inclination of F_2 is 60^0 has been described in Figure 7.4. It shows that the shear stress accumulates its maximum value near F_1 fault at $y_2 = 0$ then decrease for $|y_2| > 0$. Since after movement across F_1 fault it becomes locked then after accumulating maximum stress near F_1, shear stress decrease rapidly. Near $y_2 = D$, it attains minimum value. Since at $y_2 = D$, F_2 starts to move, it again increases and attains some positive value. After the sudden fault movement across F_2, shear stress decreases rapidly towards zero. This shear stress accumulation and release increases with increasing value of θ_1.

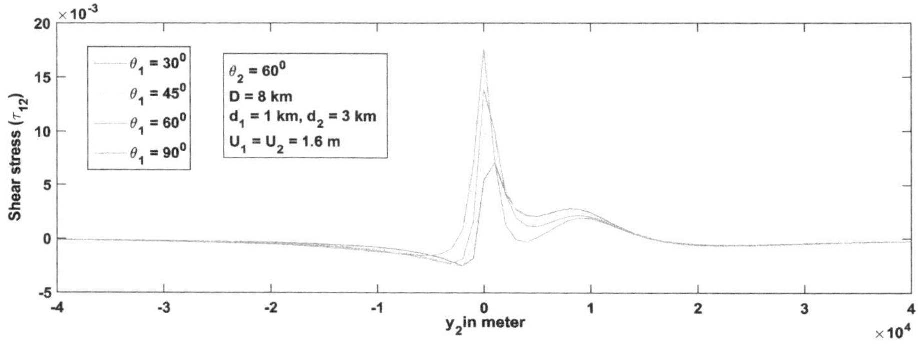

Figure 7.4: Variation of shear stress in y_2 direction for different values of inclination θ_1 of F_1 when inclination θ_2 of F_2 is 60^0.

Figure 7.5 shows the accumulation/release of shear strain per year near and away from the fault due to movement across F_1 for different θ_1. The accumulation of shear strain increases with increasing value of θ_1. Since after sudden movement across F_1 it stops moving, a large amount of strain accumulates across F_1, which then decreases rapidly towards zero.

Figure 7.6 represents the variation of surface shear strain due to the fault interaction. Near the fault F_1, surface shear strain increases for all θ_1, and this increment value depends significantly on θ_2, the higher the values of θ_2, the higher the increments. As we move away from the fault F_1 and approach towards F_2, shear strain decreases rapidly after a certain level and then starts increasing and reaches a higher value near F_2 and then it gradually decreases towards zero as $y_2 \gg D$. Shear strain accumulation increase with increasing value of θ_1 near F_1 fault and increase with decreasing value of θ_1 near F_2. After the movement of F_2 fault, there are no movement in $y_2 < 0$ direction. Since after the fault movement across both F_1 and F_2, the faults become locked, the strain rapidly

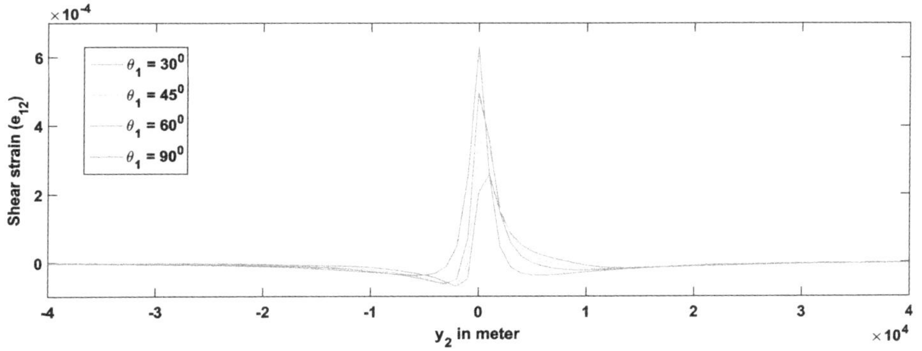

Figure 7.5: Variation of shear strain in y_2 direction due to fault movement only across F_1 (no movement across F_2) for different inclination (θ_1) of the fault F_1.

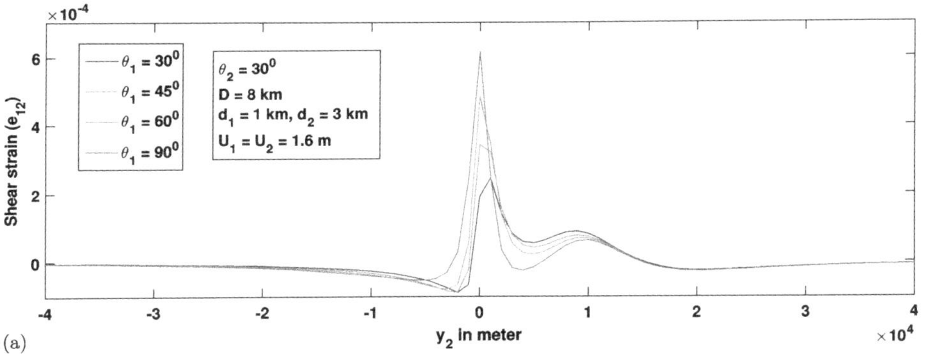

(a)

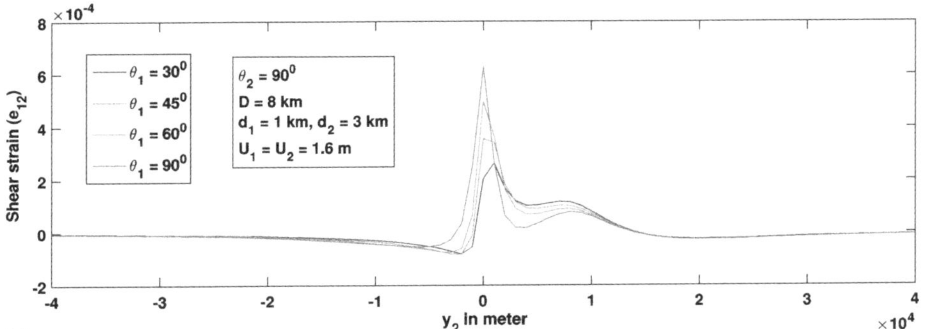

(b)

Figure 7.6: Changes of shear strain in y_2 direction due to fault movement across both F_1 and F_2 for different θ_1 when (a) $\theta_2 = 30^0$ and (b) $\theta_2 = 90^0$.

decreases towards zero after the accumulation of a huge amount of strain across F_1 and F_2.

4.1.2 Effect of slip magnitude of the faults

In Figure 7.7(a), surface displacement is plotted against y_2 for different slip magnitude due to fault movement across F_1 when there is no movement across F_2 and θ_1 is taken as 60^0. For $y_2 < 0$, it first increases and attains its maximum value on the fault F_1 and then decreases for $y_2 < 0$. For $y_2 > 0$, surface displacement decreases and after attains its minimum value, it increases. Surface displacement increases with increasing value of slip magnitude U_1. In Figure 7.7(b), the variation of surface displacement due to interaction between F_1 and F_2 has been plotted against y_2 for different slip magnitude U_1 across F_1, when the slip magnitude across F_2 is taken as 1.6 m. It is found that surface displacement increases with increasing value of slip magnitude near F_1 and F_2.

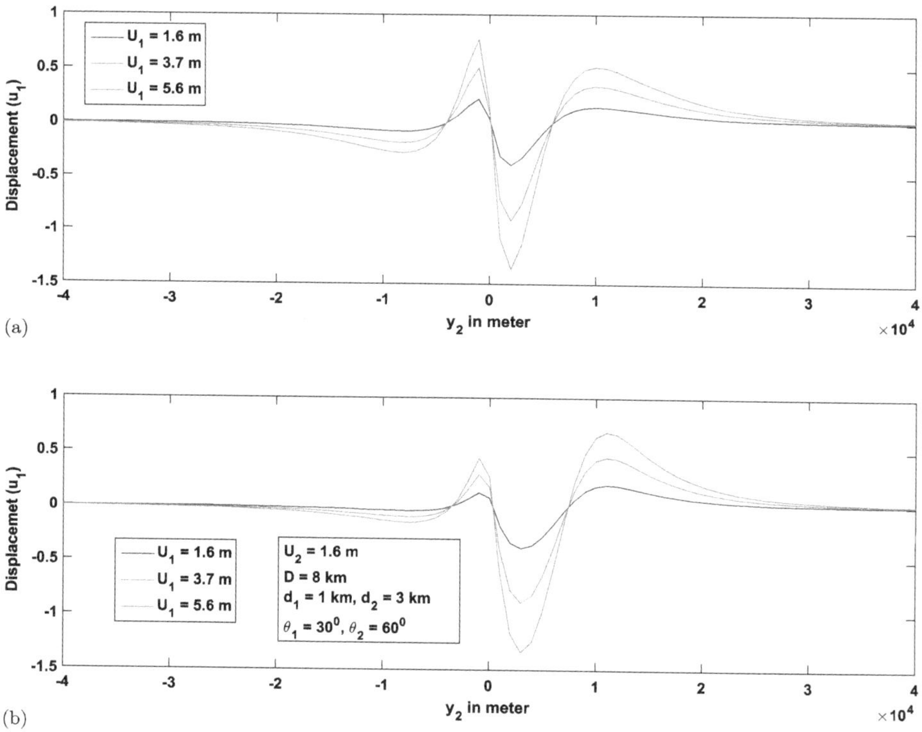

Figure 7.7: The variation of surface displacement with y_2 for different slip magnitude (U_1) (a) due to fault movement only across F_1 and (b) due to interaction between F_1 and F_2.

In Figure 7.8, the variation of shear stress (τ_{12}) after the interaction between F_1 and F_2 has been plotted against y_2 for different slip magnitude of U_1 of F_1 when U_2 is taken 1.6 m. It shows that the shear stress accumulates its maximum value near F_1 fault at $y_2 = 0$ and then decreases for $|y_2| > 0$. After decreasing to a certain value, when F_1 starts

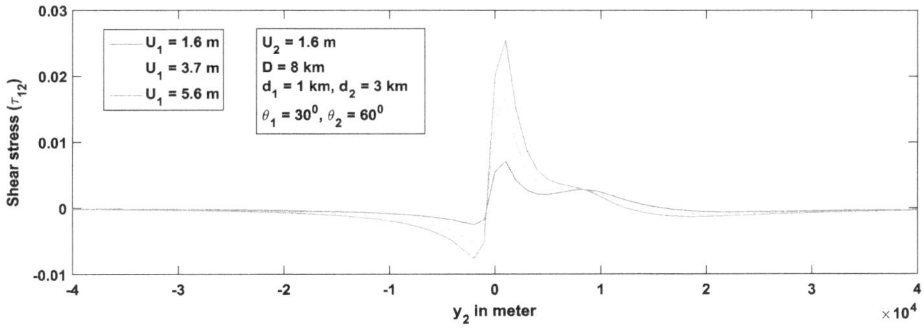

Figure 7.8: Variation of shear stress in y_2 direction for different slip magnitude U_1 of F_1, when slip magnitude U_2 of F_2 is 1.6 m.

to interact with F_2, then shear stress increases by a positive value near $y_2 = 8$ km and then decreases towards zero.

Now the nature of surface shear strain accumulation before and after the fault movement has been considered. It is noted that in the absence of any fault movement, there is a steady accumulation of surface shear strain near the fault with time. After the commence of a fault movement, this rate of strain accumulation falls off. Figure 7.9(a) shows this accumulation/release of shear strain per year near and away from the fault due to movement across F_1 for different U_1. The accumulation of shear strain increases with increasing value of slip magnitude U_1 of F_1. Since after sudden movement across F_2 the movement stops, a large amount of strain accumulates across F_1, which then releases, i. e., decreases rapidly towards zero. The variation of surface shear strain with y_2 for different slip magnitude of F_1 fault, when slip magnitude of F_2 fault is taken as 1.6 m is described in Figure 7.9(b). It shows that shear strain accumulation near both the faults increases with increasing value of slip magnitude.

4.1.3 Effect of depth from the free surface of the faults

The change in surface displacement due to sudden interaction between F_1 and F_2 for different depths of the faults from the free surface has been plotted in Figure 7.10. It is found that surface displacement decreases for increasing value of depth near F_1 and F_2.

In Figure 7.11, variation of shear strain for different depth of F_1 and F_2 has been plotted. It is found that due to sudden movement shear strain attains maximum value due to movement of F_1 fault and increases with decreasing value of surface depth of the faults.

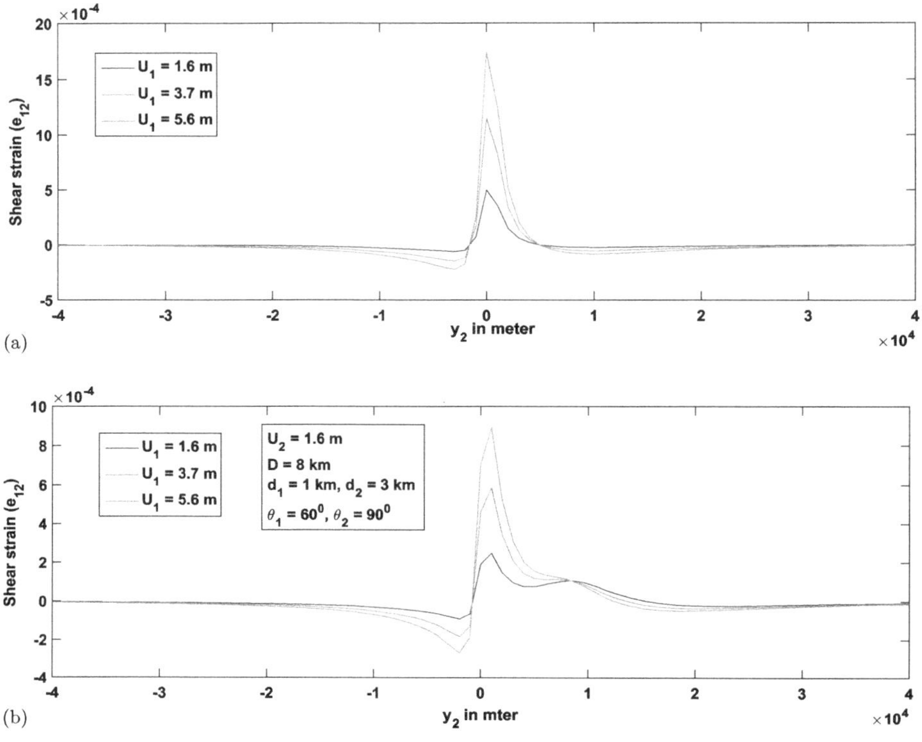

Figure 7.9: Variation of shear strain in y_2 direction for different slip magnitude (U_1) of the fault F_1 (a) only across F_1 (no movement across F_2) and (b) due to fault movement across both F_1 and F_2, when slip magnitude (U_2) across F_2 is taken as 1.6 m.

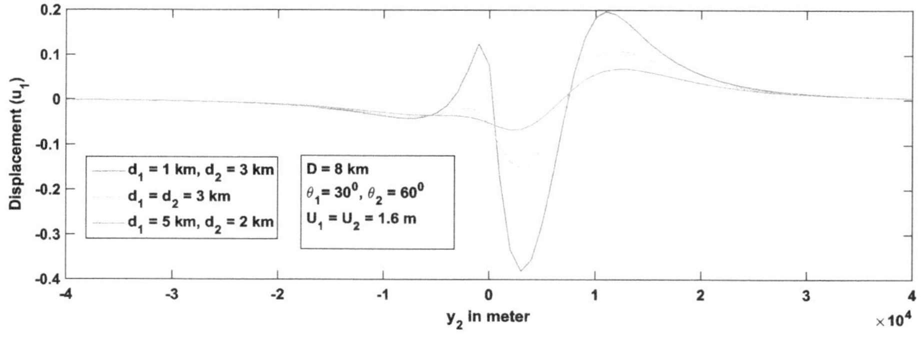

Figure 7.10: Variation of surface displacement against y_2 due to movement of both F_1 and F_2 for different depths of the faults d_1 and d_2.

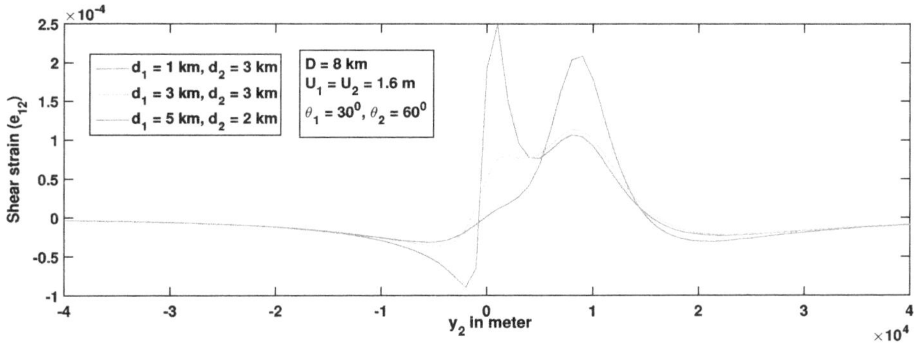

Figure 7.11: Variation of shear strain with y_2 due to fault movement across both F_1 and F_2 for different depth of the faults.

4.1.4 Effect of distance between the faults

The change of shear strain for different distances between F_1 and F_2 is described in Figure 7.12. It shows that the shear strain increase with increasing value distance between the faults. It is found that shear strain attains maximum value across F_1 fault and decreases with increasing distance and there is a change of slope of the curve at $y2 = D$, where F_1 starts to interact with F_2.

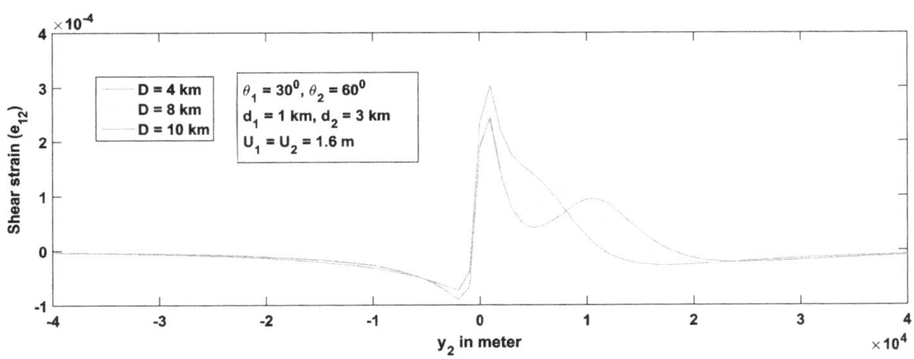

Figure 7.12: Variation of shear strain with y_2 due to interaction between F_1 and F_2 for different distance between F_1 and F_2.

4.1.5 Effect of time on shear stress

The shear stresses near the mid points of the faults F_1 and F_2 respectively

$$\left.\begin{aligned}
(T'_{12})_{F_1(\mathrm{MID})} &= \tau_{12}\sin\theta_1 - \tau_{13}\cos\theta_1 \\
&= \frac{(T'_{12})_{0F_1}}{A}[(p_1 - p_2r_1)e^{-r_1t} - (p_1 - p_2r_2)e^{-r_2t}] \\
&\quad + \left[\tau_\infty(t) - \frac{\tau_\infty(0)}{A}\{(p_1 - p_2r_1)e^{-r_1t} - (p_1 - p_2r_2)e^{-r_2t}\}\right]\sin\theta_1 \\
&\quad + \frac{v_1}{2\pi A}H(t_1)\left[\frac{q_1 - q_2r_1}{r_1}(1 - e^{-r_1t_1}) - \frac{q_1 - q_2r_2}{r_2}(1 - e^{-r_2t_1})\right] \\
&\quad [\phi'_2\sin\theta_1 - \phi'_3\cos\theta_1] + \frac{v_2}{2\pi A}H(t_2)\left[\frac{q_1 - q_2r_1}{r_1}(1 - e^{-r_1t_2})\right. \\
&\quad \left. - \frac{q_1 - q_2r_2}{r_2}(1 - e^{-r_2t_2})\right][\phi_2\sin\theta_1 - \phi_3\cos\theta_1],
\end{aligned}\right\} \tag{7.24}$$

where

$$(T'_{12})_{0F_1} = (\tau_{12})_0\sin\theta_1 - (\tau_{13})_0\cos\theta_1$$

and

$$\left.\begin{aligned}
(T''_{12})_{F_2(\mathrm{MID})} &= \tau_{12}\sin\theta_2 - \tau_{13}\cos\theta_2 \\
&= \frac{(T''_{12})_{0F_2}}{A}[(p_1 - p_2r_1)e^{-r_1t} - (p_1 - p_2r_2)e^{-r_2t}] \\
&\quad + \left[\tau_\infty(t) - \frac{\tau_\infty(0)}{A}\{(p_1 - p_2r_1)e^{-r_1t} - (p_1 - p_2r_2)e^{-r_2t}\}\right]\sin\theta_2 \\
&\quad + \frac{v_1}{2\pi A}H(t_1)\left[\frac{q_1 - q_2r_1}{r_1}(1 - e^{-r_1t_1}) - \frac{q_1 - q_2r_2}{r_2}(1 - e^{-r_2t_1})\right] \\
&\quad [\phi'_2\sin\theta_2 - \phi'_3\cos\theta_2] + \frac{v_2}{2\pi A}H(t_2)\left[\frac{q_1 - q_2r_1}{r_1}(1 - e^{-r_1t_2})\right. \\
&\quad \left. - \frac{q_1 - q_2r_2}{r_2}(1 - e^{-r_2t_2})\right][\phi_2\sin\theta_2 - \phi_3\cos\theta_2],
\end{aligned}\right\} \tag{7.25}$$

where

$$(T''_{12})_{0F_2} = (\tau_{12})_0\sin\theta_2 - (\tau_{13})_0\cos\theta_2.$$

The shear stresses $(T'_{12})_{F_1(\mathrm{MID})}$ and $(T''_{12})_{F_2(\mathrm{MID})}$ has been considered for strike slip movement in the half-space. In Figures 7.13(a) and (b), the variation of $(T'_{12})_{F_1(\mathrm{MID})}$ and $(T''_{12})_{F_2(\mathrm{MID})}$ have been described against time for different values of θ_1 taking θ_2 is 60^0 and for different values of θ_2 taking θ_1 as 90^0 respectively. From this figures it is observed that

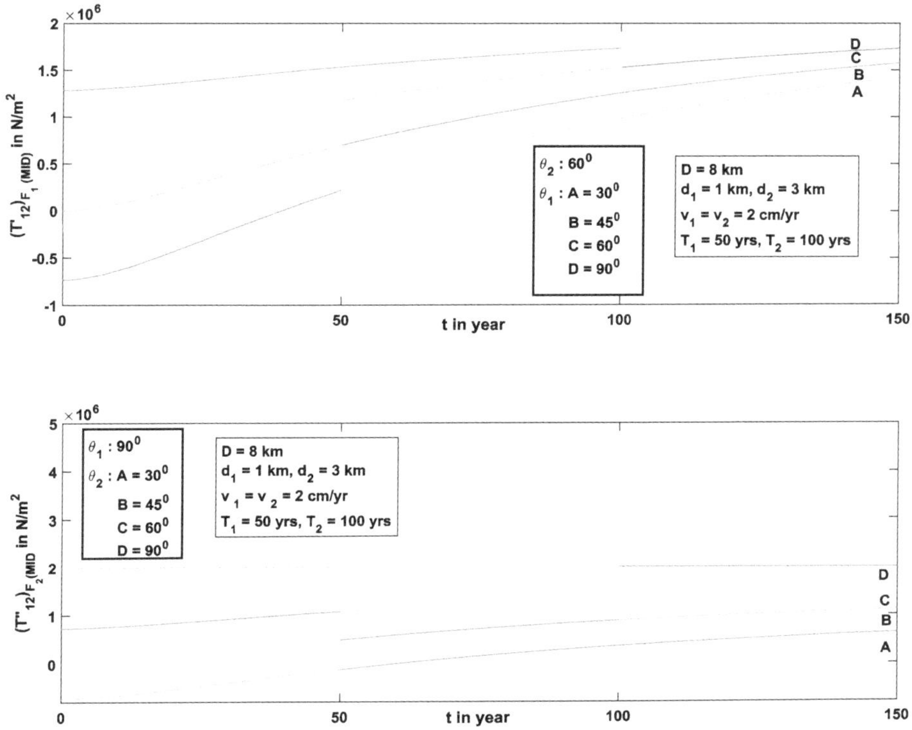

Figure 7.13: Total shear stress with time (t) for different values of (a) θ_1 at the mid point of the fault F_1 when $\theta_2 = 60^0$ and (b) θ_2 at the mid point of the fault F_2 when $\theta_1 = 90^0$.

$(T'_{12})_{F_1}$ grows gradually up to $t = T_1$ (= 50 years) and also slope of the curves changes with time. It continues up to $t = T_2$ (= 100 years). Figure 7.13(a) demonstrated that $(T'_{12})_{F_1(\mathrm{MID})}$ increases with the increasing value of θ_1 and being maximum for $\theta_1 = 90^0$. This signifies that the shear stress at the mid point of F_1 accumulates maximum value across vertical fault.

Figure 7.13(b) describes the variation of $(T''_{12})_{F_2(\mathrm{MID})}$ with time. The maximum limiting values that may be attained by $(T'_{12})_{F_1(\mathrm{MID})}$ near F_1 and $(T''_{12})_{F_2(\mathrm{MID})}$ near F_2 are $\tau_\infty \sin\theta_1$ and $\tau_\infty \sin\theta_2$ respectively. It can be observed that in the absence of any fault movement ($t < T_1$), there is gradual accumulation of shear stress near F_1, with increasing rate of accumulation depending upon the inclination of the fault, being maximum for vertical faults. But after the commencement of fault creep across $F_1 (T_1 < t < T_2)$, this rate of accumulation changes. This change depends on the inclination θ_1 and θ_2 with the horizontal. This pattern of stress accumulation near F_1 continues up to the instant $t = T_2$ at which second fault F_2 also starts creeping. At $t = T_1$ and $t = T_2$, there is a change in the rate of stress accumulation.

5 Conclusion

The development of mechanical properties of the Earth's mantle has been provided in this study. The Burger medium represents the lithosphere-asthenosphere system of the Earth. So we develop analytically and graphically the displacement, stress and strain components of finite interacting faults which are buried, inclined and strike-slip in nature. The faults are situated in a viscoelastic half-space of Burger's rheology and the movement across both the faults is sudden. The variation displacement, stress and strain components for different inclinations of the faults, different slip magnitude, depth of the faults from the free surface and distance between the faults have been studied analytically and graphically. It is concluded that due to sudden movement across the faults, the displacement, stress and strain components attain maximum value across the F_1 faults, then decreases rapidly and, when it starts to interact with F_2, it increases up to a certain level and then decreases towards zero. Since after the sudden movement across F_1 it stops moving, when F_2 fault starts to move the displacement, stress and strain components attains the values, which are less than the values attained across F_1 fault. The larger amount of accumulated shear stress and strain releases after the sudden movement across F_1, which causes the next earthquake near F_1 fault, which is why the interaction occurs. It can be observed that the displacement, stress and strain increase with increasing value of slip magnitude. That is, if the value of slipping movement increases, then the amount of accumulated stress increases and so does the release of stress. Also, the shear stress and strain decrease if the depth of the faults from the free surface increases and the distance between the faults increases, which is true from practical view. From Kundu, P. and Sarkar (Mondal), S. (2020 [12]), it is obtained that due to creeping movement most displacement occurs on the overlap zone of the interacting faults. However, in this study, the movement across the interacting faults is sudden and after the sudden movements, the faults stop moving then most displacements, shear stress and strain occur across the F_1 fault.

Such study may give some concept about the occurrence of future fault movement, which helps to save enormous life and property by predicting future earthquakes.

In the future, we can extend the concept of this paper across the creeping faults movement or one locked and other creeping. It can also be extended for non-planer fault instead of plane fault as discussed in our model.

Appendix

The boundary value problem involving $(u_i)_2$, $(\tau_{ij})_2$, $(e_{ij})_2$, where $i, j = 1, 2, 3$ can be solved by using modified Green function technique (from Maruyama ([15, 14]) and Rybicki ([23, 24])) and corresponding principle, then we find

$$(\bar{u_1})_2(Q) = \iint_{F_1} [(\bar{u}_1)_2(s)]G(P,Q)\, d\xi_3\, d\xi_1, \tag{7.26}$$

where Q is a point in half-space and a function of y_1, y_2, y_3, $P(\xi_1, \xi_2, \xi_3)$ is a point across F_1 and $[(u_1)_2(s)]$ is discontinuity of u_1. G represent Green's function and

$$G(P,Q) = \frac{\partial}{\partial \xi_2} G_1(P,Q),$$

$$G_1(P,Q) = \frac{1}{[(y_1-\xi_1)^2 + (y_2-\xi_2)^2 + (y_3-\xi_3)^2]^{\frac{1}{2}}} - \frac{1}{[(y_1+\xi_1)^2 + (y_2-\xi_2)^2 + (y_3-\xi_3)^2]^{\frac{1}{2}}}.$$

Then

$$G(P,Q) = \frac{y_2-\xi_2}{[(y_1-\xi_1)^2 + (y_2-\xi_2)^2 + (y_3-\xi_3)^2]^{\frac{3}{2}}} - \frac{y_2-\xi_2}{[(y_1+\xi_1)^2 + (y_2-\xi_2)^2 + (y_3-\xi_3)^2]^{\frac{3}{2}}}.$$

$P(\xi_1, \xi_2, \xi_3)$ is a point across the fault F_1 and $0 \le \xi_2 \le D_1 \cos\theta_1$, $0 \le \xi_3 \le D_1 \sin\theta_1$ and $\xi_2 = \xi_3 \cot\theta_1$. A change in co-ordinate from (ξ_1, ξ_2, ξ_3) to (ξ_1', ξ_2', ξ_3') connected by the relations

$$\xi_1 = \xi_1', \quad \xi_2 = \xi_2' \sin\theta_1 + \xi_3' \cos\theta_1, \quad \xi_3 = -\xi_2' \cos\theta_1 + \xi_3' \sin\theta_1 + d_1.$$

From $\xi_2 = \xi_3 \cot\theta_1$, we get $\xi_2' = 0$.

Then

$$\xi_1 = \xi_1',\ \xi_2 = \xi_3' \cos\theta_1,\ \xi = \xi_3' \sin\theta_1 + d_1 \quad \text{and} \quad d\xi_1 = d\xi_1',\ d\xi_2' = 0,\ d\xi_3 = \sin\theta_1\, d\xi_3'$$

$$(\bar{u_1})_2(Q) = \iint_{F_1} [(\bar{u}_1)_2(s)]G(P,Q) d\xi_3 d\xi_1.$$

Since the coordinates of the end points of the fault are taken w. r. t. prime co-ordinate system as A $(-L_1, 0, 0)$ and B $(L_1, 0, 0)$ respectively, the limit of the integration of ξ_1' is $-L_1$ to L_1. Also w. r. t. prime co-ordinate system, the width of the fault is D_1, which is in ξ_3' direction, the limit of ξ_3' is taken as 0 to D_1.

Then

$$\begin{aligned}(\bar{u_1})_2(Q) &= \int_{-L_1}^{L_1} \int_0^{D_1} U_1(p) f_1(\xi_1', \xi_3') G(P,Q) \sin\theta_1 d\xi_3' d\xi_1' \\ &= U_1(s)\phi'(y_1, y_2, y_3).\end{aligned}$$

Taking inverse Laplace transform, we get

$$(u_1)_2(Q) = \frac{U_1(t_1)}{2\pi}\phi' H(t_1),$$

and

$$\phi'(y_1, y_2, y_3) = \int_{-L_1}^{L_1} \int_{0}^{D_1} f_1(x_1', x_3') G(P, Q) \sin\theta_1 \, dx_3' dx_1'$$

$$= \int_{-L_1}^{L_1} \int_{0}^{D_1} \left\{ \frac{y_2 - \xi_3' \cos\theta_1}{[(y_1 - \xi_1')^2 + (y_2 - \xi_3' \cos\theta_1)^2 + (y_3 - d_1 - \xi_3' \sin\theta_1)^2]^{\frac{3}{2}}} \right.$$

$$\left. - \frac{y_2 - \xi_3' \cos\theta_1}{[(y_1 + \xi_1')^2 + (y_2 - \xi_3' \cos\theta_1)^2 + (y_3 - d_1 - \xi_3' \sin\theta_1)^2]^{\frac{3}{2}}} \right\}$$

$$\times f_1(\xi_1', \xi_3') \sin\theta_1 \, d\xi_3' d\xi_1'.$$

Similarly, the displacement components $(u_1)_3$ after the fault movement across F_2 has been obtained. (z_1, z_2, z_3) and $(\xi_1'', \xi_2'', \xi_3'')$ are taken as the field point in the half space and the fault point, respectively, and simple linear transformation of co-ordinates (z_1, z_2, z_3) and $(\xi_1'', \xi_2'', \xi_3'')$ into (y_1'', y_2'', y_3'') and $(\eta_1', \eta_2', \eta_3')$ respectively is assumed.

Then

$$(u_1)_3 = \frac{U_2(t_2)}{2\pi} H(t - T_2) \phi(y_1, y_2, y_3),$$

where

$$\phi = \int_{-L_2}^{L_2} \int_{0}^{D_2} f_2(\eta_1', \eta_3') \left[\frac{z_2 - \xi_2'}{\{(z_1 - \xi'_1)^2 + (z_2 - \xi'_2)^2 + (z_3 - \xi'_3)^2\}^{\frac{3}{2}}} \right.$$

$$\left. + \frac{z_2 - \xi_2'}{\{(z_1 + \xi'_1)^2 + (z_2 - \xi'_2)^2 + (z_3 - \xi'_3)^2\}^{\frac{3}{2}}} \right] \sin\theta_2 d\eta_1' \, d\eta_3'$$

$$z_2 = y_2 - D \quad \text{and} \quad z_3 = y_3 - d_2.$$

Now by co-ordinate transformation, we get $\xi_1' = \eta_1'$

$$\xi_2' = \eta_2' \sin\theta_2 + \eta_3' \cos\theta_2,$$
$$\xi_3' = -\eta_2' \cos\theta_2 + \eta_3' \sin\theta_2.$$

Since $\eta_2' = 0, 0 \leq \eta_3' \leq D_2$ on F_2,

$$\xi_1' = \eta_1',$$
$$\xi_2' = \eta_3' \cos\theta_2,$$
$$\xi_3' = \eta_3' \sin\theta_2.$$

Bibliography

[1] J. Attanayake, D. Sandiford, L. S. Schleicher, A. Jones, G. Gibson and M. Sandiford, Interacting intraplate fault systems in Australia: the 2012 thorpdale, Victoria, seismic sequences, *J. Geophys. Res., Solid Earth* **124** (2019). https://doi.org/10.1029/2018JB016945.

[2] L. M. Cathles III., *The Viscoelasticity of the Earth's Mantle*, Princeton University Press. Princeton N. J., 1975.

[3] P. Clift, J. Lin and U. Baarckhausen, Evidence of low flexural rigidity and low viscosity lower continental crust during continental break - up in the South China, *Mar. Pet. Geol.* **19** (2002), 951–970.

[4] P. Debnath and S. Sen, Two neighbouring strike slip fault and their interaction, *IOSR J. Appl. Geol. Geophys.* **2**(6) (2014), 44–56. e-ISSN: 2321-0990 p-ISSN: 2321-0982.

[5] S. K. Debnath, Two interacting creeping vertical rectangular strike-slip faults in a viscoelastic half-space model of the lithosphere, *Int. J. Sci. Eng. Res.* **4**(6) (2013).

[6] U. Ghosh, A. Mukhopadhyay and S. Sen, On two aseismic creeping and interacting vertical strike-slip faults- one buried and the other extended up to the surface in a two layer model of the lithosphere, *Bull. Ind. Sos. Earth. Tech.* **29**(1) (1992), 1–15, Paper No. 313.

[7] U. Ghosh, A. Mukhopadhyay and S. Sen, On two interacting creeping vertical surface breaking strike-slip faults in a two layer model of the lithosphere, *Phys. Earth Planet. Inter.* **70** (1992), 119–129.

[8] U. Ghosh and S. Sen, Stress accumulation near locked buried faults in the lithosphere-asthenosphere system, *Int. J. Comput.* **1**(4) (2011), 786–795.

[9] Y. Hu, R. Burgmann, P. Banerjee, L. Feng, E. M. Hill, T. Ito, T. Tabei and K. Wang, Asthenosphere rheology inferred from observations of the 2012 Indian Ocean Earthquake, *Nature* **538** (2016), 368–372.

[10] S. Karato, Rheology of the Earth's mantle A historical review, *Gondwana Res.* **18**(1) (2010).

[11] N. Kato and X. Lei, Interaction of parallel strike-slip faults and a characteristic distance in the spatial distribution of active faults, *Geophys. J. Int.* **144**(1) (2001).

[12] P. Kundu and S. Sarkar (Mondal), Deformation analysis of a viscoelastic half-space due to a finite and an infinite interacting faults, *Phys. Scr.* **95**(5) (2020).

[13] K. Manna and S. Sen, Interacting inclined strike-slip faults in a layered medium, *Mausam* **68**(3) (2017), 487–498.

[14] T. Maruyama, Statical elastic dislocations in an infinite and semi-infinite medium, *Bull. Earthq. Res. Inst. Tokyo Univ.* **42** (1964), 289–368.

[15] T. Maruyama, On two dimensional dislocations in an infinite and semi-infinite medium, *Bull. Earthq. Res. Inst. Tokyo Univ.* **44**(part 3) (1966), 811–871.

[16] G. M. Mavko, Fault interaction near Hollister, *Calif. J. Geophys. Res.* **87** (1982), 7807–7816.

[17] D. Mondal, S. Sarkar (Mondal) and S. Sen, Effect of movement across a long, inclined, buried, creeping, strike - slip fault in the visco-elastic medium of Burger's Rheology, *J. Eng. Appl. Sci.* **14**(3) (2019), 965–974.

[18] A. Mukhopadhyay, S. Sen and P. Maji, On the interaction between two locked strike-slip faults, in: *Proceedings of the Sixth International Symposium on Earthquake Engineering Roorkee*, Vol. 1, 1978, pp. 77–82.

[19] A. Mukhopadhyay, M. Maji and S. Sen, On stress accumulation in the lithosphere and interaction between two strike-slip faults, *Indian J. Meteol. Hudrol. Geophys. (Mausam)* **30** (1979), 359–363.

[20] A. Mukhopadhyay and P. Mukherji, On two interacting creeping vertical surface breaking strike-slip faults in the lithosphere, *Bull. ISET* **21**(4) (1984), 163–191.

[21] A. Mukhopadhyay and P. Mukherji, On two aseismically creeping and interacting buried vertical strike-slip faults in the lithosphere, *Bull. ISET* **23**(3) (1986), 91–117.

[22] T. Parsons, R. Sliter, E. L. Geist, R. C. Jachens, B. E. Jaffe, A. Foxgrover, P. E. Hart and J. McCarthy, Structure and mechanics of the Hayward-Rodgers creek fault step-over, San Francisco bay, California, *Bull. Seismol. Soc. Am.* **93**(5) (2003), 2187–2200.

[23] K. Rybicki, The elastic residual field of a very long strike-slip fault in the presence of a discontinuity, *Bull. Seismol. Soc. Am.* **61** (1971), 79–92.
[24] K. Rybicki, Static deformation of a multilayered half-space by a very long strike-slip fault, *Pure Appl. Geophys.* **110** (1973), 1955–1966.
[25] P. Segall and D. D. Pollard, Mechanics of discontinuous faults, *J. Geophys. Res.* **83**(B8) (1980), 4337–4350.
[26] B. Shao and H. Guiting, The interactions of fault patterns and stress fields during active faulting in Central North China Block: Insights from numerical simulations, *Chinese Academy of Geological Sciences* **14**(4) (2019), e0215893. https://doi.org/10.1371/journal.pone.0215893.

Melvin Varghese and G. Sheeja

On the number of bounded regions generated by a class of paths in the complex plane

Abstract: A path P in the complex plane $\mathbb{C}$ is a continuous function $P : [0,1] \longrightarrow \mathbb{C}$. This paper investigates the number of regions bounded by a path in a class of paths between two points $x, y \in \mathbb{C}$. We investigate some combinatorial properties of this class of paths.

Keywords: Paths, fibres, vertex generated by the path, regions bounded by the path, graph, regular graph

MSC 2010: 05C30, 05C10, 22E30

1 Introduction

Open-connected sets are important in the topology of the complex plane, as they are actively involved in complex integration theory, and a path in the complex plane plays a vital role in contour integrals. Experts in the field of analysis have already carried out the study of classes of paths. In this paper, we investigate some combinatorial properties of certain classes of paths. Our results are motivated by the idea of Alexander numbering introduced in [2]. A simple combinatorial rule for defining the winding number was proposed by August Ferdinand Möbius in 1865 [7] and again independently by James Waddell Alexander II in 1928. [2] Any curve partitions the plane into several connected regions, one of which is unbounded. The winding numbers of the curve around two points in the same region are equal. The winding number around (any point in) the unbounded region is zero. Finally, the winding numbers for any two adjacent regions differ by exactly 1; the region with the larger winding number appears on the left side of the curve (with respect to motion down the curve). While numbering, a large planar graph can determine the number of regions. This number of regions can be calculated by a formula, which is, in fact, the conclusion of this paper. Several authors have conducted extensive research in this area related to differential geometry and curve theory [5, 4, 8].

Acknowledgement: The first author is highly grateful to SRM Institute of Science and Technology, Kattankulathur, India, for providing financial support through a research fellowship.

Melvin Varghese, Department of Mathematics, College of Engineering and Technology, SRM Institute of Science and Technology, E 503 Travencore Residency, Potheri, Kavanur Road, Chenkelpattu district, Kattankulathur 603203, Tamil Nadu, India, e-mail: mv2121@srmist.edu.in
G. Sheeja, Department of Mathematics, College of Engineering and Technology, SRM Institute of Science and Technology, C 82/2 Chandran Enclave, Second Main Road, Thiruvengada Nagar, Ambattur, Chennai 600053, Tamil Nadu, India, e-mail: sheejag@srmist.edu.in

https://doi.org/10.1515/9783111313634-008

We introduce a graph generated by a path in the complex plane to estimate the number of bounded regions generated by a path in the complex plane. One of our major motivations is to find a connection between the winding number of a point and the number of bounded regions generated by a circuit.

2 Preliminaries

Definition 1 ([1, 6]). A path P is a continuous function $P : [0,1] \longrightarrow \mathbb{C}$. We assume that $P(0), P(1)$ are finite. The *range of* P is denoted by $P([0,1]) := \{y \in \mathbb{C} \mid y = P(x)$ for some $x \in [0,1]\}$. *Fibre* of an element v under a function P is the preimage of the singleton set $\{v\}$. And it is denoted by $v_P^{-1} := \{x \in [0,1] \mid P(x) = v\}$.

Example 1. $P(t) = t + it, 0 \le t \le 1.$

This is a path from 0 to $1 + i$. This illustrated in the Figure 8.1.

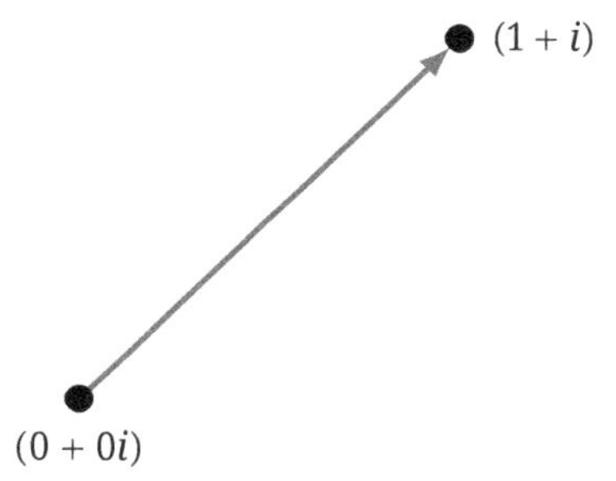

Figure 8.1: The image of $P(t) = t + it, 0 \le t \le 1$ in $\mathbb{C}$.

Definition 2 ([1]). A simple path is an injective continuous function $P : [0,1] \longrightarrow \mathbb{C}$ and a path P is non-simple if P is not injective. A simple path ans a non-simple path is given in the Figures 8.2 and 8.3 respectively.

Figure 8.2: A simple path.

Figure 8.3: A non-simple path.

Theorem 1 ([1]). *A non-empty open set in $\mathbb{C}$ is connected if and only if any two of its points can be joined by a polygon that lies in the set.*

Definition 3 ([1]). A non-empty, connected, open set is called a region.

Definition 4 ([1]). A component of a set is a connected subset that is not contained in any larger connected subset.

Theorem 2 ([1]). *Every set has a unique decomposition into components.*

Definition 5 ([3]). An undirected multigraph G is an ordered triple $G := (V, E, r)$ with
1. V is a set of vertices or nodes,
2. $\deg(v)$ denotes the number of edges incident with v.
3. E is a set of edges or lines,
4. $r : E \longrightarrow \{\{x, y\} : x, y \in V\}$ assigning to each edge an unordered pair of endpoint nodes. A multigraph is illustrated in the Figure 8.4.

Figure 8.4: A multigraph.

Definition 6 ([3]). Paths and connectedness
1. A *walk* in a graph G is a sequence of vertices such that each consecutive vertices form an edge.
2. A *trail* is a walk without repeated edges.
3. A *circuit* is a trail with its initial vertex equal to the terminal vertex.
4. A *path* is a non-empty graph $P = (V, E)$ of the form $V = \{x_0, x_1, x_2, \ldots, x_k\}$ and $E = \{x_0x_1, x_1x_2, x_2x_3, \ldots, x_{k-1}x_k\}$, where x_i are all distinct.
5. A non-empty graph G is *connected* if any two of its vertices are linked by a walk.
6. A connected graph is Eulerian if it admits an Euler circuit.

Theorem 3 (Euler, [3]).
1. *A connected graph is Eulerian if every vertex has an even degree.*
2. *A connected graph G contains an Euler path if and only if G has exactly two vertices of odd degree.*

Definition 7 ([3]). A planar graph is a pair (V, E) of finite sets with the following properties.
1. $V \subseteq \mathbb{R}^2$;
2. Every edge is an arc between two vertices;
3. Different edges have different sets of endpoints;
4. The interior of an edge contains no vertex and no point of any other edge.

Theorem 4 (Euler, [3]). *If a finite, connected, planar graph is drawn in the plane without edge intersections, and v is the number of vertices, e is the number of edges, and f is the number of faces, then* $f = e - v + 2$.

3 Main results

Definition 8. Let $P : [0,1] \longrightarrow \mathbb{C}$ be a path from $P(0)$ to $P(1)$ in the complex plane. A point $v \in \mathbb{C}$ is called a *vertex generated by the path* if $|v_P^{-1}| \geq 2$. Where v_P^{-1} is the fibre of v under P.

Remark 1. If $v \notin P([0,1])$, then $v_P^{-1} = \phi$.

Definition 9. $\langle \mathbb{P}; A\rangle := \{P : [0,1] \longrightarrow \mathbb{C} \mid v$ is a vertex generated by the path P iff $|v_P^{-1}| \in A\}$, where $A \subset \mathbb{N} - \{1\}$ such that A and V_P^A are finite, where $V_P^A := \{v \mid |v_P^{-1}| \in A\}$.

Proposition 1. $\langle \mathbb{P}; A\rangle \neq \phi$ *for every finite subset A of* $\mathbb{N} - \{1\}$.

Proof. Let $A = \{a_1, a_2, \ldots, a_k\}$. Define a continuous path in the following way. Choose $0 = P(0)$. Then take i such that $|i_P^{-1}| = a_i, 1 \leq i \leq k$ and set $P(1) = k + 1$. Denote $i_P^{-1} = \{t_1^i, t_2^i, t_3^i, \ldots, t_{a_i}^i\}$. Then we can construct a path in the following Figure 8.5 that belongs to $\langle \mathbb{P}; A\rangle$. □

Figure 8.5: The path P, constructed from A.

Example 2. In Figure 8.6, $V_P^{\{2\}} = \{a, b, c, d, e, f, g, h, k\}$.

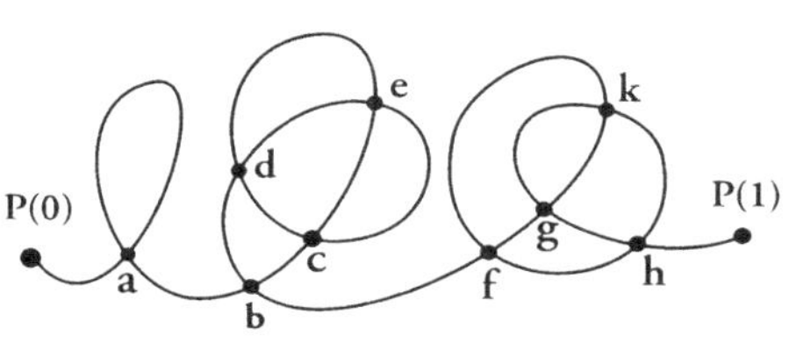

Figure 8.6: A path with $A = \{2\}$.

Let us denote fibres of the vertices in the following manner, $a_P^{-1} = \{t_1^a, t_2^a\}$, $b_P^{-1} = \{t_1^b, t_2^b\}$, $c_P^{-1} = \{t_1^c, t_2^c\}$, $d_P^{-1} = \{t_1^d, t_2^d\}$, and so on. Finally, $k_P^{-1} = \{t_1^k, t_1^k\}$.

Example 3. In Figure 8.7, $V_P^{\{2,3\}} = \{a, b, c, d, e, f, g, h\}$ also $a_P^{-1} = \{t_1^a, t_2^a, t_3^a\}$, $b_P^{-1} = \{t_1^b, t_2^b\}$, $c_P^{-1} = \{t_1^c, t_2^c\}$, $d_P^{-1} = \{t_1^d, t_2^d\}$, $e_P^{-1} = \{t_1^e, t_2^e\}$, $f_P^{-1} = \{t_1^f, t_2^f, t_3^f\}$, $h_P^{-1} = \{t_1^h, t_2^h\}$.

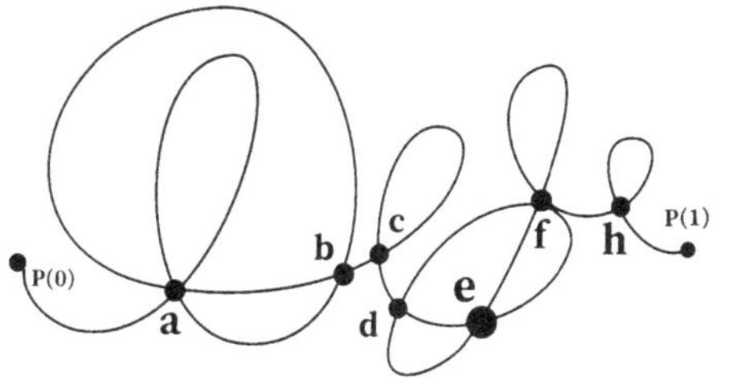

Figure 8.7: A path with $A = \{2, 3\}$.

Definition 10. Let $P : [0, 1] \longrightarrow \mathbb{C}$ be a path. A subset B of $\mathbb{C}$ is called a region generated by P if B is a component of $\mathbb{C} - P([0, 1])$.

Remark 2. Components of $\mathbb{C} - P([0, 1])$ are open, connected and bounded.

Definition 11. Define R_P^A := The set of all bounded regions generated by the path P. In Figure 8.9, dark blue shaded regions are elements of $R_Q^{\{2\}}$ and $|R_Q^{\{2\}}| = 8$. Where path Q is in Figure 8.8.

Figure 8.8: A path Q with $A = \{2\}$.

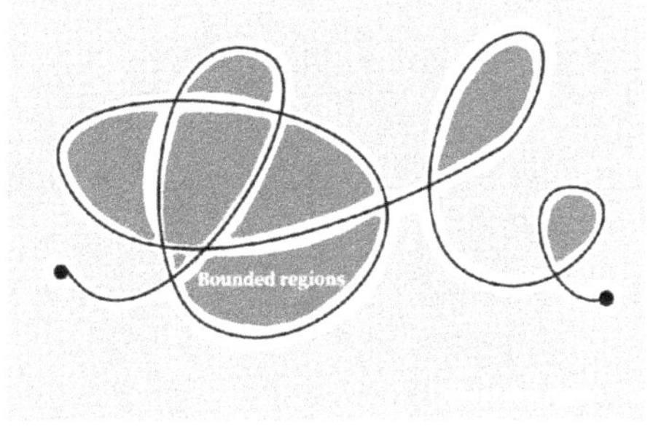

Figure 8.9: Shaded region, $\mathbb{C} - Q([0, 1])$.

Definition 12. The multigraph G_P^A is triple $\langle V_P^A, E_P^A, \eta_P^A \rangle$. Let $(V_P^A)^{-1} = \bigcup_{v \in V_P^A} v^{-1}$, then $(V_P^A)^{-1}$ is a subset of $[0,1]$ and it has an order inherited from real numbers.

For $v \in V_P^A$, let $v_P^{-1} = \{t_1^v, t_2^v, t_3^v, \ldots, t_k^v\}$, where $k = |v_P^{-1}|$. Now E_P^A defined as $E_P^A = \{\{P(t_i^u), P(t_j^v)\} \mid t_i^u \text{ and } t_j^v \text{ are consecutive in } (V_P^A)^{-1}\}$ and $\eta_P^A : E_P^A \longrightarrow \{\{x,y\} \mid x,y \in V_P^A\}$ such that $\eta_P^A(\{P(t_i^u), P(t_j^v)\}) = \{u, v\}$. Note that the bounded faces of G_P^A are in bijection with bounded regions generated by a path $P \in \langle \mathbb{P}; A \rangle$. refer the Figure 8.10.

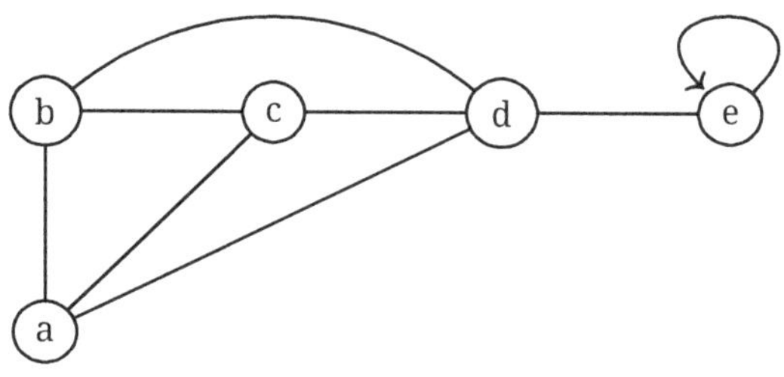

Figure 8.10: The multigraph G_P^A generated by P in Figure 8.11.

Example 4. In Figure 8.11, the ordered $(V_P^A)^{-1} = \{t_1^a, t_1^c, t_1^b, t_2^a, t_1^d, t_2^b, t_2^c, t_2^d, t_1^e, t_2^e\}$, then we have the mapping η_P^A as follows

$$\begin{aligned}
\{P(t_1^a), P(t_1^c)\} &\longrightarrow \{a, c\}, \\
\{P(t_1^c), P(t_1^b)\} &\longrightarrow \{c, b\}, \\
\{P(t_1^b), P(t_2^a)\} &\longrightarrow \{b, a\}, \\
\{P(t_2^a), P(t_1^d)\} &\longrightarrow \{a, d\}, \\
\{P(t_1^d), P(t_2^b)\} &\longrightarrow \{d, b\}, \\
\{P(t_2^b), P(t_2^c)\} &\longrightarrow \{b, c\}, \\
\{P(t_2^c), P(t_2^d)\} &\longrightarrow \{c, d\}, \\
\{P(t_2^d), P(t_1^e)\} &\longrightarrow \{d, e\}, \\
\{P(t_1^e), P(t_2^e)\} &\longrightarrow \{e, e\}.
\end{aligned}$$

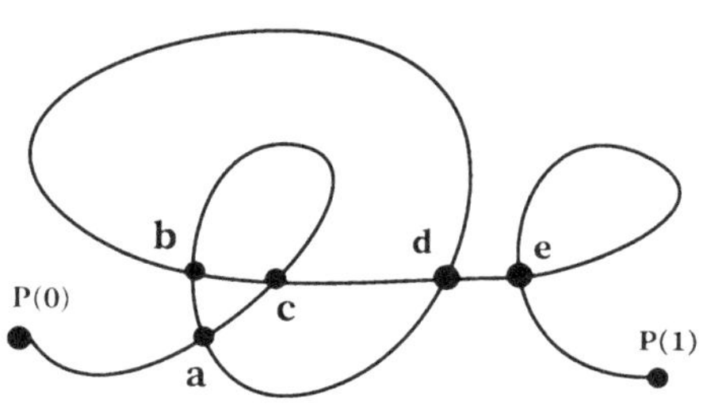

Figure 8.11: A path P with $A = \{2\}$.

Lemma 1. *If $G_P^{\{k\}}$ be a multigraph generated by $P \in \langle \mathbb{P}; \{k\}\rangle$, then*

$$|E_P^{\{k\}}| = k|V_P^{\{k\}}| - 1.$$

Proof. If $\{P(t_i^u), P(t_j^v)\} \in E_P^{\{k\}}$, then t_i^u and t_j^v are consecutive.

Note that the number of edges is exactly the number of consecutive unordered pairs in $(V_P^{\{k\}})^{-1}$. For each $v \in V_P^{\{k\}}$ we have $|v_P^{-1}| = k$. Then $(V_P^{\{k\}})^{-1}$ contains $k|V_P^{\{k\}}|$ elements. Now let B be any finite subset of real numbers containing q elements. Then the number of unordered pairs (x, y) consecutive in B is exactly $q - 1$. To see this, take $B = \{x_1, x_2, \dots, x_q\}$ such that $x_1 < x_2 < \cdots < x_q$ then the consecutive pairs are $\{(x_1, x_2), (x_2, x_3), \dots, (x_{q-1}, x_q)\}$. There are $q - 1$ of them. Take $B = (V_P^{\{k\}})^{-1}$ then the count is exactly $k|V_P^{\{k\}}| - 1$. Hence the proof. □

Theorem 5. *If $P \in \langle \mathbb{P}; \{k\}\rangle$, then $|R_P^{\{k\}}| = (k-1)|V_P^{\{k\}}|$.*

Proof. First, we will show that the multigraph $G_P^{\{k\}}$ is a connected planar graph. To see this, let $u, v \in V_P^{\{k\}}$. Then there exist t_i^u and t_j^v such that $u = P(t_i^u)$ and $v = P(t_j^v)$. Without loss of generality assume that $t_i^u < t_j^v$. We know that P is a continuous path on $[0, 1]$ so in its sub-intervals. In particular, P restricted to the interval $[t_i^u, t_j^v]$ is a continuous path. Now either $[t_i^u, t_j^v] \cap V_P^{\{k\}} = \phi$ or $[t_i^u, t_j^v] \cap V_P^{\{k\}} \neq \phi$. In the former case, $\{u, v\}$ form an edge in $G_P^{\{k\}}$ and in later case, there exists a finite ordered sequence $t_{i1}^{h1} < t_{i2}^{h2} < \cdots < t_{im}^{hm}$ (for some $m \in \mathbb{N}$) from $[0, 1]$ such that $uP(t_{i1}^{h1})P(t_{i2}^{h2})...P(t_{im}^{hm})v$ forms a path in P. That is, in both cases, we got a path between u and v in $G_P^{\{k\}}$. That is, $G_P^{\{k\}}$ is connected. The graph is planar. Since all vertex in $G_P^{\{k\}}$ corresponds to a self-intersection point of the path P. And for each self-intersection point p of the path P there is a vertex in $G_P^{\{k\}}$, since the path self intersects at a point p (if any) exactly k times. Then we observe that the regions generated by the path P are the faces of the multigraph $G_P^{\{k\}}$. Also, notice that $|R_P^{\{k\}}| = F - 1$, where F is the number of faces of $G_P^{\{k\}}$. Now, by Euler's formula, we have

$$F = |E_P^{\{k\}}| - |V_P^{\{k\}}| + 2.$$

This implies

$$F = k|V_P^{\{k\}}| - 1 - |V_P^{\{k\}}| + 2 \Longrightarrow F - 1 = (k-1)|V_P^{\{k\}}| = |R_P^{\{k\}}|.$$ □

Corollary 1. *Let $P \in \langle \mathbb{P}; \{h\}\rangle$ and $Q \in \langle \mathbb{P}; \{k\}\rangle$. If $h \leq k$ and $|V_P^{\{h\}}| \leq |V_Q^{\{k\}}|$, then*

$$|R_P^{\{h\}}| \leq |R_Q^{\{k\}}|.$$

Proof. The proof is immediately derived from the formula in Theorem 5 using given conditions. □

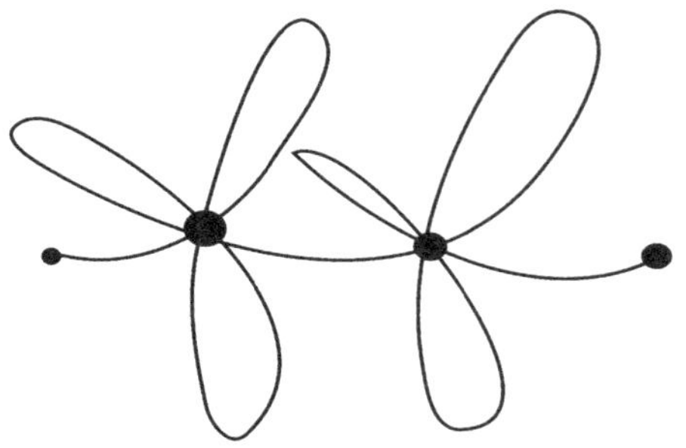

Figure 8.12: A path P with $A = \{3\}$.

Example 5. In Figure 8.12, by theorem 5, we have $k = 4$, $|V_P^{\{3\}}| = 2$, and from the picture, it is clear that there are 6 bounded regions. If we apply the formula, $|R_P^{\{k\}}| = (k-1)|V_P^{\{k\}}|$, we get $|R_P^{\{k\}}| = (4-1)2 = (3)2 = 6$, which is one and the same.

Theorem 6. *Let $P \in \langle \mathbb{P}; A\rangle$, where $A = \{k_1, k_2, k_3, \ldots, k_m\}$.*
If $a_i = |\{v \in V_P^A : |v_P^{-1}| = k_i\}|$ then $|R_P^A| = \sum_{i=1}^m a_i(k_i - 1)$.

Proof. The number of edges of G_P^A can be proved similarly as in Lemma 1, and it is found to be $\sum_{i=1}^m a_i k_i - 1$. Observe that $\sum_{i=1}^m a_i = |V_P^A|$. Now, by Euler's formula,

$$|R_P^A| = F - 1 = \sum_{i=1}^m a_i k_i - |V_P^A|.$$

Which implies,

$$|R_P^A| = \sum_{i=1}^m a_i k_i - \sum_{i=1}^m a_i.$$

That is,

$$|R_P^A| = \sum_{i=1}^m a_i(k_i - 1).$$

Hence the proof. □

Example 6. In Figure 8.13, $V_P^A = \{a, b, c, d, e, f, g\}$. $|a_P^{-1}| = 3$, $|e_P^{-1}| = 3$, $|b_P^{-1}| = 2$, $|c_P^{-1}| = 2$, $|d_P^{-1}| = 2$, $|f_P^{-1}| = 2$, $|g_P^{-1}| = 2$. That is $k_1 = 2$, $k_2 = 3$ and $a_1 = 5$, $a_2 = 2$. Then, by Theorem 6, $|R_P^A| = \sum_{i=1}^m a_i(k_i - 1)$. If we apply the formula, we get $|R_P^A| = 5(2-1) + 2(3-1) = 5 + 4 = 9$.

Corollary 2. *If $P \in \langle \mathbb{P}; A\rangle$, where A is finite subset of $\mathbb{N} - \{1\}$, then*

$$|R_P^A| \leq (\max A - 1)|V_P^A|.$$

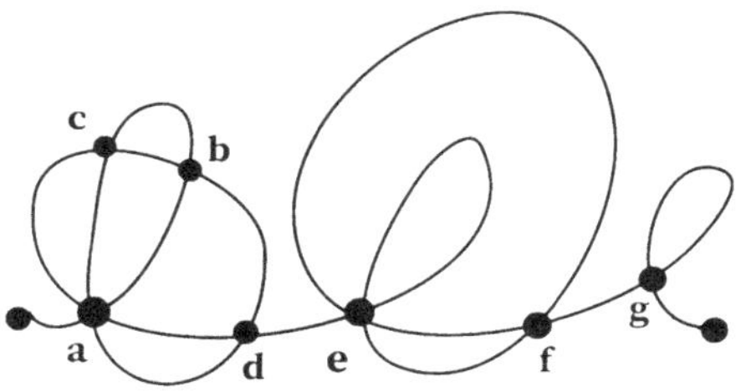

Figure 8.13: A path P with $A = \{2, 3\}$.

Proof. Using Theorem 6, we have $|R_P^A| = \sum_{i=1}^{m} a_i(k_i - 1)$. By expanding the right-hand side of the equation, we get

$$|R_P^A| = \sum_{i=1}^{m} a_i k_i - \sum_{i=1}^{m} a_i \Longrightarrow |R_P^A| = \sum_{i=1}^{m} a_i k_i - |V_P^A|.$$

$$\Longrightarrow |R_P^A| \leq \sum_{i=1}^{m} a_i \max A - |V_P^A| \Longrightarrow |R_P^A| \leq \max A\left(\sum_{i=1}^{m} a_i\right) - |V_P^A|.$$

$$\Longrightarrow |R_P^A| \leq \max A|V_P^A| - |V_P^A| \Longrightarrow |R_P^A| \leq (\max A - 1)|V_P^A|. \qquad \square$$

Lemma 2. *Let $P \in \langle \mathbb{P}; A\rangle$ and G_P^A be the multigraph associated with it. If $t \in (0,1) \cap (V_P^A)^{-1}$, then $\deg(v) = 2|v_P^{-1}|$, where $v = P(t)$.*

Proof. Let $t \in (0,1) \cap (V_P^A)^{-1}$, then $v = P(t)$ is a vertex in V_P^A and $|v_P^{-1}| \geq 2$. Assume that $|v_P^{-1}| = k$ for some positive integer $k > 1$. Then there exists a finite sequence $t_1^v, t_2^v, \ldots, t_k^v$ from $(0,1) \cap (V_P^A)^{-1}$ such that $v = P(t) = P(t_1^v) = P(t_2^v) = P(t_3^v) = \cdots = P(t_k^v)$. Each $t_i^v, 1 \leq i \leq k$ contributes 2 edges incident with v. That is, for each $t_i^v, i \leq i \leq k$, there exists $z_i = P(x)$ and $w_i = P(y)$ for some $x, y \in (V_P^A)^{-1}$ such that there are edges $\{z_i, v\}$ and $\{v, w_i\}$ in E_P^A. Then there are $2k$ edges that are incident with v. That is, $\deg(v) = 2k = 2|v_P^{-1}|$. $\square$

See the diagram 8.14 below,

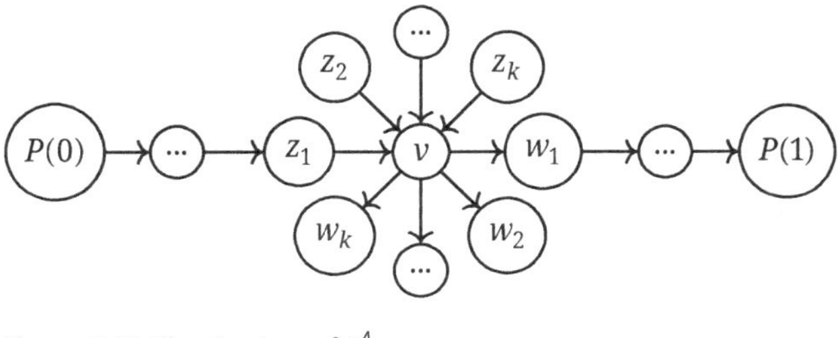

Figure 8.14: The structure of G_P^A.

Theorem 7. *Let $P \in \langle \mathbb{P}; A\rangle$ and G_P^A be the multigraph associated with it. Let $\sim$ be the equivalence relation defined on V_P^A such that $u \sim v$ if and only if $\deg(u) = \deg(v)$. Denote*

$[u]$ is the equivalence class of u under $\sim$ and $(V_P^A/\sim)$ be the set of all distinct equivalence classes under $\sim$. Let

$$M = \sum_{[u]\in(V_P^A/\sim)} |[u]|\left(\left\lceil \frac{\deg(u)}{2} \right\rceil - 1\right).$$

Then

$$|R_P^A| = \begin{cases} M & \text{if } P(0) \neq P(1), \\ 1 + M & \text{if } P(0) = P(1). \end{cases}$$

Proof. For $u \in V_P^A$, let $|u_P^{-1}| = k$ for some positive integer $k > 1$. First, let us prove for all $u \in V_P^A$, $k = \lceil \frac{\deg(u)}{2} \rceil$ in three mutually exclusive cases.

Case 1:
If $u \neq P(0)$ and $u \neq P(1)$, then by Lemma 2, we have $\deg(u) = 2k$.

In this case, $k = \frac{\deg(u)}{2} = \lceil \frac{\deg(u)}{2} \rceil$.

Case 2:
Now if $u = P(0) \neq P(1) = v$, then $P(0)$ has degree $2m+1$ and $P(1)$ has degree $2h+1$. Where $m + 1 = |P(0)^{-1}|$ and $h + 1 = |P(1)^{-1}|$. Since there are $m = \frac{|P(0)^{-1}|-1}{2}$ elements from $P(0)^{-1}$, that provides two edges and the remaining is outgoing edge. Since there are $h = \frac{|P(1)^{-1}|-1}{2}$ elements from $P(1)^{-1}$, that provides two edges and the remaining is incoming edge.

Then $|P(0)^{-1}| = m + 1 = \lceil \frac{2m+1}{2} \rceil = \lceil \frac{\deg(u)}{2} \rceil$. Also $|P(1)^{-1}| = h + 1 = \lceil \frac{2h+1}{2} \rceil = \lceil \frac{\deg(v)}{2} \rceil$.

Case 3:
Assume $w = P(0) = P(1)$. Then $q = \frac{|P(0)^{-1}|-2}{2}$ elements from $P(0)^{-1}$ provide two edges and the remaining two are one outgoing edge and one incoming edge.

$|P(0)^{-1}| = q + 1 = \lceil \frac{2q+2}{2} \rceil = \lceil \frac{\deg(w)}{2} \rceil$. That is in all three cases $|u_P^{-1}| = \frac{\deg(u)}{2}$. See the diagrams 8.15 and 8.16 for $P(0) \neq P(1)$ and $P(0) = P(1)$ respectively,

Now recall the formula from Theorem 6. Then we have, $a_i = |\{v \in V_P^A : |v_P^{-1}| = k_i\}|$. For each a_i corresponds to a unique equivalence class $[u]$ in $V_P^A/\sim$. And $k_i = \frac{\deg(u)}{2}$. Then for $P(0) \neq P(1)$, we have

$$|R_P^A| = \sum_{i=1}^{m} a_i(k_i - 1) = \sum_{[u]\in(V_P^A/\sim)} |[u]|\left(\left\lceil \frac{\deg(u)}{2} \right\rceil - 1\right).$$

When $P(0) = P(1)$ the factor $(\lceil \frac{\deg(u)}{2} \rceil - 1)$ is unchanged, but we are getting one more bounded region. Then the formula will become,

$$|R_P^A| = 1 + \sum_{[u]\in(V_P^A/\sim)} |[u]|\left(\left\lceil \frac{\deg(u)}{2} \right\rceil - 1\right).$$

Hence the proof. □

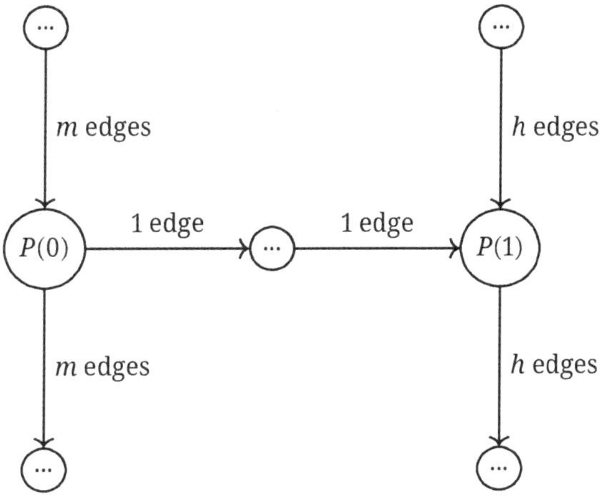

Figure 8.15: $P(0) \neq P(1)$.

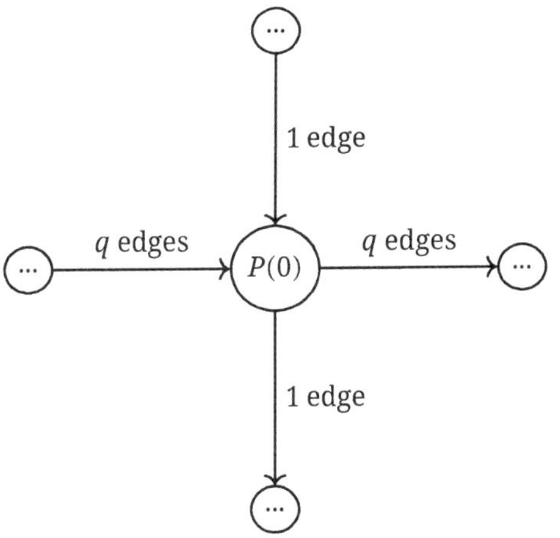

Figure 8.16: $P(0) = P(1)$.

Example 7. In Figure 8.17, $u = P(0)$ and $u \neq v = P(1)$. This path belongs to the class $\langle \mathbb{P}; \{2\} \rangle$. It is clear from the above diagram that $|R_P^A| = 7$. Let us verify the formula given by the Theorem 7. We have all vertices from V_P^A are of degree 4. Here $V_P^A = \{p, q, r, s, t, w, z\}$ and $(V_P^A / \sim) = \{[p]\}$. Then the number of regions bounded by the path is given by

$$M = \sum_{[u] \in (V_P^A/\sim)} |[u]| \left(\left\lceil \frac{\deg(u)}{2} \right\rceil - 1 \right) = |[p]| \left(\frac{\deg(p)}{2} \right) - 1 = 7(2 - 1) = 7.$$

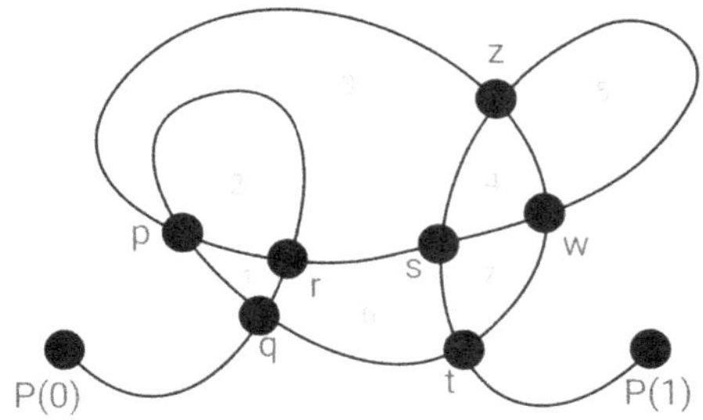

Figure 8.17: A path with $A = \{2\}$.

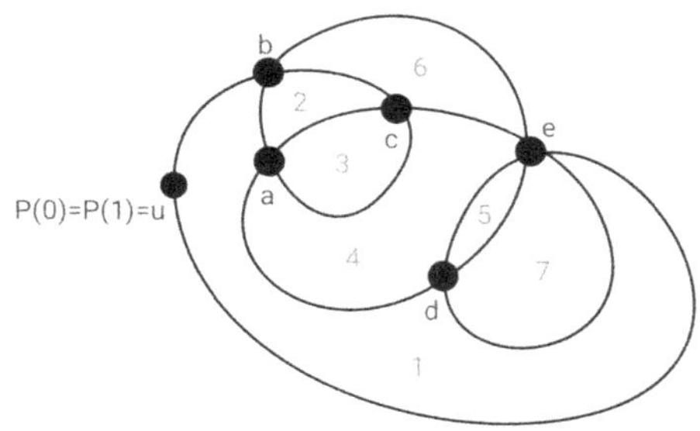

Figure 8.18: $P(0) = P(1)$ and $A = \{2, 3\}$.

Example 8. In Figure 8.18, $u = P(0) = P(1)$. This path belongs to the class $\langle \mathbb{P}; \{2,3\}\rangle$. It is clear from the above diagram that $|R_P^A| = 7$ Let us verify the formula given by the Theorem 7. Here $V_P^A = \{u, a, b, c, d, e\}$ and $(V_P^A/\sim) = \{[a], [e], [u]\}$, where $[a] = \{a, b, c, d\}$, $[e] = \{e\}$, $[u] = \{u\}$. Then the number of regions bounded by the path is given by

$$\begin{aligned}1 + M &= 1 + \sum_{[u]\in(V_P^A/\sim)} |[u]|\left(\left\lceil \frac{\deg(u)}{2}\right\rceil - 1\right)\\ &= \left(|[a]|\left(\frac{\deg(a)}{2}\right) - 1\right) + \left(|[e]|\left(\frac{\deg(e)}{2}\right) - 1\right) + \left(|[u]|\left(\frac{\deg(u)}{2}\right) - 1\right)\\ &= 1 + (4(2-1)) + (1(3-1)) + (1(1-1)) = 1 + 4 + 2 = 7.\end{aligned}$$

Theorem 8. *Let $P \in \langle \mathbb{P}; A\rangle$. Then the following statements are true.*

1. *G_P^A contains an Euler trail(or circuit).*
2. *If G is a finite graph containing an Euler trail(or circuit), then there exists a finite subset A_G of $\mathbb{N} - \{1\}$ and a path $P \in \langle \mathbb{P}; A_G\rangle$ such that $G = G_P^{A_G}$.*

Proof. 1. **Case 1:** $P(0) = P(1)$.

By Lemma 2, if $u \neq P(0)$, $v \neq P(1)$, then $\deg(v) = 2k$ for some k. Now $\deg(P(0)) = 2q + 2$. See the diagram 8.19 for details.

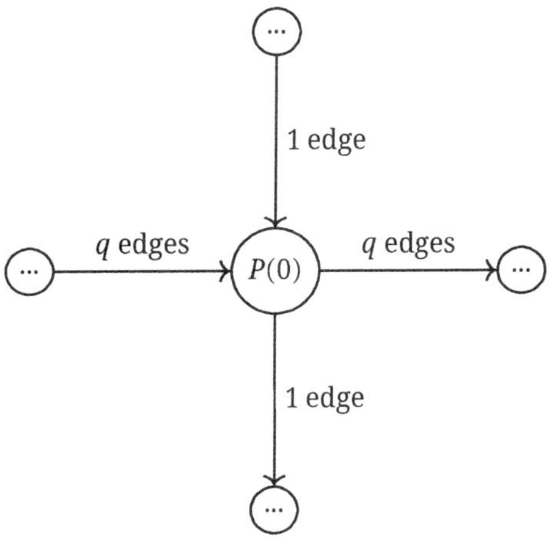

Figure 8.19: $P(0) = P(1)$.

That is every vertex of G_P^A has an even degree. Then G_P^a contains an Euler circuit.
Case 2: $P(0) \neq P(1)$.
Here we have $\deg(P(0)) = 2m + 1$ and $\deg(P(1)) = 2h + 1$. See the diagram 8.20 for details.

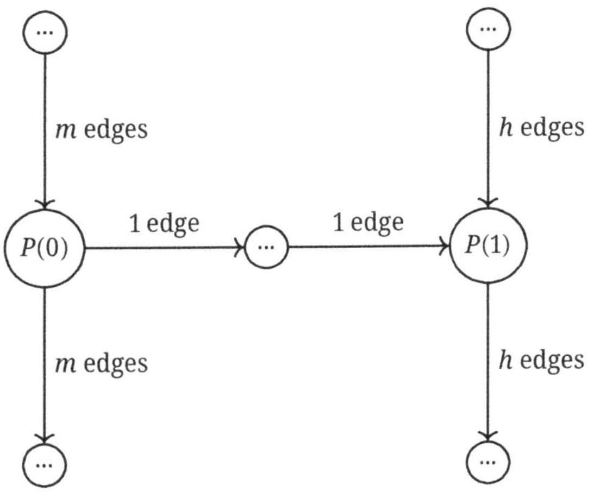

Figure 8.20: $P(0) \neq P(1)$.

That is by Lemma 2, there are exactly 2 vertices of an odd degree and then G_P^A has an Euler trail.

2. Let G contain an Euler trail(or circuit). Let this trail be $T = u_1e_1u_2e_2...e_{k-1}u_k$. Here T contains all the edges of G exactly once. Note that $k = |E(G)| + 1$. Also, observe

that each vertex $v \in V(G)$ is present at least once in T. Choose a finite subset $B := \{b_1, b_2, b_3, \ldots, b_k\}$ of $\mathbb{C}$ in bijection with $V(T) = \{u_1, u_2, \ldots, u_k\}$. $V(T)$ is a multiset. Define paths

$$P_i : [t_i, t_{i+1}] \longrightarrow \mathbb{C}$$

such that $t_i \in [0,1]$ where $1 \leq i \leq k-1$ and $P_i(t_i) = b_i$ and $P_i(t_{i+1}) = b_{i+1}$, where $t_1 = 0$, $t_2 = 1$. Notice that $\bigcup_{1 \leq i \leq k-1} [t_i, t_{i+1}] = [0,1]$. Now the join of all these paths P_i can be denoted by $P := \bigcup_{1 \leq i \leq k-1} .P_i$ is path from $[0,1]$ to $\mathbb{C}$ such that $G_P^{A_G} = G$ and $A_G = \{|(b_i)^{-1}| : 1 \leq i \leq k-1\}$. □

Bibliography

[1] L. V. Ahlfors and L. V. Ahlfors, *Complex Analysis*, Vol. 3, McGraw-Hill, New York, 1979.

[2] J. W. Alexander, Topological invariants of knots and links, *Trans. Am. Math. Soc.* **30**(2) (1928), 275–306.

[3] J. A. Bondy, U. S. R. Murty et al., *Graph Theory with Applications*, Vol. 290, Macmillan, London, 1976.

[4] M. P. Do Carmo, *Differential Geometry of Curves and Surfaces*, revised and updated 2nd edn, Courier Dover Publications, 2016.

[5] E. Feldman, Deformations of closed space curves, *J. Differ. Geom.* **2**(1) (1968), 67–75.

[6] J. Lee, *Introduction to Topological Manifolds*, Vol. 202, Springer Science & Business Media, 2010.

[7] A. Mobius, Über die bestimmung des inhaltes eines polyëders, *Math.-Phys. Kl.* **17** (1865), 31–68.

[8] D. Sunday, The crossing number, in: *Fast Winding Number Inclusion of a Point in a Polygon*, 2006.

Jitender Singh and Renu Bajaj

Binomial and de Moivre revisited through differential equations

Abstract: In this note, the classical binomial theorem, de Moivre's formula, and numerous other deductions are made using the uniqueness theorem for the initial value problems in linear ordinary differential equations.

Keywords: Linear ODEs, binomial theorem, de Moivre's theorem

MSC 2010: 34A12, 00A05, 26A06

1 Background

The binomial and de Moivre theorems are among the most important formulas in mathematics with variety of applications. The classical proofs of the binomial theorem [3, 5, 7] and de Moivre's formula [4, Ch. 8, pp. 106–107], [8] using mathematical induction are well known. In fact, several other proofs of the binomial theorem can be found in the literature using combinatorial [14], probabilistic [9], and calculus approaches [6, 10]. However, the proofs using ordinary differential equations are not known much. For instance, Staring [12] settles the Pythagoras theorem by solving a first-order nonlinear ordinary differential equation. The binomial theorem is also proved in Ungar [13, pp. 874, Eq. 6] using an addition theorem for linear ordinary differential equations. Recently, in [11], the author obtained de Moivre's formula using calculus. His proof is based on the fact that if derivative of a complex-valued function of a real variable vanishes identically on real line, then the function is constant.

In this paper, we prove these classical results using the uniqueness theorem for initial value problems in linear ODEs. The uniqueness theorem for the solution of an initial value problem in a system of first-order linear differential equations is well known (see [2, pp. 229, Th. 7.4]). An equivalent form for mth order linear problem is as follows:

Theorem A (Uniqueness). *Given a positive integer m; $a_1(t), \ldots, a_m(t)$, and $b(t)$ as continuous complex valued functions of the real variable t on an open interval I (either bounded*

Acknowledgement: The authors are indebted to the anonymous referees for useful comments leading to improvement of the article.

Jitender Singh, Department of Mathematics, Guru Nanak Dev University Amritsar, C44, Guru Nanak Dev University Campus, 143005 Amritsar, Punjab, India, e-mail: jitender.math@gndu.ac.in
Renu Bajaj, Department of Mathematics, Panjab University Chandigarh, Sector 14, Chandigarh, 160014 Chandigarh, Chandigarh UT, India, e-mail: rbajaj@pu.ac.in

https://doi.org/10.1515/9783111313634-009

or unbounded) containing a real number t_0, the mth order linear initial value problem

$$y^{(m)}(t) + a_1(t)y^{(m-1)}(t) + \cdots + a_m(t)y(t) = b(t),$$
$$y(t_0) = y_0, y'(t_0) = y_1, \ldots, y^{(m-1)}(t_0) = y_{m-1} \tag{9.1}$$

has a unique solution on the interval I.

As we shall see in this paper, in addition to the binomial and de Moivre's theorems, Theorem A can be applied to prove numerous other elementary known results by formulating an appropriate initial value problem in linear ODEs.

2 Binomial and de Moivre

Theorem 1 (Binomial). *For a positive integer n and real t, the following holds.*

$$(1 + t)^n = 1 + \sum_{k=1}^{n} \frac{n!}{k!(n-k)!} t^k. \tag{9.2}$$

Proof. Observe that if (9.2) holds for all $t > -1$, then taking limit as $t \to -1^+$ on both sides of (9.2), we get the result for $t = -1$ as well. Further, if (9.2) holds for all $t > -1$, then it holds for all $t < -1$, since if that is the case, then we can write

$$(1 + t)^n = t^n \left(1 + \frac{1}{t}\right)^n = t^n \sum_{k=0}^{n} \frac{n!}{k!(n-k)!} \frac{1}{t^k} = \sum_{k=0}^{n} \frac{n!}{k!(n-k)!} t^{n-k} = \sum_{k=0}^{n} \frac{n!}{(n-k)!k!} t^k.$$

Thus, it is sufficient to prove (9.2) for $t > -1$. So, we consider the initial value problem

$$y'(t) - \frac{n}{1+t} y(t) = 0, \quad y(0) = 0, \ t > -1, \tag{9.3}$$

which has the zero function as its unique solution for $t > -1$. Define

$$\psi(t) = (1 + t)^n - 1 - \sum_{k=1}^{n} \frac{n!}{k!(n-k)!} t^k, \quad t > -1.$$

Differentiating ψ with respect to t and multiplying throughout by $(1 + t)$, we get

$$\begin{aligned}
(1 + t)\psi'(t) &= n(1 + t)^n - (1 + t) \sum_{k=1}^{n} \frac{n!}{k!(n-k)!} k t^{k-1} \\
&= n(1 + t)^n - \sum_{k=1}^{n} \frac{n!}{(k-1)!(n-k)!} (t^{k-1} + t^k) \\
&= n(1 + t)^n - n - \sum_{k=2}^{n} \frac{n! t^{k-1}}{(k-1)!(n-k)!} - \sum_{k=1}^{n} \frac{n! t^k}{(k-1)!(n-k)!} \\
&= n(1 + t)^n - n - \sum_{k=1}^{n-1} \frac{n!}{k!(n-k-1)!} t^k - \sum_{k=1}^{n} \frac{n!}{(k-1)!(n-k)!} t^k
\end{aligned}$$

$$
\begin{aligned}
&= n(1+t)^n - n - \sum_{k=1}^{n-1} \frac{n!}{(k-1)!(n-k-1)!}\left(\frac{1}{k} + \frac{1}{n-k}\right)t^k - nt^n \\
&= n\left\{(1+t)^n - 1 - \sum_{k=1}^{n-1} \frac{n!}{k!(n-k)!}t^k - t^n\right\} \\
&= n\left\{(1+t)^n - 1 - \sum_{k=1}^{n} \frac{n!}{k!(n-k)!}t^k\right\} = n\psi(t),
\end{aligned}
$$

which together with the observation that $\psi(0) = 0$ shows that $\psi(t)$ satisfies (9.3). By Theorem A, we must have $\psi(t) = 0$, as desired. □

Theorem 2 (de Moivre). *For any integer n and a real number t,* $(\cos t + i\sin t)^n = \cos nt + i\sin nt$, *where* $i = \sqrt{-1}$.

Proof. The zero function ($y(t) = 0$) satisfies the initial value problem

$$y'(t) - iny(t) = 0, \quad y(0) = 0. \tag{9.4}$$

Let $\varphi(t) = (\cos t + i\sin t)^n - \cos nt - i\sin nt$ on real line. Observe that $\varphi(0) = 0$ and

$$
\begin{aligned}
\varphi'(t) &= n(\cos t + i\sin t)^{n-1}(-\sin t + i\cos t) + n\sin nt - in\cos nt \\
&= in(\cos t + i\sin t)^n - in(\cos nt + i\sin nt) = in\varphi(t).
\end{aligned}
$$

So, φ satisfies the initial value problem (9.4) on real line. By Theorem A, $\varphi(t) = 0$ for all $t \in \mathbb{R}$ from which the assertion follows. □

Remark 1. The proofs of binomial and de Moivre formulas in [10] and [11], respectively, can be translated to fit in the present study. More precisely, each of the functions

$$y_1(t) = \frac{1}{(1+t)^n} + \sum_{k=1}^{n} \frac{n!}{k!(n-k)!}\frac{t^k}{(1+t)^n}, \qquad t > -1$$

and

$$y_2(t) = (\cos t + i\sin t)^{-n}(\cos nt + i\sin nt)$$

satisfies the initial value problem

$$y'(t) = 0, \quad y(0) = 1.$$

By Theorem A, we have $y_1(t) = 1$ for $t > -1$ and $y_2(t) = 1$ for all t. The case $y_1(t) = 1$ for $t \leq -1$ can be treated in the same way as in the proof of Theorem 1.

Numerous other applications of the uniqueness Theorem A can be obtained, such as the following ones, which we leave to the reader for further exploration.

3 Euler's formula

The complex exponential e^{it} of a real variable t can be defined as the unique solution of the initial value problem $y' - iy = 0; y(0) = 1$. Since the function $\cos t + i \sin t$ also satisfies this initial value problem, it follows by uniqueness (Theorem A) that

$$e^{it} = \cos t + i \sin t.$$

4 Trigonometric identities

For any real numbers c and t, each of the functions $y_1(t) = \sin(t + c)$ and $y_2(t) = \sin t \cos c + \cos t \sin c$ on real line satisfy the following second-order initial value problem.

$$y''(t) + y(t) = 0; \quad y(0) = \sin c, \quad y'(0) = \cos c.$$

By Theorem A, we have $y_1(t) = y_2(t)$ for all t, and so

$$\sin(t + c) = \sin t \cos c + \cos c \sin t. \tag{9.5}$$

Many other identities about the trigonometric functions can be proved in the same way after formulating appropriate initial value problems for them.

5 Sum of geometric progression

The initial value problem

$$y' + \frac{1}{t-1} y = \frac{nt^{n-1}}{t-1}; \quad y(0) = 1, \ t < 1$$

is satisfied by both the functions $y_1(t) = 1 + \sum_{k=1}^{n-1} t^k$ as well as $y_2(t) = \frac{1-t^n}{1-t}$. Consequently, by Theorem A, we have $y_1(t) = y_2(t)$ for all $t < 1$. For $t > 1$, $\frac{1}{t} < 1$, and so, we have

$$1 + \sum_{k=1}^{n-1} t^k = t^{n-1} y_1(1/t) = t^{n-1} y_2(1/t) = \frac{1-t^n}{1-t}.$$

We then have the following:

$$1 + \sum_{k=1}^{n-1} t^k = \frac{1-t^n}{1-t} \text{ for all } t \neq 1. \tag{9.6}$$

Remark 2. By differentiating both sides of (9.6) with respect to t and then multiplying throughout by t, we get

$$\sum_{k=1}^{n}(k-1)t^{k-1} = t\frac{d}{dt}\left(\frac{1-t^n}{1-t}\right) \text{ for all } t \neq 1.$$

6 Sum of arithmetic-geometric progression

Given three real numbers a, t, and $r \neq 1$, the formula for the sum of arithmetic-geometric progression

$$\sum_{k=1}^{n}\{a+(k-1)t\}r^{k-1} = a\frac{1-r^n}{1-r} + tr\frac{1-nr^{n-1}+(n-1)r^n}{(1-r)^2}, \tag{9.7}$$

can be verified through the initial value problem

$$y''(t) = 0; \quad y(0) = a\frac{1-r^n}{1-r}, \quad y'(0) = r\frac{d}{dr}\left(\frac{1-r^n}{1-r}\right), \quad r \neq 1,$$

which is satisfied by each of the left hand side and right hand side of (9.7) for all t (see Remark 2 for the initial condition $y'(0)$).

Taking limit as $r \to 1$ on both sides of (9.7) and solving the limit in the right-hand side using L'Hopital rule [1], we get

$$\sum_{k=1}^{n}\{a+(k-1)t\} = \frac{n}{2}\{2a+(n-1)t\}, \tag{9.8}$$

which provides an alternative proof of the formula for the sum of an arithmetic progression to n terms, which otherwise is often done using the famous Gauss's clever trick.

7 Sum of sine and cosine series

Given a real number a and $0 < t < 2\pi$, one can verify the formulas

$$\begin{aligned}\sum_{k=0}^{n-1}\cos(a+kt) &= \frac{\cos\{a+(n-1)t/2\}\sin\{nt/2\}}{\sin(t/2)},\\ \sum_{k=0}^{n-1}\sin(a+kt) &= \frac{\sin\{a+(n-1)t/2\}\sin\{nt/2\}}{\sin(t/2)}\end{aligned} \tag{9.9}$$

as follows. Here, both the functions $y_1(t) = e^{ia}\sum_{k=1}^{n} e^{i(k-1)t}$ as well as $y_2 = e^{ia}\frac{e^{int}-1}{e^{it}-1}$ satisfy the initial value problem

$$y' + \frac{ie^{it}}{e^{it}-1}y = \frac{ie^{i(a+n)t}}{e^{it}-1}, \quad y(\pi) = e^{ia}\frac{1-(-1)^n}{2}, \quad 0 < t < 2\pi.$$

By the uniqueness Theorem A, we have $y_1(t) = y_2(t)$ for all $0 < t < 2\pi$, which on comparing real and imaginary parts recover the two formulas in (9.9).

The present approach highlights the importance of the uniqueness theorem of initial value problems of linear ordinary differential equations in obtaining classical results such as the present ones. The uniqueness theorem may therefore be introduced to the students at schools and colleges at the level, where the fundamental theorem of calculus and differential equations are introduced to them. The present technique may be useful for stimulating the student who is familiar with ordinary differential equations. Such an enthusiast will enjoy these entertaining applications.

Bibliography

[1] T. M. Apostol, J. Singh and S. Goyal, *Calculus: An Indian Adaptation*, Second ed, J. Singh and S. Goyal, eds, Vol. 1, Wiley India Pvt. Ltd., 2022.

[2] E. A. Coddington and R. Carlson, *Linear Ordinary Differential Equations*, SIAM, Philadelphia, 1997.

[3] J. L. Coolidge, The story of the binomial theorem, *Am. Math. Mon.* **56** (1949), 147–157.

[4] L. Euler, *Introductio in analysin infinitorum*, Bosquet, Lausanne, 1748, English translation by John Blanton, Springer, New York, 1988 and 1990.

[5] C. M. Fulton, A simple proof of the binomial theorem, *Am. Math. Mon.* **59** (1952), 243–244.

[6] L. C. Hwang, A simple proof of the binomial theorem using differential calculus, *Am. Stat.* **63** (2009), 43–44.

[7] G. Lanzo, Note on the binomial theorem, *Analyst* **1** (1874), 177–178.

[8] S. Lefschetz, A direct proof of de Moivre's formula, *Am. Math. Mon.* **23** (1916), 366–368.

[9] A. Rosalsky, A simple and probabilistic proof of the binomial theorem, *Am. Stat.* **61** (2007), 161–162.

[10] J. Singh, Another proof of the binomial theorem, *Am. Math. Mon.* **124** (2017), 658.

[11] J. Singh, A noninductive proof of de Moivre's formula, *Am. Math. Mon.* **125** (2018), 80.

[12] M. Staring, The Pythagorean proposition: A proof by means of calculus, *Math. Mag.* **69** (1996), 45–46.

[13] A. Ungar, Addition theorems in ordinary differential equations, *Am. Math. Mon.* **94** (1987), 872–875.

[14] L. M. Weiner, A direct proof of the binomial theorem, *Math. Teach.* **48** (1955), 412.

Akshay S. Rane and Kshitij Patil

Modified projection method for second kind Fredholm integral equations with trigonometric polynomials

Abstract: We consider a second kind Fredholm integral equation with a periodic kernel. Modified projection method is applied to such an integral equation with the approximating space to be the space of trigonometric polynomials, and orders of convergence are obtained. This method proves to be better than the classical Galerkin method. A numerical example which arises from a Dirichlet problem for the Laplace equation is considered to illustrate our theoretical results.

Keywords: Fredholm integral operator, trigonometric polynomials, modified projection method

MSC 2010: 11A25, 11A67

1 Introduction

Let $X = L^2[0,1]$, be the space of Lebesgue measurable square integrable functions. Consider the integral operator

$$Tx(s) = \int_0^1 k(s,t)x(t)dt, \quad s \in [0,1],$$

with $k(\cdot,\cdot) \in C([0,1]\times[0,1])$, the space of continuous functions on $[0,1]\times[0,1]$. There are many classical methods such as the Nyström method and projection methods available to solve the equation

$$u - Tu = f, \quad f \in X,$$

approximately. An account of this can be found in Atkinson [1]. Let X_n be the space of piecewise polynomials of degree $\leq (n-1)$, with respect to a uniform partition

Acknowledgement: The first author would like to acknowledge UGC Faculty Recharge Program, India.

Akshay S. Rane, Kshitij Patil, Department of Mathematics, Institute of Chemical Technology, 1 Nathalal Parekh Marg, Matunga, Mumbai 400 019, Maharastra, India, e-mails: as.rane@ictmumbai.edu.in, kshitijanandpatil@gmail.com

https://doi.org/10.1515/9783111313634-010

$$0 < \frac{1}{n} < \frac{2}{n} < \cdots < \frac{n-1}{n} < 1,$$

of $[0,1]$. Let $h = \frac{1}{n}$. Let $\pi_n : X \to X_n$ be the orthogonal projection. Let u_n^G denote the Galerkin solution and u_n^S denote the iterated Galerkin or the Sloan solution. u_n^G solves the equation

$$u_n^G - \pi_n T u_n^G = \pi_n f \tag{10.1}$$

and u_n^S solves

$$u_n^S - T\pi_n u_n^S = f.$$

A modified projection method was introduced in Kulkarni [4], which involves solving

$$u_n^M - T_n^M u_n^M = f, \tag{10.2}$$

where

$$T_n^M = \pi_n T \pi_n + \pi_n T(I - \pi_n) + (I - \pi_n) T \pi_n.$$

Define the iterated modified projection solution by

$$\tilde{u}_n^M = T u_n^M + f.$$

The above method and its iterated version have proven to be better than then the classical Galerkin and Sloan method. Let $\Delta \subset \mathbb{R}$. Let $C^r(\Delta)$ be the space of continuous functions on Δ. Assume that $k(\cdot,\cdot) \in C^r([0,1] \times [0,1])$, $f \in C^r[0,1]$, so that $u \in C^r[0,1]$. Then it is known that

$$\|u - u_n^G\| = O(h^r)$$

and

$$\|u - u_n^S\| = O(h^{2r}).$$

(See Atkinson [1].) But for the modified projection method, we have

$$\|u - u_n^M\| = O(h^{3r})$$

and

$$\|u - \tilde{u}_n^M\| = O(h^{4r}).$$

(See Kulkarni [4].)

Thus, the modified projection solution and its iterated version provide a better approximation to the actual solution than the classical Galerkin or the Sloan solution. Dirichlet problem with Laplace's equation gives rise to integral operators with periodic kernels. If the kernel $k(s,t)$ is periodic in both the variables with period 2π and f is periodic, then the solution u is 2π periodic. Instead of using local approximation by piecewise polynomials, quite often it is preferable to use global approximation in the form of trigonometric polynomials, since this leads to rapid convergence. Trigonometric polynomial approximations are used by Atkinson [1] to solve boundary integral equations on smooth planar regions. Sometimes such methods are referred as Spectral methods. Atkinson [1] has considered the Galerkin and the iterated Galerkin method, when the approximating space X_n is the space of trigonometric polynomials. We study the orders of convergence of the modified projection solution and its iterated version, when the approximating space is the space of trigonometric polynomials. In this paper, we shall see that the new method with trigonometric polynomials has an improved rate of convergence than the Galerkin method. Another important example of globally defined smooth polynomial approximations is the use of spherical polynomials to functions defined on the unit sphere in $\mathbb{R}^3$. In the future, we would like to study the modified projection method with spherical polynomial approximations to solve a boundary integral equation on smooth closed surfaces in $\mathbb{R}^3$.

The paper is organized as follows. In Section 2, we define the orthogonal projection onto the space of trigonometric polynomials. We obtain the orders of convergence of the modified and the iterated modified projection solution in the Sobolov space setting. Section 3 is devoted, in obtaining the orders of convergence of the modified projection solution in the uniform norm. A numerical example is considered in Section 4.

2 Orders of convergence

In this section, we assume that the kernel $k(s,t)$ is periodic in both the variables with period 2π. Let $X = L^2[0,2\pi]$, with the usual norm given by

$$\|x\| = \left(\int_0^{2\pi} |x(t)|^2 dt \right)^{\frac{1}{2}}, \quad x \in L^2[0,2\pi],$$

Consider the following integral operator

$$(Tx)(s) = \int_0^{2\pi} k(s,t)x(t)dt, \quad s \in [0,2\pi],$$

with $k(\cdot,\cdot) \in L^2([0,2\pi] \times [0,2\pi])$. Assume that the operator equation

$$u(s) - \int_0^{2\pi} k(s,t)u(t)dt = f(s), \quad s \in [0, 2\pi], \tag{10.3}$$

compactly written as

$$u - Tu = f$$

has a unique solution. Let r be a non negative integer. For $j = 1, 2, \dots, r$, let $x^{(j)}$ denote the jth derivative of x (distributional sense).

Define:

$$H^r[0, 2\pi] := \{x \in L^2[0, 2\pi] : x \text{ is } 2\pi \text{ periodic}, \ x^{(j)} \in L^2[0, 2\pi], j = 1, 2, \dots, r\}.$$

We consider the following norm

$$\|x\|_r := \left[\sum_{j=0}^{r} \|x^{(j)}\|^2\right]^{\frac{1}{2}},$$

on $H^r[0, 2\pi]$. In addition, assume kernel $k(\cdot,\cdot) \in H^r([0, 2\pi] \times [0, 2\pi])$ and $f \in H^r[0, 2\pi]$. Then the exact solution $u \in H^r[0, 2\pi]$. Let X_n be the space of all trigonometric polynomials of degree $\leq n$. Then the dimension of X_n is $2n + 1$ and a basis of X_n is given by

$$\varphi_j(s) = e^{ijs}, \quad j = 0, \pm 1, \pm 2, \dots, \pm n, \quad (i = \sqrt{-1}). \tag{10.4}$$

Let $\pi_n : L^2[0, 2\pi] \to X_n$ be the orthogonal projection given by

$$\pi_n x = \frac{1}{2\pi} \sum_{j=-n}^{n} < x, \varphi_j > \varphi_j. \tag{10.5}$$

Then π_n converges to the identity operator pointwise and for $x \in H^r[0, 2\pi]$, we have

$$\|x - \pi_n x\| \leq \frac{C_1}{n^r} \|x\|_r. \tag{10.6}$$

(See Chapter 3, Page 69, Atkinson [1].) The above estimate shall be crucially used in what follows. We consider the modified projection approximation given by

$$T_n^M = \pi_n T \pi_n + \pi_n T(I - \pi_n) + (I - \pi_n)T\pi_n. \tag{10.7}$$

The operator equation (10.3) is approximated by

$$u_n^M - T_n^M u_n^M = f. \tag{10.8}$$

The iterated solution is defined as

$$\tilde{u}_n^M = Tu_n^M + f. \tag{10.9}$$

In the subsequent section, we shall observe that

$$\|T_n^M - T\| = \|(I - \pi_n)T(I - \pi_n)\| \to 0.$$

As a consequence, we can use the following theorem from Kulkarni [4].

Theorem 1. *For all large n,*

$$\|u - u_n^M\| \le C_2\|(I - \pi_n)T(I - \pi_n)u\|$$

and

$$\|u - \tilde{u}_n^M\| \le C_3\|(I - T)^{-1}\|(\|T(I - \pi_n)T(I - \pi_n)u\| + \|T(I - \pi_n)T(I - \pi_n)\|\|u - u_n^M\|),$$

where C_2 and C_3 are constants independent of n.

In order to obtain the rates of convergence of $\|u - u_n^M\|$ and $\|u - \tilde{u}_n^M\|$, we need a bound for the norms of $\|(I - \pi_n)T(I - \pi_n)u\|$, $\|T(I - \pi_n)T(I - \pi_n)u\|$ and the operator norm $\|T(I - \pi_n)T(I - \pi_n)\|$. We first prove a preliminary result using standard techniques. We now set some notations. For $0 \le j \le r$, let

$$l_j(s,t) = \frac{\partial^j}{\partial s^j}k(s,t), \quad s,t \in [0,2\pi].$$

For a fixed $s \in [0,2\pi]$, we denote

$$l_{s,j}(t) = l_j(s,t), \quad t \in [0,2\pi].$$

The complex conjugate of $l_{s,j}(t)$ is denoted by $\bar{l}_{s,j}(t)$. Assume that $l_{s,j} \in H^r[0,2\pi]$, for $j = 0,1,\ldots,r$.

Proposition 1. *For $x \in H^r[0,2\pi]$, we have*

$$\|T(I - \pi_n)x\|_r \le \frac{C_1^2}{n^{2r}}\|x\|_r\left(\sum_{j=0}^{r}\int_0^{2\pi}\|\bar{l}_{s,j}\|_r ds\right)^{\frac{1}{2}}. \tag{10.10}$$

Proof. Note that

$$\|T(I - \pi_n)x\|_r = \left[\sum_{j=0}^{r}\|\,[T(I - \pi_n)x]^{(j)}\,\|^2\right]^{\frac{1}{2}}.$$

By definition,

$$T(I-\pi_n)x(s) = \int_0^{2\pi} k(s,t)(I-\pi_n)x(t)dt, \quad s \in [0,2\pi].$$

Hence for $0 \leq j \leq r$,

$$[T(I-\pi_n)x]^j(s) = \int_0^{2\pi} \frac{\partial^j}{\partial s^j} k(s,t)(I-\pi_n)x(t)dt.$$

Since $(I-\pi_n)$ is an orthogonal projection, for $s \in [0,2\pi]$,

$$\begin{aligned}
[T(I-\pi_n)x]^j(s) &= \int_0^{2\pi} l_{s,j}(t)(I-\pi_n)x(t)dt \\
&= \langle (I-\pi_n)x, \bar{l}_{s,j} \rangle \\
&= \langle (I-\pi_n)^2 x, \bar{l}_{s,j} \rangle \\
&= \langle (I-\pi_n)x, (I-\pi_n)\bar{l}_{s,j} \rangle.
\end{aligned}$$

Now using Cauchy–Schwarz inequality and the estimate (10.6), we get

$$\begin{aligned}
|[T(I-\pi_n)x]^j(s)| &\leqslant \|(I-\pi_n)x\| \|(I-\pi_n)\bar{l}_{s,j}\| \\
&\leqslant \frac{C_1}{n^r}\|x\|_r \frac{C_1}{n^r}\|\bar{l}_{s,j}\|_r \\
&\leqslant \frac{C_1^2}{n^{2r}}\|x\|_r \|\bar{l}_{s,j}\|_r, \quad s \in [0,2\pi].
\end{aligned}$$

Thus,

$$\|[T(I-\pi_n)x]^j\|^2 = \int_0^{2\pi} |[T(I-\pi_n)x]^j(s)|^2 ds \leqslant \frac{C_1^4}{n^{4r}}\|x\|_r^2 \int_0^{2\pi} \|\bar{l}_{s,j}\|_r^2 ds.$$

As a result, we have

$$\sum_{j=0}^{r} \|[T(I-\pi_n)x]^j\|^2 \leqslant \frac{C_1^4}{n^{4r}}\|x\|_r^2 \sum_{j=0}^{r} \int_0^{2\pi} \|\bar{l}_{s,j}\|_r^2 ds.$$

By taking square roots of both the sides, we obtain

$$\|T(I-\pi_n)x\|_r = \left(\sum_{j=0}^{r} \|[T(I-\pi_n)x]^j\|^2\right)^{\frac{1}{2}} \leqslant \frac{C_1^2}{n^{2r}}\|x\|_r \left(\sum_{j=0}^{r} \int_0^{2\pi} \|\bar{l}_{s,j}\|_r^2 ds\right)^{\frac{1}{2}},$$

which completes the proof. □

We now obtain the bounds for $\|(I-\pi_n)T(I-\pi_n)x\|$, $\|T(I-\pi_n)T(I-\pi_n)x\|$ and $\|T(I-\pi_n)T(I-\pi_n)\|$

Proposition 2. *For $x \in H^r[0,2\pi]$, we have*

$$\|(I-\pi_n)T(I-\pi_n)x\| \leq \frac{C_1^3}{n^{3r}}\|x\|_r\left(\sum_{j=0}^{r}\int_0^{2\pi}\|\bar{l}_{s,j}\|_r ds\right)^{\frac{1}{2}}$$

and

$$\|T(I-\pi_n)T(I-\pi_n)x\| \leq \frac{C_1^4}{n^{4r}}\|x\|_r^2\sum_{j=0}^{r}\int_0^{2\pi}\|\bar{l}_{s,j}\|_r^2 ds.$$

Also

$$\|T(I-\pi_n)T(I-\pi_n)\| = O(\|(I-\pi_n)T\|) = O\left(\frac{1}{n^r}\right).$$

Proof. From (10.6), we have

$$\|(I-\pi_n)T(I-\pi_n)x\| \leqslant \frac{C_1\|T(I-\pi_n)x\|_r}{n^r}. \tag{10.11}$$

From Proposition 1 and the above estimate, it follows that

$$\|(I-\pi_n)T(I-\pi_n)x\| \leq \frac{C_1^3}{n^{3r}}\|x\|_r\left(\sum_{j=0}^{r}\int_0^{2\pi}\|\bar{l}_{s,j}\|_r ds\right)^{\frac{1}{2}}.$$

Using (10.10), we get

$$\begin{aligned}\|T(I-\pi_n)T(I-\pi_n)x\| &\leqslant \frac{C_1^2}{n^{2r}}\|T(I-\pi_n)x\|_r\left(\sum_{j=0}^{r}\int_0^{2\pi}\|\bar{l}_{s,j}\|_r ds\right)^{\frac{1}{2}}\\ &\leqslant \frac{C_1^4}{n^{4r}}\|x\|_r^2\sum_{j=0}^{r}\int_0^{2\pi}\|\bar{l}_{s,j}\|_r ds.\end{aligned}$$

This completes the proof. Next, we need to compute an upper bound for the operator norm $\|T(I-\pi_n)T(I-\pi_n)\|$. Since π_n is a sequence of projections converging to identity pointwise. $\|\pi_n\|$ are uniformly bounded. Since $T : L^2[0,2\pi] \to L^2[0,2\pi]$ is a bounded operator, to compute an upper bound for $\|T(I-\pi_n)T(I-\pi_n)\|$ it is enough to compute an upper bound for $\|(I-\pi_n)T\|$. We have

$$\|(I-\pi_n)Tx\| \le \frac{C_1}{n^r}\|Tx\|_r = \frac{C_1}{n^r}\left(\sum_{j=0}^{r}\|(Tx)^{(j)}\|^2\right)^{\frac{1}{2}}$$

Since

$$(Tx)^{(j)}(s) = \int_0^{2\pi} l_{j,s}(t)x(t)dt$$

we have by Cauchy–Schwarz inequality,

$$|(Tx)^{(j)}(s)|^2 \le \int_0^{2\pi}|l_{j,s}(t)|^2dt\int_0^{2\pi}|x(t)|^2dt$$

and hence

$$\int_0^{2\pi}|(Tx)^{(j)}(s)|^2ds \le \int_0^{2\pi}\int_0^{2\pi}|l_{j,s}(t)|^2dtds\int_0^{2\pi}|x(t)|^2dt$$

So

$$\sum_{j=0}^{r}\|(Tx)^{(j)}\|^2 \le \sum_{j=0}^{r}\left(\int_0^{2\pi}\int_0^{2\pi}|l_{j,s}(t)|^2dtds\right)\|x\|^2.$$

Let $M^2 = \sum_{j=0}^{r}(\int_0^{2\pi}\int_0^{2\pi}|l_{j,s}(t)|^2dtds)$ and as a consequence

$$\|(I-\pi_n)Tx\| \le M\frac{C_1}{n^r}\|x\|.$$ □

We now state the main theorem.

Theorem 2. *Let $\pi_n : L^2[0,2\pi] \to X_n$ be the orthogonal projection. Assume that $k(.,.) \in H^r([0,2\pi]\times[0,\pi])$ and $f \in H^r([0,2\pi])$. Then*

$$\|u-u_n^M\| = O\left(\frac{1}{n^{3r}}\right)$$

and

$$\|u-\tilde{u}_n^M\| = O\left(\frac{1}{n^{4r}}\right).$$

Proof. The proof follows from Theorem 1 and Proposition 2. □

3 Uniform convergence

We are often interested in obtaining uniform convergence of the solution u_n^M to u. Let $C_p(2\pi)$ be the set of all 2π-periodic and continuous functions on $x(s)$, $-\infty < s < \infty$. $C_p(2\pi)$ with the uniform norm $\|.\|_\infty$ is a Banach space. We consider the solution of (10.3) that is

$$u(s) - \int_0^{2\pi} k(s,t)u(t)dt = f(s), \quad s \in [0, 2\pi].$$

It is no longer true that π_n converges to identity pointwise. We have

$$\|\pi_n\| = O(\log n).$$

See Zygmund [5]. We do not have

$$\|(I - \pi_n)T(I - \pi_n)\| \to 0,$$

and as a consequence $\|(I - \pi_n)T(I - \pi_n)\|$ has to be estimated differently. We need the following results to prove that $\|(I - \pi_n)T(I - \pi_n)\| \to 0$. Given $x \in C_p(2\pi)$, define the minimax error in approximating x by elements of X_n as follows:

$$\rho_n(x) = \inf_{z \in X_n} \|x - z\|_\infty.$$

The minimax approximation to x by elements of X_n is that element $q_n \in X_n$, for which

$$\|x - q_n\|_\infty = \rho_n(x).$$

It can be shown that there is a unique q_n and for any $x \in C_p(2\pi)$ it can be shown that $\rho_n(x) \to 0$ as $n \to \infty$. We have the following result from Section 3.3.3 of Atkinson [1].

Theorem 3. $\rho_n(x) = \inf_{z \in X_n} \|x - z\|_\infty = \|x - q_n\|_\infty$, $q_n \in X_n$

$$\|x - \pi_n x\|_\infty \le O(\log n)\rho_n(x)$$

If x is Hölder continuous of exponent α, then

$$\|x - \pi_n x\|_\infty \le c\, O\left(\frac{\log n}{n^\alpha}\right),$$

for some constant c. This converges to 0 as $n \to \infty$. Higher speeds of convergence can be obtained by assuming greater smoothness for $x(t)$. We quote the following result from Atkinson [1]

Theorem 4. *If $k(s,t)$ satisfies the Hölder condition*

$$|k(s_1,t)-k(s_2,t)| \le c(k)|s_1-s_2|^{\alpha},$$

then

$$\|(I-\pi_n)T\|_{\infty} \le C\frac{\log n}{n^{\alpha}}$$

and

$$\|u-u_n^G\|_{\infty} = O\left(\frac{\log n}{n^{\alpha}}\right),$$

where u_n^G is the Galerkin solution.

For the modified projection method, we have the following result.

Theorem 5. *If $k(s,t)$ satisfies the Hölder condition*

$$|k(s_1,t)-k(s_2,t)| \le c(k)|s_1-s_2|^{\alpha}$$

and x satisfies the Hölder condition, then

$$\|(I-\pi_n)T(I-\pi_n)x\|_{\infty} \le C\frac{(\log n)^2}{n^{2\alpha}}$$

and

$$\|u-u_n^M\|_{\infty} = O\left(\frac{(\log n)^2}{n^{2\alpha}}\right).$$

Proof. For $s \in [0,2\pi]$, we have

$$[(I-\pi_n)T(I-\pi_n)x](s) = \int_0^{2\pi}(I-\pi_n)K(s,t)[(I-\pi_n)x](t)dt$$

$$[(I-\pi_n)T(I-\pi_n)x](s) = \int_0^{2\pi}(I-\pi_n)K_t(s)[(I-\pi_n)x](t)dt.$$

We have

$$\|(I-\pi_n)T(I-\pi_n)x\|_{\infty} \le \max_{s\in[0,2\pi]}\int_0^{2\pi}|(I-\pi_n)K_t(s)|\,\|(I-\pi_n)x\|_{\infty}dt$$

If x and $K(s,t)$ satisfies the Hölder continuous of exponent α, then

$$\|(I-\pi_n)K_t\|_{\infty} \le c\ O\left(\frac{\log n}{n^{\alpha}}\right)$$

and

$$\|x - \pi_n x\|_\infty \leq c\, O\left(\frac{\log n}{n^\alpha}\right),$$

which gives

$$\|(I - \pi_n)T(I - \pi_n)x\|_\infty \leq \frac{C(\log n)^2}{n^{2\alpha}},$$

and the result follows from Theorem 4. □

In actual computations, integrations have to be replaced by numerical integration. In orthogonal projection, we generally expect iteration to improve the order of convergence. i The analysis suggests that in the uniform norm, iterated Galerkin is not an improvement over Galerkin solution and iterated modified solution is not an improvement over modified solution. Modified projection solution is better than the Galerkin solution. The next section illustrates this fact.

4 Numerical results

4.1 Example 1

We consider the example (12.15) presented in Kress [3]. Consider the Fredholm integral equation of the second kind given by

$$u(s) + \frac{ab}{\pi}\int_0^{2\pi} \frac{u(t)}{a^2 + b^2 - (a^2 - b^2)\cos(s+t)}dt = f(s), \quad s \in [0, 2\pi], a \geq b > 0. \tag{10.12}$$

This integral equation arises from the solution of the Dirichlet problem for the Laplace equation on an ellipse with semi-axes a and b. We set $(a, b) = (1, 0.5)$

The exact solution is given by:

$$u(s) = e^{\cos s}\cos(\sin s), \quad s \in [0, 2\pi]. \tag{10.13}$$

We take

$$f(s) = u(s) + e^{c\cos s}\cos(c \sin s).$$

The above integral operator in (10.12) is approximated by the modified integral operator given by equation (10.2). Computing the solution in this case involves solving a system of linear equations, which further involves computing inner products and hence integrals. These integrals are evaluated numerically using the composite trapezoidal rule for single, double and triple variables for varying grids of size N.

In our analysis, we have observed that the modified projection method has proved to be better than the Galerkin method, which reconciles with the theory. We approximate the solution using both the Galerkin method given by operator equation (10.1) and the modified projection method given by operator equation (10.2), and then compare the two with the exact solution given by equation (10.13). The tables 10.1, 10.2, 10.3 provide a comparison of errors between the Galerkin and Modified Projection errors, at points $s = 0, \frac{\pi}{2}$ and π for $n = 3, 7$ and 9. The errors are defined as follows:

$$\text{Galerkin Error} = |u(s) - u_n^G(s)|,$$
$$\text{Sloan Error} = |u(s) - u_n^S(s)|,$$
$$\text{Modified Error} = |u(s) - u_n^M(s)|,$$
$$\text{Iterated Modified Error} = |u(s) - \tilde{u}_n^M(s)|.$$

Table 10.1: Error comparison for $s = 0$.

		$s = 0$		
n	**Galerkin Error**	**Sloan Error**	**Modified Error**	**Iterated Modified Error**
3	5.2×10^{-2}	5.51×10^{-4}	5.51×10^{-4}	6.49×10^{-6}
7	2.79×10^{-5}	3.92×10^{-9}	3.92×10^{-9}	5.86×10^{-13}
9	3.03×10^{-7}	4.8×10^{-12}	4.8×10^{-12}	2.22×10^{-15}

Table 10.2: Error comparison for $s = \frac{\pi}{2}$.

		$s = \frac{\pi}{2}$		
n	**Galerkin Error**	**Sloan Error**	**Modified Error**	**Iterated Modified Error**
3	4.03×10^{-2}	5.13×10^{-4}	5.13×10^{-4}	6.35×10^{-6}
7	2.45×10^{-5}	3.78×10^{-9}	3.78×10^{-9}	5.69×10^{-13}
9	2.73×10^{-7}	4.66×10^{-12}	4.68×10^{-12}	7.99×10^{-15}

Table 10.3: Error comparison for $s = \pi$.

		$s = \pi$		
n	**Galerkin Error**	**Sloan Error**	**Modified Error**	**Iterated Modified Error**
3	3.45×10^{-2}	4.82×10^{-4}	4.82×10^{-4}	6.21×10^{-6}
7	2.23×10^{-5}	3.64×10^{-9}	3.64×10^{-9}	5.63×10^{-13}
9	2.52×10^{-7}	4.53×10^{-12}	4.53×10^{-12}	6.44×10^{-15}

4.2 Example 2

We consider example (13.2.1) presented in Atkinson and Han [2]. Consider the boundary of an ellipse, $r(t) = (a\cos t, b\sin t)$, $t \in [0, 2\pi]$. We now consider the integral equation:

$$-\pi\, u(s) + \int_0^{2\pi} u(t) K\left(\frac{s+t}{2}\right) dt = f(s), \quad s \in [0, 2\pi], \tag{10.14}$$

where

$$K(\theta) = \frac{-ab}{2(a^2\sin^2(\theta) + b^2\cos^2(\theta))} \tag{10.15}$$

and

$$f(x,y) = e^x \cos y,\ (x,y) \in S. \tag{10.16}$$

We take $(a, b) = (1, 2)$. The true solution is unknown, but we obtain a highly accurate solution by using a large value of $N = 128$ and a fixed value of $n = 9$, N and n as described above. We compute approximate errors assuming the above computed solution to be true. We see that the modified method converges faster than Galerkin, again serving as a confirmation of the theoretical results. The tables 10.4, 10.5 and 10.6 gives the errors. The errors are defined as follows:

$$\text{Galerkin Error} = |u_N(s) - u_n^G(s)|,$$
$$\text{Modified Error} = |u_N(s) - u_n^M(s)|.$$

Table 10.4: Error comparison for $s = 0$.

	$s = 0$	
N	**Galerkin Error**	**Modified Error**
16	0.136	1.58×10^{-4}
24	6.17×10^{-4}	5.61×10^{-10}
32	6.17×10^{-4}	4.61×10^{-11}
64	6.17×10^{-4}	2.92×10^{-12}

Table 10.5: Error comparison for $s = \pi/2$.

	$s = \pi/2$	
N	**Galerkin Error**	**Modified Error**
16	0.114	1.06×10^{-4}
24	2.85×10^{-3}	7.09×10^{-10}
32	2.85×10^{-3}	2.62×10^{-11}
64	2.85×10^{-3}	2.01×10^{-11}

Table 10.6: Error comparison for $s = \pi$.

	$s = \pi$	
N	**Galerkin Error**	**Modified Error**
16	0.955	2.64×10^{-3}
24	4.78×10^{-3}	9.91×10^{-8}
32	4.78×10^{-3}	4.33×10^{-11}
64	4.78×10^{-3}	5.21×10^{-11}

Bibliography

[1] K. E. Atkinson, *The Numerical Solutions of Integral Equations of the Second Kind*, Cambridge University Press, Cambridge, U.K., 1997.

[2] K. E. Atkinson and W. Han, *Theoretical Numerical Analysis, A Functional Analysis Framework*, 3rd edn, Springer, 2009.

[3] R. Kress, *Numerical Analysis*, Graduate Texts in Mathematics, Springer, 1998.

[4] R. P. Kulkarni, A superconvergence result for solutions of compact operator equations, *Bull. Aust. Math. Soc.* **68**(3) (2003), 517–528.

[5] A. Zygmund, *Trigonometric Series, Vols. I and II*, Cambridge university press, New York, 1959.

Shilpee and B. S. Bhadauria

Effect of heat and mass transfer in binary nanofluids (for silver nanoparticles) saturated in porous medium under the influence of LTNE model and different gravity modulation

Abstract: Effect of gravity modulation on binary nanofluid saturated with porous media is examined. The results are shown for a horizontal parallel plate taken at a distance of $z = d$, where lower plate is $z = 0$ and upper plate is $z = d$ heated from below. The local thermal non-equilibrium model is observed for the system where the particle, solid, and fluid phases are taken to be at different temperatures. Linear and non-linear stability analysis has been performed. Porosity and thermal diffusivity ratio shown effect on heat and mass transfer. Frequency of modulation shows unmodulated cases for its higher value, while for smaller values of (frequency of modulation) shows earlier mass and heat transfer convection rate. Day-night profile shows earlier mass heat on heat and mass transfer in all the fluid, particle and solid matrix phase.

Keywords: Nanofluid, porous medium, binary nanofluid, gravity modulation, local thermal non-equilibrium

Nomenclature

Latin symbols

(x, y, z)	Cartesian or Eucledian coordinates.
d	Dimensional layer depth.
Ra	thermal Rayleigh–Darcy number.
Rs	solute Rayleigh number.
Rm	Rayleigh number for basic density.
Rn	Concentration Rayleigh number.
N_B	Modified particle-density increment.
N_A	Modified diffusivity ratio.
Le	Lewis number.
Nu	Nusselt number.
Ls	Solute Lewis number.
ε	Porosity
g	Gravitational acceleration.
ε_s, ε_p	Modified thermal diffusivity ratio.

Shilpee, B. S. Bhadauria, Department of Mathematics, Babasaheb Bhimrao Ambedkar University, Lucknow-226025, Uttarpradesh, India, e-mails: srivastavashilpi324@gmail.com, mathsbsb@yahoo.com

https://doi.org/10.1515/9783111313634-011

γ	Amplitude of gravity modulation.
γ_p and γ_s	Modified thermal capacity ratios.
Ω	Gravity modulation (frequency of)
t	Time.
T_1	Upper wall temperature.
T_0	Lower wall temperature.
T_f	Temperature of the fluid phase.
T_s	Temperature of the solid phase.
T_p	Temperature.
$\mathbf{q}$	Binary nanofluid velocity.
D_B	Coefficient of Brownian diffusion.
a	Wave number.
D_T	Coefficient of thermophoretic diffusion.

Greek symbols

ε	Porosity ($0 < \varepsilon < 1$)
ρ	Density
ϕ	Nanoparticle-volume fraction.

Subscripts and superscripts

0	Ref value.
b	Basic value.

Operators

∇^2	$\frac{\partial^2}{\partial x^2} + \frac{\partial^2}{\partial y^2} + \frac{\partial^2}{\partial z^2}$

1 Introduction

In year 1995, the most efficient and useful fluid coined by Choi named as nanofluid [5]. Nanofluid has a very vast area of research from year 1995 to present. Because of its innovative properties comprising of high thermal conductivity and good cooling efficiency, it has many application in heat and mass transfer. The phenomenon of increase in thermal conductivity was first reported by Masuda et al. [24]. In year 2004 [7], Eastman et al. arrange a comprehensive and theoretical review that focused on thermal convection in nanofluid. Buongiorno [4] examined the nature of nanofluid based on brownian diffusion of the particle, inertia, magnus effects, etc. In 2022, authors of [11] examined the effect of rotation on Rivlin–Ericksen elastico-viscous nanofluid.

Porous medium convection has been a very popular topic among which is being studied by many researchers. Henery–Darcy (1856) was first man to proposed the law known as Darcy-law. The Darcy law explains the movement through porous medium of the fluid. Horton and Rogers (1945) used Darcy's law and examined the onset of convection. Then came a Sheikholeslami [27], who used the Darcy model under the influence of electric field. After that, in 1901, Forchheimer work on an experiment that uses the findings of coarse sand and hydraulic conductivity. Later in 1949, Brinkmann used Navier's stoke equation in the calculation for viscous incompressible fluid. In [12], the thermal instability of micropolar nanofluid with rotation was examined.

When the temperature of the particles, fluid and solid-matrix phase are at different the model is defined as local thermal non-equilibrium. Marin examined the LTNE model by taking an assumption at the wall. In year 2009, Neild and Kuznetsov [25] examined the thermal stability and instability in porous medium and later extended the study under the effect of LTNE. In year 2010, the same result using the Brinkmann model was examined [20]. Baytas 2003 [2] and Popin started examining the natural convection by considering a square enclosure using LTNE model. On the onset of Lapwood–Brinkmann convection, convection in double diffusive mode on porous layer and the hasten of convection for the anisotropic porous medium were examined by Malashetty [22]. In 2011, Bhadauria examined the repercussion of local thermal non-equilibrium for non-linear analysis and linear analysis. The very first approach to obtain the analytical solution was first observed by Vafai and Lee, who obtain the analytical type solution for temperature profiles, Nusselt number and the temperature differences between the two phases [21]. The effect of rotation nanofluid within Hele–Shaw cell saturating Casson nanofluid were examined in [17] and the magnetic field in Hele-Shaw cell in [16]. Authors of [28] have studied the gravity modulated thermal instability in Oldroyd-B nanofluids.

Gravity modulation is a case when a vessel containing a heavy liquid vibrates vertically having constant frequency and amplitude. Ismail [15] studied the three types of magnetic field modulation with different shapes of nanoparticles in (Al_2O_3) nanofluid. Again, Ismail [14] examined the thermal instability of Casson nanofluid. The first researchers to examined the effect of gravity modulation on fluid layer heated from below were Gershuni et al. [8] and Gresho with Sani [9]. Ismail and Bhadauria [13] studied the three types of gravity modulation on Rivlin–Ericksen nanofluid. Repercussion of modulation of gravity in a long cylinder oscillating around a vertical axis to know the instability of convection by Wadih et al. [31]. The repercussions of very small amplitude of modulation of gravity on the beginning of convection having porous layer is observed by Malashetty et al. [23]. Dyko [6] study the repercussion of gravity modulation in between the annulus of two horizontal coaxial cylinders. Many studies has been carried out in recent years to check the effect of gravity modulation. In present era, it has a great effect on convection rate to check the instability of thermal convection. Shilpee [29] examines the thermal instability of micropolar nanofluid under LTNE. Ismail [18] examined the effect of magnetic field modulation of micropolar nanofluid within Hele–Shaw cell.

In simple words, when for the making of nanoliquid binary liquid is used referred as binary nanofluid. According to Kim et al. [19], base fluid having binary mixture such as ammonia/water referred as *binary nanofluid*. Kuznetsov and Nield were the first to look into the repurcussion of convection on binary nanofluid saturating the porous medium [26]. Later, D. Yadav et.al starts look into the convection in binary nanofluid with saturated by porous layer [32]. In 2014 [10], the effect of thermosolutal convection in a layer having nanofluid was examined. Due to efficiency of binary nanofluid, there is a vast area for research.

With the application of binary nanofluids in many fields like solar energy, absorption the main aim of the present article is to look out the effect of different gravity modulation on the binary nanofluids using porous media under local thermal non-equilibrium in a horizontal parallel plate. Non-linear stability analysis is also performed to obtain the repurcussion of frequency and amplitude of gravity modulation.

2 Mathematical formulation of the problem

Consider an extended horizontal parallel plate (see (Fig. 11.1)), where layer having binary nanofluid saturated by porous layer b/w two plates $z = 0$ and $z = d$, where thickness d is taken. Temperature on upper lower boundary of the plane to be T_1 and T_0 such that $T_0 > T_1$. The boundaries are in such a way so that they are considered to be thermally conducting and impermeable. Homogeneity in the system and local thermal non-equilibrium with porous medium is considered.

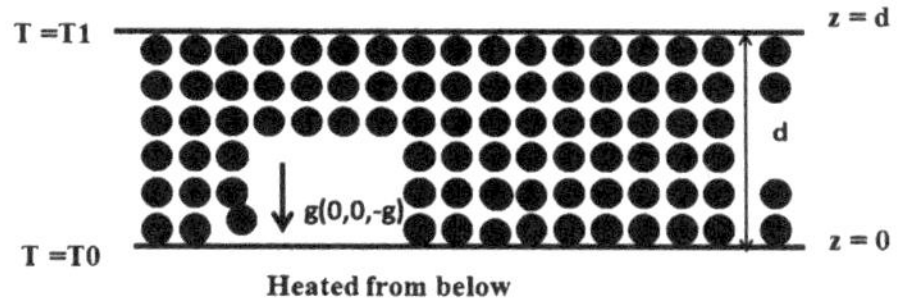

Figure 11.1: Schematic diagram.

Table 11.1: Properties of silver nanoparticles.

Nanoparticle	Ag (Silver)
Rn	95.95
N_A	0.5

The basic governing equations for the system in porous medium are Shukla et al. [30] and Bhadauria et al. [3] are written as:

$$\tilde{\nabla} \cdot \tilde{\mathbf{q}} = 0, \tag{11.1}$$

$$\frac{\rho}{\varepsilon}\left[\frac{\partial}{\partial \tilde{t}} + \frac{1}{\varepsilon}(\tilde{\mathbf{q}} \cdot \tilde{\nabla})\right]\tilde{\mathbf{q}} = -\tilde{\nabla}\tilde{p} + \mu\tilde{\nabla}^2\tilde{\mathbf{q}} + [\phi\rho_p + (1-\phi)\rho_{f_0}\{1 - \beta_T(\tilde{T}_f - \tilde{T}_c) - \beta_c(\tilde{C}_f - \tilde{C}_0)\}]\mathbf{g}, \tag{11.2}$$

$$\left[\frac{\partial}{\partial \tilde{t}} + \frac{1}{\varepsilon}(\tilde{\mathbf{q}} \cdot \tilde{\nabla})\right]\tilde{T}_f = k_f\tilde{\nabla}^2\tilde{T}_f + \frac{(\rho c)_p}{(\rho c)_f}\left[D_B\tilde{\nabla}\tilde{\phi} \cdot \tilde{\nabla}\tilde{T}_f + D_T\frac{\tilde{\nabla}\tilde{T}_f \cdot \tilde{\nabla}\tilde{T}_f}{\tilde{T}_f}\right] + \frac{h_{f_p}}{\varepsilon(1-\phi_0)(\rho c)_f}(\tilde{T}_p - \tilde{T}_f) + \frac{h_{f_s}}{\varepsilon(1-\phi_0)(\rho c)_f}(\tilde{T}_s - \tilde{T}_f), \tag{11.3}$$

$$\left[\frac{\partial}{\partial \tilde{t}} + \frac{1}{\varepsilon}(\tilde{\mathbf{q}} \cdot \tilde{\nabla})\right]\tilde{T}_p = \frac{k_p}{(\rho c)_p}\tilde{\nabla}^2\tilde{T}_p + \frac{h_{f_p}}{\varepsilon\phi_0(\rho c)_p}(\tilde{T}_f - \tilde{T}_p), \tag{11.4}$$

$$\left[\frac{\partial T_s}{\partial \tilde{t}}\right] = \frac{k_s\nabla^2\tilde{T}_s}{(\rho c)_s} + \frac{h_{f_s}}{(1-\varepsilon)(\rho c)_s}(T_f - T_s), \tag{11.5}$$

$$\left[\frac{\partial}{\partial \tilde{t}} + \frac{1}{\varepsilon}(\tilde{\mathbf{q}} \cdot \tilde{\nabla})\right]\tilde{\phi} = D_B\tilde{\nabla}^2\tilde{\phi} + \left(\frac{D_T}{\tilde{T}_c}\right)\tilde{\nabla}^2\tilde{T}_f, \tag{11.6}$$

$$\left[\frac{\partial C}{\partial \tilde{t}} + (\tilde{\mathbf{q}} \cdot \tilde{\nabla C})\right] = D_s\nabla^2 C \tag{11.7}$$

thus the boundary conditions considered for the system are:

$$\left.\begin{array}{llllll} q = 0, & T_f = 1, & T_p = 1, & T_s = 1 & C = 1, & \phi = 1 \quad \text{at } z = 0, \\ q = 0, & T_f = 0, & T_p = 0, & T_s = 0 & C = 0 & \phi = 0 \quad \text{at } z = 1. \end{array}\right\} \tag{11.8}$$

Now we non-dimensionalize the equation (11.1)–(11.7) for which non-dimensional parameters are defined below as:

$$\left.\begin{array}{r} (x,y,z) = d(x^*,y^*,z^*), \quad t^* = t\alpha_f/d^2, \quad \alpha_f = \dfrac{k_f}{(\rho c)_f}, \quad q^* = qd/\alpha_f, \quad p^* = pd^2/\mu\alpha_f, \\ \phi^* = \dfrac{\phi}{\phi_b}, \quad T^* = (T - T_0)/(T_1 - T_0), C^* = (C - C_0)/(C_1 - C_0) \\ \nabla = \nabla/d. \end{array}\right\} \tag{11.9}$$

The equation obtained after non-dimensionalizing are as follows:

$$\nabla \cdot \mathbf{q} = 0, \tag{11.10}$$

$$\frac{1}{\Pr}\left[\frac{1}{\varepsilon}\frac{\partial}{\partial t} + \frac{1}{\varepsilon^2}(\mathbf{q} \cdot \nabla)\right]\mathbf{q} = -\nabla p - \nabla^2\mathbf{q} - g_m\left[\text{Rm}\,\hat{\mathbf{e}_z} + \text{Ra}\,T_f\hat{\mathbf{e}_z} - \text{Rn}\,\phi\hat{\mathbf{e}_z} + \frac{R_s}{L_s}C\hat{\mathbf{e}_z}\right], \tag{11.11}$$

$$\begin{aligned}\left[\frac{\partial}{\partial t} + \frac{1}{\varepsilon}(\mathbf{q} \cdot \nabla)\right]T_f = \nabla^2 T_f + \left(\frac{N_B}{\text{Le}}\right)\nabla\phi \cdot \nabla T_f + \left(\frac{N_A N_B}{\text{Le}}\right)\nabla T_f \cdot \nabla T_f \\ + N_{HP}(T_p - T_f) + N_{HS}(T_s - T_f),\end{aligned} \tag{11.12}$$

$$\left[\frac{\partial}{\partial t} + \frac{1}{\varepsilon}(\mathbf{q} \cdot \nabla)\right]T_p = \varepsilon_p\nabla^2 T_p + \gamma_p N_{HP}(T_f - T_p), \tag{11.13}$$

$$\left[\frac{\partial}{\partial t} + \frac{1}{\varepsilon}(\mathbf{q} \cdot \nabla)\right]T_s = \varepsilon_s\nabla^2 T_p + \gamma_s N_{HS}(T_f - T_s), \tag{11.14}$$

$$\left[\frac{\partial}{\partial t} + \frac{1}{\varepsilon}(\mathbf{q} \cdot \nabla)\right]\phi = \left(\frac{1}{\text{Le}}\right)\nabla^2\phi + \left(\frac{N_A}{\text{Le}}\right)\nabla^2 T_f, \tag{11.15}$$

$$\left[\frac{\partial}{\partial t} + \frac{1}{\varepsilon}(\mathbf{q} \cdot \nabla)\right]C = \frac{1}{\text{Ls}}\nabla^2 C, \tag{11.16}$$

where

$N_A = \dfrac{D_T(T_1 - T_0)}{D_B T_0 \phi_0}$ is the modified diffusivity ratio,

$N_B = \dfrac{(\rho c)_p \phi_b}{(\rho c)_f}$ is the modified density particle increment,

$\text{Le} = \dfrac{\alpha_f}{D_B}$ is the Lewis number,

$N_{HP} = \dfrac{h_{fp} d^2}{\varepsilon(1-\phi_0)k_f}, \quad N_{HS} = \dfrac{h_{fp} d^2}{\varepsilon(1-\phi_0)k_s}$ are the interface heat transfer coefficients,

$\gamma_s = \dfrac{\varepsilon(1-\phi_0)(\rho c)_f}{(1-\varepsilon)(\rho c)_s}, \quad \gamma_p = \dfrac{\varepsilon(1-\phi_0)(\rho c)_f}{(1-\varepsilon)(\rho c)_s}$ are modified thermal capacity ratios.

3 Basic solution

Taking the basic state as inoperative, we assume the basic state as:

$$\begin{aligned} &\mathbf{q} = 0, \quad T_f = T_{f_b}(z), \quad T_p = T_{p_b}(z), \quad T_s = T_{sb}(z) \quad T_p = T_{p_b}(z) \quad \phi = \phi_b(z), \\ &p = p_b(z), \quad C = C_b(z). \end{aligned} \tag{11.17}$$

Using eq. (11.17) and b.cs (11.16) in (11.13)–(11.16) and using the fact that Le for most of the nanofluids is quite larger than N_A [4] and also implementing order magnitude analysis, we get:

$$\frac{d^2 T_{fb}}{d^2 z} = 0 \quad \frac{d^2 T_{pb}}{d^2 z} = 0,$$
$$\frac{d^2 T_{sb}}{d^2 z} = 0 \quad \frac{d^2 C_b}{d^2 z} = 0.$$

Using eq. (11.8) in above equations and solving, we get:

$$T_{f_b} = 1 - z, \tag{11.18}$$
$$T_{p_b} = 1 - z, \tag{11.19}$$
$$T_{sb} = 1 - z, \tag{11.20}$$
$$C_b = 1 - z. \tag{11.21}$$

4 Perturbation state solution

Applying very small perturbation to the initial solution and neglecting the quantities in product form we apply using following perturbation: $v(u, v, w) = (u', v', w')$, $T_f = T_{fb} + T_{f'}$, $T_p = T_{pb} + T_{p'}$, $T_s = T_{sb} + T_{s'}$, $\phi = \phi_b + \phi'$, $C = C_b + C'$. After applying perturbation to the non-dimensionalise equation obtained above and solving the governing equation using the basic state solution. Further, we solve the equation to check linear stability to observe the effects on convection analysis.

5 Types of modulation

In this section, we will discuss the four types of gravity. Gravity modulation is defined as the vertical oscillation in a fluid layer in a constant gravitational field. Here we have impose the gravity in four ways, i. e., sinusoidal, saw-tooth, day-night and square wave form. By applying the four types of modulation, we infer about the case where rate of convection takes place earlier. Here $G(\Omega_1, t)$ is for gravity modulation.

Four types of modulation	**Formula used**
Trigonometric sine wave (sinusoidal)	$G(\Omega_1,t) = \sin(\Omega_i t)$
saw-tooth wave	$G(\Omega_1,t) = \sum_{m=1}^{10} \frac{2(-1)^{m-1}\sin(mt\Omega_i)}{\pi m}$
day-night wave	$G(\Omega_1,t) = \sum_{m=1}^{10} \frac{2(\sin(\frac{\pi m}{2})+\frac{3}{7}\sin(\frac{5\pi m}{6}))\sin(mt\Omega_i)}{\pi^2 m^2} + \sum_{m=1}^{10} \frac{2(\cos(\frac{\pi m}{2})+\frac{3}{7}\cos(\frac{5\pi m}{6})-\frac{10}{7})\cos(mt\Omega_i)}{\pi^2 m^2} + \frac{14}{12}$

6 Stability analysis

6.1 Linear stability

The equation obtained after perturbed and neglecting pressure term in (11.10)–(11.16), along with the boundary condition and using basic state obtained in (11.18)–(11.21), constitutes the linear boundary value problem. We now apply normal mode analysis by using

$$\left.\begin{aligned} W &= W(z)\exp(st + i(my + lx)) \\ T_f &= \Theta_f(z)\exp(st + i(my + lx)) \\ T_p &= \Theta_p(z)\exp(st + i(my + lx)) \\ T_s &= \Theta_s(z)\exp(st + i(my + lx)) \\ \phi &= \psi(z)\exp(st + i(my + lx)) \\ C &= \zeta(z)\exp(st + i(my + lx)) \end{aligned}\right\} \tag{11.22}$$

After applying normal mode technique, we get the required equation of differential form:

$$(D^2 - a^2)^2 W - \frac{s}{\Pr\varepsilon}(D^2 - a^2)W + \mathrm{Rn}\, a^2\psi - \mathrm{Ra}\, a^2\Theta_f - \frac{R_s}{L_s}a^2\zeta = 0, \tag{11.23}$$

$$W + (D^2 - a^2)^2\Theta_f - s\Theta_f - \frac{2N_A N_B}{\mathrm{Le}}D\Theta_f - \frac{N_B}{\mathrm{Le}}D\psi + N_{HP}(\Theta_p - \Theta_f) = 0, \tag{11.24}$$

$$\frac{W}{\varepsilon} + \varepsilon_p(D^2 - a^2)\Theta_p + N_{HP}\gamma_p(\Theta_f - \Theta_p) - s\Theta_p = 0, \tag{11.25}$$

$$\frac{W}{\varepsilon} + \varepsilon_s(D^2 - a^2)\Theta_s + N_{HS}\gamma_p(\Theta_f - \Theta_s) - s\Theta_s = 0, \tag{11.26}$$

$$\frac{W}{\varepsilon} + \frac{1}{\text{Le}}(D^2 - a^2)\psi + \frac{N_A}{\text{Le}}(D^2 - a^2)\Theta_f - s\psi = 0, \tag{11.27}$$

$$\frac{W}{\varepsilon} + \frac{1}{\text{Ls}}(D^2 - a^2)\zeta - s\zeta = 0, \tag{11.28}$$

where s is taken to be the growth rate of disturbances, l_x and m_y are defined as the wave numbers.

$D = \frac{\partial}{\partial z}$ and $a = \sqrt{l_x^2 + m_y^2}$ is the wave number resultant (dimensionless).

To obtain the analytical solution to the system of equations, orthogonality condition is used. Accordingly for the process, the base function are chosen in the following way:

Galerkin method

Galerkin's method is a type of weighted residual numerical methods. The residual is minimized with respect to the each suitable weight function, satisfying the given boundary/initial conditions, using orthogonality property over the given region.

$$\int R(\text{Residual})(\text{weight function})dR = 0.$$

In this way, a system of linear simultaneous equations is obtained, which can be easily solved to get the solution of desired accuracy. Accordingly, for the process, we chose trial function or base function (satisfying the boundary conditions) $W_k, \Phi_k, \Theta_{f\,k}, \Theta_{sk}$, and Θ_{pk} the following way and we write:

$$W = \sum_{k=1}^{N} A_k W_k, \quad \Theta_f = \sum_{k=1}^{N} B_k \Theta_k, \quad \Theta_p = \sum_{k=1}^{N} C_k \Theta_k, \quad \Theta_s = \sum_{k=1}^{N} D_k \Theta_k,$$

$$\Psi = \sum_{k=1}^{N} E_k \Psi_k, \quad \zeta = \sum_{k=1}^{N} F_k \zeta_k.$$

$$\begin{aligned} W &= A_1 \sin \pi z, \\ \Theta_f &= B_1 \sin \pi z, \\ \Theta_p &= C_1 \sin \pi z, \\ \Theta_s &= D_1 \sin \pi z, \\ \psi &= E_1 \sin \pi z, \\ \zeta &= F_1 \sin \pi z. \end{aligned} \tag{11.29}$$

All of them satisfy boundary conditions.

$A_1, B_1, C_1, D_1, E_1, F_1$ are the unknown coefficients using the above base functions in the equation (11.23)–(11.28). The given system of algebraic equations has non-trivial solutions, which are obtained when the determinant of the coefficient matrix vanishes, resulting in the characteristic equations for the present system. Further, for solving Rayleigh number, we obtain the result for the first approximation as:

$$\begin{pmatrix} P_{11} & -\frac{a^2\,\mathrm{Ra}}{2} & 0 & 0 & \frac{a^2\,\mathrm{Rn}}{2} & -\frac{a^2\,\mathrm{Rs}}{2\,\mathrm{Ls}} \\ \frac{1}{2\varepsilon} & P_{22} & \frac{N_{HP}}{2} & \frac{N_{HS}}{2} & 0 & 0 \\ \frac{1}{2\varepsilon} & \frac{\gamma p N_{HP}}{2} & P_{33} & 0 & 0 & 0 \\ \frac{1}{2\varepsilon} & \frac{\gamma s N_{HS}}{2} & 0 & -\frac{1}{2}((a^2+\pi^2)\varepsilon s + \gamma s N_{HS} + s) & 0 & 0 \\ 0 & -\frac{(a^2+\pi^2)N_A}{2\,\mathrm{Le}} & 0 & 0 & -\frac{a^2+\pi^2}{2\,\mathrm{Le}} - \frac{s}{2} & 0 \\ \frac{1}{2\varepsilon} & 0 & 0 & 0 & 0 & -\frac{a^2+\mathrm{Ls}\,s+\pi^2}{2\,\mathrm{Ls}} \end{pmatrix} \begin{pmatrix} A_1 \\ B_1 \\ C_1 \\ D_1 \\ E_1 \\ F_1 \end{pmatrix} = \begin{pmatrix} 0 \\ 0 \\ 0 \\ 0 \\ 0 \\ 0 \end{pmatrix}, \tag{11.30}$$

where $P_{11} = \frac{1}{2\,\mathrm{Ls}\,\mathrm{Pr}\,\varepsilon^2}(a^2 s L_s + \mathrm{Ls}\,\pi^2 s\varepsilon + a^4\,\mathrm{Ls}\,\mathrm{Pr}\,\varepsilon^2 + 2a^2\,\mathrm{Ls}\,\pi^2\,\mathrm{Pr}\,\varepsilon^2 + \mathrm{Ls}\,\pi^4\,\mathrm{Pr}\,\varepsilon^2)$, $P_{22} = \frac{\varepsilon(-N_{HP}-N_{HS}-s-\pi^2)-a^2\varepsilon}{2\varepsilon}$, $P_{33} = \frac{1}{2}(-(a^2+\pi^2)\delta - \gamma p N_{HP} - s)$.

For stationary convection, we put $s = 0$. For the assumption, we obtain the Rayleigh number for local thermal equilibrium, putting $N_{HS} = 0$ and $N_{HP} = 0$.

$$\mathrm{Ra} = \frac{(a^2+\pi^2)^3}{a^2} - \mathrm{Rs} - N_A\,\mathrm{Rn}. \tag{11.31}$$

7 Non-linear stability

In non-linear stability, we introduce the stream function and the equations obtained are as follows: Taking $u = -\frac{\partial\Psi}{\partial z}$ and $w = \frac{\partial\Psi}{\partial x}$

$$\frac{1}{\mathrm{Pr}\,\varepsilon}\partial_t(\nabla^2\Psi) - \nabla^4\Psi - \mathrm{Rn}\,\partial_x\phi g_m + \mathrm{Ra}\,\partial_x T_f g_m + \frac{\mathrm{Rs}}{\mathrm{Ls}}\partial_x C g_m = 0, \tag{11.32}$$

$$\partial_t\phi + \frac{1}{\varepsilon}\left(-\frac{\partial\Psi}{\partial x}\frac{\partial\phi_b}{\partial z}\right) - \frac{1}{\mathrm{Le}}\nabla^2\phi - \frac{N_A}{\mathrm{Le}}\nabla^2 T_f - \frac{1}{\varepsilon}\left(\frac{\partial\Psi}{\partial x}\frac{\partial\phi}{\partial z} - \frac{\partial\Psi}{\partial z}\frac{\partial\phi}{\partial x}\right) = 0, \tag{11.33}$$

$$\begin{aligned} \partial_t T_f &+ \frac{1}{\varepsilon}\left(-\frac{\partial\Psi}{\partial x}\frac{\partial T_{fb}}{\partial z}\right) - \nabla^2 T_f - \frac{N_B}{\mathrm{Le}}\left(\frac{\partial T_f}{\partial z}\frac{\partial\phi_b}{\partial z} + \frac{\partial\phi}{\partial z}\frac{\partial T_{fb}}{\partial z}\right) \\ &- \frac{1}{\varepsilon}\left(\frac{\partial\Psi}{\partial x}\frac{\partial T_f}{\partial z} - \frac{\partial\Psi}{\partial z}\frac{\partial T_f}{\partial x}\right) - N_{HP}(T_p - T_f) - N_{HS}(T_s - T_f) = 0, \end{aligned} \tag{11.34}$$

$$\partial_t T_p + \frac{1}{\varepsilon}\left(-\frac{\partial\Psi}{\partial x}\frac{\partial T_{pb}}{\partial z}\right) - \delta\nabla^2 T_p - \frac{1}{\varepsilon}\left(\frac{\partial\Psi}{\partial x}\frac{\partial T_p}{\partial z} - \frac{\partial\Psi}{\partial z}\frac{\partial T_p}{\partial x}\right) - \gamma_p\,N_{HP}(T_f - T_p) = 0, \tag{11.35}$$

$$\partial_t T_s - \varepsilon\nabla^2 T_s - \gamma_s\,N_{HS}(T_f - T_s) = 0, \tag{11.36}$$

$$\partial_t C + \frac{1}{\varepsilon}\left(-\frac{\partial\Psi}{\partial x}\frac{\partial C_b}{\partial z}\right) - \frac{1}{\mathrm{Ls}}\nabla^2 C - \frac{1}{\varepsilon}\left(\frac{\partial\Psi}{\partial x}\frac{\partial C}{\partial z} - \frac{\partial\Psi}{\partial z}\frac{\partial C}{\partial x}\right) = 0, \tag{11.37}$$

where $g_m = (1 + \gamma \, \mathrm{Cos}[\Omega t])$.

Since physical quantities are unchanged by y, we apply the Fourier series expansions. In order to found non-linear stability analysis, we take Fourier series expression as follows [1]:

$$\Psi = \sum_{i=1}^{\infty}\sum_{j=1}^{\infty} A_{ij}(t)\sin(j\pi z)\sin(i\pi ax),$$

$$\phi = \sum_{i=1}^{\infty}\sum_{j=1}^{\infty} B_{ij}(t)\sin(j\pi z)\cos(i\pi ax),$$

$$T_f = \sum_{i=1}^{\infty}\sum_{j=1}^{\infty} C_{ij}(t)\sin(j\pi z)\cos(i\pi ax),$$

$$T_s = \sum_{i=1}^{\infty}\sum_{j=1}^{\infty} D_{ij}(t)\sin(j\pi z)\cos(i\pi ax),$$

$$T_p = \sum_{i=1}^{\infty}\sum_{j=1}^{\infty} E_{ij}(t)\sin(j\pi z)\cos(i\pi ax),$$

$$C = \sum_{i=1}^{\infty}\sum_{j=1}^{\infty} F_{ij}(t)\sin(j\pi z)\cos(i\pi ax).$$

Here for stream function, we consider (1, 1) mode and for nanoparticle concentration, fluid, solid, and particle phase mode (1, 1) and (0, 2). In the following section, to implement the non-linear stability analysis, we take the following modes in expansion of Fourier series as follows:

$$\Psi = A_{11}(t)\sin(\pi z)\sin(ax), \tag{11.38}$$

$$\phi = B_{11}(t)\sin(\pi z)\cos(ax) + B_{02}(t)\sin(2\pi z)(-N_a), \tag{11.39}$$

$$T_f = C_{11}(t)\sin(\pi z)\cos(ax) + C_{02}(t)\sin(2\pi z), \tag{11.40}$$

$$T_p = D_{11}(t)\cos(\pi z)\cos(ax) + D_{02}(t)\cos(2\pi z), \tag{11.41}$$

$$T_s = E_{11}(t)\cos(\pi z)\cos(ax) + E_{02}(t)\cos(2\pi z), \tag{11.42}$$

$$C = F_{11}(t)\cos(\pi z)\cos(ax) + F_{02}(t)\cos(2\pi z). \tag{11.43}$$

Substituting (11.38)–(11.43) into obtained non-linear equations of perturbed state and using orthogonalization technique of the Galerkin's technique, we get:

$$\begin{aligned} A_{11}'(t) = &-\frac{\varepsilon \,\mathrm{Pr}}{(a^2+\pi^2)\,\mathrm{Ls}}(a^4 \,\mathrm{Ls}\, A_{11}(t) + 2\pi^2 a^2 \,\mathrm{Ls}\, A_{11}(t) + a\gamma \,\mathrm{Ls}\, N_a \,\mathrm{Rn}\, B_{11}(t)\cos(t\Omega) \\ &+ a \,\mathrm{Ls}\, N_A \,\mathrm{Rn}\, B_{11}(t) + a\gamma \,\mathrm{Ls}\,\mathrm{Ra}\, C_{11}(t)\cos(t\Omega) + a \,\mathrm{Ls}\,\mathrm{Ra}\, C_{11}(t) + a\gamma \,\mathrm{Rs}\, E_{11}(t)\cos(t\Omega) \\ &+ a \,\mathrm{Rs}\, E_{11}(t) + \pi^4 \,\mathrm{Ls}\, A_{11}(t)(a^2+\pi^2)\,\mathrm{Ls}), \end{aligned} \tag{11.44}$$

$$B'_{11}(t) = \frac{1}{\varepsilon \operatorname{Le} N_A}(a^2\varepsilon(-N_A)B_{11}(t) + a^2\varepsilon N_A C_{11}(t) - \pi a \operatorname{Le} N_A A_{11}(t)B_{02}(t)$$
$$- a \operatorname{Le} A_{11}(t) - \pi^2\varepsilon N_A B_{11}(t) + \pi^2\varepsilon N_A C_{11}(t)), \tag{11.45}$$

$$B'_{02}(t) = -\frac{\pi(-a \operatorname{Le} A_{11}(t)B_{11}(t) + 8\pi\varepsilon B_{02}(t) - 8\pi\varepsilon C_{02}(t))}{2\varepsilon \operatorname{Le}}, \tag{11.46}$$

$$C'_{11}(t) = \frac{1}{\varepsilon}(a^2(-\varepsilon)C_{11}(t) - \pi a A_{11}(t)C_{02}(t) - aA_{11}(t) - \varepsilon N_{HP}C_{11}(t)$$
$$- \varepsilon N_{HS}C_{11}(t) - \pi^2\varepsilon C_{11}(t) + \varepsilon N_{HP}D_{11}(t) + \varepsilon N_{HS}F_{11}(t)), \tag{11.47}$$

$$C'_{02}(t) = \frac{1}{2\varepsilon}(\pi a A_{11}(t)C_{11}(t) - 2\varepsilon N_{HP}C_{02}(t) - 2\varepsilon N_{HS}C_{02}(t)$$
$$- 8\pi^2\varepsilon C_{02}(t) + 2\varepsilon N_{HP}D_{02}(t) + 2\varepsilon N_{HP}F_{02}(t)), \tag{11.48}$$

$$D'_{11}(t) = \frac{1}{\varepsilon}(a^2(-\delta)\varepsilon D_{11}(t) - \pi a A_{11}(t)D_{02}(t) - aA_{11}(t) \tag{11.49}$$
$$+ \gamma_p\varepsilon N_{HP}C_{11}(t) - \gamma_p\varepsilon N_{HP}D_{11}(t) - \pi^2\delta\varepsilon D_{11}(t)), \tag{11.50}$$

$$D'_{02}(t) = \frac{1}{2\varepsilon}(\pi a A_{11}(t)D_{11}(t) + 2\gamma_p\varepsilon N_{HP}C_{02}(t) - 2\gamma_p\varepsilon N_{HP}D_{02}(t) - 8\pi^2\delta\varepsilon D_{02}(t)), \tag{11.51}$$

$$E'_{11}(t) = \frac{a^2(-\varepsilon)E_{11}(t) - \pi a \operatorname{Ls} A_{11}(t)E_{02}(t) - a \operatorname{Ls} A_{11}(t) - \pi^2\varepsilon E_{11}(t)}{\varepsilon \operatorname{Ls}}, \tag{11.52}$$

$$E'_{02}(t) = -\frac{\pi(8\pi\varepsilon E_{02}(t) - a \operatorname{Ls} A_{11}(t)E_{11}(t))}{2\varepsilon \operatorname{Ls}}, \tag{11.53}$$

$$F'_{11}(t) = a^2(-\varepsilon s)D_{11}(t) + \gamma_s N_{HS}C_{11}(t) - \pi^2\varepsilon_s D_{11}(t) - \gamma_s N_{HS}F_{11}(t), \tag{11.54}$$

$$F'_{02}(t) = \gamma s N_{HS}C_{02}(t) - 4\pi^2\varepsilon_s D_{02}(t) - \gamma_s N_{HS}F_{02}(t). \tag{11.55}$$

The introduced system of simultaneously ordinary differential equations above is answer with the use of an built-in tool ND solve of MATHEMATICA numerically.

8 Heat transport and nanoparticle concentration

The heat transport for fluid, solid, and particle phase is given by Nusselt number, $\mathrm{Nu}_{T_f}(t)$, $\mathrm{Nu}_{T_s}(t)$, $\mathrm{Nu}_{T_p}(t)$ is described as:

$$\mathrm{Nu}_{T_f}(t) = \frac{\text{Heat transport by (conduction + convection)}}{\text{Heat transport by conduction}},$$
$$\mathrm{Nu}_{T_f} = 1 + \left[\frac{\int_0^{2\pi/a}(\frac{\partial T_f}{\partial z})dx}{\int_0^{2\pi/a}(\frac{\partial T_{fb}}{\partial z})dx}\right]_{z=0}. \tag{11.56}$$

By using the values of T_{fb}, T_{sb}, T_{pb} and C_b from Eqs. (11.18)–(11.21) into the Eq. (11.56) and similar equations, we obtain the Nusselt number for solid, fluid, and particle phase as follows:

$$\mathrm{Nu}_{T_f}(t) = 1 - 2\pi C_{02}(t), \tag{11.57}$$

$$\mathrm{Nu}_{T_s}(t) = 1 - 2\pi D_{02}(t), \tag{11.58}$$

$$\mathrm{Nu}_{T_p}(t) = 1 - 2\pi E_{02}(t). \tag{11.59}$$

In the same way, for the nanoparticle concentration, Nusselt number, $\mathrm{Nu}_\phi(t)$, can be given as:

$$\mathrm{Nu}_\phi(t) = \frac{\text{Mass transport by (diffusion + advection)}}{\text{mass transport by diffusion}}.$$

Similarly, equation of nanoparticle concentration Nusselt number is

$$\mathrm{Nu}_\phi(t) = 1 + 2\pi(-B_{02}(t) + C_{02}(t)). \tag{11.60}$$

9 Results and discussion

9.1 Linear stability analysis

In case for the assumption of local thermal equilibrium, we obtain the Rayleigh number as follows [30]:

$$\mathrm{Ra} = \frac{(a^2 + \pi^2)^3}{a^2} - \mathrm{Rs} - N_A\,\mathrm{Rn}. \tag{11.61}$$

In this section, we discuss the linear stability of the problem. Here, the Marginal stability curve is drawn for Ra against a for the fixed values of the parameters. For values of $N_{HP} = N_{HS} = 10$, $\varepsilon = 0.4$, $\varepsilon_s = 0.7$, $\delta = 0.4$, $\gamma_s = \gamma_p = 5$, $R_s = 50$, (for silver nanoparticle Rn = 95.95 and Na = 0.5 from Table 11.1) figures are plotted. In Fig. 11.2, the graph of binary nanofluid and nanofluid is observed in both the LTE and LTNE (local thermal-non equilibrium) case. The rate of convection rate in binary nanofluid is faster than the rate of nanofluid. This is due to the fact that as binary nanofluid having base fluid a binary

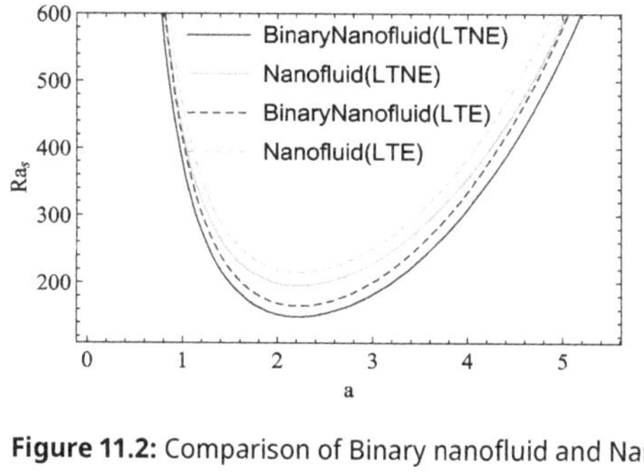

Figure 11.2: Comparison of Binary nanofluid and Nanofluid in LTE and LTNE.

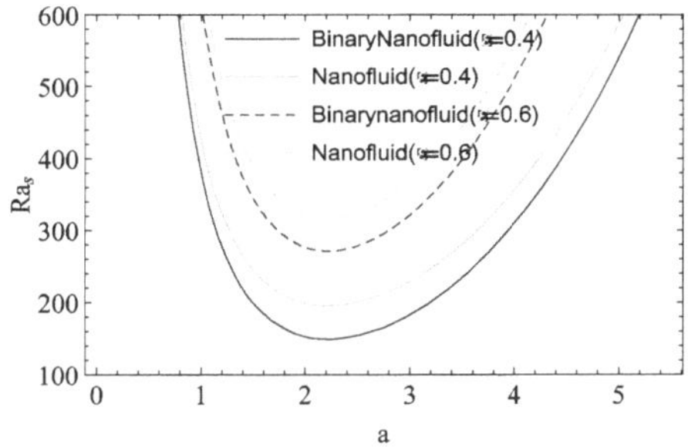

Figure 11.3: Effect of porosity in Binary nanaofluid and nanofluid (LTE and LTNE).

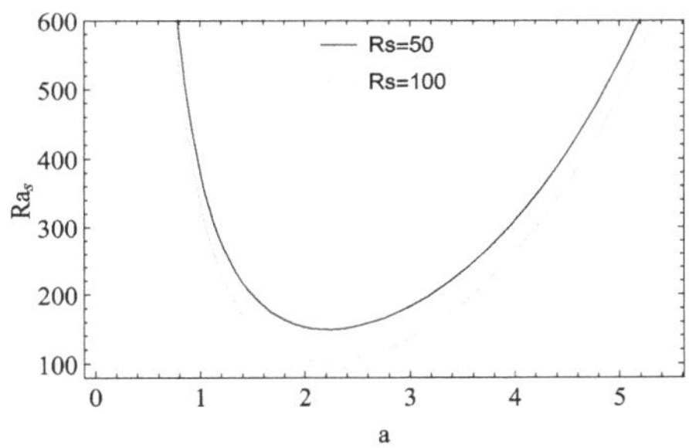

Figure 11.4: Effect of solute Rayleigh number.

mixture, so with presence of binary nanofluid rate of convection increases, thereby increasing the thermal conductivity. Therefore, binary nanofluid has a stabilizing effect. In Fig. 11.3, the repercussion of increasing porosity has been observed. From graph, it is clear that with increase in value of porosity convection cell increases, which decreases the rate of convection. Hence increase in value of porosity has a stabilizing effect on convection rate. One thing is also noticed that convection rate in nanofluid is slower in comparison to convection rate in binary nanofluid. Fig. 11.4 describes the repurcussion of solute Rayleigh number (R_s). With increase in value of solute Rayleigh number, convection rate increases. This is observed because increasing value of R_s decreases the Rayleigh number and hence convection occurs faster. Therefore, R_s has destabilizing effect on rate of convection. In Fig. 11.5, repercussion of N_A is observed. Figures depict that on increasing the value of N_A convection rate increases. This is because on increasing N_A diffusion rate increases and hence convection get faster. Hence Na has stabilizing effect on convection rate.

Figure 11.6–11.9 shows the effect of R_s, ε, ε_s, ε_p, where the graph of R_{ac} vs $\log_{NHP}$ is drawn. In Fig. 11.6, we examine that increasing the value of R_s increases the rate of convection in LTNE region. When the value of N_{HS} increases, the system attains thermal state. From Fig. 11.7, we observed the effect of porosity in LTE and LTNE region. In the range from −2 to 2, we depict that rising the value of porosity has a effect on convection

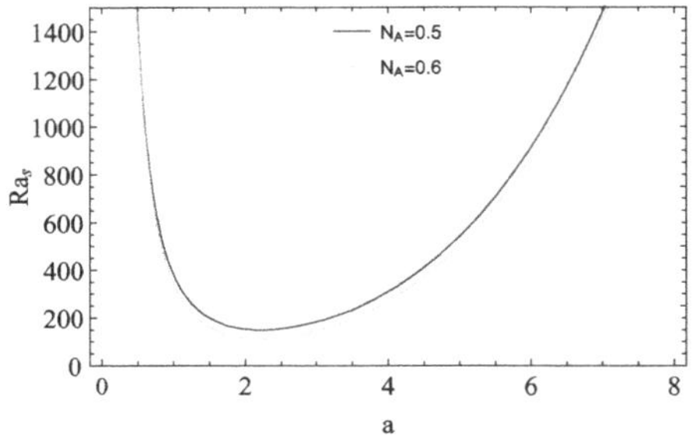

Figure 11.5: Effect of N_A.

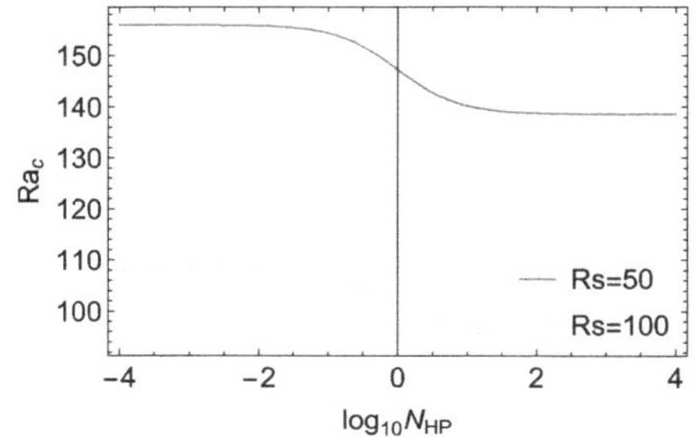

Figure 11.6: Graph of Ra$_c$ vs log N_{HP} with variation in Rs.

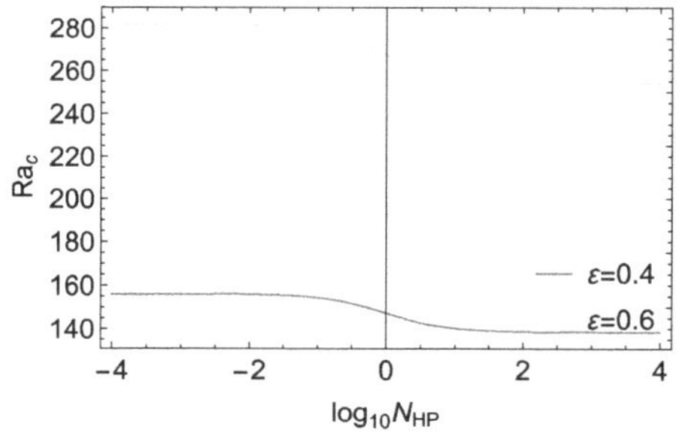

Figure 11.7: Graph of Ra$_c$ vs log N_{HP} with variation in ε.

rate with increase in critical Rayleigh number. But as the value of interface heat transfer increases, system attains thermal state. Fig. 11.8 shows the effect of ε_s, where in LTNE region, we can observe the effect of increasing value of ε_s. Rate of convection decreases with increasing value of ε_s because as thermal diffusivity ratio increases, it increases the value of a_c, the critical Rayleigh number. In Fig. 11.9, we observe the repercussion of thermal diffusivity of particle on thermal convection. We notice from the figure that as value of thermal diffusivity increases convection rate decreases. For very small-scale values of N_{HS}, the system reaches to thermal equilibrium, but as the value of N_{HS} in-

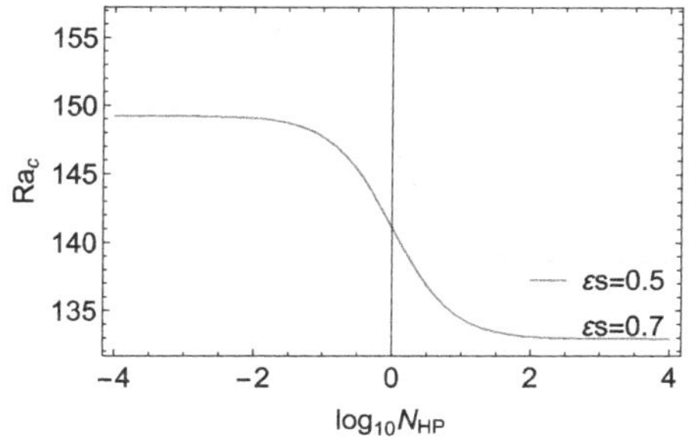

Figure 11.8: Graph of Ra_c vs log N_{HP} with variation in εs.

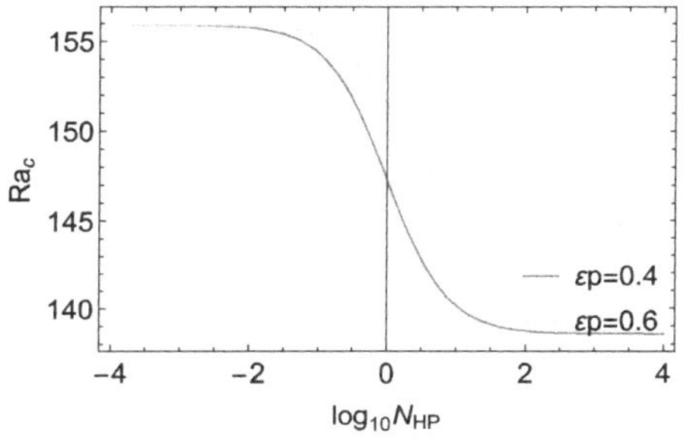

Figure 11.9: Graph of Ra_c vs log N_{HP} with variation in εp.

creases, the heat-transfer between the phases is too early, which again after some time maintains the thermal equilibrium in the system. While in case of LTE region no effect is observed in convection rate and value remains constant.

Figures 11.10–11.13 depicts the graph of a_c vs $\log_{NHP}$. From Fig. 11.10 we observed that increasing the value of R_s rate of convection decreases thereby decreasing the value of a_c. In LTNE region, we can see that for very small range convection is getting delayed on increasing value of solute Rayleigh number. But as the value of N_{HS} increases, it reaches to its thermal equilibrium state. Figure 11.11 depicts the effect of porosity. We can see

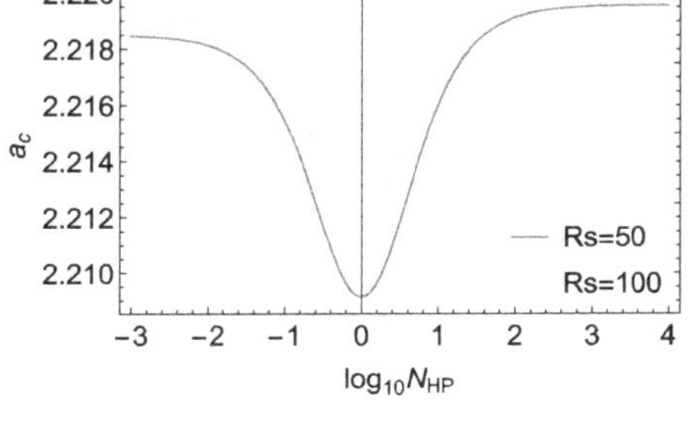

Figure 11.10: Graph of a_c vs log N_{HP} with variation in Rs.

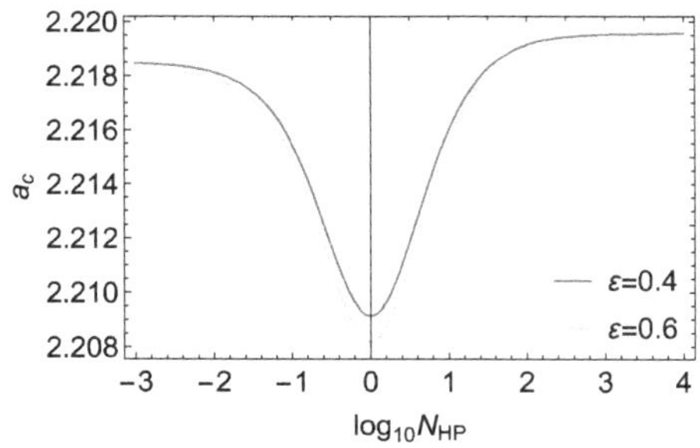

Figure 11.11: Graph of a_c vs log N_{HP} with variation in ε.

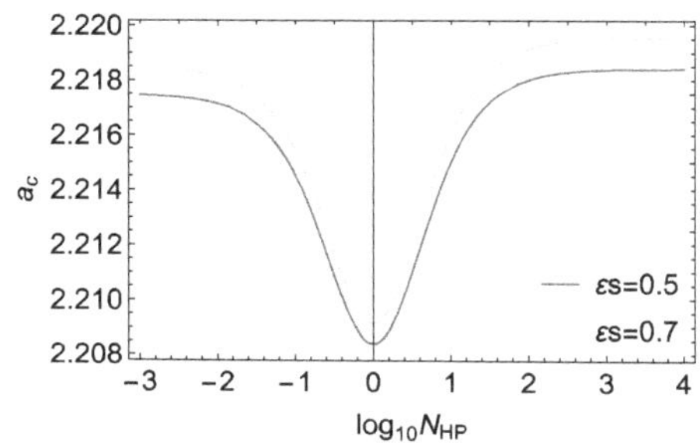

Figure 11.12: Graph of a_c vs log N_{HP} with variation in εs.

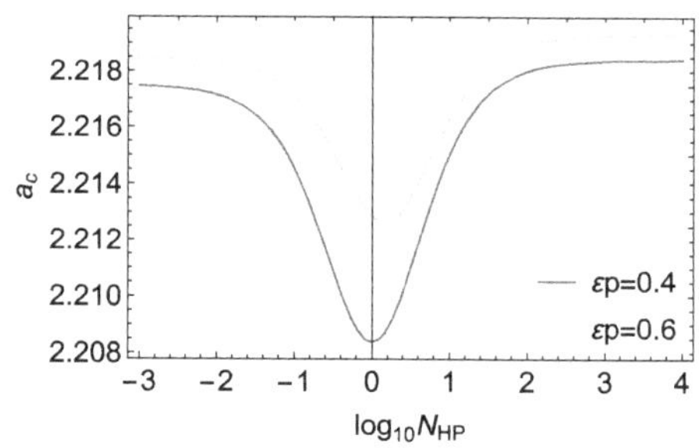

Figure 11.13: Graph of a_c vs log N_{HP} with variation in εp.

from figure that on increasing value of porosity critical wave number decreases thereby increasing wavelength of the cells. Hence convection get faster on increasing porosity, which can be observed in LTNE region. Figures 11.12 and 11.13 show the repurcussion of thermal diffusivity ratio ε_s and ε_p on thermal convection rate. With rise in value of diffusivity ratio critical wave number increases, decreasing the wavelength and reduce convection rate. ε_s and ε_p decrease the rate of convection.

9.2 Non-linear stability analysis

In this section, we look into the repercussion of frequency and (amplitude of) gravity modulation on mass and heat transfer rate in fluid, solid, particle, and concentration Nusselt number for non-linear stability.

Figures 11.14–11.17 show the effect of frequency of gravity modulation in non-linear stability. In Figure 11.14, we observe that for very small value of Ω, large number of periods are formed, which cover the entire volume of the fluid and hence heat transfer is slower. On increasing the value of Ω, less number of periods formed, which increases

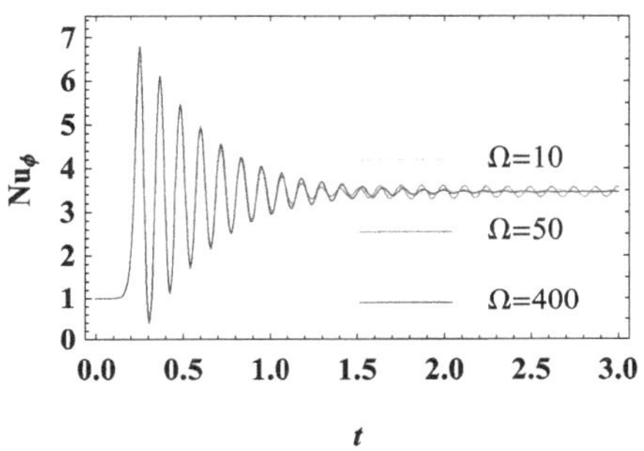

Figure 11.14: Graph of Nu_ϕ vs t with variation in Ω.

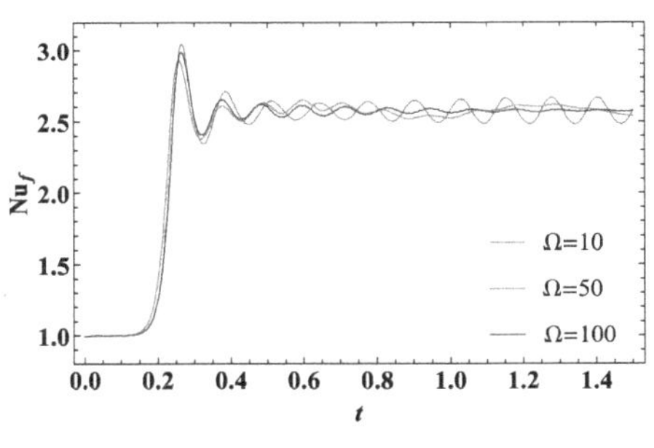

Figure 11.15: Graph of Nu_f vs t with variation in Ω.

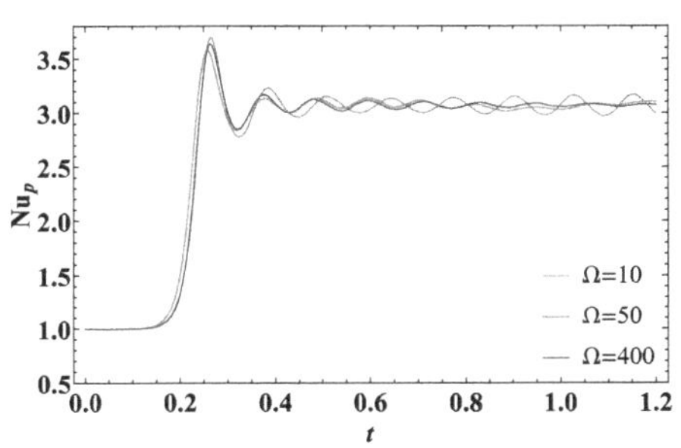

Figure 11.16: Graph of Nu_p vs t with variation in Ω.

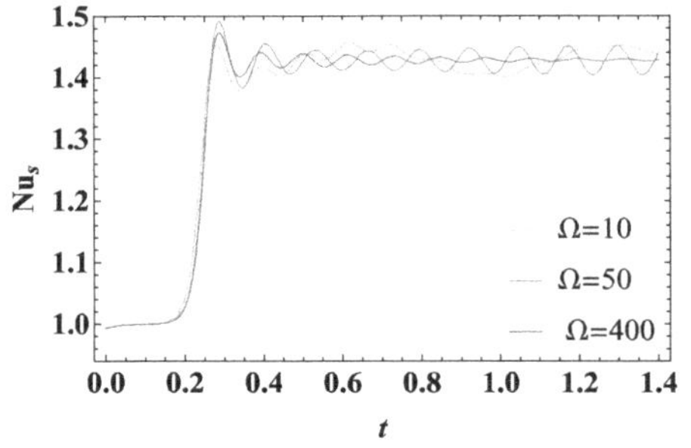

Figure 11.17: Graph of Nu_S vs t with variation in Ω.

the mass transfer and after some time reaches to steady state. Figures 11.15–11.17 exhibit the effect of frequency of gravity modulation on Nusselt number for fluid, solid, and particle phase. We observe from these figures that for very small values of Ω, frequency of gravity modulation, due to formation of large number of periods convection rate, is slower, but as the value of Ω increases, it increases the rate of heat transfer. For values of $\Omega = 400$ and above, we observed the case that is similar to unmodulated case.

Figures 11.18–11.21 exhibit the effect of (amplitude of) gravity modulation. In Fig. 11.18, we observe that increasing value of amplitude has neglecting effect on mass-

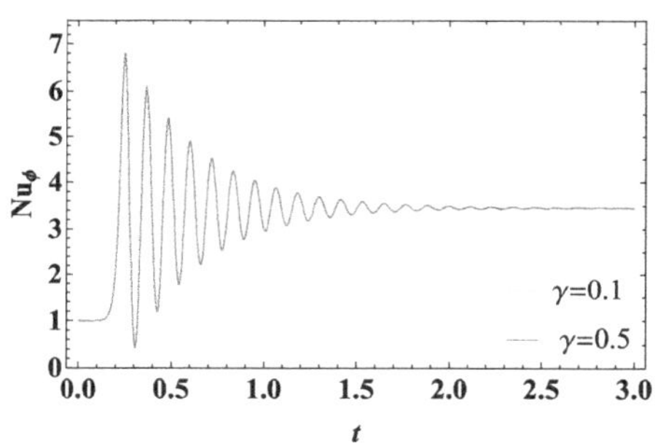

Figure 11.18: Graph of Nu_ϕ vs t with variation in γ.

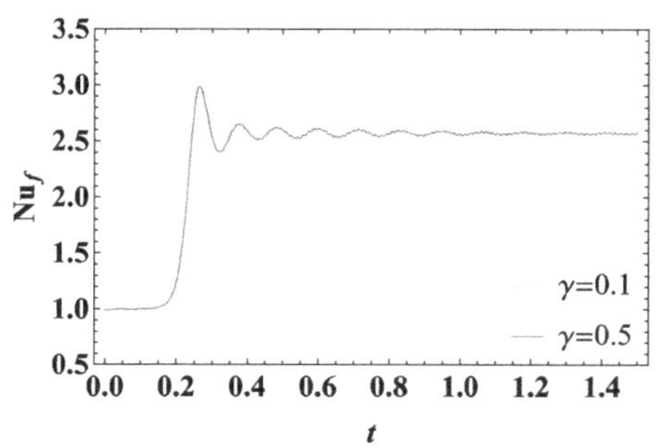

Figure 11.19: Graph of Nu_f vs t with variation in γ.

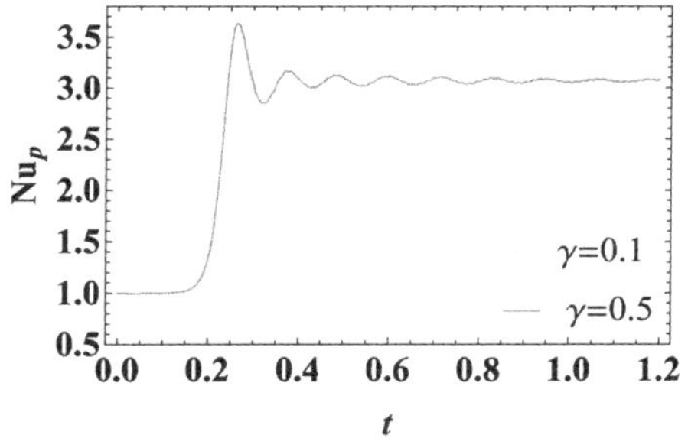

Figure 11.20: Graph of Nu_p vs t with variation in γ.

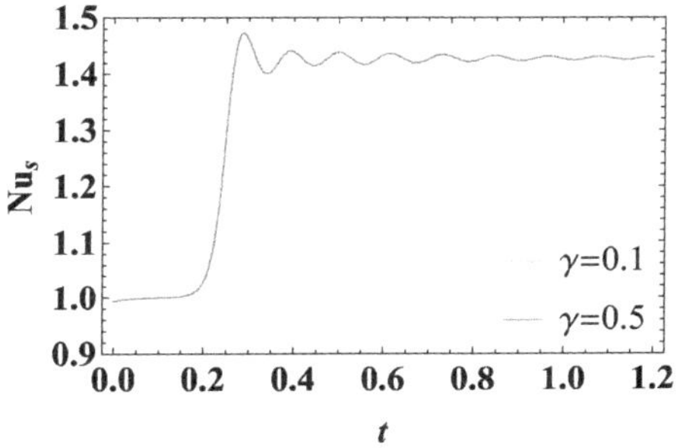

Figure 11.21: Graph of Nu_s vs t with variation in γ.

transfer rate. Similarly, in Figures 11.19–11.21, we observe that increasing value of amplitude of gravity modulation has neglecting effect on heat transfer rate on fluid, solid, and particle phase.

Figure 11.22 exhibit the effect of porosity on rate of mass transfer. Hence the figure clearly shows that increasing value of porosity delays the rate of convection. For smaller value of porosity, conduction starts initially for short interval of time and starts the convection at faster rate. Hence mass transfer rate decreases as the value of porosity

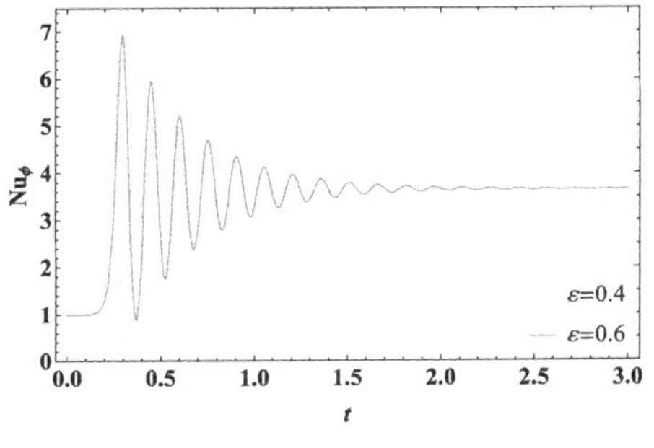

Figure 11.22: Graph of Nu_ϕ vs t with variation in ε.

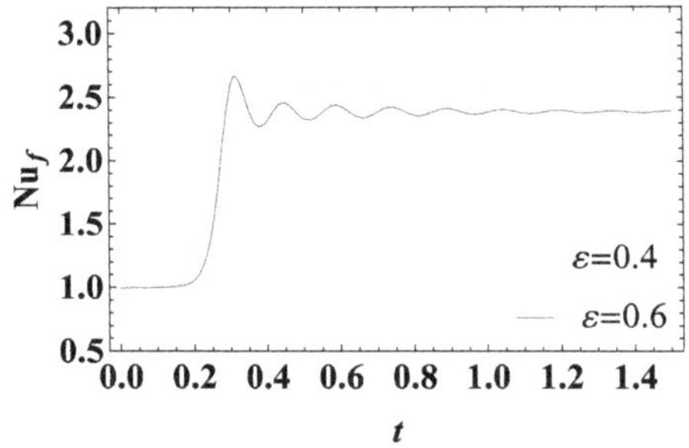

Figure 11.23: Graph of Nu_f vs t with variation in ε.

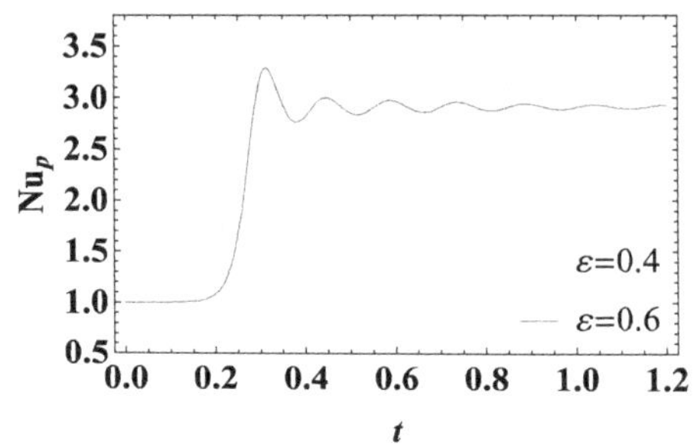

Figure 11.24: Graph of Nu_p vs t with variation in ε.

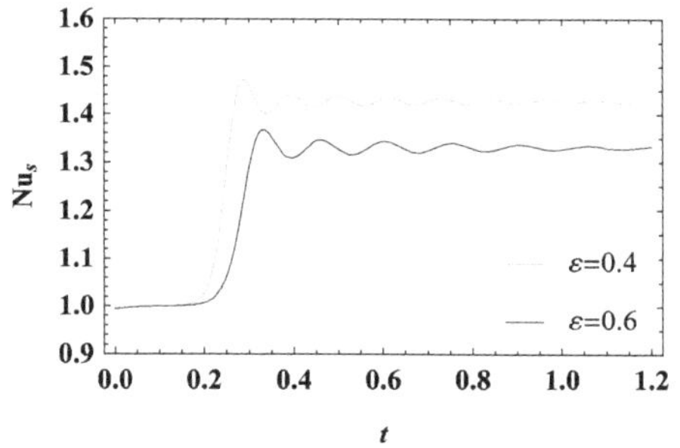

Figure 11.25: Graph of Nu_s vs t with variation in ε.

increases. Similarly in case of heat transfer rate for fluid, solid, and particle phase on increasing value of porosity convection is getting delayed, hence rate of heat and mass transfer occurs slower (see 11.23, 11.24, 11.25).

10 Effect of different type of gravity modulation

We have observed the effects of different types of gravity modulation, i. e., day-night, sinusoidal and, sawtooth, on Nusselt number for concentration and Nusselt number for fluid, particles and solid matrix. It is clearly observed from the Figures 11.26–11.29 that day-night gravity modulation starts mass transfer and heat transfer earlier than a sinusoidal and saw-tooth.

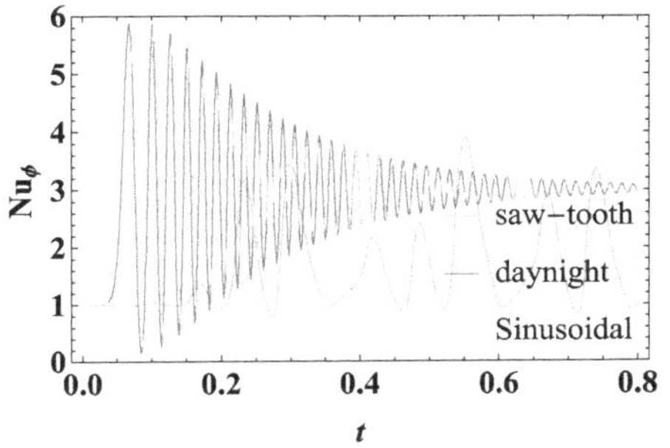

Figure 11.26: Graph of Nu_ϕ vs t with different types of modulation.

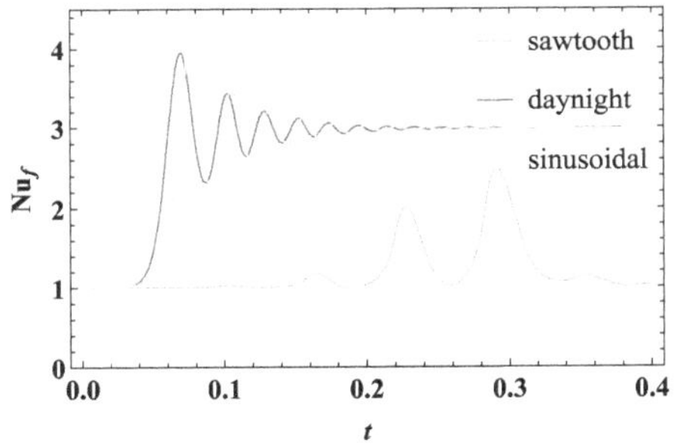

Figure 11.27: Graph of Nu_f vs t with different types of modulation.

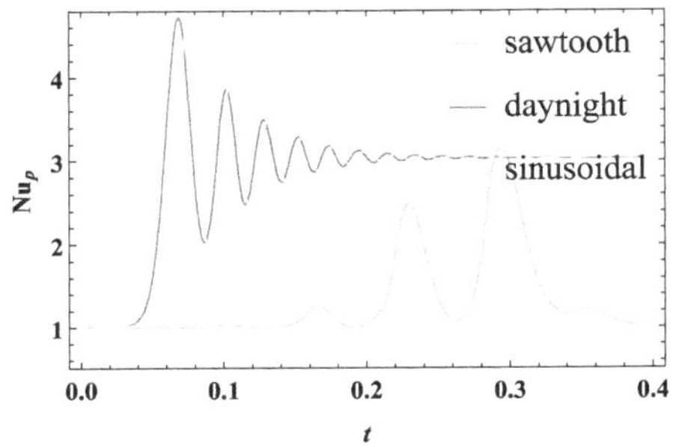

Figure 11.28: Graph of Nu_p vs t with different types of modulation.

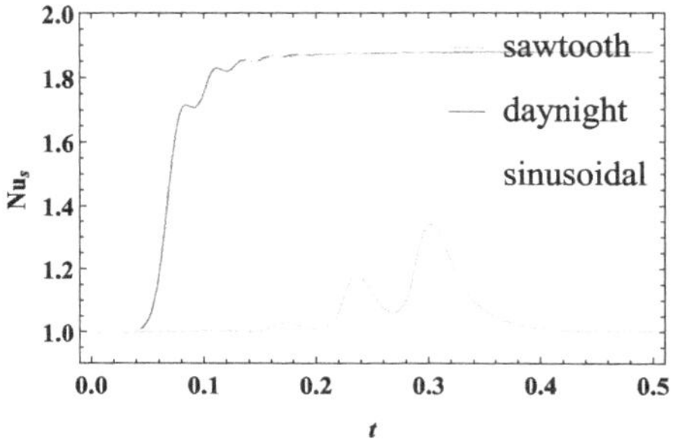

Figure 11.29: Graph of Nu_s vs t with different types of modulation.

11 Conclusion

In this paper, we have looked into the merge repercussion of gravity modulation and local thermal non-equilibrium having binary nanofluid saturated by porous media for the two and linear and non-linear stability analysis, respectively. Different parameters have different effects on linear (stationary) and non-linear thermal stability analysis. We notice the succeeding effect which are as follows:

1. From Figure 11.2(a) it is observed that rate of convection hastens faster in binary nanofluid than in nanofluid.
2. Porosity has a stabilizing effect in both LTE and LTNE case.
3. With increase in solute Rayleigh number convection hastens faster and hence has a destabilizing effect on rate of convection.
4. In graph of Ra_c vs. $\log N_{HP}$, porosity has a stabilizing effect.
5. ε_p, ε_s has stabilizing effect on convection.
6. In case of non-linear analysis, Nusselt number for concentration, fluid, particle and solid-matrix phase, with increasing value of frequency of modulation, higher rate of heat transfer is observed.
7. For $\Omega = 400$, the effect of frequency of gravity modulation is same as unmodulated case.
8. Low porosity increases the heat and mass transfer, thereby delaying in convection rate as value of porosity increase.
9. Day-night profile show earlier rate of heat and mass transfer in all the concentration, fluid, particles and solid matrix phase.

Bibliography

[1] S. Agarwal, N. C. Sacheti, P. Chandran, B. S. Bhadauria and A. K. Singh, Non-linear convective transport in a binary nanofluid saturated porous layer, *Transp. Porous Media* **93** (2012), 29–49.

[2] A. C. Baytas, Thermal non-equilibrium natural convection in a square enclosure filled with a heat-generating solid phase, non-Darcy porous medium, *Int. J. Energy Res.* **27**(10) (2003), 975–988.

[3] B. S. Bhadauria and A. Srivastava, Combined effect of internal heating and through-flow in a nanofluid saturated porous medium under local thermal nonequilibrium, *J. Porous Media* **25**(2) (2022).

[4] J. Buongiorno, Convective transport in nanofluids, *ASME J. Heat Transf.* **128**(3) (2006), 240–250.

[5] S. U. Choi and J. A. Eastman, Enhancing thermal conductivity of fluids with nanoparticles (No. ANL/MSD/CP-84938; CONF-951135-29). Argonne National Lab. (ANL), Argonne, IL (United States), 1995.

[6] M. P. Dyko and K. Vafai, Effects of gravity modulation on convection in a horizontal annulus, *Int. J. Heat Mass Transf.* **50**(1–2) (2007), 348–360.

[7] J. A. Eastman, S. R. Phillpot, S. U. S. Choi and P. Keblinski, Thermal transport in nanofluids, *Annu. Rev. Mater. Res.* **34** (2004), 219.

[8] G. Z. Gershuni, E. M. Zhukhovitskii and I. S. Iurkov, On convective stability in the presence of periodically varying parameter, *J. Appl. Math. Mech.* **34**(3) (1970).

[9] P. M. Gresho and R. L. Sani, The effects of gravity modulation on the stability of a heated fluid layer, *J. Fluid Mech.* **40**(4) (1970), 783–806.

[10] U. Gupta, J. Sharma and R. K. Wanchoo, Thermosolutal convection in a horizontal nanofluid layer: introduction of oscillatory motions, in: *2014 Recent Advances in Engineering and Computational Sciences (RAECS)*, IEEE, March 2014, pp. 1–6.

[11] Ismail and B. S. Bhadauria, Thermal instability of Rivlin-Ericksen elastico-viscous nanofluid saturated by a porous medium with rotation, in: *International Workshop of Mathematical Modelling, Applied Analysis and Computation*, Springer Nature, Switzerland, Cham, August 2022, pp. 436–455.

[12] Ismail and B. S. Bhadauria, Effect of thermal instability on heat transfer of micropolar nanofluid saturated in porous medium with rotation, *Numer. Heat Transf., Part B, Fundam.* (2023), 1–26.

[13] Ismail and B. S. Bhadauria, Effect of three types of gravity modulation on Rivlin Ericksen nanofluid saturated in porous medium with internal heating, *J. Nanofluids* **12**(8) (2023), 2245–2262.

[14] Ismail, B. S. Bhadauria, A. Kumar, S. K. Rawat and M. Yaseen, Thermal instability of tri-hybrid Casson nanofluid with thermal radiation saturated porous medium in different enclosures, *Chin. J. Phys.* **87** (2024), 710–727.

[15] Ismail, B. S. Bhadauria, A. Kumar and A. Srivastava, Effect of three types of magnetic-field modulation and different shapes of nanoparticles on the instability of nanofluid (Al_2O_3+H_2O) saturated porous medium with internal heating, *J. Nanofluids* **12** (2023), 2115–2128, https://doi.org/10.1166/jon.2023.2077.

[16] Ismail, B. S. Bhadauria and Shilpee, Effect of magnetic-field modulation on the instability of micropolar nanofluid filled within Hele-Shaw cell, in: *International Conference on Mathematical Modelling, Applied Analysis and Computation*, Springer Nature, Switzerland, Cham, 2024, pp. 83–104.

[17] Ismail, B. S. Bhadauria and A. Srivastava, Effect of rotation on thermal instability under Hele-Shaw cell saturated by Casson nanofluid, *Int. J. Mod. Phys. B* (2023), 2450405.

[18] Ismail, B. S. Bhadauria and A. Srivastava, Effect of three types of magnetic field modulation on the instability of heat transfer in micropolar nanofluid filled within Hele-Shaw cell, *Numer. Heat Transf., Part A, Appl.* (2023), 1–25.

[19] J. K. Kim, J. Y. Jung and Y. T. Kang, The effect of nano-particles on the bubble absorption performance in a binary nanofluid, *Int. J. Refrig.* **29**(1) (2006), 22–29.

[20] A. V. Kuznetsov and D. Nield, Thermal instability in a porous medium layer saturated by a nanofluid: Brinkman model, *Transp. Porous Media* **81**(3) (2010), 409–422.

[21] D. Y. Lee and K. Vafai, Analytical characterization and conceptual assessment of solid and fluid temperature differentials in porous media, *Int. J. Heat Mass Transf.* **42**(3) (1999), 423–435.

[22] M. S. Malashetty, Anisotropic thermoconvective effects on the onset of double diffusive convection in a porous medium, *Int. J. Heat Mass Transf.* **36**(9) (1993), 2397–2401.

[23] M. S. Malashetty and M. S. Swamy, Effect of gravity modulation on the onset of thermal convection in rotating fluid and porous layer, *Phys. Fluids* **23**(6) (2011), 064108.

[24] H. Masuda, A. Ebata and K. Teramae, Alteration of thermal conductivity and viscosity of liquid by dispersing ultra-fine particles. Dispersion of Al_2O_3, SiO_2 and TiO_2 ultra-fine particles, *Netsu Bussei* (1993).
[25] D. A. Nield and A. V. Kuznetsov, Thermal instability in a porous medium layer saturated by a nanofluid, *Int. J. Heat Mass Transf.* **52**(25–26) (2009), 5796–5801.
[26] D. A. Nield and A. V. Kuznetsov, The onset of double-diffusive convection in a nanofluid layer, *Int. J. Heat Fluid Flow* **32**(4) (2011), 771–776.
[27] M. Sheikholeslami and H. B. Rokni, Simulation of nanofluid heat transfer in presence of magnetic field: a review, *Int. J. Heat Mass Transf.* **115** (2017), 1203–1233.
[28] Shilpee and B. S. Bhadauria, Gravity-modulated thermal instability in Oldroyd-B nanofluids, in: *Computational Fluid Flow and Heat Transfer*, CRC Press, 2024.
[29] Shilpee and B. S. Bhadauria, Combined effect of local thermal nonequilibrium and gravity modulation on thermal instability in micropolar nanofluid saturated porous media, *J. Porous Media* **27**(2) (2024).
[30] S. Shukla, U. Gupta and M. Devi, LTNE effects on binary nanofluid convection: A modified model, *Mater. Today Proc.* **45** (2021), 4753–4759.
[31] M. Wadih and B. Roux, Natural convection in a long vertical cylinder under gravity modulation, *J. Fluid Mech.* **193** (1988), 391–415.
[32] D. Yadav, G. S. Agrawal and R. Bhargava, The onset of convection in a binary nanofluid saturated porous layer, *Int. J. Theor. Appl. Multiscale Mech.* **2**(3) (2012), 198–224.

Dhriti Raj Talukdar and H. Imo Mani Singha

A comparative analysis of standard Knapsack and Legendre–Knapsack cryptosystem

Abstract: In this paper, we do a comparative analysis between the Knapsack cryptosystem using Legendre symbol and the standard Knapsack cryptosystem. It has been observed that the addition of the Legendre symbol enhances the Knapsack cryptosystem's decryption time. And we know that the longer the decryption time taken, the longer the time necessary to crack the cryptosystem if any feasible assaults exist. As a result, we can observe that the Legendre–Knapsack cryptosystem is better than the standard Knapsack cryptosystem in terms of security and is more effective.

Keywords: Knapsack cryptosystem, Legendre symbol, public and private keys

MSC 2010: 94A60, 68P25, 11Z05

1 Introduction

Data transmission is becoming a matter of concern in today's world of internet, as it is very vulnerable to becoming public and can be misused by others. As a result, the privacy and any other source of online communications are under major threat of being used by unknown people. Therefore, cryptography is used to transform the plaintext into ciphertext in order to prevent the transmitted data from being exploited. Cryptography is a secure way of sharing informations and communication between two individuals using mathematical rules called algorithms so that any third person cannot intercept the information shared between the two individuals. Using deterministic algorithms, messages can be transformed in ways that cannot be decoded. These algorithms are used to generate cryptographic keys, verify digital signatures, browse the web on the internet and communicate confidentially, such as credit card transactions. In addition to cryptography, cryptanalysis and cryptology share many similarities. Information can be hidden in storage or transit through microdots, merging words with images, and other techniques. Modern cryptography is most often associated with scrambling plaintext (ordinary text) into ciphertext (encryption), then restoring it to plaintext (decryption). The people who practice this profession are known as cryptographers.

Acknowledgement: We would like to sincerely thank the Department of Mathematics, The Assam Royal Global University, Guwahati, Assam, India, for providing all the necessary assistance in conducting this work successfully.

Dhriti Raj Talukdar, H. Imo Mani Singha, Department of Mathematics, The Assam Royal Global University, B-Block 2nd Floor Room No. B-208, NH-37, Opp. Tirupati Balaji Temple, Betkuchi, Guwahati-781035, Assam, India, e-mails: dhritirajtalukdar43@gmail.com, hmsingha@rgu.ac

https://doi.org/10.1515/9783111313634-012

The objectives of modern cryptography are as follows:
1. The information is not accessible to anyone who was not intended.
2. No alterations can be made to the data in storage or transit between the sender and the intended receiver without being detected.
3. It is impossible to deny at a later stage the intentions of the person who created the information, or who transmitted it, once the information has been transmitted.
4. Sender and receiver can identify each other and verify origin/destination of the information.

It is known as a cryptosystem when it meets some or all of the criteria above.

Knapsack cryptosystem is a public key cryptosystem that uses public and private keys to encrypt and decrypt data. Merkle–Hellman knapsack cryptosystem is one of the earliest public-key cryptosystems, which was invented by Ralph Merkle and Martin Hellman in 1978. They used subset sum problem [5–8] to make it difficult for anyone to hack it. However, it was broken by Adi Shamir [9] in 1982. Many researchers have been trying to improve the security of this cryptosystem, for example, using modular Knapsack formula [1], elliptic curve [4] and shift Knapsack problem [10], by developing a new Knapsack cryptosystem [12], using the fact of permutation combination algorithm [3], by combining the linear transformation of the secret sequences with Chinese remainder theorem [6] and by converting Knapsack cryptosystem to 3CNF [11].

In this work, we do a comparative analysis of the Legendre–Knapsack cryptosystem mentioned in [2] with the standard Knapsack cryptosystem using three different words. Because of the unpredictability based on the Legendre symbol, the Legendre–Knapsack cryptosystem is safe from the LLL algorithm and Shamir's attacks. Furthermore, a comparison of the Legendre–Knapsack and standard cryptosystems reveals that the encryption and decryption times of the Legendre–Knapsack cryptosystem are longer than those of the standard Knapsack cryptosystem. Longer decryption time means that it will take more time to break the system if any other possible attack is applied. Therefore, the results reveal that the Legendre–Knapsack cryptosystem is more safe and efficient compared to the standard Knapsack cryptosystem. Many other improvements of Knapsack cryptosystems are also available, like the one in [7], etc.

1.1 Definitions

Congruence

Let n be a fixed positive integer. Two Integers a and b are said to be congruent modulo n, written as $a \equiv b \pmod{n}$ if $n \mid (a - b)$, i. e., if $a - b = nk$, for some integer k.

Public key cryptography (PKC)

Public key cryptography is an encryption method that uses an asymmetric key algorithm or a pair of public and private keys for secure data transmission. A message sender uses a recipient's public key to encrypt a message. To decrypt the sender's message, only the recipient's private key may be used. PKC encryption evolved to meet the growing secure communication demands of multiple sectors and industries, such as the military.

Private key cryptography

Private key cryptography is a type of encryption where a single key is both utilised to encrypt and decrypt the cypher text that corresponds to the plain text. Both sender and receiver are agreed upon sharing the same key for encryption and decryption process. Private key cryptography algorithm is used in payment applications, such as card transactions.

Some examples of public key cryptography algorithms are *RSA*, which is an acronym named after this algorithm's inventors: Rivest, Shamir and Adelman. *DSA*, or Digital Signature Algorithm, *ElGamal*, *Elliptic Curve Cryptography*. Some examples of private key cryptography algorithms are *DES*, or Data Encryption Standard, *AES*, or Advanced Encryption Standard, *IDEA*, or International Data Encryption Algorithm, *RC4*, or Rivest Cipher 4.

Fibonacci sequence

It is a sequence of numbers, starting with zero and one, that steadily increases and where each number is equal to the sum of the preceding two numbers, e. g., 0, 1, 1, 2, 3, 5, 8, 13, 21, 34, 55... and so on.

Super-increasing sequence

A sequence $r = \{r_1, r_2, \dots, r_n\}$ of positive integers is super-increasing if $r_{i+1} \geq r_i + r_{i-1} + \cdots + r_1$, for all i with $1 \geq i \geq n - 1$.

2 Legendre symbol and its properties

Definition. Let p be an odd prime and let $\gcd(a, p) = 1$. If the quadratic congruence $x^2 \equiv a (\text{mod } p)$ has a solution, then a is said to be a *quadratic residue* of p. Otherwise, a is said to be a *quadratic non-residue* of p.

Definition. Let p be an odd prime and let $\gcd(a, p) = 1$. *The Legendre symbol* $(\frac{a}{p})$ is defined by,

$$\left(\frac{a}{p}\right) = \begin{cases} 1, & \text{if } a \text{ is quadratic residue of } p, \\ -1, & \text{if } a \text{ is quadratic non-residue of } p. \end{cases}$$

Theorem 1 (Euler's Criterion). *Let p be an odd prime and $\gcd(a, p) = 1$. Then a is a quadratic residue of p if and only if $a^{\frac{p-1}{2}} \equiv 1 (\text{mod } p)$.*

Theorem 2. *Let p be any odd prime and let a and b be two integers such that* $\gcd(a, p) = 1$ *and* $\gcd(b, p) = 1$. *Then the Legendre symbol has following properties:*

1. *If* $a \equiv b(\text{mod } p)$, *then* $(\frac{a}{p}) \equiv (\frac{b}{p})$.
2. $(\frac{a^2}{p}) = 1$.
3. $(\frac{ab}{p}) = (\frac{a}{p}).(\frac{b}{p})$.
4. $(\frac{-1}{p}) = (-1)^{\frac{p-1}{2}} = \begin{cases} 1, & \text{if } p \equiv 1(\text{mod } 4), \\ -1, & \text{if } p \equiv 3(\text{mod } 4). \end{cases}$
5. $(\frac{1}{p}) = 1$.

3 Methodology

3.1 The standard Knapsack cryptosystem

Suppose Alice and Bob want to communicate using standard Knapsack cryptosystem, then the following algorithm must be followed.

1. *Key generation process*
   The key generation process followed by Alice is given below
   - A super-increasing sequence, $S = (r_i)_{i=1}^k$ is selected.
   - A large integer n is selected, such that, $n \geq \sum_i^k r_i$.
   - An integer u is chosen, such that $\gcd(u, n) = 1$. Thus $[S, n, u]$ will be treated as the private key, which is kept secret.
   - She then calculates $M_i \equiv u * r_i(\text{mod } n)$, where $1 \leq i \leq n$, then the sequence $\vec{M} = (M_i)_{i=1}^k$ is the public key.
2. *The encryption process*
   In order for Bob to encrypt any plaintext, he follows the following steps.
   - He converts the characters of the plaintext to a binary form $\vec{x} = (x_1, x_2, \ldots, x_n)$.
   - For each x_i, he calculates and publishes the ciphertext $C = \vec{x}.\vec{M}$ or, $C = \sum_{i=1}^k x_i.M_i$.
3. *The decryption process*
   Once the cipher text is received by Alice, she applies the decryption process. She requires to know the private key $[S, u, n]$ in this process. Initially, Alice calculates the modular multiplicative inverse of $u^{-1}(\text{mod } n)$. Then, she multiplies C by u^{-1}, i. e., $C' \equiv u^{-1}.C(\text{mod } n)$
   $\therefore C' \equiv u^{-1} \sum_{i=1}^k x_i.M_i(\text{mod } n)$
   $\therefore C' \equiv u^{-1} \sum_{i=1}^k x_i.u.r_i(\text{mod } n)$
   $\therefore C' \equiv \sum_{i=1}^k x_i.r_i(\text{mod } n)$

3.1.1 The standard Knapsack problem

Example. Let us consider the following example.

1. Suppose Alice has a secret key that is the superincreasing knapsack $(2, 3, 7, 14, 30, 57, 120)$.
2. To convert the superincreasing knapsack into general knapsack, Alice chooses a multiplier u and a large integer as modulus n, such that $\gcd(u, n) = 1$ where $n \geq$ sum of all elements of the superincreasing Knapsack $(2, 3, 7, 14, 30, 57, 120)$.
3. To create the public key, Alice chooses $u = 41$ and $n = 491$.
   Now Alice computes $M_i \equiv u.r_i(\bmod n)$, $0 \leq M_i \leq n$
   Now, for each r_i,
   $2 : 2.u = 2.41 = 82 \equiv 82(\bmod 491)$
   $3 : 3.u = 3.41 = 123 \equiv 123(\bmod 491)$
   $7 : 7.u = 7.41 = 287 \equiv 287(\bmod 491)$
   $14 : 14.u = 14.41 = 547 \equiv 83(\bmod 491)$
   $30 : 30.u = 30.41 = 1230 \equiv 248(\bmod 491)$
   $57 : 57.u = 57.41 = 2337 \equiv 373(\bmod 491)$
   $120 : 10.u = 10.41 = 4920 \equiv 10(\bmod 491)$
   $\therefore$ The general Knapsack or the public key is $(82, 123, 287, 83, 248, 373, 10)$, which is available for everyone.
4. To encrypt the plaintext: Suppose Bob wants to send the message M = *"HELP"*
   Now, he converts it to binary form
   $72 \equiv (1\,0\,0\,1\,0\,0\,0)_2 = \vec{x}$ (say)
   $\therefore$ the Ciphertext, $C = \sum_{i=1}^{k} x_i.M_i$
   $C = (82 * 1 + 123 * 0 + 287 * 0 + 83 * 1 + 248 * 0 + 373 * 0 + 10 * 0)$
   $\therefore\ C = 165$
5. *At receiver's side*, Alice received $C = 165$.
   Alice uses the ciphertext together with the modular inverse of the conversion factor u, i. e., $C' \equiv C.u^{-1}(\bmod n)$
   $\therefore C' \equiv 165.41^{-1}(\bmod 491)$
   $\therefore 16 \equiv 165.41^{-1}(\bmod 491)$
   or, $C' \equiv u^{-1}.C(\bmod n)$
   or, $C' \equiv 41^{-1}.165(\bmod 491)$
   Now, $165.u^{-1} = u^{-1}. \sum_{i=1}^{k} x_i.M_i$
   $165.u^{-1} = (82 * 1 + 123 * 0 + 287 * 0 + 83 * 1 + 248 * 0 + 373 * 0 + 10 * 0).u^{-1}$
   $165.u^{-1} = (2 * u + 14 * u).u^{-1}$
   $165.u^{-1} = 2 + 14 = 16$
   Now, Alice solves the superincreasing Knapsack for $C' = 16$ with the help of the superincreasing sequence (or, the private key).
   Private key: $(2, 3, 7, 14, 30, 57, 120)$
   Alice recovers the binary message $(1\,0\,0\,1\,0\,0\,0) = 72$ = *"HELP"*.

3.2 Knapsack cryptosystem using the Legendre symbol

Here, the Legendre symbol is incorporated to the standard Knapsack cryptosystem. Firstly, Alice and Bob both agree on a secret large prime number p. That is,

$$\left(\frac{a}{p}\right) = \begin{cases} 1, & \text{Key Generating process 1;} \\ -1, & \text{Key Generating process 2.} \end{cases} \tag{12.1}$$

Keeping the above formula secret, Alice sends the public key 1 and public key 2 to Bob and use them based on Legendre symbol.

3.2.1 The Legendre–Knapsack problem

Now, to understand it properly, let us look at a functioning example below.

Assume Alice and Bob agreed on a prime $p = 19$, and the agreed, randomly sorted set of quadratic residues and quadratic nonresidues is given as $(4, 2, 7, 3, 8, 13, 5, \ldots)$. That is, they agreed-upon matching set of Legendre symbols is $(1, -1, 1, -1, -1, -1, 1, \ldots)$.

3.2.1.1 Key generating process

- *Step-1:*
  Suppose that Alice generates $S_1 = (3, 5, 11, 20, 41)$ and she chooses $n_1 = 85$ and $u_1 = 44$. Therefore, the first private key is (S_1, u_1, n_1).
  Now, Alice calculates the public key by using the formula, $q_i = u_1 * s_i \pmod{n_1}$.
  So, the first public key is $Q_1 = (47, 50, 59, 30, 19)$.
- *Step-2:*
  Suppose Alice generates $S_2 = (2, 3, 7, 13, 27)$ and selects $n_2 = 60$ and $u_2 = 7$. Thus the second private key is (S_2, u_2, n_2).
  Now, Alice calculates the second public key by using the formula, $q_i = u_2 * s_i \pmod{n_2}$.
  So, the second public key is $Q_2 = (14, 21, 49, 31, 9)$

3.2.1.2 Encryption process

Suppose Bob wants to send the plaintext "HELP" to Alice. To do this, he first converts the plaintext to binary form. He then calculates $en_i = \sum_{j=1}^{k} q_j * b_{ij}$ based on the resulting public key where the agreed set of Legendre symbol is $(1, -1, 1, -1, -1, -1, 1, \ldots)$. The encryption method is depicted in Table 12.1.

As a result, the ciphertext is $E_n = (50, 58, 109, 14)$ and which he sends to Alice.

Table 12.1: Encryption process.

Letter	b_i	$(\frac{a}{p})$	Public key	en_i
H	01000	1	$Q_1 = (47, 50, 59, 30, 19)$	50
E	00101	−1	$Q_2 = (14, 21, 49, 31, 9)$	58
L	01100	1	$Q_1 = (47, 50, 59, 30, 19)$	109
P	10000	−1	$Q_2 = (14, 21, 49, 31, 9)$	14

3.2.1.3 Decryption process

Alice receives $E_n = (50, 58, 109, 14)$ as the ciphertext, then she first computes the inverse of both $u_1(\text{mod } n_1)$ and $u_2(\text{mod } n_2)$, which are $u_1^{-1} = 29$ and $u_2^{-1} = 43$ respectively. Then, based on the agreed set of Legendre symbols $(1, -1, 1, -1, -1, -1, 1, \ldots)$, she calculates $M_i = en_i * u_j^{-1}(\text{mod } n_j)$, where $1 \leq j \leq 2$, $1 \leq i \leq k$. Then, b_i is computed by subtracting the largest term in S_j from M_i and by repeating the subtraction step with the remaining terms in S_j. The process is shown in Table 12.2.

Table 12.2: Decryption process.

en_i	$(\frac{a}{p})$	n_j	u_j^{-1}	M_i	b_i
50	1	85	29	5	01000
58	−1	60	43	34	00101
109	1	85	29	16	01100
14	−1	60	43	2	10000

Thus, $B_i = (01000, 00101, 01100, 10000)$.

After converting B_i to the numerical form, we obtain the plaintext “HELP”.

4 Results and discussions

We compare the run time of encryption and decryption processes for different words using the standard and Legendre–Knapsack cryptosystems. Table 12.3 and Table 12.4 shows the results respectively. The calculations are performed by Mathematica on a computer with Intel(R) Platinum(R) CPU 3825U @1.90 GHz 2.43 and 4 GB RAM.

Tables 12.3 and 12.4 are highlighted in Figure 12.1 and Figure 12.2, respectively. With increasing text length characters, Legendre–Knapsacks encryption time increases compared to the standard Knapsack cryptosystem, as shown in Figure 12.1.

Also, it can be seen from Figure 12.2, Legendre–Knapsack decryption time is much longer than standard Knapsack decryption time.

Table 12.3: Encryption time comparison.

Text	Standard Knapsack	Legendre-Knapsack
"HELP"	0.147 s	0.148 s
"HELLOW"	0.147 s	0.269 s
"MODERN"	0.101 s	0.251 s

Table 12.4: Decryption time comparison.

Text	Standard Knapsack	Legendre-Knapsack
"HELP"	0.032 s	0.154 s
"HELLOW"	0.035 s	0.168 s
"MODERN"	0.023 s	0.124 s

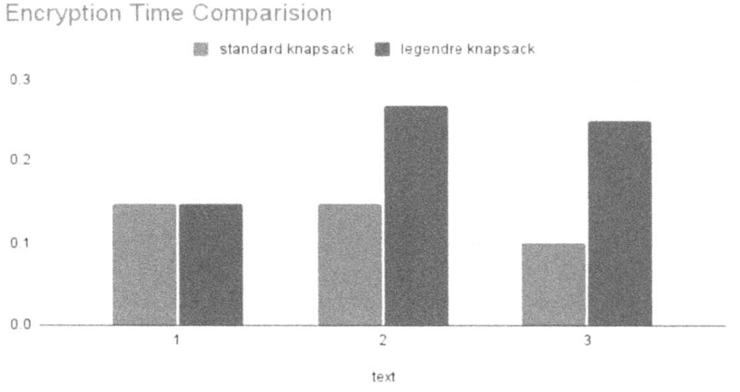

Figure 12.1: Encryption time for the both cryptosystems.

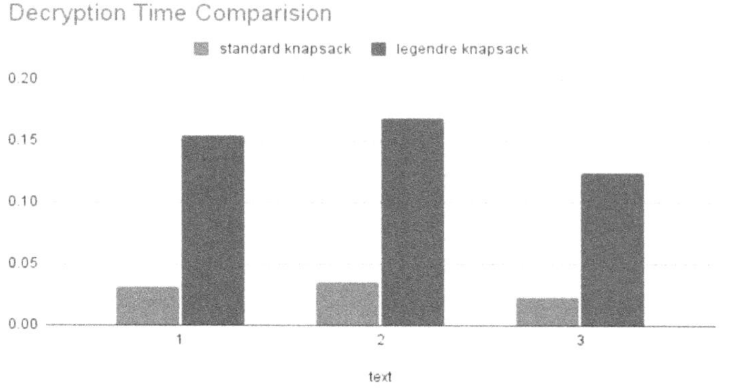

Figure 12.2: Decryption time for the both cryptosystems.

According to the figures, we observe that the Legendre–Knapsack algorithm encryption and decryption times are longer than standard Knapsack algorithm encryption and decryption times. As a result, the Legendre–Knapsack cryptosystem is more secure with an increasing in the decryption times.

5 Conclusion

In this work, we have compared the encryption and decryption time of the standard Knapsack and Legendre–Knapsack cryptosystems. It has been observed that the decryption time in the Legendre–Knapsack cryptosystem is longer than the time taken in the standard Knapsack cryptosystem. The longer decryption time is due to the two different private keys used. And as the longer the time it takes to decrypt, the longer it takes to breach the system, and hence the system is secured. Thus, the Legendre–Knapsack cryptosystem is a highly secure and efficient system compared to the standard Knapsack cryptosystem.

Bibliography

[1] R. M. Goodman and A. J. McAuley, A new trapdoor knapsack public key cryptosystem, in: *Advances in Cryptology: Proceedings of EUROCRYPT 84 A Workshop on the Theory and Application of Cryptographic Techniques Paris, France, April 9–11, 1984 3*, Springer, 1985, pp. 150–158.

[2] H. B. Habib et al., Improving the security of the knapsack cryptosystem by using legendre symbol, *Turk. J. Comput. Math. Educ. (TURCOMAT)* **12**(11) (2021), 2249–2255.

[3] M. S. Hwang, C. C. Lee and S. F. Tzeng, A new knapsack public-key cryptosystem based on permutation combination algorithm, *Inf. J. Appl. Math. Comput. Sci.* **5**(1) (2009), 33–38.

[4] N. Koblitz and N. Koblitz, Elliptic curves, in: *A Course in Number Theory and Cryptography*, 1987, pp. 150–179.

[5] R. Markle and M. E. Hellman, Hiding information and signatures in trapdoor knapsack, *IEEE Inf. Theory* **24** (1978), 525–530.

[6] Y. Murakami, A new construction method of knapsack pkc using linear transformation and chinese remainder theorem, in: *2010 International Symposium On Information Theory & Its Applications*, IEEE, 2010, pp. 433–436.

[7] T. A. Pham, The improvement of the knapsack cipher, *Comput. Commun.* **34**(3) (2011), 342–343.

[8] K. Sachdeva, Public key cryptography with knapsack systems, *Int. J. Eng. Adv. Technol.* **3**(2) (2013).

[9] A. Shamir, A polynomial time algorithm for breaking the basic merkle-hellman cryptosystem, in: *23rd Annual Symposium on Foundations of Computer Science (sfcs 1982)*, IEEE, 1982, pp. 145–152.

[10] P. C. Su and C. H. Tsai, New cryptosystems design based on hybrid-mode problems, *Comput. Electr. Eng.* **35**(3) (2009), 478–484.

[11] J. Thomas and N. Chaudhari, Knapsack cryptosystem and its reduction to 3cnf, in: *Twenty Fifth National Convention of Computer Engineers and National Seminar on Networked Home Systems and Services*, 2011, pp. 23–26.

[12] W. Zhang, B. Wang and Y. Hu, A new knapsack public-key cryptosystem, in: *2009 Fifth International Conference on Information Assurance and Security*, Vol. 2, IEEE, 2009, pp. 53–56.

Tania Bose, Renu Bala, and Madhu Aneja

Study of creep in composite functionally graded disc with non-linear thickness

Abstract: In the present study, a mathematical model is constructed to compare the creep response in a spinning FGM disc of $AlSiC_p$ having non-linear thickness geometry and SiC particles distributed non-linearly along radial direction with a spinning non-FGM disc with uniform thickness subjected to constant temperature. The threshold stress based law describes the creep behavior of the disc material, while the yielding is assumed to follow Tresca criterion. The study shows the FGM disc with thickness profile shows superior creep in terms of stress and strain rates in comparison to non-FGM disc with constant thickness. The use of thickness profile may help to reduce the creep-induced deformations in rotating disc.

Keywords: Creep, rotating, functionally graded, thickness, temperature, stress, strain rate

MSC 2010: 74A10, 74E30, 74G15, 74Q15

1 Introduction

A disc that is rotating at a high angular speed and subjected to high temperature is a crucial structural element in many different systems, including jet engines, ships, turbine rotors, automobile brakes in sports cars, and many others ([12, 18, 1, 2]). The disc is sensitive to creep in certain of these applications, such as turbine rotors and disc brakes, where it is subjected to extreme heat and mechanical loadings. To reduce the stresses and deformations in these application caused by creep, the researchers have used ceramic-reinforced aluminum matrix composites, such as Al-SiC ([16] and [13]) and functionally graded material (FGM), a novel type of metal matrix composite have been created ([6, 14, 8, 11, 4, 3]). FGM is capable of withstanding extremely harsh thermo-mechanical loading conditions. FGMs are employed in modern aircraft and aerospace engine components, computer circuit boards, thermal barrier coatings, and other applications because of their exceptional characteristics. Researchers have used various disc profiles to further enhance the performance of rotating disc ([15] and [21]).

Acknowledgement: The authors thank the anonymous referee for helpful comments.

Tania Bose, Renu Bala, Madhu Aneja, Chitkara University Institute of Engineering and Technology, Chitkara University, Chandigarh-Patiala National Highway, 140401 Rajpura, Punjab, India, e-mails: tania.bose@chitkara.edu.in, renu.bala@chitkara.edu.in, madhu.aneja@chitkara.edu.in

https://doi.org/10.1515/9783111313634-013

According to the literature, there has been some study done on the analysis of creep in rotating discs with varied thickness. Singh et al. [22] has compared rotating Al–SiC discs with varied thickness profiles for their steady state creep response. It was seen the variable thickness disc was found to exhibit a better creep response in terms of loads and strain rates in comparison to a constant thickness disc. Garg et al. [10] investigated on the steady state creep behavior of a spinning FGM disc with linearly variable thickness. The impact of changing the disc thickness gradient on the stresses and strain rates in the FGM disc has been studied. With the disc's thickness gradient increasing, it was seen that the radial and tangential stresses that were created in the disc diminished across the thickness profile. Singh et al. [20] estimated the creep response of a revolving Al-SiCp disc with an exponential SiC_p gradient. With the same amount of reinforcement, the findings were calculated for an exponential FGM disc and compared to those for hyperbolic FGM and non-FGM discs. The highest variation was seen for the exponential FGM disc. The imposition of reinforcement gradient raised the radial stress throughout, the tangential stress close to the inside radius, but lowered the tangential stress towards the outside radius.

In the present framework, a mathematical model is constructed to compare the creep response in a rotating FGM disc of $AlSiC_p$ with non-linear thickness geometry and SiC particles distributed non-linearly along radial direction with a rotating non-FGM disc having uniform thickness subjected to constant temperature. The threshold stress based law describes the creep behavior of the disc material, while the yielding is assumed to follow Tresca criterion.

2 Disc geometry

The creep analysis has been performed on two different disc profiles, i. e., Disc 1 is a rotating composite disc with uniform thickness and Disc 2 is a functionally graded disc having non-linearly distributed SiC particles along radial direction with non-linear thickness. Both the disc are subjected to uniform temperature of 623 K. The inner and outer radii of both the disc are taken similar to those reported in literature [5], i. e., inner radius, $r_a = 0.03175$ m and outer radius, $r_b = 0.1524$ m.

For Disc 1, volume fraction, V, of the reinforcement is considered to be 20 vol.% fraction of the constituent material $AlSiC_p$.

For Disc 2, volume profile is given by:

$$\mathrm{vol}(r) = \alpha + \beta/r, \quad r_a \le r \le r_b, \tag{13.1}$$

where

$$\alpha = \frac{r_b.\mathrm{vol}(r_b) - r_a.\mathrm{vol}(r_a)}{b - a} \tag{13.2}$$

and

$$\beta = r_a.r_b.\left(\frac{\mathrm{vol}(r_a) - \mathrm{vol}(r_b)}{b - a}\right). \tag{13.3}$$

Here $\mathrm{vol}(r_a) = 30\mathrm{vol}\%$ at r_a and $\mathrm{vol}(r_b) = 10\mathrm{vol}\%$ at r_b respectively.

The density of composites as per law of mixture:

$$\rho(r) = \rho_{\mathrm{Al}} + (\rho_{\mathrm{SiC}} - \rho_{\mathrm{Al}})\frac{\mathrm{vol}(r)}{100}, \tag{13.4}$$

where $\rho_{\mathrm{SiC}} = 3210\ \mathrm{kg/m^3}$ and $\rho_{\mathrm{Al}} = 2713\ \mathrm{kg/m^3}$ are the respective densities [9]. Using (13.1) and (13.4),

$$\rho(r) = \rho_{\mathrm{Al}} + (\rho_{\mathrm{SiC}} - \rho_{\mathrm{Al}})\frac{\alpha + \beta/r}{100}. \tag{13.5}$$

The thickness profile of Disc 2, $t(r)$ varies non-linearly with radius as

$$t(r) = t_{r_b}\left(\frac{r}{r_b}\right)^k, \tag{13.6}$$

where t_{r_b} and k are thickness at r_b and thickness gradient of disc respectively.

The average particle content in Disc 2 with $t(r)$ as variable thickness is given as

$$\int_a^b 2\pi r.t(r).\mathrm{vol}(r)dr = \mathrm{vol}_{\mathrm{avg}}\pi.t(r)(b^2 - a^2). \tag{13.7}$$

Eqs. (13.1) and (13.6) gives

$$\mathrm{vol}_{\mathrm{avg}} = 2\frac{b^k}{r^k}\left[\frac{\alpha}{k+2}(b^{k+2} - a^{k+2}) + \frac{\beta}{k+1}(b^{k+1} - a^{k+1})\right]. \tag{13.8}$$

3 Analysis of creep and creep parameters

The steady state creep of Al-SiC$_p$ composite is given by [19] law as:

$$\dot{\varepsilon} = [M(\bar{\sigma} - \sigma_0)]^8, \tag{13.9}$$

where $\dot{\varepsilon}$, $\bar{\sigma}$ and σ_0 have usual meaning as in [7].

Values of M and σ_0 (13.10), (13.11) are procured from [17] results for Al-SiC$_p$ composite under uniaxial loading.

In this study, $p = 1.7\ \mu\mathrm{m}$ and regression equations taking M, σ_0 as functions of p, $V(r)$ and T are achieved using Data Fit software given as:

$$M(r) = p^{0.2112}.T^{4.89}.\mathrm{vol}(r)^{-0.591}.e^{-34.91}, \tag{13.10}$$

$$\sigma_0(r) = p^{-0.02050}.T^{0.0378}.\mathrm{vol}(r)^{1.033}.e^{-4.9695}. \tag{13.11}$$

4 Mathematical modeling of creep

Creep modeling is performed by making the following assumptions by taking principal stresses in r, θ and z directions.

(a) Steady state condition of stress exists.
(b) Elastic deformations are ignored as they are small in magnitude.
(c) At any radius of the disc, biaxial stress prevails.

The generalized constitutive equations for biaxial stress are:

$$\dot{\epsilon}_r = \frac{\dot{\bar{\epsilon}}}{2\bar{\sigma}}[2\sigma_r - \sigma_\theta], \tag{13.12}$$

$$\dot{\epsilon}_\theta = \frac{\dot{\bar{\epsilon}}}{2\bar{\sigma}}[2\sigma_\theta - \sigma_r], \tag{13.13}$$

$$\dot{\epsilon}_z = -\frac{\dot{\bar{\epsilon}}}{2\bar{\sigma}}[\sigma_r + \sigma_\theta], \tag{13.14}$$

where

$$\bar{\sigma} = \left[\sigma_r^2 + \sigma_\theta^2 - \sigma_r.\sigma_\theta\right]^{1/2} \tag{13.15}$$

and $\dot{\epsilon}_r$, $\dot{\epsilon}_\theta$, $\dot{\epsilon}_z$ are the strain rates and σ_r, σ_θ are stresses. These equations get transformed into:

$$\dot{\epsilon}_r = \frac{d\dot{u}_r}{dr} = \frac{\dot{\bar{\epsilon}}.(2z-1)}{2(z^2-z+1)^{1/2}}, \tag{13.16}$$

$$\dot{\epsilon}_\theta = \frac{\dot{u}_r}{r} = \frac{\dot{\bar{\epsilon}}.(2-z)}{2(z^2-z+1)^{1/2}}, \tag{13.17}$$

$$\dot{\epsilon}_z = -(\dot{\epsilon}_r + \dot{\epsilon}_\theta), \tag{13.18}$$

where $z = \frac{\sigma_r(r)}{\sigma_\theta(r)}$ and u_r is the radial displacement.

For rotating FGM disc 2, equation of motion is given as

$$\frac{d}{dr}[t(r).r.\sigma_r(r)] - t(r).\sigma_\theta(r) + \rho(r)\omega^2.r^2.t(r) = 0, \tag{13.19}$$

where angular velocity of the disc is $\omega = 15000\text{rpm}$.

The above equations are solved to get $\sigma_\theta(r)$ as:

$$\sigma_\theta(r) = \left[\frac{\int_b^a M(r).\sigma_{\theta_{\text{avg}}}dr - \int_b^a M(r).\psi_2(r)dr}{\int_b^a \psi_1(r)}\right]\frac{\psi_1(r)}{M(r)} + \psi_2(r), \tag{13.20}$$

where

$$\psi_1(r) = \frac{\psi(r)}{(z^2 - z + 1)^{1/2}}, \quad (13.21)$$

$$\psi_2(r) = \frac{\sigma_0(r)}{(z^2 - z + 1)^{1/2}}, \quad (13.22)$$

$$[\psi(r)]^8 = \frac{2(z^2 - z + 1)^{1/2}}{r(2 - z)} \exp \int_a^r \frac{\Psi(r)}{r} dr, \quad (13.23)$$

and

$$\Psi(r) = \frac{(2z - 1)}{(2 - z)}. \quad (13.24)$$

After tangential stress distribution, values of $\sigma_r(r)$ are obtained from eq. (13.20):

$$\sigma_r(r) = \frac{\sigma_\theta.A_c - \rho(r).\omega^2.I_m}{t(r).r}, \quad (13.25)$$

where A_c is the area of cross-section of disc with radius r given by,

$$A_c = \int_r^a t(r)dr, \quad (13.26)$$

I_m is the moment of inertia of disc with radius r given by

$$I_m = \int_r^a t(r).r^2 dr, \quad (13.27)$$

σ_θ and σ_r are calculated by (13.20) and (13.25) at any radius of rotating disc. Equations (13.12), (13.13) and (13.14) are used to evaluate strain rates in r, θ and z directions.

5 Results and discussion

5.1 Validation of program

The results for a revolving steel disc were compared with the experimental data from [23] for the same type of disc working under the same conditions so as to validate the generated MATLAB program. The same graph is obtained as in the work of [7].

Figure 13.1 shows the distribution of radial stress along radial distance. As r increases from inner to outer radius, σ_r follow a parabolic path. However, σ_r in Disc 2 are lower in comparison to σ_r of Disc 1. This is due to the difference in profile of the discs taken in the study. Disc 1 assumes maximum value of σ_r as 37.22 MPa and Disc 2 attains 36.84 MPa value at a radial distance of 0.0800 m.

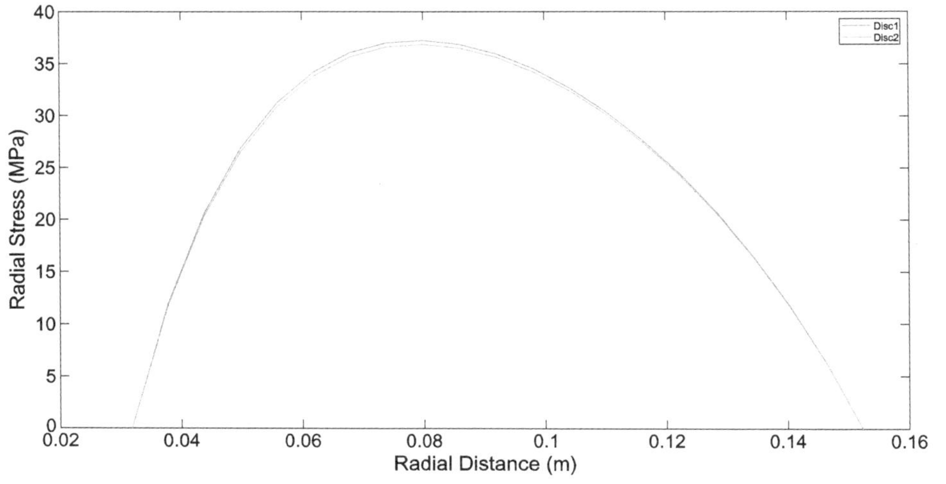

Figure 13.1: Distribution of radial stress in different discs.

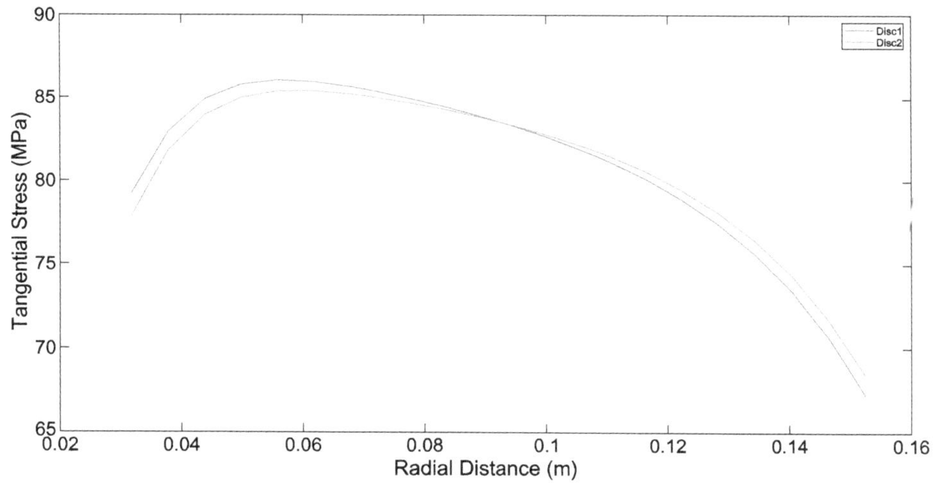

Figure 13.2: Distribution of tangential stress in different discs.

The distribution of tangential stress along radial distance is represented in Figure 13.2. σ_θ values of Disc 2 are lower in comparison to Disc 1 till the middle value and then the trend changes. At 0.03175 m, σ_θ value is 79.21 MPa, the value increases to 86.04 MPa at 0.0559 m and then decreases to 67.24 MPa at 0.1524 m for Disc 1. However, for Disc 2, the σ_θ is 77.85 MPa at 0.03175 m, then increases to 85.39 MPa at 0.0619 m and then decreases to 68.43 MPa at 0.1524 m.

Figure 13.3 shows the distribution of radial strain rate along radial distance. The graph clearly shows that Disc 1 has lower radial strain values in comparison to Disc 2.

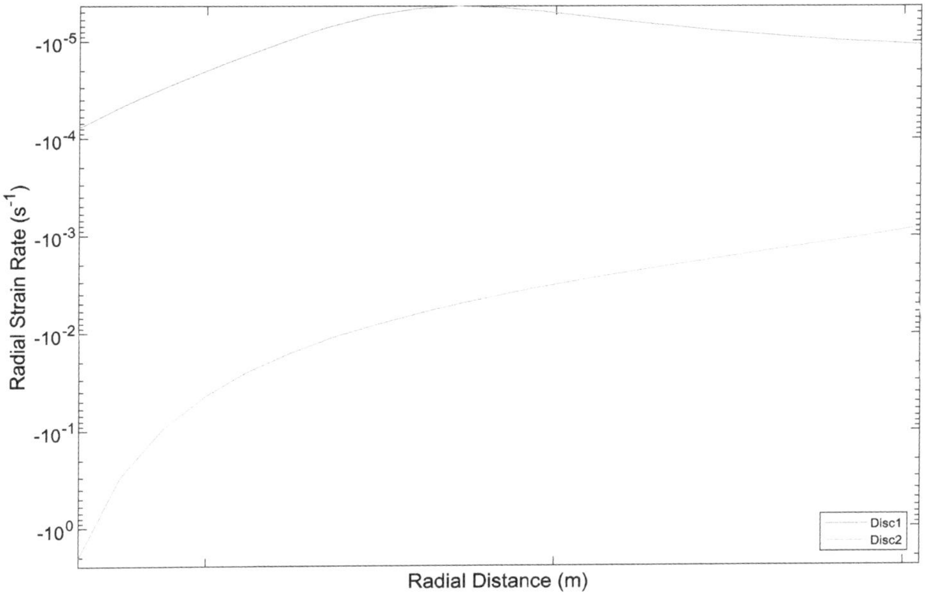

Figure 13.3: Distribution of radial strain rate in different discs.

At $r = 0.03175$ m, the value of $\epsilon_r = -7.74 \times 10^{-5}\,\mathrm{s}^{-1}$ for Disc 1, while for Disc 2, the value of $\epsilon_r = -2.011\,\mathrm{s}^{-1}$. At the other radial end, the value of $\epsilon_r = -1.086 \times 10^{-5}\,\mathrm{s}^{-1}$ for Disc 1, while for Disc 2, the value of $\epsilon_r = -8.1875\,\mathrm{s}^{-1}$.

Figure 13.4 shows the distribution of tangential strain rate along radial distance. The graph clearly shows that Disc 1 has lower tangential strain values in comparison to Disc 2. At $r = 0.03175$ m, the value of $\epsilon_\theta = -1.54 \times 10^{-4}\,\mathrm{s}^{-1}$ for Disc 1, while for Disc 2, the value of $\epsilon_\theta = 4.02\,\mathrm{s}^{-1}$. At the other radial end, the value of $\epsilon_\theta = 2.17 \times 10^{-5}\,\mathrm{s}^{-1}$ for Disc 1, while for Disc 2, the value of $\epsilon_\theta = 1.63 \times 10^{-3}\,\mathrm{s}^{-1}$.

6 Conclusion

Radial stress in non-linear FGM disc decreases as compared to uniform composite disc. The tangential stress decreases in inner part for non-linear FGM disc, but follows opposite trend for outer part of the disc. The radial and tangential strain rates in FGM disc with variable thickness are more in comparison to composite disc with uniform thickness. Thus, by employing FGM disc having variable thickness, the starin rate distributions become more gradual, thus leading to lesser distortion of FGM disc.

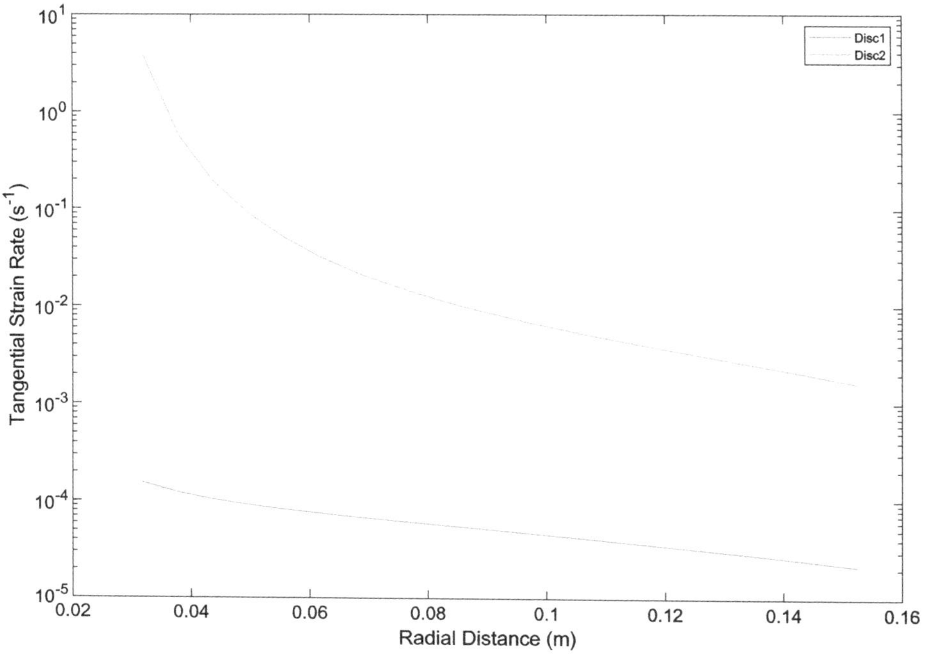

Figure 13.4: Distribution of tangential strain rate in different discs.

Bibliography

[1] G. Altan, M. Topcu, N. B. Bektas and B. D. Altan, Elastic-plastic thermal stress analysis of an aluminum composite disc under parabolic thermal load distribution, *J. Mech. Sci. Technol.* **22** (2008), 2318–2327.

[2] M. Bayat, M. Saleem, B. B. Sahari, A. M. S. Hamouda and E. Mahdi, Mechanical and thermal stresses in a functionally graded rotating disk with variable thickness due to radially symmetry loads, *Int. J. Press. Vessels Piping* **86** (2009), 357–372.

[3] T. Bose, N. Chamoli, M. Rattan and K. Gupta, Effect of anisotropy on creep of rotating functionally graded disc with varying thickness, *IOP Conf. Ser., Mater. Sci. Eng.* **1033**(1) (2021), 012061.

[4] T. Bose, K. Gupta, M. Rattan and N. Chamoli, Study of thermal gradation on creep deformation of non-linear varying functionally graded rotating disc, *IOP Conf. Ser., Mater. Sci. Eng.* **1033**(1) (2021), 012084.

[5] T. Bose and M. Rattan, Effect of thermal gradation on steady state creep of functionally graded rotating disc, *Eur. J. Mech. A, Solids* **67** (2018), 169–176.

[6] T. Bose, M. Rattan and N. Chamoli, Creep analysis of an isotropic linearly decreasing functionally graded rotating disc at linearly increasing temperature, *Int. J. Appl. Sci.-Res. Rev.* **3**(1) (2016), 122–129.

[7] T. Bose, M. Rattan and N. Chamoli, Steady state creep of isotropic rotating composite disc under thermal gradation, *Int. J. Appl. Mech.* **9**(6) (2017), 1750077.

[8] T. Bose, M. Rattan, N. Chamoli and K. Gupta, Investigation into steady state creep of AlSiC cylinder on account of residual stress with internal and external pressure, *AIP Conf. Proc.* **2357**(1) (2022), 110001.

[9] T. W. Clyne and P. J. Withers, *An Introduction to Metal Matrix Composites*, Cambridge University Press, p. 479, 1993.

[10] M. Garg, B. S. Salaria and V. K. Gupta, Effect of disc geometry on the steady state creep in a rotating disc made of functionally graded materials, *Mater. Sci. Forum* **736** (2013), 183–191.

[11] K. Gupta, T. Bose, M. Rattan and N. Chamoli, Modeling creep on account of residual stress of thermally graded rotating disc, *ECS Trans.* **107**(1) (2022), 9805.
[12] V. K. Gupta, S. B. Singh, H. N. Chandrawat and S. Ray, Creep behavior of a rotating functionally graded composite disc operating under thermal gradient, *Metall. Mater. Trans. A* **35** (2004), 1381–1391.
[13] V. K. Gupta, S. B. Singh, H. N. Chandrawat and S. Ray, Modeling of creep behavior of a rotating disc in the presence of both composition and thermal gradients, *J. Eng. Mater. Technol.* **127** (2005), 97–105.
[14] H. Jahed, B. Farshi and J. Bidabadi, Minimum weight design of inhomogeneous rotating discs, *Int. J. Press. Vessels Piping* **82** (2005), 35–41.
[15] K. Khanna, V. K. Gupta and S. P. Nigam, Creep analysis in functionally graded rotating disc using Tresca criterion and comparison with von-mises criterion, *Mater. Today Proc.* **4** (2017), 2431–2438.
[16] Z. Y. Ma and S. C. Tjong, Creep deformation characteristics of discontinuously reinforced aluminium-matrix composites, *Compos. Sci. Technol.* **61** (2001), 771–786.
[17] A. B. Pandey, R. S. Mishra and Y. R. Mahajan, Steady state creep behavior of silicon carbide particulate reinforced aluminum composites, *Acta Metall. Mater.* **40**(8) (1992), 2045–2052.
[18] M. Rattan, T. Bose and N. Chamoli, Effect of linear thermal gradient on steady-state creep behavior of isotropic rotating disc, *Int. J. Mech. Mechatron. Eng.* **11**(5) (2017), 1074–1080.
[19] O. D. Sherby, R. H. Klundt and A. K. Miller, Flow stress, subgrain size and subgrain stability at elevated temperature, *Metall. Trans. A* **8** (1977), 843–850.
[20] R. Singh, R. K. Saxena, K. Khanna and V. K. Gupta, Modelling creep response of a functionally graded rotating disc with exponential reinforcement gradient, *Mater. Today Proc.* **18** (2019), 3891–3899.
[21] R. Singh, R. K. Saxena, K. Khanna and V. K. Gupta, Assessment of creep in composite disc having exponential, hyperbolic and uniform thickness profiles, *Mater. Today Proc.* **26**(2) (2020), 1972–1976.
[22] R. Singh, R. K. Saxena, K. Khanna and V. K. Gupta, Creep response of rotating composite discs having exponential, hyperbolic, linear and constant thickness profiles, *Def. Sci. J.* **70**(3) (2020), 292–298.
[23] A. M. Wahl, G. O. Sankey, M. J. Manjoine and E. Shoemaker, Creep tests of rotating disks at elevated temperature and comparison with theory, *J. Appl. Mech.* **76** (1954), 225–235.

M. K. Gupta and Abha Sahu

On the $h(hv)$-torsion tensor and the v-curvature tensor of Matsumoto change with an h-vector

Abstract: In the present paper, we have considered a Finsler space characterized by h-Matsumoto change and obtained condition for C-reducibility. We have derived certain geometric properties of v-curvature tensor for such change. We have also obtained the T-tensor for the transformed space and studied the Finsler space with T'-condition and semi-T'-condition.

Keywords: Matsumoto change, Finsler space, T-tensor, torsion tensor, curvature tensor

MSC 2010: 53B40, 53C60

1 Introduction

M. Matsumoto first established the concept of (α,β)-metrics in 1972. Many authors [2, 1, 8] have investigated Finsler space with such metric. Although it is an old concept, but an important part of Finsler geometry and its applications in physics. The (α,β)-metric, which emerge iteratively in physical sciences and is written in the form of $F = \alpha\,\psi(s)$, $s = \frac{\beta}{\alpha}$, where α is a Riemannian metric and $\beta = b_i(x)y^i$ is differential 1-form on manifold M, are one major type of Finsler metrics. Matsumoto metric (or slope metric) was developed by M. Matsumoto to determine how long it takes to climb or drop any particular path on a slope [7]. Understanding the geometric meanings of various quantities and their effects on global geometric structures is one of the most challenging problems we can see in Finsler geometry.

The h-vector field in the Finsler geometry was first introduced by H. Izumi [4] and is defined as
(i) $b_i|_j = 0$,
(ii) $LC^h_{ij}b_h = \rho h_{ij}$; $\rho \neq 0$.

Recently, we [3] have considered Finsler space F^n which undergoes Matsumoto change with an h-vector; that is, $L(x,y) \to \overline{L}(x,y) = \frac{L^2(x,y)}{L(x,y)-\beta(x,y)}$, where $\beta(x,y) = b_i(x,y)y^i$ and b_i is an h-vector. We have obtained some geometric properties for this transformation.

In this paper, we make some observations regarding the v-curvature tensor and some special type of Finsler spaces like $S3$-like, $S4$-like, quasi-reducible, C-reducible,

M. K. Gupta, Abha Sahu, Department of Mathematics, Guru Ghasidas Vishwavidyalaya, 495009 Bilaspur, C.G., India, e-mails: mkgiaps@gmail.com, abhasahu118@gmail.com

https://doi.org/10.1515/9783111313634-014

$C2$-like Finsler spaces. We have also obtained the transformed T-tensor under this change. This paper is arranged as follows: Section 2, contains some essential geometric objects of Finsler metrics, as well as notations and conventions. We have dealt with v-curvature tensor for the transformed Finsler space in Section 3. In Section 4, we have computed the T-tensor of the associated Finsler space.

2 Preliminaries

Consider $F^n = (M^n, L)$ be an n-dimensional smooth Manifold with Finsler metric function $L(x,y)$. Let the following change of Finsler space $\overline{F}^n = (M^n, \overline{L})$, which will be referred to as h-Matsumoto change:

$$L(x,y) \to \overline{L}(x,y) = \frac{L^2(x,y)}{L(x,y) - \beta(x,y)}. \tag{14.1}$$

The normalized supporting element, metric tensor, angular metric tensor and Cartan tensor are defined by $l_i = \dot{\partial}_i L$, $g_{ij} = \frac{1}{2}\dot{\partial}_i\dot{\partial}_j L^2$, $h_{ij} = L\dot{\partial}_i\dot{\partial}_j L$ and $C_{ijk} = \frac{1}{2}\dot{\partial}_k g_{ij}$ respectively. The Cartan connection for the Finsler space F^n is given by $(F^i_{jk}, N^i_j, C^i_{jk})$. The h-covariant derivative $T_{ij|k}$ and v-covariant derivative $T_{ij}|_k$ of the tensor T_{ij} with respect to the Cartan connection, are given as

$$T_{ij|k} = \partial_k T_{ij} - N^r_k \dot{\partial}_k T_{ij} - T_{ir}F^r_{jk} - T_{jr}F^r_{ik},$$
$$T_{ij}|_k = \dot{\partial}_k T_{ij} - T_{ir}C^r_{jk} - T_{jr}C^r_{ik},$$

respectively.

The normalized supporting element and the angular metric tensor of the Finsler space $\overline{F}^n$, are calculated as [3]

$$\overline{l}_i = \frac{\tau}{(\tau-1)} l_i + \frac{\tau^2}{(\tau-1)^2} m_i, \tag{14.2}$$

and

$$\overline{h}_{ij} = \frac{\tau^2(\tau + \rho\tau - 2)}{(\tau-1)^3} h_{ij} + \frac{2\tau^4}{(\tau-1)^4} m_i m_j, \tag{14.3}$$

where $m_i = b_i - \frac{l_i}{\tau}$, $\tau = \frac{L}{\beta}$.

The metric tensor $\overline{g}_{ij}$ is given by

$$\overline{g}_{ij} = p\, g_{ij} + p_1\, l_i l_j + p_2\, (m_i l_j + m_j l_i) + p_3\, m_i m_j, \tag{14.4}$$

where

$$p = \frac{\tau^2(\tau + \rho\tau - 2)}{(\tau - 1)^3}, \quad p_1 = \frac{\tau^2(1 - \rho\tau)}{(\tau - 1)^3}, \quad p_2 = \frac{\tau^3}{(\tau - 1)^3}, \quad p_3 = \frac{3\tau^4}{(\tau - 1)^4}.$$

The inverse metric g^{ij} of the metric g_{ij} is computed as

$$\overline{g}^{ij} = q\, g^{ij} + q_1\, l^i l^j + q_2\,(l^i m^j + m^i l^j) + q_3\, m^i m^j, \tag{14.5}$$

where

$$q = \frac{1}{p}, \quad q_1 = \frac{-1}{2}\left[\frac{p_1 p_3 - p_2^2}{(p_1 + p)p_3 - p_2^2} + \frac{2p^2 p_2^2 p_3}{(3p + 2p_3 m^2)\{(p_1 + p)p_3 - p_2^2\}^2}\right],$$
$$q_2 = \frac{-2p_2 p_3}{(3p + 2p_3 m^2)\{(p_1 + p)p_3 - p_2^2\}}, \quad q_3 = \frac{-2p_3}{p(3p + 2p_3 m^2)}.$$

The Cartan tensor $\overline{C}_{ijk}$ of the transformed Finsler space $\overline{F}^n$ can be written as

$$\overline{C}_{ijk} = p\, C_{ijk} + V_{ijk}, \tag{14.6}$$

where

$$V_{ijk} = K_1(h_{ij} m_k + h_{jk} m_i + h_{ki} m_j) + K_2 m_i m_j m_k$$

and

$$K_1 = \frac{(\tau + 3\rho\tau - 4)\tau^3}{2L(\tau - 1)^4}, \quad K_2 = \frac{6\tau^4}{\beta(\tau - 1)^5}.$$

The associated Cartan tensor is written as [3]

$$\overline{C}^i_{jk} = C^i_{jk} + M^i_{jk}, \tag{14.7}$$

where

$$M^i_{jk} = q\, K_1(m_k h^i_j + m_j h^i_k) + (q_2\, l^i + q_3\, m^i)\left\{2K_1 m_j m_k + \frac{p}{L}\rho\, h_{jk}\right\}$$
$$+ \{q\, m^i + (q_2\, l^i + q_3\, m^i) m^2\}(K_2 m_j m_k + K_1 h_{jk}).$$

3 *C*-Reducibility

The concept of C-reducibility in Finsler geometry was introduced by M. Matsumoto. He gave two examples of (α, β)-metric, first is the Finsler metric for electromagnetic field and second is the Kropina metric, and both the spaces are C-reducible [5]. Here, we will study about the C-reducibility of h-Matsumoto change.

Contracting i and k in equation (14.7), we obtain

$$\overline{C}_j = C_j + \theta m_j, \tag{14.8}$$

where

$$\theta = \left\{(n+1)q K_1 + 3m^2 q_3 K_1 + K_2 m^2 (q + q_3 m^2) + \frac{p\rho}{L} q_3\right\}. \tag{14.9}$$

Remark. (i) $C_i m^i = \frac{\rho}{L}(n-1)$ and (ii) $C_{ijk} m^k = \frac{\rho}{L} h_{ij}$.

Contracting equation (14.8) with $\overline{g}^{ij}$, we obtain

$$\overline{C}^i = q C^i + \left\{(q + q_3 m^2)\theta + \frac{\rho}{L}(n-1) q_3\right\} m^i + \left\{\frac{\rho}{L}(n-1) q_3 + \theta q_3 m^2\right\} l^i. \tag{14.10}$$

The magnitude $\overline{C}$ of $\overline{C}^i$ is given by

$$\overline{C}^2 = q C^2 + \phi, \tag{14.11}$$

where

$$\phi = \left\{\frac{2\rho}{L}(n-1)\theta + \theta^2 m^2\right\}(q + q_3 m^2) + \frac{2\rho^2}{L^2}(n-1)^2 q_3.$$

Definition 1 ([6]). If the Cartan tensor C_{ijk} can be written as $C_{ijk} = W_{ij} C_k + W_{jk} C_i + W_{ik} C_j$, for some symmetric and indicatory tensor W_{ij}, then the Finsler space F^n $(n \geq 3)$ is called quasi-C-reducible.

From equation (14.6), we can easily see

$$\overline{C}_{ijk} = p\, C_{ijk} + \frac{1}{3} \sum_{ijk} \{(3K_1 h_{ij} + K_2 m_i m_j) m_k\},$$

where $\sum_{ijk}$ denotes cyclic permutation of indices i, j, k and their addition.

Now, let us rewrite above equation as

$$\overline{C}_{ijk} = \frac{1}{3\theta} \sum_{ijk} \{(3K_1 h_{ij} + K_2 m_i m_j)\overline{C}_k\} + \frac{1}{3\theta} \sum_{ijk} \{\theta p C_{ijk} - (3K_1 h_{ij} + K_2 m_i m_j)\}. \tag{14.12}$$

Proposition 1. *Under the h-Matsumoto change, the Cartan tensor of transformed space can be written as*

$$\overline{C}_{ijk} = \sum_{ijk} \{\overline{W}_{ij} \overline{C}_k\} + w_{ijk},$$

where

$$\overline{W}_{ij} = \frac{1}{3\theta}\{(3K_1 h_{ij} + K_2 m_i m_j)\} \quad \textit{is symmetric and indicatory,}$$

and

$$w_{ijk} = \frac{1}{3\theta}\sum_{ijk}\{\theta p C_{ijk} - (3K_1 h_{ij} + K_2 m_i m_j)\}.$$

Thus, we have

Theorem 1. *If the tensor w_{ijk} vanishes, then the Finsler space $\overline{F}^n$ has the property of quasi C-reducible.*

Definition 2. A Finsler space of dimension $n \geq 2$ is said to be C-reducible, if the Cartan tensor can be written in the form

$$C_{ijk} = \frac{1}{(n+1)}(h_{ij}C_k + h_{jk}C_i + h_{ki}C_j). \tag{14.13}$$

Let us define a tensor $K_{ijk} = C_{ijk} - \frac{1}{(n+1)}(h_{ij}C_k + h_{jk}C_i + h_{ki}C_j)$, then K_{ijk} is symmetric and indicatory. Now,

$$\overline{K}_{ijk} = \overline{C}_{ijk} - \frac{1}{(n+1)}(\overline{h}_{ij}\overline{C}_k + \overline{h}_{jk}\overline{C}_i + \overline{h}_{ki}\overline{C}_j),$$

which on compilation gives us

$$\overline{K}_{ijk} = p\,K_{ijk} + d_{ijk}, \tag{14.14}$$

where

$$d_{ijk} = \frac{1}{(n+1)}\sum_{ijk}\Bigg[\bigg\{(K_1(n+1) - p\,\theta)h_{ij} + \bigg(K_2(n+1) - \frac{2p_3}{3}\theta\bigg)m_i m_j\bigg\}\, m_k - \frac{2p_3}{3}m_i m_j C_k\Bigg].$$

Thus, we can state that

Theorem 2. *If the tensor d_{ijk} vanishes from equation* (14.14), *then we have a C-reducible Finsler space $\overline{F}^n$.*

Definition 3 ([6]). If the Cartan tensor C_{ijk} can be written as $C^2 C_{ijk} = C_i C_j C_k$, then the Finsler space $\overline{F}^n$ of dimension $n > 2$ is called $C2$-like.

Let us define a tensor $\eta_{ijk} = C^2 C_{ijk} - C_i C_j C_k$, which is symmetric and indicatory. Now,

$$\overline{\eta}_{ijk} = \overline{C}^2\,\overline{C}_{ijk} - \overline{C}_i\,\overline{C}_j\,\overline{C}_k,$$

which on solving gives us

$$\overline{\eta}_{ijk} = \eta_{ijk} + I_{ijk}, \tag{14.15}$$

where

$$\begin{aligned} I_{ijk} &= p\,\phi\, C_{ijk} + (q\,C^2 + \phi)V_{ijk} - \theta(m_i C_j C_k + m_j C_i C_k + m_k C_j C_i) \\ &\quad - \theta^2(m_i m_j C_k + m_j m_k C_i + m_i m_k C_i) - \theta^3 m_i m_j m_j. \end{aligned}$$

Theorem 3. *If the tensor I_{ijk} in equation* (14.15) *vanishes, then $\overline{F}^n$ is C2-like.*

4 *S*-Tensor

In a Finsler space, the v-curvature tensor S_{hijk} with respect to the Cartan connection is defined by [6]

$$S_{hijk} = C^r_{hk} C_{ijr} - C^r_{hj} C_{ikr}.$$

From equation (14.2) and (14.3)

$$\begin{aligned} \overline{C}^r_{hk}\overline{C}_{ijr} &= (C^r_{hk} + M^r_{hk})(p\,C_{ijr} + V_{ijr}) \\ &= p\,C^r_{hk} C_{ijr} + K_1(C_{ijk} m_h + C_{jkh} m_i + C_{khi} m_j + C_{jkh} m_i) \\ &\quad + I_1(h_{ij} m_h m_k + h_{hk} m_i m_j) + I_2 m_i m_j m_h m_k + J h_{ij} h_{hk} \\ &\quad + A K_1^2 (h_{jh} m_i m_k + h_{ih} m_j m_k + h_{jk} m_i m_h + h_{ik} m_j m_h), \end{aligned} \tag{14.16}$$

where

$$\begin{aligned} I_1 &= \left(2K_1^2 + K_1 K_2 m^2 + \frac{K_2}{L} p\rho\right)(q + q_3 m^2) + \frac{2p\,\rho}{L} K_1 q_3, \\ I_2 &= (4K_1 K_2 + K_2 m^2)(q + q_3 m^2) + 4K_1^2 q_3, \\ J &= \left(\frac{2}{L} K_1 p\rho + {K_1}^2 m^2\right)(q + q_3 m^2) + \frac{\rho^2}{L^2} p^2\, q_3. \end{aligned}$$

The v-curvature tensor of Cartan connection of the transformed space can be written as

$$\begin{aligned} \overline{S}_{hijk} &= \overline{C}^r_{hk}\overline{C}_{ijr} - \overline{C}^r_{hj}\overline{C}_{ikr} \\ &= pS_{hijk} + (I_1 - q\,K_1^2)(h_{ij} m_h m_k + h_{hk} m_i m_j - h_{ik} m_h m_j - h_{hj} m_i m_k) \\ &\quad + J(h_{ij} h_{kh} - h_{ik} h_{hj}). \end{aligned} \tag{14.17}$$

Theorem 4. *The relation between the v-curvature tensor S_{hijk} of Finsler space F^n and $\overline{F}^n$ is given by* (14.17).

After re-solving equation (14.17), we have

$$\overline{S}_{hijk} = A_{hijk} + J\,q^2\,(\overline{h}_{ij}\overline{h}_{kh} - \overline{h}_{ik}\overline{h}_{hj}), \tag{14.18}$$

where

$$A_{hijk} = p\,S_{hijk} + q\,(\overline{h}_{ij}m_h m_k + \overline{h}_{hk}m_i m_j - \overline{h}_{ik}m_h m_j - \overline{h}_{hj}m_i m_k)\left(I_1 - q\,K_1^2 - \frac{2}{3}p_3\,qJ\right).$$

Definition 4 ([6]). A Finsler space is said to be $S3$-like, if the S-Tensor of a Finsler space of dimension $n > 3$ has the form $L^2 S_{hijk} = S(h_{ij}h_{hk} - h_{ik}h_{jh})$.

Therefore, we have

Theorem 5. *Under the h-Matsumoto transformation of the associated Finsler space,* $\overline{F}^n$ *is S3-like if the tensor* A_{hijk} *vanishes in equation* (14.18).

Equation (14.17) can be rewritten as

$$\begin{aligned}\overline{S}_{hijk} = p\,S_{hijk} + \Big[h_{kh}\Big\{(I_1 - q\,K_1^2)m_i m_j + \frac{J}{2}h_{ij}\Big\} \\ + h_{ij}\Big\{(I_1 - q\,K_1^2)m_h m_k + \frac{J}{2}h_{hk}\Big\} - j/k\Big],\end{aligned} \tag{14.19}$$

where $-j/k$ means the interchanging the indices j, k and substract within the bracket.

Definition 5 ([6]). A non-Riemannian space $F^n(n > 4)$ is called $S4$-like if the v-curvature tensor S_{hijk} is in the form

$$L^2 S_{hijk} = h_{ij}Y_{hk} + h_{hk}Y_{ij} - h_{ik}Y_{jk} - h_{jh}Y_{ik}, \tag{14.20}$$

where Y_{ij} is a symmetric and indicatory tensor.

In view of equation (14.3), the equation (14.19) becomes

$$\overline{L}^2\,\overline{S}_{hijk} = \frac{p\,\tau^2 L^2}{(\tau-1)^2}S_{hijk} + M_{ij}\overline{h}_{hk} + M_{kh}\overline{h}_{ij} - M_{jh}\overline{h}_{ik} - M_{ik}\overline{h}_{jh}, \tag{14.21}$$

where

$$M_{ij} = \frac{\tau^2 L^2}{(\tau-1)^2}\left[\frac{q^2 J}{2}\,\overline{h}_{ij} + q\left(I_1 - q\,K_1^2 - \frac{2}{3}p_3\,qJ\right)m_i m_j\right].$$

Now, if we consider the Finsler space F^n to be $S4$-like. Then from equation (14.21), we can get

$$\overline{L}^2\overline{S}_{hijk} = H_{ij}\overline{h}_{kh} + H_{hk}\overline{h}_{ij} - H_{ik}\overline{h}_{jh} - H_{jh}\overline{h}_{ik} + B_{hijk} \tag{14.22}$$

where

$$H_{ij} = \frac{\tau^2}{(\tau-1)^2} Y_{ij} + M_{ij},$$

and

$$B_{hijk} = \frac{2p_3\tau^2}{3(\tau-1)^2}\{K_{ik}m_jm_h + K_{jh}m_hm_i - K_{hk}m_im_j - K_{ij}m_hm_k\}.$$

Therefore, we obtain

Theorem 6. *If the Finsler space F^n is S4-like, then the h-Matsumoto changed Finsler space $\overline{F}^n$ is S4-like provided the tensor B_{hijk} vanishes.*

5 *T*-tensor

The T-tensor is constructed by the derivatives of Finsler metric function L with respect to y^i upto the forth order and is defined by Matsumoto [6] as

$$T_{hijk} = LC_{hij}|_k + C_{hij}l_k + C_{hik}l_j + C_{hjk}l_i + C_{ijk}l_h. \tag{14.23}$$

In the theory of Finsler space, the T-tensor is introduced as symmetry and indicatory tensor. The vanishing of the T-tensor is known as T-condition. Transvecting equation (14.23) with tensor g^{hk}, we get

$$T_{ij} = LC_i|_j + C_il_j + C_jl_i, \tag{14.24}$$

where T_{ij} is called the T'-tensor.

Next, we shall consider the T-tensor with h-Matsumoto change. On differentiating Cartan tensor (14.6) with respect to y^k, we obtain

$$\begin{aligned}\dot\partial_k\overline{C}_{hij} &= p\dot\partial_kC_{hij} + 2K_1(C_{hik}m_j + C_{ijk}m_h + C_{hjk}m_i + C_{hij}m_k) + \alpha_1m_im_jm_km_h\\ &- \frac{K_1}{L}\left(\rho-\frac{1}{\tau}\right)(h_{hk}h_{jk} + h_{hj}h_{ik} + h_{ij}h_{hk}) + \frac{K_2}{L}\left(\rho-\frac{1}{\tau}\right)(h_{hi}m_jm_k\\ &+ h_{ij}m_hm_k + h_{hj}m_im_k + h_{hk}m_im_j + h_{jk}m_im_h + h_{ik}m_jm_h)\\ &- \frac{K_1}{L}(h_{hi}f_{kj} + h_{hk}f_{ij} + h_{ik}f_{hj} + h_{ij}f_{hk} + h_{jk}f_{ih} + h_{hj}f_{ik})\\ &+ \frac{K_2}{L}(m_jm_km_hl_i + m_jm_km_il_h + m_jm_im_hl_k + m_im_km_hl_j),\end{aligned} \tag{14.25}$$

where

$$f_{ij} = m_il_j + l_im_j \quad\text{and}\quad \alpha_1 = \frac{30\tau^4}{\beta^2(\tau-1)^6}.$$

Now, we are in a position to find the T-tensor of the transformed space. First, we shall calculate $\overline{C}_{hij}\bar{|}_k$ as follows:

$$\overline{C}_{hij}\bar{|}_k = \dot{\partial}_k\overline{C}_{hij} - \overline{C}^r_{hk}\overline{C}_{ijr} - \overline{C}^r_{jk}\overline{C}_{ihr} - \overline{C}^r_{ik}\overline{C}_{hjr}. \tag{14.26}$$

Using equation (14.16) and (14.25) in above equation, we have

$$\begin{aligned}\overline{C}_{hij}\bar{|}_k &= p(\dot{\partial}_k C_{hij} - C^r_{hk}C_{ijr} - C^r_{jk}C_{ihr} - C^r_{ik}C_{hjr}) - K_1(C_{hik}m_j + C_{ijk}m_h \\ &\quad + C_{hjk}m_i + C_{hij}m_k) + \left\{\frac{K_2}{L}\left(\rho - \frac{1}{\tau}\right) - I_1 - 2qK_1^2\right\}(h_{hi}m_jm_k + h_{ij}m_hm_k \\ &\quad + h_{hj}m_im_k + h_{hk}m_im_j + h_{jk}m_im_h + h_{ik}m_jm_h) - \left[\frac{K_1}{L}\left(\rho - \frac{1}{\tau}\right) - J\right] \\ &\quad \times (h_{hk}h_{jk} + h_{hj}h_{ik} + h_{ij}h_{hk}) + \frac{K_2}{L}(m_jm_km_hl_i + m_jm_km_il_h \\ &\quad + m_jm_im_hl_k + m_im_km_hl_j) + (\alpha_1 - 3I_2)m_im_jm_km_h \\ &\quad - \frac{K_1}{L}(h_{hi}f_{kj} + h_{hk}f_{ij} + h_{ik}f_{hj} + h_{ij}f_{hk} + h_{jk}f_{ih} + h_{hj}f_{ik}).\end{aligned} \tag{14.27}$$

After solving above equation, the T-tensor of the transformed Finsler space can be obtained as

$$\begin{aligned}\overline{T}_{hijk} &= p\,t\,T_{hijk} + T_1(h_{ij}h_{hk} + h_{ih}h_{jk} + h_{ik}h_{jh}) + T_2(C_{hij}m_k + C_{ijk}m_h \\ &\quad + C_{hik}m_j + C_{jhk}m_i) + T_3m_im_jm_km_h + T_4(h_{hi}m_jm_k + h_{ij}m_hm_k \\ &\quad + h_{hj}m_im_k + h_{hk}m_im_j + h_{jk}m_im_h + h_{ik}m_jm_h),\end{aligned} \tag{14.28}$$

where

$$\mu = \frac{\tau}{(\tau-1)}, \quad T_1 = \mu K_2\left(\rho - \frac{1}{\tau}\right) - \mu L(I_1 + 2qK_1^2),$$

$$T_2 = p\mu^2 - \mu K_1 L, \quad T_3 = 9\mu^2 K_2 - 3\mu L I_2, \quad \text{and} \quad T_4 = -\mu K_1\left(\rho - \frac{1}{\tau}\right) - J.$$

Transvecting equation (14.28) by $\overline{g}^{hk}$ and using equation (14.3) gives us

$$\overline{T}_{ij} = \mu\,T_{ij} + \lambda_1\overline{h}_{ij} + \lambda_2 m_im_j + q\,T_2(C_im_j + C_jm_i), \tag{14.29}$$

where

$$\lambda_1 = \left(\frac{2\rho}{L}T_2 + T_1m^2 + T_4\right)(q + q_3m^2) + nAT_4 + \frac{p\rho q}{\tau L},$$

$$\lambda_2 = (q + q_3m^2)T_3m^2 + (n+3)qT_1 + 5T_1q_3m^2 + 2q_3T_4 + \frac{3\rho}{L}q_3T_2 - \frac{2}{3}p_3\,q\lambda_1.$$

Equation (14.29) can be rewritten as

$$\overline{T}_{ij} = \mu\, T_{ij} + \mu_{ij}, \tag{14.30}$$

where

$$\mu_{ij} = \lambda_1 \overline{h}_{ij} + \lambda_2 m_i m_j + q\, T_2(C_i m_j + C_j m_i).$$

"The Finsler space is said to satisfy T'-condition, if the T'-tensor of the Finsler space F^n vanishes identically". Therefore, from equation (14.30), we have

Theorem 7. *Consider the Finsler space F^n with T'-condition, then the transformed Finsler space $\overline{F}^n$ also satisfies the T'-condition if and only if μ_{ij} vanishes.*

Definition 6. If the tensor has the form $T_{ij} = \alpha\, h_{ij}$, where $\alpha = \frac{LC^i|_i}{(n-1)}$, then the Finsler space is called semi-T'-condition [9].

Now, let us assume that the Finsler space F^n satisfies the semi-T'-condition. Then, equation (14.29) becomes

$$\overline{T}_{ij} = (q\,\alpha + \lambda_1)\overline{h}_{ij} + \lambda_{ij}, \tag{14.31}$$

where

$$\lambda_{ij} = \left(\lambda_2 - \frac{2}{3} p_3 q \alpha\right) m_i m_j + q\, T_2(C_i m_j + C_j m_i).$$

Theorem 8. *Let the Finsler space F^n with the semi-T'-condition, then the associated Finsler space $\overline{F}^n$ also satisfies the semi-T'-condition if and only if λ_{ij} vanishes in equation* (14.31).

Bibliography

[1] M. K. Gupta and P. N. Pandey, On hypersurface of a Finsler space with Randers conformal metric, *Tensor* **70**(3) (2008), 229–240.

[2] M. K. Gupta and P. N. Pandey, Finsler space subjected to a Kropina change with an *h*-vector, *Facta Univ., Ser. Math. Inform.* **30**(4) (2014), 513–525.

[3] M. K. Gupta, A. Sahu and S. Sharma, Cartan connection for *h*-Matsumoto change, arXiv:2204.07298.

[4] H. Izumi, Conformal transformations of Finsler spaces II, *Tensor* **33** (1980), 337–359.

[5] M. Matsumoto, On *C*-reducible Finsler spaces, *Tensor* **24** (1972), 29–37.

[6] M. Matsumoto, *Foundations of Finsler Geometry and Special Finsler Spaces*, Kaiseisha Press, Saikawa Otsu, Japan, 1986.

[7] M. Matsumoto, A slope of a mountain is a Finsler surface with respect to a time measure, *J. Math. Kyoto Univ.* **29**(1) (1989), 17–25.

[8] P. N. Pandey and M. K. Gupta, On subspaces of a Finsler space with a special metric, *Bull. Allahabad Math. Soc.* **23**(2) (2008), 263–272.

[9] B. Tiwari, On S-4 like Finsler spaces with semi-T'-condition, *Bull. Calcutta Math. Soc.* **96**(5) (2004), 389–394.

Anup Kumar Behera and Priyanka Agrawal

Software reliability growth model considering power-law testing effort using optimization on hybrid PSO-GWO algorithm

Abstract: Software reliability is a crucial aspect when it comes to assessing the quality of software. To ensure that software is free of errors before its release in the market, it undergoes thorough testing multiple times. Software reliability growth models (SRGMs) are valuable tools for software developers and testers as they help in analyzing stochastic failures and ensure software quality. Over the past four decades, several SRGMs have been proposed, but there is a constant need for upgraded versions in the software industry and research organizations. To address this need, it is essential to incorporate more updated testing effort functions and various fault detection rates into SRGMs to achieve more suitable software than ever before. This study presents a novel SRGM incorporating a power-law testing effort function (PL-TEF) and a non-linear fault detection rate (FDR), which are more practical in a realistic context. This research discusses a model that is compared by optimization techniques, using hybrid PSO-GWO algorithms conducted using a real dataset of failures to validate the model. The findings of the study reveal that the suggested model is the best fit for the data, and the ideal release time for the proposed model is also determined. These outcomes provide favorable results for software developers, offering valuable insights for decision-making processes.

Keywords: NHPP, software reliability, power-law TEF, non-linear FDR, hybrid PSO-GWO

MSC 2010: 94C12

1 Introduction

Software reliability refers to the capacity of a software system to operate as intended and produce accurate outcomes consistently over a specific duration. Several factors influence the reliability of software, including its design, coding, testing, and maintenance. This aspect becomes particularly critical for software systems deployed in aerospace, medical devices, and other safety-critical domains. It is crucial to thoroughly test the software to detect and address any defects before its release, ensuring it remains dependable and devoid of vulnerabilities through ongoing maintenance and support. De-

Acknowledgement: The authors thank the anonymous referee for helpful comments.

Anup Kumar Behera, Priyanka Agrawal, Department of Mathematics, SRM Institute of Science and Technology, Delhi-NCR, Ghaziabad, utter pradesh, India

https://doi.org/10.1515/9783111313634-015

velopers must design stable, secure, and easily maintainable software to enhance software reliability. The reliability of software plays a vital role in meeting user requirements and delivering expected outcomes [1].

Over the past forty years, numerous software reliability growth models (SRGMs) have been discussed, and they play a significant role in software testing [5]. Testing requires substantial resources, including manpower and resource allocation, depending on the software's complexity. In earlier studies it was found that the Weibull distribution fits well for estimating the resource expenditure in software reliability modeling. In 1987, Yamada et al. proposed an SRGM that considered the Weibull curve as a testing effort function and analyzed expenditure patterns [18]. Yamada and Ohtera (1990) used Rayleigh and exponential curves to forecast the necessary expenditures for achieving the desired reliability [17]. Following these developments, several researchers explored different testing efforts functions such as logistic TEF, log-logistic TEF, and Gompertz TEF under various scenarios. They applied these models in software industries to optimize cost and reliability. Some notable studies include Huang and Kuo (2002) [4], Jha et al. (2009) [7], Rafi et al. (2010) [16], Ahmad and Imam (2013) [2], and Jin and Jin (2016) [8]. In 2016, Li and Yi introduced a novel SRGM that adopted a power-law testing effort function and considered multi-generation faults [12]. Dwivedi and Kumar (2018) further explained a testing effort-based reliability growth model that incorporated the concept of testers learning from previous experiences [3]. Jain et al. (2020) studied a new way to predict how reliable the software will be by using the FRF and Gompertz methods to measure the testing effort [6]. According to Kassaymeh et al. (2021), the conventional Backpropagation Neural Network (BPNN) has a limitation where its performance in making accurate estimation is influenced by the initial parameter values, impacting prediction [11]. John et al. (2023) analyzing reliability of multi-system, considering different types of failure interactions. It assumes an exponential distribution for the failure and repair and use differential [9].

This paper uses the NHPP to calculate the number of faults and then uses them to estimate parameter. Traditional estimation methods have difficulty estimating software reliability models because they are often nonlinear and involve multiple parameters. The metaheuristic algorithm is excellent at navigating complex search space and finding the optimal parameters. Section 2 of this study briefs the power-law testing-effort function (PL-TEF). In Section 3, a new framework is deliberated upon to obtain the SRGM consideration of power-law TEF and non-linear FDR. The optimization technique Hybrid PSO-GWO is discussed in detail in Section 4. Section 5 discusses parameter estimation in HPSGWO algorithm. Finally, Section 6 concludes the paper.

2 Testing-effort function

Testing is essential before launching any software into the market, requiring significant resources and time. The testing effort function (TEF) is a crucial metric measuring the amount of effort invested in CPU time, labor, human resources, and the number of test cases used for the software. Various functions can be employed to model the expenditure on resources and the efficiency of the software detection and removal process. These functions help to understand and optimize resource allocation, identify bottlenecks, and improve the overall testing process to ensure the software's reliability and quality. Numerous functions can model resource expenditure and software detection/removal process efficiency [10].

Let $W(t)$ be the cumulative number of testing effort expenditures in the testing time interval $(0, t]$.

$$W(t) = \int_0^t w(x)\,dx \tag{15.1}$$

where $w(t)$ represents the amount of testing currently being done at time t, and it is adapted according to the severity of the software. This paper will study power-law TEF (PL-TEF) and compare it with the existing TEF-based SRGM.

2.1 Power-law TEF

The power-law curve is flexible to estimate the reliability and its cumulative testing effort consumed in the time $(0, t]$ is

$$W(t) = \frac{at^{k+1}}{k+1} \tag{15.2}$$

and its current testing effort function is

$$W(t) = at^k. \tag{15.3}$$

3 SRGM with PL-TEF and non-linear FDR

Software reliability growth models (SRGMs) are mathematical models that predict the reliability of software systems over time based on the number of failures and the testing effort [15]. PL-TEF stands for power law testing effort function, which is a function that describes how the testing effort changes over time. Non-linear FDR stands for non-linear fault detection rate, which is a function that describes how the rate of finding faults changes over time.

Our proposed model is based on the following assumptions:

- The software system experiences failures occurring at unpredictable intervals due to underlying faults that remain present in the system. The process of eliminating this fault follows NHPP.
- when a failure arises, it is resolved immediately, and the testing process is flawless because no new faults are detected during the debugging phase.
- The FDR exhibits non-linear changes and becomes more efficient over time. This improvement results from testers gaining knowledge and expertise through their experience.

Based on the assumptions, the mean value function $m(t)$ as:

$$\frac{dm(t)}{dt} \cdot \frac{1}{w(t)} = b(t)[a - m(t)], \tag{15.4}$$

where the function $a(t)$ represents the initial fault content at a specific time t, $b(t)$ represents the rate at which faults are detected over time, and $m(t)$ represents the mean failure function.

$$b(t) = \begin{cases} b: & \text{constant,} \\ b(1 + \beta t^d): & \text{non-linear.} \end{cases} \tag{15.5}$$

On putting the values of $b(t)$ in equation (15.4), we get the expected removed faults by time t

$$m(t) = \begin{cases} a[1 - e^{-ab\frac{t^{k+1}}{k+1}}], & \text{when } b \text{ constant,} \\ a[1 + e^{-ab(\frac{t^{k+1}}{k+1} + \beta\frac{t^{k+d+1}}{k+d+1})}], & \text{when } b \text{ is non-linear.} \end{cases} \tag{15.6}$$

4 Hybrid PSO-GWO algorithm

A hybrid PSO-GWO algorithm combines particle swarm optimization (PSO) and grey wolf optimizer (GWO) to improve the exploration and exploitation abilities of the original algorithms [13]. PSO is a swarm-based algorithm that mimics the social behavior of birds or fish, while GWO is a nature-inspired algorithm that simulates the hunting behavior of gray wolves. The hybrid algorithm uses PSO to update the position of the particles and GWO to update the position of the best particle in each iteration. The mathematical equations for the hybrid PSO-GWO algorithm are as follows:

Let X_i be the position of the i-th particle in a D-dimensional search space, V_i be its velocity, X_i be its personal best position, and G be the global best position. Then, the velocity and position update equations for PSO are

$$V_i = wV_i + c_1r_1(P_1 - X_i) + c_2r_2(G - X_i), \tag{15.7}$$

$$X_i = X_i + V_i, \tag{15.8}$$

where w is the inertia weight, c_1 and c_2 are the acceleration coefficients, and r_1 and r_2 are random numbers between 0 and 1.

Let α, β, and δ be the three best particles in the swarm, representing the alpha, beta, and delta wolves in GWO. Then, the position update equation for GWO is

$$X_i = \frac{1}{3}(\alpha + \beta + \gamma), \tag{15.9}$$

where α, β, and δ are calculated as

$$\alpha = G - A - D^{\alpha}, \tag{15.10}$$

$$\beta = G - A - D^{\beta}, \tag{15.11}$$

$$\delta = G - A - D^{\delta}, \tag{15.12}$$

where A and D are coefficient vectors defined as

$$X_i = X_i + V_i, \tag{15.13}$$

$$D = |2r_4 G - X_i|. \tag{15.14}$$

The hybrid PSO-GWO algorithm combines these two methods by replacing a particle of PSO with a small probability of a partially improved particle with GWO.

5 Parameter estimation in Hybrid PSO-GWO algorithm

Parameter estimation is the process of finding the optimal values of the unknown parameters of a model that best fits the observed data. In this study, we are using a hybrid PSO-GWO algorithm to optimize the parameters proposed and existing models.

This section aims to determine the optimal values that minimize the faults between the actual faults and the estimated faults obtained using the parameters estimated through PSO, GWO, and HPSGWO. We have taken an existing data set of 25 measurements as discussed in Chin-Yu Huang et al. [14] and applied three methods: PSO, GWO, and Hybrid PSO-GWO, to estimate a software project parameter using four SRGMs: Goel–Okumoto model, Inflection S-shaped model, Delay s-shaped model, and the Proposed model [Table 15.1]. The parameter range is the same for all models. We use a population size of 30 and a maximum iteration number of 500 for PSO, GWO, and HPSGWO, respectively.

From Table 15.2, we analyze that HPSGWO has a lower mean and best solutions than PSO and GWO in all cases. This shows that HPSGWO is powerful and that it can avoid the drawbacks of PSO and GWO, such as getting stuck in the local best and converging too early. Figure 15.1 shows the best convergence graph for the evolutionary process

Table 15.1: Models of software reliability and mean value function.

NHPP Models		
Serial No.	**Name of the model**	**Mean value function**
1	Goel–Okumoto Model (1979) (GOM)	$m(t) = a(1 - e^{-bt})$
2	Inflection S-shaped Model (1984) (ISSM)	$m(t) = \frac{a(1-e^{-bt})}{(1-\beta e^{-bt})}$
3	Delay S-shaped Model (1984) (DSSM)	$m(t) = a(1 - (1 + bt)e^{-bt})$
4	Proposed model (PM)	equation (15.6)

Table 15.2: SRGMs with parameters estimated using HPSGWO-25 weeks data set.

Model Name	PSO			GWO			HPSGWO		
	Mean	**SD**	**Best**	**Mean**	**SD**	**Best**	**Mean**	**SD**	**Best**
Goel–Okumoto Model (GOM)	13.904	1.23	11.6	14.883	8.15	10.09	13.730	2.85	10.66
Inflection S-shaped Model (ISSM)	15.5529	1.39	14.26	13.554	1.61	11.51	12.7	2.9	10.28
Delay S-shaped Model (DSSM)	15.45	2.44	12.04	13.35	2.04	10.65	12.4	3.09	10.51
Proposed Model (PM)	12.72	1.21	11	12.01	1.5	10.02	11.71	1.39	8.67

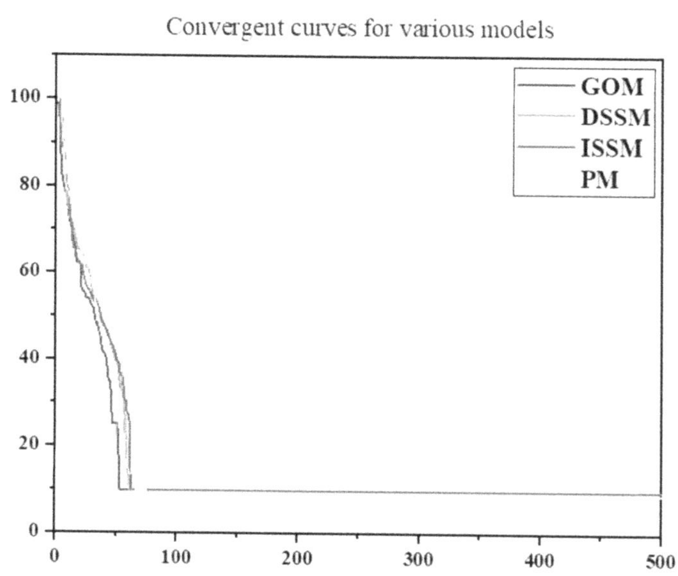

Figure 15.1: Best convergence for HPSGWO algorithm applying various models for 25 measurements.

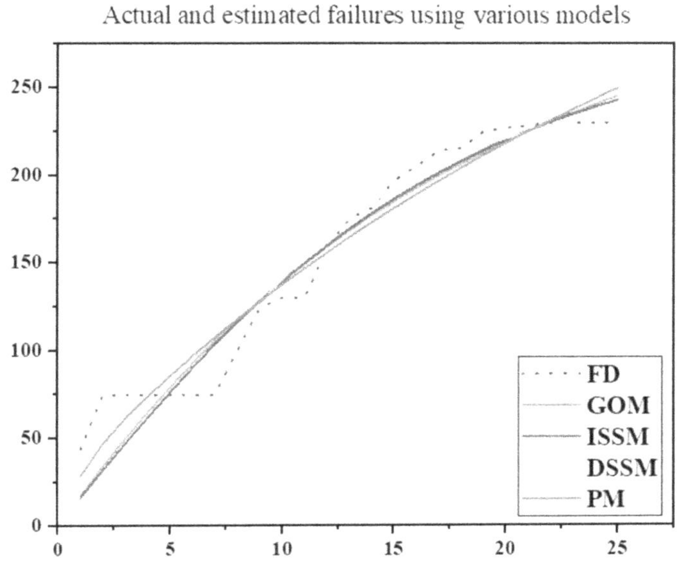

Figure 15.2: Actual and estimated failure for the HPSGWO algorithm using various models for 25 measurements.

of the four models Goel–Okumoto model, Inflection S-shaped model, Delay S-shaped model, and the proposed model using the HPSGWO algorithm. Figure 15.2 shows a graph that compares the actual and estimated failures curve for the four development models. These studies have shown that the hybrid PSO-GWO algorithm can achieve better results than the conventional PSO or GWO algorithms in terms of accuracy, convergence speed, and robustness.

6 Conclusion

In this study, we studied the SRGM that includes power-law testing effort and a non-linear FDR function. We analyzed the proposed model with three optimization algorithms PSO, GWO, and HPSGWO, and it was observed then HPSGWO converges faster than the other two optimization methods. The HPSGWO algorithm has proven its effectiveness in many studies, where it has performed better than both PSO and GWO in terms of convergence speed, solution quality, and robustness. The suggested approach will aid software developers by allowing them to release software at the appropriate time while maintaining the required degree of dependability. The Non-linear FDR function and PL-TEF may be combined with change-point phenomena, coverage function, and delay effects in future studies to get more dignified outcomes that would undoubtedly.

Bibliography

[1] A. G. Aggarwal, A. Tandon and H. Pham, *Optimization Models in Software Reliability*, Springer, 2022.

[2] N. Ahmad, M. Z. Imam et al., Software reliability growth models with log-logistic testing-effort function: A comparative study, *Int. J. Comput. Appl.* **75**(12) (2013), 6–11.

[3] A. Dwivedi and D. Kumar, Fault dependency srgm with testing effort using learning function, *Int. J. Manag. Concepts Philos.* **11**(1) (2018), 1–10.

[4] C.-Y. Huang and S.-Y. Kuo, Analysis of incorporating logistic testing-effort function into software reliability modeling, *IEEE Trans. Reliab.* **51**(3) (2002), 261–270.

[5] C.-Y. Huang, C.-T. Lin, S.-Y. Kuo, M. R. Lyu and C.-C. Sue, Software reliability growth models incorporating fault dependency with various debugging time lags, in: *Proceedings of the 28th Annual International Computer Software and Applications Conference, 2004. COMPSAC 2004*, IEEE, 2004, pp. 186–191.

[6] M. Jain, P. Agarwal and R. Solanki, Nhpp-based srgm using time-dependent fault reduction factors (frf) and gompertz tef, in: *Decision Analytics Applications in Industry*, 2020, pp. 81–89.

[7] P. Jha, D. Gupta, B. Yang and P. Kapur, Optimal testing resource allocation during module testing considering cost, testing effort and reliability, *Comput. Ind. Eng.* **57**(3) (2009), 1122–1130.

[8] C. Jin and S.-W. Jin, Parameter optimization of software reliability growth model with s-shaped testing-effort function using improved swarm intelligent optimization, *Appl. Soft Comput.* **40** (2016), 283–291.

[9] Y. M. John, A. Sanusi, I. Yusuf and U. M. Modibbo, Reliability analysis of multi-hardware–software system with failure interaction, *J. Comput. Cogn. Eng.* **2**(1) (2023), 38–46.

[10] P. Kapur, H. Pham, A. Gupta, P. Jha et al., *Software Reliability Assessment With OR Applications*, Vol. 364, Springer, 2011.

[11] S. Kassaymeh, S. Abdullah, M. Al-Laham, M. Alweshah, M. A. Al-Betar and Z. Othman, Salp swarm optimizer for modeling software reliability prediction problems, *Neural Process. Lett.* **53** (2021), 4451–4487.

[12] F. Li, Z.-L. Yi et al., A new software reliability growth model: multigeneration faults and a power-law testing-effort function, *Math. Probl. Eng.* **2016** (2016).

[13] D.-T. Nguyen, J.-R. Ho, P.-C. Tung and C.-K. Lin, A hybrid pso–gwo fuzzy logic controller with a new fuzzy tuner, *Int. J. Fuzzy Syst.* (2022), 1–19.

[14] K. Okumoto and A. L. Goel, Optimum release time for software systems, in: *COMPSAC 79. Proceedings. Computer Software and The IEEE Computer Society's Third International Applications Conference, 1979*, IEEE, 1979, pp. 500–503.

[15] H. Pham, *System Software Reliability*, Springer Science & Business Media, 2007.

[16] S. M. Rafi and S. Akthar, Software reliability growth model with logistic-exponential testing effort function and analysis of software release policy, in: *Proceedings of International Conference on Advances in Computer Science*, Citeseer, 2010.

[17] S. Yamada and H. Ohtera, Software reliability growth models for testing-effort control, *Eur. J. Oper. Res.* **46**(3) (1990), 343–349.

[18] S. Yamada, H. Ohtera and H. Narihisa, A testing-effort dependent software reliability model and its application, *Microelectron. Reliab.* **27**(3) (1987), 507–522.

Flora Jeba S, Anirban Roy, and Manjil P. Saikia

On near-perfect numbers with five prime factors

Abstract: Let n be a positive integer and $\sigma(n)$ the sum of all the positive divisors of n. We call n a near-perfect number with redundant divisor d if $\sigma(n) = 2n + d$. Let n be an odd near-perfect number of the form $n = p_1^{a_1} \cdot p_2^{a_2} \cdot p_3^{a_3} \cdot p_4^{a_4} \cdot p_5^{a_5}$ where p_i's are odd primes and a_i's ($1 \le i \le 5$) are positive integers. In this article, we prove that $3 \mid n$ and one of $5, 7, 11 \mid n$. We also show that there exists no odd near-perfect number when $n = 3^{a_1} \cdot 7^{a_2} \cdot p_3^{a_3} \cdot p_4^{a_4} \cdot p_5^{a_5}$ with $p_3 \in \{17, 19\}$ and when $n = 3^{a_1} \cdot 11^{a_2} \cdot p_3^{a_3} \cdot p_4^{a_4} \cdot p_5^{a_5}$.

Keywords: Near-perfect numbers, perfect numbers, divisor sum function

MSC 2010: 11A25, 11A67

1 Introduction

Let $n = p_1^{a_1} p_2^{a_2} \cdots p_i^{a_i}$ be the canonical representation of a positive integer where p_i's are prime numbers and a_i's are natural numbers. If $\sigma(n)$ denotes the sum of the positive divisors of n, then

$$\sigma(n) = \frac{p_1^{a_1+1} - 1}{p_1 - 1} \frac{p_2^{a_2+1} - 1}{p_2 - 1} \frac{p_3^{a_3+1} - 1}{p_3 - 1} \frac{p_4^{a_4+1} - 1}{p_4 - 1} \cdots \frac{p_i^{a_i+1} - 1}{p_i - 1}. \tag{16.1}$$

If $\sigma(n) < 2n$, then n is a deficient number, and n is an abundant number if $\sigma(n) > 2n$. n is a perfect number if $\sigma(n) = 2n$.

Pollock and Shevelev introduced near-perfect numbers in 2012 [2]. A positive integer n is near-perfect if $\sigma(n) = 2n + d$, where d is a proper divisor of n, known as redundant divisor of n. They also gave a method for the construction of near-perfect numbers with two prime divisors, and they introduced the concept of k-near-perfect numbers. A positive integer n is a k-near-perfect number when at most, k proper divisors are removed from the summation to make the sum equal to $2n$. When $\sigma(n) = 2n - d$, n is called a deficient perfect number with deficient divisor d; also, when $d = 1$, n is called an

Acknowledgement: The authors thank the anonymous referee for helpful comments.

Flora Jeba S, Department of Mathematics, CHRIST (Deemed to be University), Kanmanike, Kumbalgodu, Mysore Road, 560074 Bengaluru, Karnataka, India, e-mail: flora.jeba@res.christuniversity.in
Anirban Roy, Department of Sciences and Humanities, CHRIST (Deemed to be University), Kanmanike, Kumbalgodu, Mysore Road, 560074 Bengaluru, Karnataka, India, e-mail: anirban.roy@christuniversity.in
Manjil P. Saikia, Mathematical and Physical Sciences division, School of Arts and Sciences, Ahmedabad University, Commerce Six Roads, Navrangpura, Ahmedabad 380009, Gujarat, India, e-mail: manjil.saikia@ahduni.edu.in

https://doi.org/10.1515/9783111313634-016

almost perfect number. Ren and Chen [3] demonstrated that all near-perfect numbers with two prime factors besides 40 fall within one of the three constructions provided by Pollock and Shevelev [2]. In the same year, Tang, Ren and Li [6] proved that there are no odd near-perfect numbers with three prime factors and determined all deficient-perfect numbers with at most two distinct prime factors. Tang and Feng [4] proved that there are no odd deficient perfect numbers with three distinct prime factors. Later in 2016, Tang, Ma and Feng [5] demonstrated that the only odd near-perfect number with four distinct prime factors is $173369889 = 3^4 \times 7^2 \times 11^2 \times 19^2$ whose redundant divisor is $d = 3^2 \times 7 \times 11^2 \times 19^2$. Subsequently, Dutta and Saikia [1] proved a few properties of odd deficient perfect numbers with four different prime factors. Recently, Yang and Togbe [7] showed that the only deficient perfect number with four distinct prime factors is $3^2 \times 7^2 \times 11^2 \times 13^2$ whose deficient divisor is $d = 3^2 \times 7 \times 13$.

Inspired by the findings of Tang, Ma and Feng, we investigate near-perfect numbers with five prime factors. Throughout this paper, we assume that

$$n = p_1^{a_1} \cdot p_2^{a_2} \cdot p_3^{a_3} \cdot p_4^{a_4} \cdot p_5^{a_5}, \quad 3 \le p_1 < p_2 < p_3 < p_4 < p_5 \tag{16.2}$$

is an odd near-perfect number whose redundant divisor is

$$d = p_1^{b_1} \cdot p_2^{b_2} \cdot p_3^{b_3} \cdot p_4^{b_4} \cdot p_5^{b_5}, \quad \text{with } b_1 + b_2 + b_3 + b_4 + b_5 < a_1 + a_2 + a_3 + a_4 + a_5,$$

where $p_i (1 \le i \le 5)$ are distinct primes, $a_i \in \mathbb{N}$, $b_i \in \{0\} \bigcup \mathbb{N}, 1 \le i \le 5$. Since n is a near-perfect number, we have

$$\sigma(n) = 2n + d \tag{16.3}$$

$$\Rightarrow 2n < \sigma(n)$$

$$\Rightarrow 2 < \frac{\sigma(n)}{n}. \tag{16.4}$$

However,

$$\sigma(n) = \prod_{i=1}^{5} \frac{p_i^{a_i+1} - 1}{p_i - 1} < \prod_{i=1}^{5} \frac{p_i^{a_i+1}}{p_i - 1} = \prod_{i=1}^{5} \frac{p_i^{a_i} p_i}{p_i - 1} = n \times \prod_{i=1}^{5} \frac{p_i^{a_i}}{p_i - 1}$$

$$\Rightarrow \frac{\sigma(n)}{n} < \prod_{i=1}^{5} \frac{p_i^{a_i}}{p_i - 1}. \tag{16.5}$$

Thus combining equations (16.4) and (16.5), we get

$$2 < \frac{\sigma(n)}{n} < \prod_{i=1}^{5} \frac{p_i}{p_i - 1}. \tag{16.6}$$

Additionally, since n is an odd near-perfect number, $\sigma(n) \equiv 1 \pmod 2$, therefore

$$\sigma(n) = \prod_{i=1}^{5} (1 + p_i + p_i^2 + p_i^3 + \cdots + p_i^{a_i})$$

is odd. This implies $1 + p_i + p_i^2 + p_i^3 + \cdots + p_i^{a_i}$ being odd, $p_i + p_i^2 + p_i^3 + \cdots + p_i^{a_i}$ must be even for each $1 \le i \le 5$, consequently $a_i \equiv 0 \pmod 2$ for each $1 \le i \le 5$.

We define a function

$$f(a_1, a_2, a_3, a_4, a_5) = \left(1 - \frac{1}{p_1^{a_1+1}}\right)\left(1 - \frac{1}{p_2^{a_2+1}}\right)\left(1 - \frac{1}{p_3^{a_3+1}}\right)\left(1 - \frac{1}{p_4^{a_4+1}}\right)\left(1 - \frac{1}{p_5^{a_5+1}}\right). \tag{16.7}$$

This function (16.7) is strictly less than 1. We will be using (16.7) often in the paper.

Theorem 1. *For any odd near-perfect number of the form* (16.2), *we have the following bounds:*

1. $p_1 = 3$,
2. $5 \le p_2 \le 11$,
3. *If* $p_2 = 7$, *then* $11 \le p_3 \le 19$,
4. *If* $p_2 = 7$ *and* $p_3 = 17$, *then* $19 \le p_4 \le 23$,
5. *If* $p_2 = 7$ *and* $p_3 = 19$, *then* $p_4 = 23$, *and*
6. *If* $p_2 = 11$, *then* $p_3 = 13$,
7. *If* $p_2 = 11$ *and* $p_3 = 13$, *then* $p_4 = 17$.

Proof. Assuming $p_1 \ge 5$, we get from (16.6),

$$2 < \frac{\sigma(n)}{n} < \frac{5.7.11.13.17}{4.6.10.12.16} = 1.84646 < 2.$$

This is not possible, therefore, $p_1 = 3$. Similarly, by using (16.6), it is easy to find out that the only possible p_2's in n are 5, 7 and 11. From (16.6), by setting $p_1 = 3$, $p_2 = 7$ and $p_3 \ge 23$, we get

$$2 < \frac{\sigma(n)}{n} < \frac{3.7.23.29.31}{2.6.22.28.30} = 1.95805 < 2.$$

This is not possible, therefore, $p_3 \in \{11, 13, 17, 19\}$. By setting $p_1 = 3$, $p_2 = 7$, $p_3 = 17$, and $p_4 \ge 29$ from (16.6), we get

$$2 < \frac{\sigma(n)}{n} < \frac{3.7.17.29.31}{2.6.16.28.30} = 1.98997 < 2.$$

This is not possible, therefore, $p_4 \le 23$. Similarly, we obtain the values of p_4 for the other cases of p_3, when $p_2 = 7$ and $p_2 = 11$. They are listed in Table 16.1. □

Theorem 1 shows that to study near-perfect numbers of the form (16.2) it is enough to consider $p_1 = 3$ and $p_2 = 5, 7$ or 11. The main results of this paper are the following:

Theorem 2. *There is no odd near-perfect number* $n = 3^{a_1} \cdot 7^{a_2} \cdot p_3^{a_3} \cdot p_4^{a_4} \cdot p_5^{a_3}$, *when* $p_3 \in \{17, 19\}$.

Table 16.1: Possible values of p_4's.

p_1	p_2	p_3	p_4
3	7	17	≤ 23
3	7	19	23
3	11	13	17

Theorem 3. *There is no odd near-perfect number* $n = 3^{a_1} \cdot 11^{a_2} \cdot p_3^{a_3} \cdot p_4^{a_4} \cdot p_5^{a_3}$.

This paper is organized as follows, by fixing $p_1 = 3$, $p_2 = 7$ and $p_3 = 17$,

- Section 2 discusses the various possibilities of p_5 when p_4 is fixed to 19.
- Section 3 discusses the various possibilities of p_5 when p_4 is fixed to 23.

By fixing $p_1 = 3$, $p_2 = 7$ and $p_3 = 19$,

- Section 4 discusses the various possibilities of p_5 when p_4 is fixed to 23.

By fixing $p_1 = 3$, $p_2 = 11$ and $p_3 = 13$,

- Section 5 discusses the various possibilities of p_5 when p_4 is fixed to 17.

2 Odd near-perfect number of the form $3^{a_1}7^{a_2}17^{a_3}19^{a_4}p_5^{a_5}$

In this section, we prove the following result.

Proposition 1. *There exists no odd near-perfect number of the form* $n = 3^{a_1}7^{a_2}17^{a_3}19^{a_4}p_5^{a_5}$, *where a_i's are natural numbers.*

Proof. Let $n = 3^{a_1}7^{a_2}17^{a_3}19^{a_4}p_5^{a_5}$ be an odd near-perfect number with redundant divisor $d = 3^{b_1}7^{b_2}17^{b_3}19^{b_4}p_5^{b_5}$, where $b_1 + b_2 + b_3 + b_4 + b_5 < a_1 + a_2 + a_3 + a_4 + a_5$ and $b_i \leq a_i$, $i = 1, 2, 3, 4, 5$. Then by (16.1) and (16.3), we have

$$\begin{aligned} \sigma(n) &= \frac{3^{a_1+1}-1}{2}\frac{7^{a_2+1}-1}{6}\frac{17^{a_3+1}-1}{16}\frac{19^{a_4+1}-1}{18}\frac{p_5^{a_5+1}-1}{p_5-1} \\ &= 2.3^{a_1}7^{a_2}17^{a_3}19^{a_4}p_5^{a_5} + 3^{b_1}7^{b_2}17^{b_3}19^{b_4}p_5^{b_5}. \end{aligned} \tag{16.8}$$

And (16.7) becomes

$$f(a_1, a_2, a_3, a_4, a_5) = \left(1 - \frac{1}{3^{a_1+1}}\right)\left(1 - \frac{1}{7^{a_2+1}}\right)\left(1 - \frac{1}{17^{a_3+1}}\right)\left(1 - \frac{1}{19^{a_4+1}}\right)\left(1 - \frac{1}{p_5^{a_5+1}}\right). \tag{16.9}$$

Using (16.6), we find $p_5 \leq 53$, and from (16.8), we get the following bounds as laid down in Table 16.2 for the powers of the primes in n.

Table 16.2: The lower bounds of a_i's for $3^{a_1}7^{a_2}17^{a_3}19^{a_4}p_5^{a_5}$.

p_5	a_i
$23 \le p_5 \le 37$	$a_1 \ge 4$
$41 \le p_5 \le 43$	$a_1 \ge 4, a_2 \ge 4$
$p_5 = 47$	$a_1 \ge 6, a_2 \ge 4$
$p_5 = 53$	$a_1 \ge 8, a_2 \ge 6, a_3 \ge 4$

Now we will discuss the cases for $p_5 \in \{23, 29, 31, 37, 41, 43, 47, 53\}$.

We define the function,

$$g(a_1, a_2, a_3, a_4, a_5) = \frac{2^8 \times 3^2 \times (p_5 - 1)}{7 \times 17 \times 19 \times p_5} + \frac{2^7 \times 3^2 \times (p_5 - 1)}{D}, \tag{16.10}$$

where

$$D = 3^{a_1 - b_1} \times 7^{a_2 - b_2 + 1} \times 17^{a_3 - b_3 + 1} \times 19^{a_4 - b_4 + 1} \times p_5^{a_5 - b_5 + 1}. \tag{16.11}$$

From (16.8) and (16.9), it is clear that

$$g(a_1, a_2, a_3, a_4, a_5) = f(a_1, a_2, a_3, a_4, a_5) < 1.$$

Since $17 \nmid (3^{a_1+1}-1)\times(7^{a_2+1}-1)\times(17^{a_3+1}-1)\times(19^{a_4+1}-1)\times(p_5^{a_5+1}-1)$ for $p_5 = 23, 29, 31, 37$ and 43, by (16.8), we get $b_3 = 0$.

We now consider the following six cases:

Case 1. $p_5 \in \{23, 29, 31\}$.

Now by Table 16.2 and equations (16.10), (16.9), we have for $p_5 = 23$,

$$g(a_1, a_2, a_3, a_4, a_5) \le \frac{2^8 \times 3^2 \times 22}{7 \times 17 \times 19 \times 23} + \frac{2^7 \times 3^2 \times 22}{7 \times 17^3 \times 19 \times 23} = 0.976399,$$

which contradicts $f(a_1, a_2, a_3, a_4, a_5) \ge 0.992553$.

For $p_5 = 29$,

$$g(a_1, a_2, a_3, a_4, a_5) \le \frac{2^8 \times 3^2 \times 28}{7 \times 17 \times 19 \times 29} + \frac{2^7 \times 3^2 \times 28}{7 \times 17^3 \times 19 \times 29} = 0.985582$$

and $f(a_1, a_2, a_3, a_4, a_5) \ge 0.992594$. This is a contradiction.

For $p_5 = 31$,

$$g(a_1, a_2, a_3, a_4, a_5) \le \frac{2^8 \times 3^2 \times 30}{7 \times 17 \times 19 \times 31} + \frac{2^7 \times 3^2 \times 30}{7 \times 17^3 \times 19 \times 31} = 0.987853$$

and $f(a_1, a_2, a_3, a_4, a_5) \ge 0.992601$, which is a contradiction.

Case 2. $p_5 = 37$.

By Table 16.2 and equation (16.9), we find that

$$f(a_1, a_2, a_3, a_4, a_5) \geq 0.992615. \tag{16.12}$$

It is easy to see that $37 \mid \sigma(7^{a_2})$, whenever $a_2 + 1 \equiv 9 \pmod{18}$. Therefore, we have the following two sub-cases depending on a_2.

Sub-case 2.1. $a_2 + 1 \not\equiv 9 \pmod{18}$.

Since $a_2 + 1 \not\equiv 9 \pmod{18}$, therefore $37 \nmid \sigma(7^{a_2})$ and thus by (16.8), we get $b_5 = 0$. Therefore by (16.10),

$$g(a_1, a_2, a_3, a_4, a_5) \leq \frac{2^8 \times 3^2 \times 36}{7 \times 17 \times 19 \times 37} + \frac{2^7 \times 3^2 \times 36}{7 \times 17^3 \times 19 \times 37^3} \leq 0.991478.$$

Since the functions f and g are equal, the above bound contradicts (16.12).

Sub-case 2.2. $a_2 + 1 \equiv 9 \pmod{18}$.

If $D \leq 3\times7\times17^2\times19\times37$, by (16.10), we get $g(a_1, a_2, a_3, a_4, a_5) = 1.0012 > 1$, which is not possible. Again, if $D \geq 3\times7^4\times17\times19\times37$, from (16.10), we get $g(a_1, a_2, a_3, a_4, a_5) = 0.991959$, which contradicts (16.12). Since $b_3 = 0$, the only D's that satisfies (16.12) and the inequality

$$3 \times 7 \times 17^2 \times 19 \times 37 < D < 3 \times 7^4 \times 17 \times 19 \times 37 \tag{16.13}$$

are given below in Table 16.3 along with the corresponding results for $\sigma(n)$. In Table 16.3, l_i's, indicate the indices of prime factors 3, 7, 17, 19, 37 of n in D respectively and t_i's, refer to the powers of primes 3, 7, 17, 19, 37 respectively in $\sigma(n)$, where $1 \leq i \leq 5$, M denotes the co-efficient in $\sigma(n)$, obtained in (16.8). In the following, l_i's and t_i's denote the indices of each prime factor of D and $\sigma(n)$, respectively.

Table 16.3: Calculations of $\sigma(3^{a_1}7^{a_2}17^{a_3}19^{a_4}37^{a_5})$.

D $[3^{l_1}\times7^{l_2}\times17^{l_3}\times19^{l_4}\times37^{l_5}]$					$\sigma(n)$ $[M \times 3^{t_1} \times 7^{t_2} \times 17^{t_3} \times 19^{t_4} \times 37^{t_5}]$					
l_1	l_2	l_3	l_4	l_5	M	t_1	t_2	t_3	t_4	t_5
0	1	3	1	1	193	$a_1 + 1$	a_2	$a_3 - 2$	a_4	a_5

Since $37 \mid \sigma(7^{a_2})$ whenever $a_2 + 1 \equiv 9 \pmod{18}$, $1063 \mid \sigma(n)$. But none of the $\sigma(n)$ values found in Table 16.3 is divisible by 1063. This is a contradiction.

Case 3. $p_5 = 41$.

Using Table 16.2 and equation (16.9), we find that $f(a_1, a_2, a_3, a_4, a_5) \geq 0.995463$. But, $(7\times17) \nmid (3^{a_1+1} - 1) \times (7^{a_2+1} - 1) \times (17^{a_3+1} - 1) \times (19^{a_4+1} - 1) \times (41^{a_5+1} - 1)$. Then by (16.8) we get, $b_2 = b_3 = 0$ and thus by (16.10),

$$g(a_1, a_2, a_3, a_4, a_5) \leq \frac{2^8 \times 3^2 \times 40}{7 \times 17 \times 19 \times 41} + \frac{2^7 \times 3^2 \times 40}{7^5 \times 17^3 \times 19 \times 41} \leq 0.994165.$$

This is not possible.

Case 4. $p_5 = 43$.

By Table 16.2 and equation (16.9), we get

$$f(a_1, a_2, a_3, a_4, a_5) \geq 0.995465. \tag{16.14}$$

It is easy to see that $43 \mid \sigma(17^{a_3})$ whenever $a_3 + 1 \equiv 21 \pmod{42}$. Therefore, we have the following two sub-cases depending on a_3.

Sub-case 4.1. $a_3 + 1 \not\equiv 21 \pmod{42}$.

In this case, $43 \nmid \sigma(17^{a_3})$ and thus by (16.8) we get, $b_5 = 0$. Therefore, by (16.10), we get

$$g(a_1, a_2, a_3, a_4, a_5) \leq \frac{2^8 \times 3^2 \times 42}{7 \times 17 \times 19 \times 43} + \frac{2^7 \times 3^2 \times 42}{7 \times 17^3 \times 19 \times 43^3} \leq 0.995321.$$

Since the functions f and g are equal, the above bound contradicts (16.14).

Sub-case 4.2. $a_3 + 1 \equiv 21 \pmod{42}$.

If $D \leq 3 \times 7 \times 17^2 \times 19 \times 43$, by (16.10), we get $g(a_1, a_2, a_3, a_4, a_5) = 1.00508 > 1$, which is not possible. If $D \geq 3 \times 7 \times 17^4 \times 19 \times 43$, by (16.10), we get $g(a_1, a_2, a_3, a_4, a_5) = 0.995354 < f(a_1, a_2, a_3, a_4, a_5)$. Since the functions f and g are equal, the above bound contradicts (16.14). Since $b_3 = 0$, the D's that satisfy (16.12) and the inequality

$$3 \times 7 \times 17^2 \times 19 \times 43 < D < 3 \times 7 \times 17^4 \times 19 \times 43$$

are given below in Table 16.4 along with the corresponding results for $\sigma(n)$.

Table 16.4: Calculations of $\sigma(3^{a_1} 7^{a_2} 17^{a_3} 19^{a_4} 43^{a_5})$.

D $[3^{l_1} \times 7^{l_2} \times 17^{l_3} \times 19^{l_4} \times 43^{l_5}]$					$\sigma(n)$ $[M \times 3^{t_1} \times 7^{t_2} \times 17^{t_3} \times 19^{t_4} \times 43^{t_5}]$					
l_1	l_2	l_3	l_4	l_5	M	t_1	t_2	t_3	t_4	t_5
0	1	3	1	1	193	$a_1 + 1$	$a_2 + 1$	$a_3 - 2$	a_4	a_5
0	2	3	1	1	229	$a_1 + 1$	$a_2 - 1$	$a_3 - 2$	a_4	a_5
1	1	3	1	1	1735	$a_1 - 1$	a_2	$a_3 - 2$	a_4	a_5
2	1	3	1	1	121	$a_1 - 2$	a_2	$a_3 - 2$	a_4	$a_5 + 1$

Since $43 \mid \sigma(17^{a_3})$ whenever $a_3 + 1 \equiv 21 \pmod{42}$, $940143709 \mid \sigma(n)$. But none of the $\sigma(n)$ values, given in Table 16.4, is divisible by 940143709. This leads to a contradiction.

Case 5. $p_5 = 47$.

We observe that, $(7\times17) \nmid (3^{a_1+1}-1)\times(7^{a_2+1}-1)\times(17^{a_3+1}-1)\times(19^{a_4+1}-1)\times(47^{a_5+1}-1)$, therefore by (16.8), we get $b_2 = b_3 = 0$. Then using Table 16.2 and by (16.9) and (16.10), we get

$$g(a_1,a_2,a_3,a_4,a_5) \le \frac{2^8\times3^2\times46}{7\times17\times19\times47} + \frac{2^7\times3^2\times46}{7^5\times17^3\times19\times47} \le 0.997338$$

and $f(a_1,a_2,a_3,a_4,a_5) \ge 0.999125$. This is not possible.

Case 6. $p_5 = 53$.

Here, $(17\times53) \mid (3^{a_1+1}-1)\times(7^{a_2+1}-1)\times(17^{a_3+1}-1)\times(19^{a_4+1}-1)\times(53^{a_5+1}-1)$ and thus by (16.8), $b_3 = b_5 = 0$. Then by using Table 16.2, $f(a_1,a_2,a_3,a_4,a_5) \ge 0.999795$ and

$$g(a_1,a_2,a_3,a_4,a_5) \le \frac{2^8\times3^2\times52}{7\times17\times19\times53} + \frac{2^7\times3^2\times52}{7\times17^5\times19\times53^3} \le 0.999791.$$

This is a contradiction.

This proves that there is no odd near-perfect number of the form $n = 3^{a_1}7^{a_2}17^{a_3}19^{a_4}p_5^{a_5}$.

□

3 Odd near-perfect number of the form $3^{a_1}7^{a_2}17^{a_3}23^{a_4}p_5^{a_5}$

In this section, we prove the following result.

Proposition 2. *There exists no odd near-perfect number of the form* $n = 3^{a_1}7^{a_2}17^{a_3}23^{a_4}p_5^{a_5}$, *where a_i's are natural numbers.*

Proof. Let $n = 3^{a_1}7^{a_2}17^{a_3}23^{a_4}p_5^{a_5}$ be an odd near-perfect number with redundant divisor $d = 3^{b_1}7^{b_2}17^{b_3}23^{b_4}p_5^{b_5}$, where $b_1 + b_2 + b_3 + b_4 + b_5 < a_1 + a_2 + a_3 + a_4 + a_5$ and $b_i \le a_i$, $i = 1, 2, 3, 4, 5$. Then by (16.1) and (16.3), we have

$$\begin{aligned}\sigma(n) &= \frac{3^{a_1+1}-1}{2}\,\frac{7^{a_2+1}-1}{6}\,\frac{17^{a_3+1}-1}{16}\,\frac{23^{a_4+1}-1}{22}\,\frac{p_5^{a_5+1}-1}{p_5-1}\\ &= 2\cdot 3^{a_1}7^{a_2}17^{a_3}23^{a_4}p_5^{a_5} + 3^{b_1}7^{b_2}17^{b_3}23^{b_4}p_5^{b_5} \qquad (16.15)\end{aligned}$$

and (16.7) becomes

$$f(a_1,a_2,a_3,a_4,a_5)=\left(1-\frac{1}{3^{a_1+1}}\right)\left(1-\frac{1}{7^{a_2+1}}\right)\left(1-\frac{1}{17^{a_3+1}}\right)\left(1-\frac{1}{23^{a_4+1}}\right)\left(1-\frac{1}{p_5^{a_5+1}}\right). \tag{16.16}$$

Using (16.6), we find that $p_5 \le 31$, and from (16.15), we get the following bounds as given in Table 16.5 for the powers of the primes in n.

Table 16.5: The lower bounds of a_i's for $3^{a_1}7^{a_2}17^{a_3}23^{a_4}p_5^{a_5}$.

p_i	a_i
$p_5 = 29$	$a_1, a_2 \ge 4$
$p_5 = 31$	$a_1, a_2, a_3 \ge 4$

Now we will discuss the cases for $p_5 = \{29, 31\}$. We define the function,

$$g(a_1.a_2,a_3,a_4,a_5)=\frac{2^8\times 11\times(p_5-1)}{7\times 17\times 23\times p_5}+\frac{2^7\times 11\times(p_5-1)}{D}, \tag{16.17}$$

where

$$D=3^{a_1-b_1}\times 7^{a_2-b_2+1}\times 17^{a_3-b_3+1}\times 23^{a_4-b_4+1}\times p_5^{a_5-b_5+1}. \tag{16.18}$$

Then from (16.15) and (16.16), it is clear that

$$g(a_1,a_2,a_3,a_4,a_5)=f(a_1,a_2,a_3,a_4,a_5)<1.$$

Since, $17 \nmid (3^{a_1+1}-1)\times(7^{a_2+1}-1)\times(17^{a_3+1}-1)\times(23^{a_4+1}-1)\times(p_5^{a_5+1}-1)$ for $p_5 = 29$ and $p_5 = 31$, by (16.15), we get $b_3 = 0$.

By using (16.17) and Table 16.5 for $p_5 = 29$, we have

$$g(a_1,a_2,a_3,a_4,a_5)\le\frac{2^8\times 11\times 28}{7\times 17\times 23\times 29}+\frac{2^7\times 11\times 28}{7\times 17^3\times 23\times 29}\le 0.995104,$$

which contradicts $f(a_1,a_2,a_3,a_4,a_5) \ge 0.9955$. Therefore, $p_5 = 29$ is not possible. Similarly, for $p_5 = 31$, we have

$$g(a_1,a_2,a_3,a_4,a_5)\le\frac{2^8\times 11\times 30}{7\times 17\times 23\times 31}+\frac{2^7\times 11\times 30}{7\times 17^5\times 23\times 31}\le 0.995681,$$

which contradicts $f(a_1,a_2,a_3,a_4,a_5) \ge 0.99571$. Therefore $p_5 = 31$ is not possible.

This proves that there is no odd near-perfect number of the form $n=3^{a_1}7^{a_2}17^{a_3}23^{a_4}p_5^{a_5}$. □

4 Odd near-perfect number of the form $3^{a_1}7^{a_2}19^{a_3}23^{a_4}p^{a_5}$

In this section, we prove the following result.

Proposition 3. *There exists no odd near-perfect number of the form* $n = 3^{a_1}7^{a_2}19^{a_3}23^{a_4}p^{a_5}$, *where* a_i*'s are natural numbers.*

Proof. Let $n = 3^{a_1}7^{a_2}19^{a_3}23^{a_4}p_5^{a_5}$ be an odd near-perfect number with redundant divisor $d = 3^{b_1}7^{b_2}19^{b_3}23^{b_4}p_5^{b_5}$, where $b_1 + b_2 + b_3 + b_4 + b_5 < a_1 + a_2 + a_3 + a_4 + a_5$ and $b_i \leq a_i$, $i = 1, 2, 3, 4, 5$. Then by (16.1) and (16.3), we have

$$\begin{aligned} \sigma(n) &= \frac{3^{a_1+1}-1}{2}\frac{7^{a_2+1}-1}{6}\frac{19^{a_3+1}-1}{18}\frac{23^{a_4+1}-1}{22}\frac{p_5^{a_5+1}-1}{p_5-1} \\ &= 2\times 3^{a_1}7^{a_2}19^{a_3}23^{a_4}p_5^{a_5} + 3^{b_1}7^{b_2}19^{b_3}23^{b_4}p_5^{b_5} \end{aligned} \tag{16.19}$$

and (16.7) becomes

$$f(a_1,a_2,a_3,a_4,a_5) = \left(1-\frac{1}{3^{a_1+1}}\right)\left(1-\frac{1}{7^{a_2+1}}\right)\left(1-\frac{1}{19^{a_3+1}}\right)\left(1-\frac{1}{23^{a_4+1}}\right)\left(1-\frac{1}{p_5^{a_5+1}}\right). \tag{16.20}$$

Using (16.6), we find $p_5 = 29$, and from (16.19), we get the following bounds as given in Table 16.6 for the powers of the primes in n.

Table 16.6: The lower bounds of a_i's for $3^{a_1}7^{a_2}19^{a_3}23^{a_4}p_5^{a_5}$.

p_i	a_i
$p_5 = 29$	$a_1 \geq 8$, $a_2 \geq 6$, $a_i \geq 4$ where $i = 3, 4, 5$.

We define

$$g(a_1.a_2,a_3,a_4,a_5) = \frac{2^7\times 3^2\times 11}{19\times 23\times 29} + \frac{2^6\times 3^2\times 11}{D}, \tag{16.21}$$

where

$$D = 3^{a_1-b_1}\times 7^{a_2-b_2}\times 19^{a_3-b_3+1}\times 23^{a_4-b_4+1}\times 29^{a_5-b_5+1}. \tag{16.22}$$

From (16.19) and (16.20), it is clear that

$$g(a_1,a_2,a_3,a_4,a_5) = f(a_1,a_2,a_3,a_4,a_5) < 1.$$

By (16.20), we find that

$$f(a_1,a_2,a_3,a_4,a_5) \geq 0.999947. \tag{16.23}$$

If $D \leq 3 \times 7^3 \times 19 \times 23 \times 29$, by (16.21), $g(a_1, a_2, a_3, a_4, a_5) = 1.00041 > 1$, this is not possible. If $D \geq 3 \times 7^5 \times 19 \times 23 \times 29$, by (16.21), we get $g(a_1, a_2, a_3, a_4, a_5) = 0.999931$, which is a contradiction by (16.23). Therefore, the only D's that satisfy the (16.23) and inequality

$$3 \times 7^3 \times 19 \times 23 \times 29 < D < 3 \times 7^5 \times 19 \times 23 \times 29$$

are given in Table 16.7.

Table 16.7: Calculation of $\sigma(3^{a_1}7^{a_2}19^{a_3}23^{a_4}29^{a_5})$.

D $[3^{l_1} \times 7^{l_2} \times 19^{l_3} \times 23^{l_4} \times 29^{l_5}]$					$\sigma(n)$ $[M \times 3^{t_1} \times 7^{t_2} \times 19^{t_3} \times 23^{t_4} \times 29^{t_5}]$					
l_1	l_2	l_3	l_4	l_5	M	t_1	t_2	t_3	t_4	t_5
0	0	2	1	3	3551	a_1+2	a_2	a_3-1	a_4	a_5-2
0	0	2	2	2	1207	a_1+1	a_2+1	a_3-1	a_4-1	a_1
0	0	2	2	1	125	a_1	a_2+1	a_3-1	a_4-1	a_5
0	0	3	1	2	20939	a_1	a_2	a_3-2	a_4	a_5-1
0	0	3	2	1	16607	a_1	a_2	a_3-2	a_4-1	a_5
0	0	4	1	1	4573	a_1+1	a_2	a_3-3	a_4	a_5
0	2	3	1	1	3931	a_1+2	a_2-2	a_3-2	a_4	a_5
0	3	1	1	2	865	a_1	a_2-3	a_3	a_4+1	a_5-1
0	3	1	2	1	15779	a_1	a_2-3	a_3	a_4-1	a_5
0	3	2	1	1	4345	a_1	a_2-3	a_3-1	a_4	a_5
0	5	1	1	1	514	a_1+4	a_2-5	a_3	a_4	a_5
2	0	1	1	3	15139	a_1-2	a_2	a_3	a_4	a_5-2
3	0	1	2	2	36019	a_1-3	a_2	a_3	a_4-1	a_5-1
3	0	1	3	1	583	a_1-3	a_2+2	a_3	a_4-2	a_5
3	0	2	1	2	29755	a_1-3	a_2	a_3-1	a_4	a_5-1
3	0	2	2	1	23599	a_1-3	a_2	a_3-1	a_4-1	a_5
3	0	3	1	1	2785	a_1-3	a_2	a_3-2	a_4	a_5
5	0	1	1	2	14095	a_1-5	a_2	a_3	a_4	a_5-1
6	0	1	2	1	1765	a_1-6	a_2	a_3	a_4-1	a_5
6	0	2	1	1	277035	a_1-6	a_2	a_3-1	a_4	a_5
8	0	1	1	1	13123	a_1-8	a_2	a_3	a_4	a_5
1	1	1	1	3	35323	a_1-1	a_2-1	a_3	a_4+1	a_5-2
1	1	1	2	2	28015	a_1-1	a_2-1	a_3	a_4-1	a_5-1
1	1	1	3	1	22219	a_1-1	a_2-1	a_3	a_4-2	a_5
1	1	2	1	2	23143	a_1-1	a_2-1	a_3-1	a_4	a_5-1
1	1	2	2	1	18355	a_1-1	a_2-1	a_3-1	a_4-1	a_5
1	1	3	1	1	15163	a_1-1	a_2-1	a_3-2	a_4	a_5
1	4	1	1	1	14407	a_1-1	a_2-4	a_3	a_4	a_5
2	2	1	1	2	25579	a_1-2	a_2-2	a_3	a_4	a_5-1
2	2	1	2	1	20287	a_1-2	a_2-2	a_3	a_4-1	a_5
2	2	2	1	1	16759	a_1-2	a_2-2	a_3-1	a_4	a_5
3	3	1	1	1	18523	a_1-3	a_2-3	a_3	a_4	a_5
4	1	1	1	2	32887	a_1-4	a_2-1	a_3	a_4	a_5-1
4	1	1	2	1	26083	a_1-4	a_2-1	a_3	a_4-1	a_5
4	1	2	1	1	743	a_1-4	a_2-1	a_3-1	a_4	a_5+1
5	2	2	1	1	23815	a_1-5	a_2-2	a_3-1	a_4	a_5
7	1	1	1	1	30619	a_1-7	a_2-1	a_3	a_4	a_5

Now $23 \mid \sigma(3^{a_1})$ and $23 \mid \sigma(29^{a_5})$, when $a_1 + 1 \equiv 11 \pmod{22}$ and $a_5 + 1 \equiv 11 \pmod{22}$ respectively. Therefore, $18944890940537 \mid \sigma(29^{a_5})$ and $3851 \mid \sigma(3^{a_1})$ which implies $18944890940537 \mid \sigma(n)$ and $3851 \mid \sigma(n)$. But $\sigma(n)$ is not divisible by 3851 or 18944890940537 in Table 16.7. This is a contradiction.

This proves that there is no odd near-perfect number of the form $n = 3^{a_1}7^{a_2}19^{a_3}23^{a_4}29^{a_5}$. □

5 Odd near-perfect number of the form $3^{a_1}11^{a_2}13^{a_3}17^{a_4}p_5^{a_5}$

In this section, we prove the following result.

Proposition 4. *There exists no odd near-perfect number of the form $n = 3^{a_1}11^{a_2}13^{a_3}17^{a_4}p_5^{a_5}$, where a_i's are natural numbers.*

Proof. Let $n = 3^{a_1}11^{a_2}13^{a_3}17^{a_4}p_5^{a_5}$ be an odd near-perfect number with redundant divisor $d = 3^{b_1}11^{b_2}13^{b_3}17^{b_4}p_5^{b_5}$, where $b_1 + b_2 + b_3 + b_4 + b_5 < a_1 + a_2 + a_3 + a_4 + a_5$ and $b_i \le a_i$, $i = 1, 2, 3, 4, 5$. Then by (16.1) and (16.3), we have

$$\begin{aligned}\sigma(n) &= \frac{3^{a_1+1}-1}{2}\,\frac{11^{a_2+1}-1}{10}\,\frac{13^{a_3+1}-1}{12}\,\frac{17^{a_4+1}-1}{16}\,\frac{p_5^{a_5+1}-1}{p_5-1}\\ &= 2\times 3^{a_1}11^{a_2}13^{a_3}17^{a_4}p_5^{a_5} + 3^{b_1}11^{b_2}13^{b_3}17^{b_4}p_5^{b_5}\end{aligned} \tag{16.24}$$

and (16.7) becomes

$$f(a_1,a_2,a_3,a_4,a_5) = \left(1-\frac{1}{3^{a_1+1}}\right)\left(1-\frac{1}{11^{a_2+1}}\right)\left(1-\frac{1}{13^{a_3+1}}\right)\left(1-\frac{1}{17^{a_4+1}}\right)\left(1-\frac{1}{p_5^{a_5+1}}\right). \tag{16.25}$$

By using (16.6), we find that $p_5 = 19$ and from (16.24) and (16.6), we get the following bounds as given in Table 16.8 for the powers of the primes in n.

Table 16.8: The lower bounds of a_i's for $3^{a_1}11^{a_2}13^{a_3}17^{a_4}p_5^{a_5}$.

p_i	a_i
$p_5 = 19$	$a_1 \ge 6$

We define the function

$$g(a_1.a_2,a_3,a_4,a_5) = \frac{2^{10}\times 3^2\times 5}{11\times 13\times 17\times 19} + \frac{2^9\times 3^2\times 5}{D}, \tag{16.26}$$

where

$$D = 3^{a_1-b_1} \times 11^{a_2-b_2+1} \times 13^{a_3-b_3+1} \times 17^{a_4-b_4+1} \times 19^{a_5-b_5+1}. \quad (16.27)$$

By using (16.24) and (16.25), it is clear that

$$g(a_1, a_2, a_3, a_4, a_5) = f(a_1, a_2, a_3, a_4, a_5) < 1.$$

By (16.24) and Table 16.8, we find that

$$f(a_1, a_2, a_3, a_4, a_5) \geq 0.997988. \quad (16.28)$$

If $D \leq 3 \times 11^2 \times 13 \times 17 \times 19$, by (16.26), we get $g(a_1, a_2, a_3, a_4, a_5) = 1.01276 > 1$, this is not possible. If $D \geq 3 \times 11^4 \times 13 \times 17 \times 19$, from (16.26), we get $g(a_1, a_2, a_3, a_4, a_5) = 0.997765$, which contradicts (16.28). Therefore,

$$3 \times 11^2 \times 13 \times 17 \times 19 < D < 3 \times 11^4 \times 13 \times 17 \times 19. \quad (16.29)$$

The D's that satisfy (16.29) and (16.26) are given in Table 16.9.

Table 16.9: Calculation of $\sigma(3^{a_1} 11^{a_2} 13^{a_3} 17^{a_4} 19^{a_5})$.

D $[3^{l_1} \times 11^{l_2} \times 13^{l_3} \times 17^{l_4} \times 19^{l_5}]$					$\sigma(n)$ $[M \times 3^{t_1} \times 11^{t_2} \times 13^{t_3} \times 17^{t_4} \times 19^{t_5}]$					
l_1	l_2	l_3	l_4	l_5	M	t_1	t_2	t_3	t_4	t_5
0	1	1	1	3	241	$a_1 + 1$	a_2	a_3	a_4	$a_5 - 2$
0	1	1	2	2	647	$a_1 + 1$	a_2	a_3	$a_4 - 1$	$a_5 - 1$
0	1	2	1	2	5	$a_1 + 2$	$a_2 + 1$	$a_3 - 1$	a_4	$a_5 - 1$
0	1	2	2	1	443	a_1	a_2	$a_3 - 1$	$a_4 - 1$	a_5
0	4	1	1	1	2663	a_1	$a_2 - 3$	a_3	a_4	a_5
1	1	1	1	3	197	$a_1 - 1$	$a_2 + 1$	a_3	a_4	$a_5 - 2$
1	1	1	2	2	1939	$a_1 - 1$	a_2	a_3	$a_4 - 1$	$a_5 - 1$
1	1	1	3	1	1735	$a_1 - 1$	a_2	a_3	$a_4 - 2$	a_5
1	1	2	1	2	1483	$a_1 - 1$	a_2	$a_3 - 1$	a_4	$a_5 - 1$
1	1	2	2	1	1327	$a_1 - 1$	a_2	$a_3 - 1$	$a_4 - 1$	a_5
1	1	3	1	1	1015	$a_1 - 1$	a_2	$a_3 - 2$	a_4	a_5
1	2	1	1	2	1255	$a_1 - 1$	$a_2 - 1$	a_3	a_4	$a_5 - 1$
1	2	1	2	1	1123	$a_1 - 1$	$a_2 - 1$	a_3	$a_4 - 1$	a_5
1	2	2	1	1	859	$a_1 - 1$	$a_2 - 1$	$a_3 - 1$	a_4	a_5
1	3	1	1	1	727	$a_1 - 1$	$a_2 - 2$	a_3	a_4	a_5
2	2	2	1	1	2575	$a_1 - 2$	$a_2 - 1$	$a_3 - 1$	a_4	a_5
2	3	1	1	1	2179	$a_1 - 2$	$a_2 - 2$	a_3	a_4	a_5
3	1	1	1	2	79	$a_1 - 3$	a_2	$a_3 + 1$	a_4	$a_5 - 1$
3	1	1	2	1	919	$a_1 - 3$	a_2	a_3	$a_4 - 1$	a_5
3	1	2	1	1	37	$a_1 - 3$	a_2	$a_3 - 1$	a_4	$a_5 + 1$
3	2	1	1	1	35	$a_1 - 3$	$a_2 - 1$	a_3	$a_4 + 1$	a_5
4	1	1	2	1	145	$a_1 - 4$	a_2	a_3	$a_4 - 1$	$a_5 + 1$
4	1	2	1	1	2107	$a_1 - 4$	a_2	$a_3 - 1$	a_4	a_5
4	2	1	1	1	1783	$a_1 - 4$	$a_2 - 1$	a_3	a_4	a_5
5	1	1	1	1	487	$a_1 - 5$	a_2	a_3	a_4	a_5
6	1	1	1	1	1459	$a_1 - 6$	a_2	a_3	a_4	a_5

Since $3 \mid \sigma(13^{a_3})$ and $3 \mid \sigma(19^{a_5})$, when $a_3 + 1 \equiv 3 \pmod 6$ and $a_5 + 1 \equiv 3 \pmod 6$ respectively, therefore $61 \mid \sigma(13^{a_3})$ and $127 \mid \sigma(19^{a_5})$, which implies $61 \mid \sigma(n)$ and $127 \mid \sigma(n)$, but the above relations in Table 16.9 is neither divisible by 127, nor by 61. This is not possible. □

6 Proof of the main Theorems 2 and 3

From Propositions 1, 2 and 3 it is clear that there exists no odd near-perfect number of the form $n = 3^{a_1} \cdot 7^{a_2} \cdot p_3^{a_3} \cdot p_4^{a_4} \cdot p_5^{a_5}$ when $p_3 \in \{17, 19\}$ and from Proposition 4, it is clear that there exists no odd near-perfect number of the form $n = 3^{a_1} \cdot 11^{a_2} \cdot p_3^{a_3} \cdot p_4^{a_4} \cdot p_5^{a_5}$.

Bibliography

[1] P. Dutta and M. P. Saikia, Deficient perfect numbers with four distinct prime factors, *Asian-Eur. J. Math.* **13**(7) (2020), 13, Id/No 2050126.
[2] P. Pollack and V. Shevelev, On perfect and near-perfect numbers, *J. Number Theory* **132**(12) (2012), 3037–3046.
[3] X.-Z. Ren and Y.-G. Chen, On near-perfect numbers with two distinct prime factors, *Bull. Aust. Math. Soc.* **88**(3) (2013), 520–524.
[4] M. Tang and M. Feng, On deficient-perfect numbers, *Bull. Aust. Math. Soc.* **90**(2) (2014), 186–194.
[5] M. Tang, X. Ma and M. Feng, On near-perfect numbers, *Colloq. Math.* **144**(2) (2016), 157–188.
[6] M. Tang, X.-Z. Ren and M. Li, On near-perfect and deficient-perfect numbers, *Colloq. Math.* **133**(2) (2013), 221–226.
[7] S. Yang and A. Togbé, Odd deficient-perfect numbers with four distinct prime factors, *Integers* **23** (2023), 20, paper a5.

Neeti Ghiya and S Manjula

Eulerian integrals associated with the product of multivariable *H*-function and generalized *M*-series

Abstract: Eulerian integrals are used in mathematical and applied sciences to have enhanced comprehensions of wide variety of applications. The intent of this paper is to ascertain Eulerian integral formulae which involve the product of multivariable H-function and a generalized M-series. One of the results is expressed as fractional integral, which is an application. These integrals provide the key formulae that can be used to derive numerous other useful results. Some special cases are also mentioned.

Keywords: Eulerian integral, fractional integral, multivariable H-function, generalized M-series

MSC 2010: 33C45, 33C60

1 Introduction

Eulerian integrals play an important role in many areas of mathematics, science and engineering like algebraic geometry, data aggregation, signal processing, image processing, etc. In the process of study of the screening properties of a charged impurity located near and inside the surface of the metal subjected to a magnetic field, an stimulating and useful class of Eulerian integral containing the Bessel function $J_\nu(z)$ or $J_0(z)$ came to light, which were written in compact forms by M. L. Glasser and further L. T. Wille presented Eulerian integrals pertaining to Meijer's G-function. A number of researchers Raina and Srivastava [5], Saxena and Nishimoto [9], Srivastava and Hussain [13], Srivastava and Singh [16], Srivastava and Garg [11], Saigo and Raina [6], Chaurasia and Gupta [1] worked on these types of integrals. In an attempt to unify the aforementioned results and several other results, we evaluate general Eulerian integral formulae involving a product of multivariable H-function and a generalized M-series in compact form. These results provide key formulae from which many integrals can be deduced.

The multivariable H-function familiarized by Srivastava and Panda ([14, 15]) in the following manner:

Acknowledgement: The authors thank the anonymous referee for helpful comments.

Neeti Ghiya, Department of Mathematics, RV College of Engineering, R. V. Vidyaniketan Post, Mysore Road, 560059 Bengaluru, Karnataka, India, e-mail: neethighiya@rvce.edu.in
S Manjula, Department of Mathematics , Government PU College, Uppinangady, 574241 Dakshina Kannada, Karnataka, India, e-mail: manjula.siddesh.banakar@gmail.com

https://doi.org/10.1515/9783111313634-017

$$H[x_1,\dots,x_t] = H^{0,N;\{M_t,N_t\}}_{P,Q;\{P_t,Q_t\}}\left[\begin{matrix} x_1 \\ \vdots \\ x_t \end{matrix}\left|\begin{matrix}(s_j;S_k^{(1)},\dots,S_k^{(t)})_{1,P};(a_k^{(1)},A_k^{(1)})_{1,P_1};\dots;(a_k^{(t)},A_k^{(t)})_{1,P_t} \\ (r_j;R_k^{(1)},\dots,R_k^{(t)})_{1,Q};(b_k^{(1)},B_k^{(1)})_{1,Q_1};\dots;(b_k^{(t)},B_k^{(t)})_{1,Q_t}\end{matrix}\right.\right]$$

$$= \frac{1}{(2\pi i)^t}\int_{L_1}\cdots\int_{L_t}\phi(\xi_1,\dots,\xi_t)\prod_{i=1}^{t}(\psi_i(\xi_i)\,x_i^{\xi_i}\,d\xi_i), \tag{17.1}$$

where

$$\phi(\xi_1,\dots,\xi_t) = \frac{\prod_{k=1}^{N}\Gamma(1-s_k+\sum_{j=1}^{t}S_k^{(j)}\xi_j)}{\prod_{k=N+1}^{P}\Gamma(s_k-\sum_{j=1}^{t}S_k^{(j)}\xi_j)\prod_{k=1}^{Q}\Gamma(1-r_k+\sum_{j=1}^{t}R_k^{(j)}\xi_j)}, \tag{17.2}$$

$$\psi_j(\xi_j) = \frac{\prod_{k=1}^{N_j}\Gamma(1-a_k^{(j)}+A_k^{(j)}\xi_j)\prod_{k=1}^{M_j}\Gamma(b_k^{(j)}-B_k^{(j)}\xi_j)}{\prod_{k=N_j+1}^{P_j}\Gamma(a_k^{(j)}-A_k^{(j)}\xi_j)\prod_{k=M_j+1}^{Q_j}\Gamma(1-b_k^{(j)}+B_k^{(j)}\xi_j)}; \quad (j=1,\dots,t). \tag{17.3}$$

$\{M_t,N_t\}$ stands for $M_1,N_1;\dots;M_t,N_t$ and $(a_k^{(t)},A_k^{(t)})_{1,P_t}$ represents the arrangement of t-ordered pairs $(a_k^{(1)},A_k^{(1)})_{1,P_1};\dots;(a_k^{(t)},A_k^{(t)})_{1,P_t}$.

A new generalization of M-series initiated by Faraj, Salim, Sadek and Ismail [8] is defined as follows:

$$\underset{m',n';p',q'}{\overset{\gamma',\delta'}{M}}(y) = \underset{m',n';p',q'}{\overset{\gamma',\delta'}{M}}(c_1,\dots,c_{p'};d_1,\dots,d_{q'};y)$$

$$= \sum_{r=0}^{\infty}\frac{(c_1)_{rp'},\dots,(c_{m'})_{rp'}}{(d_1)_{rq'},\dots,(d_{n'})_{rq'}}\frac{y^r}{\Gamma(\gamma' r+\delta')}, \tag{17.4}$$

where y, γ', $\delta' \in \mathbb{C}$, $\mathrm{Re}(\gamma') > 0$ and p', q' are non-negative real numbers. For more details, refer the paper given by Sharma and Jain [10].

a. When $m' = n' = 1$, we have

$$\underset{1,1;p',q'}{\overset{\gamma',\delta'}{M}}(y) = E^{c_1,d_1,p'}_{\gamma',\delta',q'}(y) = \sum_{r=0}^{\infty}\frac{(c_1)_{rp'}}{(d_1)_{rq'}}\frac{y^r}{\Gamma(\gamma' r+\delta')}. \tag{17.5}$$

Thus the M-series reduced to the Mittag–Leffler function as mentioned by Salim and Faraz [7].

b. When $\gamma' = \delta' = 1$ and $p' = q' = 1$, we have

$$\underset{m',n';1,1}{\overset{1,1}{M}}(y) = \sum_{r=0}^{\infty}\frac{(c_1)_r,\dots,(c_{m'})_r}{(d_1)_r,\dots,(d_{n'})_r}\frac{y^r}{r!} = {}_{m'}F_{n'}\left[(c_i)_1^{m'};(d_i)_1^{n'};y\right]. \tag{17.6}$$

Thus the M-series reduced to the generalized hypergeometric function as given by Mathai and Saxena [3].

There are two types of Euler's integral:

1. Euler's integral of first kind, which is also known as Beta function and is defined as

$$B(x, y) = \int_0^1 t^{x-1}(1-t)^{y-1}dt, \tag{17.7}$$

for all $x, y \in \mathbb{C}$ such that $\operatorname{Re}(x), \operatorname{Re}(y) > 0$.

2. Euler's integral of second kind, which is also known as Gamma function, is defined as

$$\Gamma(z) = \int_0^\infty t^{z-1}e^{-t}dt, \tag{17.8}$$

for all $z \in \mathbb{C}$, such that $\operatorname{Re}(z) > 0$.

Both of the Euler's integral can be related to each other by the following formula

$$B(x, y) = \frac{\Gamma(x)\,\Gamma(y)}{\Gamma(x+y)}, \tag{17.9}$$

can be written in its equivalent form

$$\int_a^b (t-a)^{x-1}(b-t)^{y-1}dt = (b-a)^{x+y-1}B(x, y), \quad [\operatorname{Re}(x) > 0;\ \operatorname{Re}(y) > 0;\ a \neq b] \tag{17.10}$$

since

$$(vt+w)^{\gamma} = (av+w)^{\gamma}\sum_{n=0}^{\infty}\frac{(-\gamma)_n}{n!}\left\{-\frac{(t-a)v}{av+w}\right\}^n, \tag{17.11}$$

$[|(t-a)v| < |av+w|; t \in [a, b]]$.

We readily find from (17.10) that

$$\int_a^b (t-a)^{x-1}(b-t)^{y-1}(vt+w)^{\gamma}(gt+h)^{\eta}dt = (b-a)^{x+y-1}(av+w)^{\gamma}(bg+h)^{\eta} \times B(x,y)F_3\left(x, y, -\gamma, -\eta;\ x+y;\ -\frac{(b-a)v}{av+w}, \frac{(b-a)g}{bg+h}\right), \tag{17.12}$$

where for convergence

$$\max\left\{\left|\frac{(b-a)v}{av+w}\right|, \left|\frac{(b-a)g}{bg+h}\right|\right\} < 1; \quad b \neq a, \min\{\operatorname{Re}(x), \operatorname{Re}(y)\} > 0.$$

2 Main result

In this section, we shall obtain the following two Eulerian integral formulas associated with the product of multivariable H-function and generalized M-series.

Integral I

$$\int_c^d (u-c)^{\mu-1}(d-u)^{\varsigma-1}(eu+f)^{\nu}(gu+h)^{\kappa}$$
$$\times \underset{m',n';p',q'}{\overset{\alpha',\beta'}{M}}\left[T(u-c)^{\alpha}(d-u)^{\beta}(eu+f)^{\delta}(gu+h)^{\tau}\right]$$
$$\times H\left[z_1\frac{(u-c)^{p_1}(d-u)^{q_1}}{(eu+f)^{r_1}(gu+h)^{s_1}},\ldots,z_t\frac{(u-c)^{p_t}(d-u)^{q_t}}{(eu+f)^{r_t}(gu+h)^{s_t}}\right]du$$
$$=(d-c)^{\mu+\varsigma-1}(ec+f)^{\nu}(dg+h)^{\kappa}$$
$$\sum_{l=0}^{\infty}\frac{(a_1)_{lp'},\ldots,(a_{m'})_{lp'}}{(b_1)_{lq'},\ldots,(b_{n'})_{lq'}}\frac{T^l}{\Gamma(\alpha' l+\beta')}(d-c)^{(\alpha+\beta)l}(ce+f)^{\delta l}(dg+h)^{\tau l}$$
$$\times H^{0,N+4;M_1,N_1;\ldots;M_t,N_t;1,0;1,0}_{P+4,Q+3;P_1,Q_1;\ldots;P_t,Q_t;0,1;0,1}\left[I_1,\ldots,I_t,\frac{(d-c)e}{ce+f},\frac{-(d-c)g}{dg+h}\,\middle|\,\begin{matrix}A:C;\;-\;;\;-\\B:D;(0,1);(0,1)\end{matrix}\right],\tag{17.13}$$

where

$$I_j=z_j\frac{(d-c)^{p_j+q_j}}{(ce+f)^{r_j}(dg+h)^{s_j}},\quad(j=1,\ldots,t)$$
$$A=(1+\nu+\delta l;r_1,\ldots,r_t,1,0),(1+\kappa+\tau l;s_1,\ldots,s_t,0,1),(1-\mu-\alpha l;p_1,\ldots,p_t,1,0),$$
$$(1-\varsigma-\beta l;q_1,\ldots,q_t,0,1),(s_k;S_k^{(1)},\ldots,S_k^{(t)},0,0)_{1,P}$$
$$B=(1+\nu+\delta l;r_1,\ldots,r_t,0,0),(1+\kappa+\tau l;s_1,\ldots,s_t,0,0),(r_k;R_k^{(1)},\ldots,R_k^{(t)},0,0)_{1,Q},$$
$$(1-\mu-\varsigma-\beta l-\alpha l;p_1+q_1,\ldots,p_t+q_t,1,1)$$
$$C=(a_k^{(1)},A_k^{(1)})_{1,P_1};\ldots;(a_k^{(t)},A_k^{(t)})_{1,P_t}$$

and

$$D=(b_k^{(1)},B_k^{(1)})_{1,Q_1};\ldots;(b_k^{(t)},B_k^{(t)})_{1,Q_t}$$

provided that the following conditions are satisfied

(i) $\mathrm{Re}(\mu,\varsigma)>0$

(ii) $\min(p_j,q_j,r_j,s_j)\geq 0\ (j=1,\ldots,t)$ (not all zero simultaneously)

(iii) $T,u,c,d,e,f,g,h,\alpha,\beta,\delta,\tau,\alpha',\beta'\in\mathbb{C}$, $\mathrm{Re}(\alpha')>0$, p',q' are non-negative real numbers

(iv) $\mathrm{Re}[\mu] + \alpha + \sum_{j=1}^{t} p_j \min_{1\le k\le M_j} \mathrm{Re}(\frac{b_k^{(j)}}{B_k^{(j)}}) > 0,$

$\mathrm{Re}[\varsigma] + \beta + \sum_{j=1}^{t} q_j \min_{1\le k\le M_j} \mathrm{Re}(\frac{b_k^{(j)}}{B_k^{(j)}}) > 0$

(v) $\max\{|\frac{(d-c)e}{ce+f}|, |\frac{(d-c)g}{dg+h}|\} < 1; d \neq c.$

Proof. Let us denote the left-hand side of integral-I as L_λ.

$$L_\lambda = \int_c^d (u-c)^{\mu-1}(d-u)^{\varsigma-1}(eu+f)^{\nu}(gu+h)^{\kappa}$$
$$\underset{m',n';p',q'}{\overset{\alpha',\beta'}{M}} [T(u-c)^{\alpha}(d-u)^{\beta}(eu+f)^{\delta}(gu+h)^{\tau}]$$
$$\times H\left[z_1 \frac{(u-c)^{p_1}(d-u)^{q_1}}{(eu+f)^{r_1}(gu+h)^{s_1}}, \dots, z_t \frac{(u-c)^{p_t}(d-u)^{q_t}}{(eu+f)^{r_t}(gu+h)^{s_t}}\right] du.$$

Express the multivariable H-function in integral form and the generalized M-series function in series form using the equations (17.1) and (17.4) respectively.

$$L_\lambda = \int_c^d (u-c)^{\mu-1}(d-u)^{\varsigma-1}(eu+f)^{\nu}(gu+h)^{\kappa}$$
$$\sum_{l=0}^{\infty} \frac{(a_1)_{lp'}, \dots, (a_{m'})_{lp'}}{(b_1)_{lq'}, \dots, (b_{n'})_{lq'}} \frac{T^l (u-c)^{\alpha l}(d-u)^{\beta l}(eu+f)^{\delta l}(gu+h)^{\tau l}}{\Gamma(\alpha' l + \beta')}$$
$$\times \left\{ \frac{1}{(2\pi\omega)^t} \int_{L_1}, \dots, \int_{L_t} z_1^{\xi_1} \frac{(u-c)^{p_1\xi_1}(d-u)^{q_1\xi_1}}{(eu+f)^{r_1\xi_1}(gu+h)^{s_1\xi_1}}, \dots, \right.$$
$$\left. \times z_t^{\xi_t} \frac{(u-c)^{p_t\xi_t}(d-u)^{q_t\xi_t}}{(eu+f)^{r_t\xi_t}(gu+h)^{s_t\xi_t}} d\xi_1, \dots, d\xi_t \right\} du. \tag{17.14}$$

Collect the powers of $(eu+f)$ and $(gu+h)$ and apply the following binomial expansions.

$$(eu+f)^{\eta_o} = (ec+f)^{\eta_o} \sum_{m_1=0}^{\infty} \frac{(-\eta_o)_{m_1}}{m_1!} \left\{\frac{-e(u-c)}{(ec+f)}\right\}^{m_1}, \tag{17.15}$$

where $|e(u-c)| < |ec+f|$ and

$$(gu+h)^{\omega_0} = (gd+h)^{\omega_0} \sum_{m_2=0}^{\infty} \frac{(-\omega_0)_{m_2}}{m_2!} \left\{\frac{g(d-u)}{(gd+h)}\right\}^{m_2}, \tag{17.16}$$

where $|g(d-u)| < |gd+h|$.

Then, the equation (17.14) becomes as follows

$$L_\lambda = \sum_{l=0}^{\infty} \frac{(a_1)_{lp'}, \ldots, (a_{m'})_{lp'}}{(b_1)_{lq'}, \ldots, (b_{n'})_{lq'}} \frac{T^l}{\Gamma(\alpha' l + \beta')} \left\{ \frac{1}{(2\pi\omega)^t} \int_{L_1}, \ldots, \int_{L_t} z_1^{\xi_1}, \ldots, z_t^{\xi_t} \right.$$

$$\times (ec+f)^{\nu+\delta l - \sum_{i=1}^t r_i \xi_i} \sum_{\xi_{t+1}=0}^{\infty} \frac{(-\nu - \delta l + \sum_{i=1}^t r_i \xi_i)_{\xi_{t+1}}}{\xi_{t+1}!} \left\{ \frac{-e(u-c)}{(ec+f)} \right\}^{\xi_{t+1}}$$

$$\times (gd+h)^{\kappa+\tau l - \sum_{i=1}^t s_i \xi_i} \sum_{\xi_{t+2}=0}^{\infty} \frac{(-\kappa - \tau l + \sum_{i=1}^t s_i \xi_i)_{\xi_{t+2}}}{\xi_{t+2}!} \left\{ \frac{-g(u-d)}{(gd+h)} \right\}^{\xi_{t+2}}$$

$$\times \left\{ \left\{ \int_c^d (u-c)^{\mu+\alpha l + \sum_{i=1}^t p_i \xi_i - 1} (d-u)^{\varsigma + \beta l + \sum_{i=1}^t q_i \xi_i - 1} \, du \right\} d\xi_1, \ldots, d\xi_t \right\}.$$

Now, collect the powers of $(u-c)$ and $(d-u)$, we get

$$L_\lambda = \sum_{l=0}^{\infty} \frac{(a_1)_{lp'}, \ldots, (a_{m'})_{lp'}}{(b_1)_{lq'}, \ldots, (b_{n'})_{lq'}} \frac{T^l}{\Gamma(\alpha' l + \beta')} \left\{ \frac{1}{(2\pi\omega)^t} \int_{L_1}, \ldots, \int_{L_t} z_1^{\xi_1}, \ldots, z_t^{\xi_t} \right.$$

$$\times (ec+f)^{\nu+\delta l - \sum_{i=1}^t r_i \xi_i} \sum_{\xi_{t+1}=0}^{\infty} \frac{(-\nu - \delta l + \sum_{i=1}^t r_i \xi_i)_{\xi_{t+1}}}{\xi_{t+1}!} \left\{ \frac{-e}{(ec+f)} \right\}^{\xi_{t+1}}$$

$$\times (gd+h)^{\kappa+\tau l - \sum_{i=1}^t s_i \xi_i} \sum_{\xi_{t+2}=0}^{\infty} \frac{(-\kappa - \tau l + \sum_{i=1}^t s_i \xi_i)_{\xi_{t+2}}}{\xi_{t+2}!} \left\{ \frac{g}{(gd+h)} \right\}^{\xi_{t+2}}$$

$$\times \left\{ \left\{ \int_c^d (u-c)^{\mu+\alpha l + \sum_{i=1}^t p_i \xi_i + \xi_{t+1} - 1} (d-u)^{\varsigma + \beta l + \sum_{i=1}^t q_i \xi_i + \xi_{t+2} - 1} \, du \right\} d\xi_1, \ldots, d\xi_t \right\}.$$

Apply the result of equation (17.10) and interpreting the obtained Mellin–Barnes contour integral as H-function of $(t+2)$ variables, then we attain the result of Integral-I after a little simplification. □

Integral II

$$\int_c^d (u-c)^{\mu-1} (d-u)^{\varsigma-1} (eu+f)^{\nu} (gu+h)^{\kappa}$$

$$\times \underset{m',n';p',q'}{\overset{\alpha',\beta'}{M}} \left[T(u-c)^{\alpha} (d-u)^{\beta} (eu+f)^{\delta} (gu+h)^{\tau} \right]$$

$$\times H^* \left[z_1 \frac{(eu+f)^{r_1} (gu+h)^{s_1}}{(u-c)^{p_1} (d-u)^{q_1}}, \ldots, z_t \frac{(eu+f)^{r_t} (gu+h)^{s_t}}{(u-c)^{p_t} (d-u)^{q_t}} \right] du$$

$$= (d-c)^{\mu+\varsigma-1}(ec+f)^{\nu}(dg+h)^{\kappa}$$

$$\sum_{l=0}^{\infty}\frac{(a_1)_{lp'},\ldots,(a_{m'})_{lp'}}{(b_1)_{lq'},\ldots,(b_{n'})_{lq'}}\frac{T^l}{\Gamma(\alpha' l+\beta')}(d-c)^{(\alpha+\beta)l}(ce+f)^{\delta l}(dg+h)^{\tau l}$$

$$\times H^{0,4;N_1,M_1:\ldots:N_t,M_t;1,0;1,0}_{Q+4,P+3;Q_1,P_1;\ldots;Q_t,P_t;0,1;0,1}\left[J_1,\ldots,J_t,\ \frac{(d-c)e}{ec+f},\ \frac{-(d-c)g}{dg+h}\ \middle|\ \begin{matrix}A^*: C^*;\ -\ ;\ -\\ B^*: D^*; (0,1); (0,1)\end{matrix}\right], \tag{17.17}$$

where $H^*[z_1,\ldots,z_t]$ stands for $H[z_1,\ldots,z_t]$ with $N=0$,

$$I_j = \frac{1}{z_j}\frac{(d-c)^{p_j+q_j}}{(ec+f)^{r_j}(dg+h)^{s_j}}, \qquad (j=1,\ldots,t)$$

$$A^* = (1+\nu+\delta l; r_1,\ldots,r_t, 1, 0), (1+\kappa+\tau l; s_1,\ldots,s_t, 0, 1), (1-\mu-\alpha l; p_1,\ldots,p_t, 1, 0),$$
$$(1-\varsigma-\beta l; q_1,\ldots,q_t, 0, 1), (1-r_k; R_k^{(1)},\ldots,R_k^{(t)}, 0, 0)_{1,Q}$$
$$B^* = (1+\nu+\delta l; r_1,\ldots,r_t, 0, 0), (1+\kappa+\tau l; s_1,\ldots,s_t, 0, 0), (1-s_k; S_k^{(1)},\ldots,S_k^{(t)}, 0, 0)_{1,P},$$
$$(1-\mu-\varsigma-\beta l-\alpha l; p_1+q_1,\ldots,p_t+q_t, 1, 1)$$
$$C^* = (1-b_k^{(1)}, B_k^{(1)})_{1,Q_1};\ldots;(1-b_k^{(t)}, B_k^{(t)})_{1,Q_t}$$

and

$$D^* = (1-a_k^{(1)}, A_k^{(1)})_{1,P_1};\ldots;(1-a_k^{(t)}, A_k^{(t)})_{1,P_t}$$

provided that the following conditions are satisfied

(i) $\mathrm{Re}(\mu,\varsigma) > 0$,

(ii) $\min(p_j, q_j, r_j, s_j) \geq 0$ $(j=1,\ldots,t)$ (not all zero simultaneously),

(iii) $T, u, c, d, e, f, g, h, \alpha, \beta, \delta, \tau, \alpha', \beta' \in \mathbb{C}$,
$\mathrm{Re}(\alpha') > 0, p', q'$ are non-negative real numbers

(iv) $\mathrm{Re}[\mu] + \alpha + \sum_{j=1}^{t} p_j \max_{1\leq k\leq N_j} \mathrm{Re}(\frac{a_k^{(j)}-1}{A_k^{(j)}}) > 0$,
$\mathrm{Re}[\varsigma] + \beta - \sum_{j=1}^{t} q_j \max_{1\leq k\leq N_j} \mathrm{Re}(\frac{a_k^{(j)}-1}{A_k^{(j)}}) > 0$

(v) $\max\{|\frac{(d-c)e}{ce+f}|, |\frac{(d-c)g}{dg+h}|\} < 1;\ d \neq c.$

Proof. The proof of Integral-II is omitted since it is same as Integral-I (page no. 14 in [12]). □

3 Special cases

A. Set $m' = n' = 1$ in Integral-I and Integral-II, the generalized M-series reduces to generalized Mittag–Leffler function.

$$\text{(i)}\quad \int_c^d (u-c)^{\mu-1}(d-u)^{\varsigma-1}(eu+f)^{\nu}(gu+h)^{\kappa}$$
$$\times E^{a_1,b_1,p'}_{\alpha',\beta',q'}[T(u-c)^{\alpha}(d-u)^{\beta}(eu+f)^{\delta}(gu+h)^{\tau}]$$
$$\times H\left[z_1\frac{(u-c)^{p_1}(d-u)^{q_1}}{(eu+f)^{r_1}(gu+h)^{s_1}},\ldots,z_t\frac{(u-c)^{p_t}(d-u)^{q_t}}{(eu+f)^{r_t}(gu+h)^{s_t}}\right]du$$
$$=(d-c)^{\mu+\varsigma-1}(ec+f)^{\nu}(dg+h)^{\kappa}$$
$$\sum_{l=0}^{\infty}\frac{(a_1)_{lp'}}{(b_1)_{lq'}}\frac{T^l}{\Gamma(\alpha'l+\beta')}(d-c)^{(\alpha+\beta)l}(ec+f)^{\delta l}(dg+h)^{\tau l}$$
$$\times H^{0,N+4;M_1,N_1:\ldots:M_t,N_t;1,0;1,0}_{P+4,Q+3;P_1,Q_1;\ldots;P_t,Q_t;0,1;0,1}\left[I_1,\ldots,I_t,\frac{(d-c)e}{ce+f},\frac{-(d-c)g}{dg+h}\,\Big|^{A:\,C;\ -\ ;\ -}_{B:\,D;\,(0,1);\,(0,1)}\right],\tag{17.18}$$

where I_j, A, B, C and D are same mentioned in integral-I.

$$\text{(ii)}\quad \int_c^d (u-c)^{\mu-1}(d-u)^{\varsigma-1}(eu+f)^{\nu}(gu+h)^{\kappa}$$
$$\times E^{a_1,b_1,p'}_{\alpha',\beta',q'}[T(u-c)^{\alpha}(d-u)^{\beta}(eu+f)^{\delta}(gu+h)^{\tau}]$$
$$\times H^*\left[z_1\frac{(eu+f)^{r_1}(gu+h)^{s_1}}{(u-c)^{p_1}(d-u)^{q_1}},\ldots,z_t\frac{(eu+f)^{r_t}(gu+h)^{s_t}}{(u-c)^{p_t}(d-u)^{q_t}}\right]du$$
$$=(d-c)^{\mu+\varsigma-1}(ec+f)^{\nu}(dg+h)^{\kappa}$$
$$\sum_{l=0}^{\infty}\frac{(a_1)_{lp'}}{(b_1)_{lq'}}\frac{T^l}{\Gamma(\alpha'l+\beta')}(d-c)^{(\alpha+\beta)l}(ec+f)^{\delta l}(dg+h)^{\tau l}$$
$$\times H^{0,4;N_1,M_1:\ldots:N_t,M_t;1,0;1,0}_{Q+4,P+3;Q_1,P_1;\ldots;Q_t,P_t;0,1;0,1}\left[J_1,\ldots,J_t,\frac{(d-c)e}{ec+f},\frac{-(d-c)g}{dg+h}\,\Big|^{A^*:\,C^*;\ -\ ;\ -}_{B^*:\,D^*;\,(0,1);\,(0,1)}\right],\tag{17.19}$$

where J_j, A^*, B^*, C^* and D^* are same arrays in integral-II.

B. Set $\alpha'=\beta'=1$ and $p'=q'=1$ with arbitrary m', n' in integral-I and Integral-II, the generalized M-series reduces to generalized hypergeometric function.

$$(i)\quad \int_c^d (u-c)^{\mu-1}(d-u)^{\varsigma-1}(eu+f)^{\nu}(gu+h)^{\kappa}$$
$$\times\, {}_{m'}F_{n'}[(a_i)_1^{m'};(b_i)_1^{n'};T(u-c)^{\alpha}(d-u)^{\beta}(eu+f)^{\delta}(gu+h)^{\tau}]$$
$$\times H\left[z_1\frac{(u-c)^{p_1}(d-u)^{q_1}}{(eu+f)^{r_1}(gu+h)^{s_1}},\ldots,z_t\frac{(u-c)^{p_t}(d-u)^{q_t}}{(eu+f)^{r_t}(gu+h)^{s_t}}\right]du$$
$$=(d-c)^{\mu+\varsigma-1}(ec+f)^{\nu}(dg+h)^{\kappa}$$

$$\sum_{l=0}^{\infty}\frac{\prod_{i=1}^{m'}(a_i)_l}{\prod_{j=1}^{n'}(b_j)_l}\frac{T^l}{l!}(d-c)^{(\alpha+\beta)l}(ec+f)^{\delta l}(dg+h)^{\tau l}$$
$$\times H^{0,N+4;M_1,N_1;\ldots;M_t,N_t;1,0;1,0}_{P+4,Q+3;P_1,Q_1;\ldots;P_t,Q_t;0,1;0,1}\left[I_1,\ldots,I_t,\frac{(d-c)e}{ce+f},\frac{-(d-c)g}{dg+h}\,\middle|\begin{matrix}A:C;\ -\ ;\ -\\ B:D;(0,1);(0,1)\end{matrix}\right], \quad (17.20)$$

where I_j, A, B, C and D are same mentioned in integral-I.

(ii)
$$\int_c^d (u-c)^{\mu-1}(d-u)^{\varsigma-1}(eu+f)^{\nu}(gu+h)^{\kappa}$$
$$\times\, {}_{m'}F_{n'}\left[(a_i)_1^{m'};(b_i)_1^{n'};T(u-c)^{\alpha}(d-u)^{\beta}(eu+f)^{\delta}(gu+h)^{\tau}\right]$$
$$\times H^*\left[z_1\frac{(eu+f)^{r_1}(gu+h)^{s_1}}{(u-c)^{p_1}(d-u)^{q_1}},\ldots,z_t\frac{(eu+f)^{r_t}(gu+h)^{s_t}}{(u-c)^{p_t}(d-u)^{q_t}}\right]du$$
$$=(d-c)^{\mu+\varsigma-1}(ec+f)^{\nu}(dg+h)^{\kappa}$$
$$\sum_{l=0}^{\infty}\frac{\prod_{i=1}^{m'}(a_i)_l}{\prod_{j=1}^{n'}(b_j)_l}\frac{T^l}{l!}(d-c)^{(\alpha+\beta)l}(ec+f)^{\delta l}(dg+h)^{\tau l}$$
$$\times H^{0,4;N_1,M_1;\ldots;N_t,M_t;1,0;1,0}_{Q+4,P+3;Q_1,P_1;\ldots;Q_t,P_t;0,1;0,1}\left[J_1,\ldots,J_t,\frac{(d-c)e}{ec+f},\frac{-(d-c)g}{dg+h}\,\middle|\begin{matrix}A^*:C^*;\ -\ ;\ -\\ B^*:D^*;(0,1);(0,1)\end{matrix}\right], \quad (17.21)$$

where J_j, A^*, B^*, C^* and D^* are same arrays as given in Integral-II.

4 Application

Eulerian integrals can be expressed as fractional integrals ([2, 4]) involving the operator ${}_cD_z^{-\varsigma}$ defined by

$${}_cD_z^{-\varsigma}[f(z)]=\frac{1}{\Gamma(\varsigma)}\int_c^z (z-t)^{\varsigma-1}f(t)\,dt;\quad \mathrm{Re}(\varsigma)>0. \quad (17.22)$$

1. When c=0, we have

$${}_0D_z^{-\varsigma}[f(z)]=\frac{1}{\Gamma(\varsigma)}\int_0^z (z-t)^{\varsigma-1}f(t)\,dt;\quad \mathrm{Re}(\varsigma)>0. \quad (17.23)$$

Thus the fractional integral operator (17.22) reduces to the Reimann–Liouville fractional integral of order ς.

2. When $c \to \infty$, we have

$$ {}_zD_{\infty}^{-\varsigma}[f(z)] = \frac{1}{\Gamma(\varsigma)} \int_z^{\infty} (t-z)^{\varsigma-1} f(t)\,dt; \quad \operatorname{Re}(\varsigma) > 0. \tag{17.24} $$

Thus the fractional integral operator (17.22) reduces to the Weyl fractional integral of order ς.

Under the conditions of integral-I, we express the integral-I as fractional integral formula by setting $d = z, \beta = 0 = q_j$ $(j = 1, \dots, t)$.

$$ \begin{aligned} &{}_cD_z^{-\varsigma}\Big\{(z-c)^{\mu-1}(ez+f)^{\nu}(gz+h)^{\kappa} \, M_{m',n';p',q'}^{\alpha',\beta'}\left[T(z-c)^{\alpha}(ez+f)^{\delta}(gz+h)^{\tau}\right] \\ &\times H\left[z_1 \frac{(z-c)^{p_1}}{(ez+f)^{r_1}(gz+h)^{s_1}}, \dots, z_t \frac{(z-c)^{p_t}}{(ez+f)^{r_t}(gz+h)^{s_t}}\right]\Big\} = \frac{1}{\Gamma(\varsigma)}(z-c)^{\mu+\varsigma-1} \\ &\times (ec+f)^{\nu}(gz+h)^{\kappa} \sum_{l=0}^{\infty} \frac{(a_1)_{lp'}, \dots, (a_{m'})_{lp'}}{(b_1)_{lq'}, \dots, (b_{n'})_{lq'}} \frac{T^l}{\Gamma(\alpha' l + \beta')} (z-c)^{\alpha l}(ec+f)^{\delta l}(zg+h)^{\tau l} \\ &\times H_{P+3,Q+3;P_1,Q_1;\dots;P_t,Q_t;0,1;1,1}^{0,N+3;M_1,N_1:\dots:M_t,N_t;1,0;1,1}\left[V_1, \dots, V_t, \frac{(z-c)e}{ec+f}, \frac{-(z-c)g}{gz+h} \,\middle|\, \begin{matrix} A^{**}: C^{**}; -; (1-\varsigma,1) \\ B^{**}: D^{**}; (0,1); (0,1) \end{matrix}\right], \end{aligned} \tag{17.25} $$

where

$$ V_j = z_j \frac{(z-c)^{p_j}}{(ec+f)^{r_j}(zg+h)^{s_j}}; \quad (j = 1, \dots, t) $$

$$ \begin{aligned} A^{**} &= (1+\nu+\delta l; r_1, \dots, r_t, 1, 0), (1+\kappa+\tau l; s_1, \dots, s_t, 0, 1), (1-\mu-\alpha l; p_1, \dots, p_t, 1, 0), \\ &\quad (s_k; S_k^{(1)}, \dots, S_k^{(t)}, 0, 1)_{1,P} \\ B^{**} &= (1+\nu+\delta l; r_1, \dots, r_t, 0, 0), (1+\kappa+\tau l; s_1, \dots, s_t, 0, 0), (1-\mu-\varsigma-\alpha l; p_1, \dots, p_t, 1, 1), \\ &\quad (r_k; R_k^{(1)}, \dots, R_k^{(t)}, 0, 0)_{1,Q} \\ C^{**} &= (a_k^{(1)}, A_k^{(1)})_{1,P_1}; \dots; (a_k^{(t)}, A_k^{(t)})_{1,P_t} \end{aligned} $$

and

$$ D^{**} = (b_k^{(1)}, B_k^{(1)})_{1,Q_1}; \dots; (b_k^{(t)}, B_k^{(t)})_{1,Q_t}. $$

Bibliography

[1] V. B. L. Chaurasia and N. Gupta, Some finite integrals involving a product of general polynomials, Fox's H-function with the multivariable H-function, *Acta Cienc. Indica* **XXIVM**(2) (1998), 173–178.

[2] M. Garg and M. Purohit, Eulerian integral formulas and associated fractional integration, *Tamkang J. Math.* **31**(3) (2000), 203–212.

[3] A. M. Mathai and R. K. Saxena, *The H-Function With Applications in Statistics and Other Disciplines*, John Wiley and sons, New York, 1978.

[4] K. S. Miller and B. Ross, *An Introduction to the Fractional Calculus and Fractional Differential Equations*, John Wiley and sons, New York, 1993.

[5] R. K. Raina and H. M. Srivastava, Evaluation of a certain class of Eulerian integrals, *J. Phys. A, Math. Gen.* **26** (1993), 691–696.

[6] M. Saigo and R. K. Raina, Evaluation of certain integrals in terms of generalized Kampe de Feriet function and Lauricella function, *Tamkang J. Math.* **26** (1995), 41–47.

[7] T. Salim and A. Faraj, A generalization of Mittag-Leffler function and integral operator associated with fractional calculus, *J. Fract. Calc. Appl.* **3**(5) (2012), 1–13.

[8] A. Faraj, T. Salim, S. Sadek and J. Ismail, A generalization of *M*-series and integral operator associated with fractional calculus, *Asian Online J.* **2** (2014), 142–155.

[9] R. K. Saxena and K. Nishimoto, Fractional integral formula for the *H*-function, *J. Fract. Calc.* **6** (1994), 64–75.

[10] M. Sharma and R. Jain, A note on a generalized *M*-series as a special function of fractional calculus, *Fract. Calc. Appl. Anal.* **4** (2009), 449–452.

[11] H. M. Srivastava and M. Garg, Some integrals involving a general class of polynomials and the multivariable *H*-function, *Rev. Roum. Phys.* **32** (1987), 685–692.

[12] H. M. Srivastava, K. C. Gupta and S. P. Goyal, *The H-Functions of One and Two Variables With Applications*, South Asian Publishers, 1982.

[13] H. M. Srivastava and M. A. Hussain, Fractional integration of the *H*-function of several variables, *Comput. Math. Appl.* **30** (1995), 73–85.

[14] H. M. Srivastava and R. Panda, Some expansion theorems and generating relations for the *H*-function of several complex variables I, *Comment. Math. Univ. St. Pauli* **24** (1975), 119–137.

[15] H. M. Srivastava and R. Panda, Some expansion theorems and generating relations for the *H*-function of several complex variables II, *Comment. Math. Univ. St. Pauli* **25** (1976), 167–197.

[16] H. M. Srivastava and N. P. Singh, The integration of multivariable *H*-function with a general class of polynomials, *Rend. Circ. Mat. Palermo* **32** (1983), 157–187.

R. Bala, T. Bose, and M. Aneja

Upper bounds for the complex growth rates in the Darcy–Brinkman convection in a binary viscoelastic fluid in a saturated porous layer

Abstract: In the present paper, we mathematically establish that the complex growth rate (σ_r, σ_i) of an arbitrary neutral or unstable oscillatory perturbation of growing amplitude, in the Darcy–Brinkman convection in a binary viscoelastic fluid in a saturated porous layer heated from below, lie inside a semicircle in the right half of the (σ_r, σ_i)-plane whose center is origin and radius equals $|\sigma| < \min[(\frac{\lambda R_T P_r}{1+\epsilon})^2, \frac{R_S}{\chi 4\pi^2 \epsilon D_a}]$, where R_T and R_S are the thermal and solute Rayleigh number and P_r is the Prandtl number. The bounds obtained herein, in particular, yield a sufficient condition for the validity of 'the principle of the exchange of stability'. Further, it is proved that above result is uniformly valid for quite general nature of the bounding surfaces.

Keywords: Darcy–Brinkman convection, binary viscoelastic fluid, double-diffusive convection, complex growth rate, porous layer

PACS: 47.10.ab, 47.10.ad, 47.20.-k

1 Introduction

Certainly, the investigation into the dynamics of viscoelastic fluid flow is paramount importance in diverse scientific, engineering, and technological domains. This encompasses crucial applications in geophysics, bioengineering, material processing, and various industries such as petroleum, chemical, nuclear industries, and so on. Several researchers, including Benard [1], Rayleigh [15], Rajagopal et al. [14], and Sekhar and Jayalatha [16] have carried out extensive theoretical and experimental research on Rayleigh–Benard convection within Newtonian fluids, delving into the complexities of this fluid dynamic phenomenon. Double diffusive convection has evolved in to a field with fundamental distinctions from its counterpart, namely single diffusive convection. Chandrasekhar [4], Drazin and Reid [5] and Bejan [3] did detailed theoretical and experimental studies of Rayleigh–Benard convection in Newtonian fluids. Double diffusive convection has matured into a subject possessing fundamental departure from its counterpart, namely single diffusive convection. The onset of convection of a horizontal fluid layer superposed on a porous layer when the system is heated from below has been

R. Bala, T. Bose, M. Aneja, Chitkara University Institute of Engineering and Technology, Chitkara University, Chandigarh-Patiala, National Highway, 140401 Rajpura, Punjab, India, e-mails: renu.bala@chitkara.edu.in, tania.bose@chitkara.edu.in, madhu.aneja@chitkara.edu.in

https://doi.org/10.1515/9783111313634-018

considered, firstly, by Sun [19], who showed that the critical Rayleigh number in the porous layer decreases continuously as the thickness of the fluid layer increases. Nield [12] used Darcy's law in formulating the equations of porous layer, while Somerton and Catton [17] used Brinkman's extension of Darcy's law to study the thermal instability of superposed porous and fluid layers when internal heating is allowed in the porous medium. For reviews of this subject, one may refer to Horton and Rogers [7], Lapwood [10], Katto and Masuoka [8], Khaled and Vafai [9], Prakash et al. [13], Bertola et al. [2] and Vadasz [21]. They studied various features of thermal convection of Newtonian fluids with porous medium. Govender [6] and Nield and Bejan [11] also gives the outline of thermal convection in a porous medium. Straughan [18] analyzed that the occurrence of overstabilities in porous layers heated from below. Yin et al. [22] studied the thermal instability of a viscoelastic fluid in a fluid-porous system with a plane Poiseuille flow. He found that thermal instability of the viscoelastic fluid is found to be more unstable than that of the Newtonian fluid in a two-layer system.

The present paper addresses the challenge of determining the maximum rate at which an oscillating motion with increasing intensity can grow in a thermal instability scenario, specifically in Darcy–Brinkman convection within a porous layer of a dual-viscosity fluid which is heated at the base. This becomes particularly tricky when both boundaries are not freely moving, making it impossible to find exact solutions in a straight forward manner. As a result, one has to rely on numerical simulations, which can be time-consuming. In this paper, such bounds for the linear growth rate of any oscillatory motion with increasing amplitude are determined. These bounds, particularly, provide conditions ensuring the principle of the exchange of stabilities, which holds for this specific configuration. The condition herein obtained is not dependent on the wave number and represents an improvement over existing results, such as those from Swamy et al. [20], which depend upon the wave number.

2 Mathematical formulation and analysis

The physical configuration for the present problem is taken exactly the same as described by Swamy et al. [20]. Here we consider an infinite horizontal sparsely packed porous layer. This layer is filled with Oldroyd-B fluid, a type of viscoelastic fluid, and is situated between two planes defined by the vertical positions $z = 0$ and $z = d$, with vertically downward gravity g acting on it. A constant adverse temperature difference $T = T_0 - T_1$ and a stabilizing concentration gradient $S = S_0 - S_1$. The lower surface is exposed to higher temperatures $T_0 > T_1$ and there is a higher concentration at the lower surface $S_0 > S_1$. This arrangement creates conditions that may lead to thermal and concentration-driven fluid motions within the layer, as illustrated in Fig. 18.1 This coordinate system provides a standardized reference frame for describing and analyzing the properties and dynamics of the system.

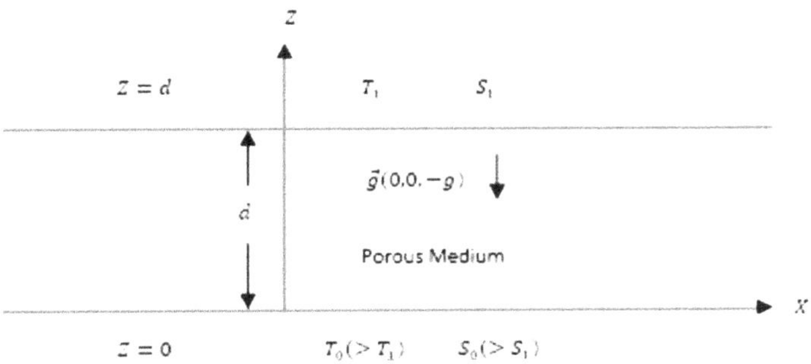

Figure 18.1: Physical configuration.

Swamy et al. [20] have presented the mathematical equations that describe how disturbances in a special fluid setup unfold. This particular system involves a twofold viscoelastic material flowing through a porosity layer, following the Darcy–Brinkman model. Additionally, the system is subject to the Oldroyd-B model and is heated from below. The equations include boundary conditions that account for the unique characteristics of this fluid system under consideration.

$$\left[\frac{\sigma}{p_r}(1+\lambda\sigma)+(1+\epsilon\sigma)\right](D^2-a^2)w-(1+\epsilon\sigma)(D^2-a^2)^2D_aw$$
$$=-(1+\lambda\sigma)R_Ta^2\theta+(1+\lambda\sigma)R_Sa^2\phi, \tag{18.1}$$
$$(D^2-a^2-\sigma)\theta=-w, \tag{18.2}$$
$$\left[\frac{1}{L_e}(D^2-a^2)-\chi\sigma\right]=-w \tag{18.3}$$

and

$$w=0=\theta=D^2w \quad \text{at } z=0 \text{ and } z=1, \tag{18.4}$$

when both the boundaries are free and

$$w=0=\theta=Dw \quad \text{at } z=0 \text{ and } z=1, \tag{18.5}$$

when both the boundaries are rigid.

The symbols in equations (18.1)–(18.5) holds specific physical meanings as follows: z is the vertical coordinate, D indicates the differentiation w. r. t. z, a^2 is square of wave number, $P_r > 0$ is the Prandtl number, l_e is the Lewis number, R_T is the thermal Rayleigh number, R_s is the concentration Rayleigh number, $\sigma = \sigma_r + i\sigma_i$ is the complex growth rate, where σ_r and σ_i are real constants, w is the vertical velocity, θ represents the temperature and ϕ is the concentration of the component. Additionally, It should be observed that equations (18.1)–(18.5) collectively describe an eigenvalue problem for σ and govern double diffusive convection, considering various combinations of flexible and fixed boundaries. The following conversation revolves around determining the maximum limits for the complex growth rates of oscillatory motions, whether they remain neutral or display increasing amplitudes, and a theorem is presented to validate the derived results.

Theorem 1. *If* $R_T > 0$, $R_S > 0$, $L_e > 0$, $\lambda > 0$, $\sigma_r \geq 0$, $\sigma_i \neq 0$, *then a prerequiste for the existence of a nontrivial outcome* $(w, \theta, \phi, \sigma)$ *of equation* (18.1)–(18.3) *alongwith the boundary conditions* (18.4)–(18.5) *is that*

$$|\sigma| < \min.\left[\left(\frac{\lambda R_T P_r}{1+\epsilon}\right)^2, \frac{R_s}{\chi 4\pi^2 \epsilon D_a}\right].$$

Proof. By taking the product of equation (18.1) by w^* ($*$ denotes the complex conjugation) and performing integration by parts over the range of z, we derive the following expression:

$$\left[\frac{\sigma}{p_r}(1+\lambda\sigma) + (1+\epsilon\sigma)\right]\int_0^1 w^*(D^2 - a^2)w dz - (1+\epsilon\sigma)D_a\int_0^1 w^*(D^2-a^2)^2 w dz$$
$$= -(1+\lambda\sigma)R_T a^2\int_0^1 \theta w^* dz + (1+\lambda\sigma)R_S a^2\int_0^1 \phi w^* dz. \tag{18.6}$$

By substituting the suitable expressions for w^* appropriately from eqs. (18.2)–(18.3) in to the right-hand side of equation (18.6), we get

$$\left[\frac{\sigma}{p_r}(1+\lambda\sigma) + (1-\epsilon\sigma)\right]\int_0^1 w^*(D^2 - a^2)w dz - (1+\epsilon\sigma)D_a\int_0^1 w^*(D^2-a^2)^2 w dz$$
$$= (1+\lambda\sigma)R_T a^2\int_0^1 \theta^*(D^2 - a^2 - \sigma)\theta dz - (1+\lambda\sigma)R_S a^2\int_0^1 \phi\left[\frac{1}{L_e}(D^2-a^2) - \chi\sigma^*\right]\phi dz. \tag{18.7}$$

After integrating equation (18.7) by successive steps and incorporating the boundary conditions (18.4) or (18.5) along with the given equality, we obtain

$$\int_0^1 \Psi^* D^2 n\psi dz = (-1)^n \int_0^1 |D^n\Psi|^2 dz \quad \psi = w(n = 1, 2) \text{ or } \psi = \theta(n = 1), \tag{18.8}$$

we get

$$\left[\frac{\sigma}{p_r}(1+\lambda\sigma) + (1+\epsilon\sigma)\right]\int_0^1 (|Dw|^2 + a^2|w|^2)dz$$

$$- (1+\epsilon\sigma)D_a\int_0^1(|D^2w|^2+2a^2|Dw|^2+a^4|w|^2)dz$$

$$= (1+\lambda\sigma)R_Ta^2\int_0^1(|D\theta|^2+a^2|\theta|^2+\sigma^*|\theta|^2)dz$$

$$- (1+\lambda\sigma)R_Sa^2\left[\frac{1}{L_e}\int_0^1(|D\phi|^2+a^2)|\phi|^2dz+\chi\sigma^*\int_0^1|\phi|^2\right]dz. \tag{18.9}$$

Now equating the imaginary parts of both sides of eq. (18.9) and cancelling $\sigma_i(\neq 0)$, we get

$$\frac{(1+2\lambda\sigma_r+\epsilon)}{P_r}\int_0^1(|Dw|^2+a^2|w|^2)dz+\epsilon D_a\int_0^1(|D^2w|^2+2a^2|Dw|^2+a^4|w|^2)dz$$

$$+R_Ta^2\int_0^1|\theta|^2dz+\frac{\lambda R_sa^2}{L_e}\int_0^1(|D\phi|^2+a^2|\phi|^2)dz$$

$$=\lambda R_Ta^2\int_0^1(|D\theta|^2+a^2|\theta|^2)dz+R_sa^2\chi\int_0^1|\phi|^2dz. \tag{18.10}$$

Now multiplying eq. (18.2) by θ^* (complex conjugate of θ), integrating the resulting equation over the vertical range of z, we get

$$\int_0^1\theta^*(D^2-a^2-\sigma)\theta dz=\int_0^1 w\theta^*dz,$$

$$\int_0^1(|D\theta|^2+a^2|\theta|^2+\sigma_r|\theta|^2)dz=\text{Real part of}\int_0^1(w\theta^*)dz$$

$$\leq\left|\int_0^1 w\theta^*dz\right|\leq|w\theta^*|dz$$

$$\leq\int_0^1|w||\theta^*|dz\leq\int_0^1|w||\theta|$$

$$\leq[|w|^2]^{1/2}[|\theta|^2]^{1/2}\quad\text{(using Schwartz inequality).}$$

Since $\sigma_r\geq 0$, we have from the above inequality

$$\int_0^1(|D\theta|^2+a^2|\theta|^2+\sigma^*|\theta|^2)dz\leq[|w|^2]^{1/2}[|\theta|^2]^{1/2}.$$

Now multiplying eq. (18.2) by its complex conjugate and integrating over the vertical range of z, we get

$$\int_0^1 (D^2 - a^2 - \sigma)\theta(D^2 - a^2 - \sigma^*)\theta^* dz = \int_0^1 ww^* dz,$$

$$\int_0^1 (|D^2\theta|^2 + 2a^2|D\theta|^2 + a^4|\theta|^2)dz + 2\sigma_r \int_0^1 (|D\theta|^2 + a^2|\theta|^2 + |\sigma|^2 \int_0^1 |\theta|^2 dz = \int_0^1 |w|^2 dz \quad (18.11)$$

from above, we have

$$\int_0^1 |\theta|^2 dz \le \frac{1}{|\sigma|^2} \int_0^1 |w|^2 dz,$$

$$[|\theta|^2]^{1/2} \le \frac{1}{|\sigma|} [|w|^2]^{1/2} \quad (18.12)$$

Combining (18.11) and (18.12), we get

$$\int_0^1 (|D\theta|^2 + a^2|\theta|^2)dz \le \frac{1}{|\sigma|} |w|^2 dz. \quad (18.13)$$

Now multiplying eq. (18.3) by its complex conjugate and integrating over the vertical range of z, we get

$$\int_0^1 \left(\frac{1}{L_e}(D^2 - a^2) - \chi\sigma\right)\phi\left(\frac{1}{L_e}(D^2 - a^2) - \chi\sigma^*\right)\phi^* = \int_0^1 ww^* dz,$$

$$\int_0^1 \frac{1}{L_e^2}(|D^2\phi|^2 + 2a^2|D\phi|^2 + a^4|\phi|^2 dz) + 2\chi\sigma_r \int_0^1 (|D\phi|^2 + a^2|\phi|^2 dz)$$

$$+ \chi^2|\sigma|^2 \int_0^1 |D\phi|^2 dz = \int_0^1 |w|^2 dz.$$

From above, we have

$$\int_0^1 |\phi|^2 dz \le \frac{1}{\chi^2|\sigma|^2} \int_0^1 |w|^2 dz. \quad (18.14)$$

Using eqs. (18.13) and (18.14) in eq. (18.10), we get

$$\frac{(1+2\lambda\sigma_r+\epsilon)}{P_r}\int_0^1(|Dw|^2+a^2|w|^2)dz+\epsilon D_a\int_0^1(|D^2w|^2+2a^2|Dw|^2+a^4|w|^2)dz$$

$$+R_Ta^2\int_0^1|\theta|^2dz+\frac{\lambda R_sa^2}{L_e}\int_0^1(|D\phi|^2+a^2|\phi|^2dz)\le\frac{\lambda R_Ta^2}{|\sigma|}\int_0^1|w|^2dz+\frac{R_sa^2\chi}{\chi^2|\sigma|^2}\int_0^1|w|^2dz,$$

$$\left(\frac{1+\epsilon}{P_r}\right)a^2\int_0^1|w|^2dz+\left(\frac{1+\epsilon}{P_r}\right)\int_0^1|Dw|^2+\frac{2\lambda\sigma_r}{P_r}\int_0^1(|Dw|^2+a^2|w|^2)dz$$

$$+\epsilon D_a(\pi^2+a^2)^2\int_0^1|w|^2+R_Ta^2\int_0^1|\theta|^2+\frac{\lambda R_sa^2}{L_e}\int_0^1(|D\phi|^2+a^2|\phi|^2)dz$$

$$<\frac{\lambda R_Ta^2}{|\sigma|}\int_0^1|w|^2dz+\frac{R_sa^2\chi}{\chi^2|\sigma|^2}\int_0^1|w|^2dz\left(\frac{1+\epsilon}{P_r}-\frac{\lambda R_T}{|\sigma|}\right)a^2\int_0^1|w|^2dz$$

$$+\left(\frac{\epsilon D_a(\pi^2+a^2)^2}{a^2}-\frac{R_s}{\chi|\sigma|^2}\right)a^2\int_0^1|w|^2dz+\left(\frac{1+\epsilon}{P_r}\right)\int_0^1|Dw|^2$$

$$+\frac{2\lambda\sigma_r}{P_r}\int_0^1(|Dw|^2+a^2|w|^2)dz+R_Ta^2\int_0^1|\theta|^2+\frac{\lambda R_sa^2}{L_e}\int_0^1(|D\phi|^2+a^2|\phi|^2dz)<0,$$

which can be rearranged as

$$\left(\frac{1+\epsilon}{P_r}-\frac{\lambda R_T}{|\sigma|}\right)a^2\int_0^1|w|^2dz+\left(\epsilon D_a4\pi^2-\frac{R_s}{\chi|\sigma|}\right)a^2\int_0^1|w|^2dz$$

$$+\frac{2\lambda\sigma_r}{P_r}\int_0^1(|Dw|^2+a^2|w|^2)dz+R_Ta^2\int_0^1|\theta|^2+\frac{\lambda R_sa^2}{L_e}\int_0^1(|D\phi|^2+a^2|\phi|^2dz)<0,$$

where minimum value of $(\frac{(\pi^2+a^2)^2}{a^2})$ is $4\pi^2$ from above, we have

$$|\sigma|<\left(\left(\frac{\lambda R_TP_r}{1+\epsilon}\right)^2,\frac{R_s}{\chi4\pi^2\epsilon D_a}\right).$$

This establishes the desired result. □

The above theorem may be stated in an equivalent form as: the complex growth rate of an arbitrary, neutral or unstable oscillatory perturbation of growing amplitude in the Darcy–Brinkman convection in a Binary viscoelastic fluid saturated porous layer heated from below must lie inside a semicircle in the right-half of the (σ_r, σ_i)-plane, whose center is origin and radius equals $|\sigma| < [(\frac{\lambda R_T P_r}{1+\epsilon})^2, \frac{R_S}{\chi 4\pi^2 \epsilon D_a}]$ where R_T and R_s are the thermal and solute Rayleigh number and P_r is the Prandtl number. The bounds obtained herein, in particular, yield a sufficient condition for the validity of the principle of the exchange of stability. Further, it is proved that above result is uniformly valid for quite general nature of the bounding surfaces.

3 Conclusions

The investigation employs linear stability analysis as a method to ascertain the highest rates of complex growth in Darcy–Brinkman convection within a porous layer saturated with a binary viscoelastic fluid. This technique is valuable for understanding the stability and behavior of the system under specific conditions, shedding light on the potential rates of growth for disturbances in the fluid. These upper bounds become essential, especially when both boundaries are constrained and do not move freely, making it impractical to find exact solutions in a closed mathematical form. Moreover, the results obtained through this analysis hold true uniformly across all combinations of rigid and free boundaries, offering a broad applicability to different boundary conditions. This emphasizes the importance of the study in situations where obtaining precise solutions is a challenging task.

Bibliography

[1] H. Benard, Les tourbillions cellulaires dans une nappe liquid, *Rev. Gén. Sci. Pures Appl.* **11** (1900), 1261.

[2] V. Bertola and E. Cafaro, Thermal instability of viscoelastic fluids in horizontal porous layers as initial problem, *Int. J. Heat Mass Transf.* **49** (2006), 4003–4012.

[3] A. Bezan, *Convection Heat Transfer*, 3rd edn, John Wiley and Sons, New Jersey, 2004.

[4] S. Chandrasekhar, *Hydrodynamic and Hydromagnetic Stability*, Clarendon, Oxford, 1961.

[5] P. Drazin and W. Reid, *Hydrodynamic Stability*, Cambridge University Press, Cambridge, 1981.

[6] S. Govender, Coriolis effect on the linear stability of convection in a porous layer placed far away from the axis of rotation, *Transp. Porous Media* **51** (2003), 315–326.

[7] C. Horton and F. Rogers, Convection currents in a porous medium, *J. Appl. Phys.* **16**(6) (1945), 367–370.

[8] Y. Katto and T. Masuoka, Criterion for the onset of convective flow in a fluid in a porous medium, *Int. J. Heat Mass Transf.* **10**(3) (1967), 297–309.

[9] A. R. A. Khaled and K. Vafai, The role of porous media in modeling flow and heat transfer in biological tissues, *Int. J. Heat Mass Transf.* **46**(26) (2003), 4989–5003.

[10] E. R. Lapwood, Convection of a fluid in a porous medium, *Proc. Camb. Philol. Soc.* **44** (1946), 508–521.

[11] D. Nield and A. Bejan, *Convection in Porous Media*, 3rd edn, Springer, Berlin, 2006.

[12] D. A. Nield, Onset of convection in a fluid layer overlying a layer of a porous medium, *J. Fluid Mech.* **81** (1977), 513–522.

[13] J. Prakash, K. Kumari and R. Kumar, Triple diffusive convection in a Maxwell fluid saturated porous layer: Darcy Brinkman-Maxwell model, *J. Porous Media* **108**(5) (2016), 871–883.

[14] K. R. Rajagopal, G. Saccomandi and L. Vergori, Stability analysis of the Rayleigh– Bénard convection for a fluid with temperature and pressure dependent viscosity, *J. Appl. Math. Phys. (ZAMP)* **60**(4) (2009).

[15] L. Rayleigh, On the convective currents in a horizontal layer of fluid when the higher temperature is on the upper side, *Philos. Mag.* **32** (1916), 529–546.

[16] G. N. Sekhar and G. Jayalatha, Elastic effects on Rayleigh-Bénard convection in liquids with temperature-dependent viscosity, *Int. J. Therm. Sci.* **49**(1) (2010), 67–75.

[17] C. W. Somerton and I. Catton, On the thermal instability of superposed porous and fluid layers, *ASME J. Heat Transf. A* **104** (1982), 160–165.

[18] B. Straughan, Global nonlinear stability in porous convection with a thermal non–equilibrium model, *Proc. R. Soc. A.* **462** (2006), 409–418.

[19] W. J. Sun, Convective instabilities in superposed porous and free layers, PhD diss., University of Minnesota, 1973.

[20] S. Swamy, N. B. Mahantesh Naduvinamani and W. Sidram, Onset of Darcy–Brinkman convection in a binary viscoelastic fluid saturated porous layer, *Transp. Porous Media* **94** (2012), 339–357.

[21] P. Vadasz, Coriolis effect on gravity-driven convection in a rotating porous layer heated from below, *J. Fluid Mech.* **376** (1998), 351–375.

[22] C. Yin and S. Wang, Thermal instability of a viscoelastic fluid in a fluid-porous system with a plane Poiseuille flow, *Appl. Math. Mech.* **41**(11) (2020), 1631–1650.

Kamal Wadhwa, Sanjay Choudhary, and Arvind Kumar Gupta

Some result of fixed-point theorems in complete *M*-fuzzy metric space

Abstract: The preeminant purpose of this work is to prove some fixed point theorems for self-mappings in complete M-fuzzy metric space, via a new concept of A-class functions without using non-decreasing conditions of M, and presented some examples in backing of our outcomes in this paper. Our result generalize of many results existing in literature. We utilize our result to find the bounded solution of functional equations arising in dynamic programming.

Keywords: Complete M-fuzzy metric space, A-class contractive mapping

MSC 2010: 47H10, 54H25

1 Introduction

In the present scenario, fixed-point theory is an extremely dynamic and vast area of research because of its applications in various fields, such as mathematics, engineering, physics, game theory, and approximation theory, etc. The notion of fuzzy sets was established by Zadeh [21] in 1965. Fuzzy set theory comes under artificial intelligence (A.I.) or soft computing methods. In 1975, Kramosil and Michelek [9] introduced the concept of fuzzy metric space. After a decade, in 1994, George and Veeramani [5] defined continuous t-norms and modified the concept of fuzzy metric space. In the next decade, Jungck and Rhoades [7, 8] presented the concept of weak compatibility. Fixed-point theory in the fuzzy metric space has been developed by Grabiee [6]. In 2011, W. Situnavarat and P. Kumam [18] gave the idea of (C.L.R.) property, which removes the condition of closedness.

On the other hand, the concept of 2-metric space presented by Gahler [4] as a generalization of metric space (X, d) in 6th decade. In 1992, Dhage [3] introduced D-metric space and prove fixed point theorems in bounded and complete D-metric space. Afterward, in 2006, Mustafa and Sim [12] gave more pertinent notion of generalized metric space. A revised definition of D-metric space was given by Sedghi and N. Shobe [15] as D^*-metric space. Recently, before this work, Sedghi et al. [14] introduced the new concept of M-fuzzy metric space which is generalized form of fuzzy metric space. Later in this sequence many authors [15–17, 19] have been proved several results about existence and

Kamal Wadhwa, Sanjay Choudhary, Department of Mathematics, Govt. Narmada College, Hoshangabad m.p. 461005, India, e-mails: wadhwakamal68@gmail.com, sanjay_choudhari123@rediffmail.com
Arvind Kumar Gupta, Department of Mathematics, Govt. Polytechnic College, Qtr. No. F-6, Government Polytechnic, Campus Rajiv Colony, Mandla m.p. 481661, India, e-mail: arvindkumargupta74@gmail.com

https://doi.org/10.1515/9783111313634-019

uniqueness of the fixed-point theorem in M-fuzzy metric space using non-decreasing condition of the degree of nearness M.

In this paper, we prove some fixed-point theorems for self-mapping without using non-decreasing condition of $M(x,y,z)$ via A-class functions in complete M-fuzzy metric space. And we use our results as an application for the existence and uniqueness bounded solution of functional equations arising in dynamic programming.

2 Preliminaries

Definition 2.1 ([18]). A binary operation $*:[0,1]\times[0,1]\to[0,1]$ is continuous t-norm if for all $p,q,r,s\in[0,1]$ the following conditions are satisfied.

N_t-1: $(p*q)=(q*p)$ commutative.
N_t-2: $(p*q)*r=p*(q*r)$ associative.
N_t-3: $p*1=p$ identity element.
N_t-4: $*$ is continuous.
N_t-5: $p*q\le r*s$, whenever $p\le r$ and $q\le s$ for all $p,q,r,s\in[0,1]$.

Example 2.2 ([18]). There are some examples of t-norms.

1 (The product t-norm) T.P.: A mapping $T_p:[0,1]\times[0,1]\to[0,1]$ is defined by

$$T_P(p,q)=pq.$$

2 (The minimum t-norm) T.M.: A mapping $T_m:[0,1]\times[0,1]\to[0,1]$ is defined by

$$T_M(p,q)=\min\{p,q\}.$$

3 (Lukasiewicz t-norm) T.L.: A mapping $T_l:[0,1]\times[0,1]\to[0,1]$ is defined by

$$T_l(p,q)=\max\{p+q-1,0\}.$$

Definition 2.3 ([5]). A3-tuple $(X,M,*)$ is called a GV fuzzy metric space, where X is nonempty set, $*$ is continuous t-norm, and if M is a fuzzy set on $X\times X\times(0,\infty)$, it satisfies the following conditions.

For all $x,y,z\in X$ and $s,t>0$

FM-1: $MG(x,y,t)>0$.
FM-2: $MG(x,y,t)=1\Leftrightarrow x=y$.
FM-3: $MG(x,y,t)=MG(y,x,t)$.
FM-4: $MG(x,y,t)*MG(y,z,s)\le MG(x,z,t+s)$.
FM-5: $MG(x,y,.):[0,\infty)\to[0,1]$ is continuous for all $x,y,z\in X$ and $t,s>0$,

where $MG(x, y, t)$ is the degree of nearness between x and y with respect to t, and axiom FM-1 and FM-2 we see that if $x \neq y$ and $\forall t > 0$

$$0 < MG(x, y, t) < 1.$$

Example 2.4 ([16]). Let (R, d) be a metric space, $p * q = T_p(p, q)$ and for all $x, y \in R$, where $t > 0$.

$$MG(x, y, t) = \frac{t}{t + d(x, y)}.$$

Then $(R, M, *)$ is a GV-fuzzy metric space.

Example 2.5 ([16]). Let $\vartheta : R_+ \to (0, 1)$ be an increasing and continuous function such that $\lim_{x\to\infty} \vartheta(t) = 1$, and X be an arbitrary nonempty set. Below are some examples of these functions.

$$\vartheta(x) = \frac{x}{x+1} \quad \text{and} \quad \vartheta(x) = \sin\frac{\pi(x)}{2x+1}$$

and $T_P(p, q) = pq$ for all $p, q, \in [0, 1]$. For each $t > 0$, define

$$MG(x, y, t) = \vartheta(t)^{d(x,y)}.$$

For all $x, y \in X$, where $d(x, y)$ is an ordinary metric. Clearly $(X, MG, *)$ is a fuzzy metric space.

Definition 2.6 ([20]). Let $(X, M, *)$ be a fuzzy metric space.
(1) Sequence $\{x_n\}$ in X is called convergent to x if $\lim_{n\to\infty} M(x_n, x, t) = 1$.
(2) Sequence $\{x_n\}$ in X is called Cauchy sequence if $\lim_{n\to\infty} M(x_n, x_{n+m}, t) = 1 \, \forall t > 0$ and $n, m \in N$.

Definition 2.7 ([14]). Let $D^* : X \times X \times X \to R^+$ be a generalized D metric defined on nonempty set X and which satisfies the following conditions for each $x, y, z, a \in X$.
(a) $D^*(x, y, z) \geq 0$.
(b) $D^*(x, y, z) = 0$ if and only if $x = y = z$.
(c) $D^*(x, y, z) = D^*(p\{x, y, z\})$, where p is a permutation function.
(d) $D^*(x, y, z) \leq D^*(x, y, a) + D^*(a, z, z)$.

Then the pair (X, D^*) is called a generalized metric space. There are some examples of such function as follows.
(1) If $X = R^+$, then we define

$$D^*(x, y, z) = \begin{cases} 0 & \text{if } x = y = z, \\ \max\{x, y, z\} & \text{otherwise.} \end{cases}$$

(2) $D^*(x,y,z) = d(x,y) + d(y,z) + d(z,x)$, where d is the ordinary metric on X.
(3) If $X = R^n$, then we define

$$D^*(x,y,z) = (\|x-y\|^p + \|y-z\|^p + \|z-x\|^p)^{\frac{1}{p}}.$$

For every $p \in R^+$.

Definition 2.8. [14] A3-tuple $(X, M, *)$ is called a M-fuzzy metric space, where X is nonempty set, $*$ is a continuous t-norm, and if M is a fuzzy set on $X^3 \times (0, \infty)$, it satisfies the following conditions.

For all $x, y, z, a \in X$ and $s, t > 0$

(1) $M(x,y,z,t) > 0$.
(2) $M(x,y,z,t) = 1 \Leftrightarrow x = y = z$.
(3) $M(x,y,z,t) = M(p\{x,y,z\}t)$, where p is a permutation function.
(4) $M(x,y,a,t) * M(a,z,z,s) \leq M(x,y,z,t+s)$.
(5) $M(x,y,z,.) : (0,\infty) \to [0,1]$ is continuous for $t > 0$.

Clearly,

$$0 < M(x,y,z,t) < 1.$$

Example 2.9 ([15]). Let X be a nonempty set and D^* be the D-metric on X, if $T_p(p,q) = pq$ for all $p,q \in [0,1]$ and for $t > 0$.

$$M(x,y,z,t) = \frac{t}{t + D^*(x,y,z)}$$

for all $x,y,z \in X$, then $(X, M, *)$ is a M-fuzzy metric space.

Lemma 2.10 ([15]). *Let $(X, MG, *)$ be a fuzzy metric space. Now we define $M : X^3 \times (0,\infty) \to [0,1]$ by*

$$M(x,y,z,t) = MG(x,y,t) * MG(y,z,t) * MG(z,x,t)$$

for all $x,y,z \in X$, then $(X, M, *)$ is a M-fuzzy metric space.

Example 2.11. Let $(X, MG, *)$ be a fuzzy metric space, with

$$MG(x,y,t) = \left(\sin\frac{\pi(t)}{(2t+1)}\right)^{d(x,y)},$$

where $T_p(p,q) = pq$ for all $p,q \in [0,1]$ and for $t > 0$.

Now we define $M : X^3 \times (0,\infty) \to [0,1]$ by

$$M(x,y,z,t) = \left(\sin\frac{\pi(t)}{(2t+1)}\right)^{D^*(x,y,z)}$$

for each $t > 0$, and for all $x, y, z \in X$, then

$$\begin{aligned} M(x,y,z,t) &= \left(\sin \frac{\pi(t)}{(2t+1)}\right)^{d(x,y)+d(y,z)+d(z,x)} \\ &= \left(\sin \frac{\pi(t)}{(2t+1)}\right)^{d(x,y)} . \left(\sin \frac{\pi(t)}{(2t+1)}\right)^{d(y,z)} . \left(\sin \frac{\pi(t)}{(2t+1)}\right)^{d(z,x)} \\ &= MG(x,y,t) * MG(y,z,t) * MG(z,x,t). \end{aligned}$$

Hence $(X, M, *)$ is a M-fuzzy metric space, and M is of second type.

Example 2.12. Let $(X, M, *)$ be a fuzzy metric space with

$$M(x,y,t) = e^{-}\frac{d(x,y)}{t},$$

where $T_p(p,q) = pq$ for all $p, q \in [0,1]$ and for $t > 0$, and for all $x, y \in X$.

Now we define $M : X^3 \times (0,\infty) \to [0,1]$ by

$$M(x,y,z,t) = e^{-}\frac{D^*(x,y,z)}{t}$$

for each $t > 0$, and for all $x, y, z \in X$, then

$$\begin{aligned} M(x,y,z,t) &= e^{-}\frac{D^*(x,y,z)}{t} \\ &= e^{-}\frac{[d(x,y) + d(y,z) + d(z,x)]}{t} \\ &= e^{-}\frac{d(x,y)}{t}.e^{-}\frac{d(y,z)}{t}.e^{-}\frac{d(z,x)}{t} \\ &= M(x,y,t).M(y,z,t).M(z,x,t) \\ &= M(x,y,t) * M(y,z,t) * M(z,x,t). \end{aligned}$$

Hence $(X, M, *)$ is a M-fuzzy metric space,

Lemma 2.13 ([14]). *Let $(X, M, *)$ be a M-fuzzy metric space, for all $x, y, z \in X$ and for each $t > 0$,*

(1) *$M(x,y,z,t)$ is continuous.*
(2) *$M(x,y,z,t)$ is non-decreasing with respect to t.*
(3) *$M(x,x,y,t) = M(x,y,y,t)$.*

Definition 2.14 ([20]). Let φ be a class of all mappings $\phi : [0,1] \to [0,1]$, which satisfies following condition.

(1) ϕ is continuous, and non-decreasing on $[0,1]$.
(2) $\phi(x) > x$ for all $x \in (0,1)$.
(3) $\phi(0) = 0$ and $\phi(1) = 1$.

Definition 2.15 ([20]). Let $\psi : [0,1] \to [0,1]$ be monotonic increasing function with $\psi(t) > t$ and

$$\psi(0) = 0 \quad \text{and} \quad \psi(1) = 1.$$

Definition 2.16 ([20]). A mapping $\mathcal{F} : (0.1] \times (0,1] \to R^+$ is called A-class function, if it is continuous and satisfies following conditions.
(1) $\mathcal{F}(x,y) \geq x$
(2) $\mathcal{F}(x,y) = x \Rightarrow x = 1$ or $y = 1$ for all $x, y \in (0,1]$.

For some $\mathcal{F}$, $\mathcal{F}(1,1) = 1$.

We denote set of A-class functions as $\mathcal{A}$.

3 Main result

Lemma 3.1. *Let $(X, M, *)$ be a M-fuzzy metric space, then for each $n \in N$ and $t > 0$,*
(1) $M(x,y,z,t) \geq M(x,y,z,\frac{t}{n})$,
(2) $M(x,y,z,t) \geq M(x,y,z,t)^n$,

for all $x, y, z \in X$.

Proof. Let $(X, M, *)$ be a M-fuzzy metric space with

$$M(x,y,z,t) = \left(\frac{t}{t+1}\right)^{D^*(x,y,z)},$$

where $x, y, z \in X$ and $t > 0$

Now for each $n \in N$

$$\begin{aligned} M(x,y,z,t) &\geq M\left(x,y,a,\frac{t}{n}\right) * M\left(a,z,z,\frac{(n-1)t}{n}\right) \\ &= M\left(x,y,z,\frac{t}{n}\right) * 1. \end{aligned}$$

So

$$M(x,y,z,t) \geq M\left(x,y,z,\frac{t}{n}\right)$$

we know that

$$\left(\frac{\frac{t}{n}}{\frac{t}{n}+1}\right) \geq \left(\frac{t}{t+1}\right)^n.$$

So

$$\begin{aligned} M\left(x,y,z,\frac{t}{n}\right) &= \left(\frac{\frac{t}{n}}{\frac{t}{n}+1}\right)^{D^*(x,y,z,)} \\ &\geq \left[\left(\frac{t}{t+1}\right)^{n}\right]^{D^*(x,y,z)} \\ &= \left[\left(\frac{t}{t+1}\right)^{D^*(x,y,z)}\right]^{n} \\ &= M(x,y,z,t)^n. \end{aligned}$$

Hence,

$$M(x,y,z,t) \geq M(x,y,z,t)^n. \qquad \square$$

Theorem 3.2. *Let $(X, M, *)$ be a complete M-fuzzy metric space and $T : X \to X$, be self-mapping with A-class contraction.*

$$\psi\{M(Tx,Ty,Tz,t)\} \geq \mathcal{F}\{\psi\{M_g(x,y,z,t)\},\phi\{M_g(x,y,z,t)\}\},$$

where

$$M_g(x,y,z,t) \geq \min\left\{\begin{array}{c} M(z,Ty,Tz,t), \\ M(z,z,Tx,t), \\ M(z,Tx,Ty,t), \\ M(z,z,Tz,t), \\ \frac{M(z,Tx,Tz,t)+M(z,Ty,Tz,t)}{2M(z,Tx,Ty,t)} \end{array}\right\}.$$

Then T has a unique fixed point.

Proof. Suppose that $x_0 \in X$ is any point. Let $\{x_n\}$ be a sequence, which is constructed by self-mapping $T : X \to X$ and defined as

$$Tx_n = x_{n-1}$$

for $n = 1, 2, 3 \ldots$.

Case I: If $x_n = x_{n-1}$, then x_n is a frivolous fixed point of T.

Case II: Suppose that $x_n \neq x_{n-1}$, we have

$$\begin{aligned} \psi\{M(x_{n+1},x_{n+1},x_n,t)\} &= \psi\{M(Tx_{n+2},Tx_{n+2},Tx_{n+1},t)\} \\ &\geq \mathcal{F}\{\psi\{M_g(x_{n+2},x_{n+2},x_{n+1},t)\},\phi\{M_g(x_{n+2},x_{n+2},x_{n+1},t)\}\}, \end{aligned} \tag{19.1}$$

$$M_g(x_{n+2}, x_{n+2}, x_{n+1}, t) \geq \min \left\{ \begin{array}{c} M(x_{n+1}, Tx_{n+2}, Tx_{n+1}, t), \\ M(x_{n+1}, x_{n+1}, Tx_{n+2}, t), \\ M(x_{n+1}, Tx_{n+2}, Tx_{n+2}, t), \\ M(x_{n+1}, x_{n+1}, Tx_{n+1}, t), \\ \frac{M(x_{n+1}, Tx_{n+2}, Tx_{n+1}, t) + M(x_{n+1}, Tx_{n+2}, Tx_{n+1}, t)}{2M(x_{n+1}, Tx_{n+2}, Tx_{n+2}, t)} \end{array} \right\}$$

$$\geq \min \left\{ \begin{array}{c} M(x_{n+1}, x_{n+1}, x_n, t), \\ M(x_{n+1}, x_{n+1}, x_{n+1}, t), \\ M(x_{n+1}, x_{n+1}, x_{n+1}, t), \\ M(x_{n+1}, x_{n+1}, x_n, t), \\ \frac{M(x_{n+1}, x_{n+1}, x_n, t) + M(x_{n+1}, x_{n+1}, x_n, t)}{2M(x_{n+1}, x_{n+1}, x_{n+1}, t)} \end{array} \right\}$$

$$\geq \min \left\{ \begin{array}{c} M(x_{n+1}, x_{n+1}, x_n, t), \\ 1, \quad 1, \\ M(x_{n+1}, x_{n+1}, x_n, t), \\ M(x_{n+1}, x_{n+1}, x_n, t) \end{array} \right\}$$

$$> M(x_{n+1}, x_{n+1}, x_n, t).$$

From equation (19.1)

$$\begin{aligned} \psi\{M(x_{n+1}, x_{n+1}, x_n, t)\} &\geq \mathcal{F}\{\psi\{M(x_{n+1}, x_{n+1}, x_n, t)\}, \phi\{M(x_{n+1}, x_{n+1}, x_n, t)\}\} \\ &\geq \psi\{M(x_{n+1}, x_{n+1}, x_n, t)\}. \end{aligned}$$

So

$$\begin{aligned} &\mathcal{F}\{\psi\{M(x_{n+1}, x_{n+1}, x_n, t)\}, \phi\{M(x_{n+1}, x_{n+1}, x_n, t)\}\} = \psi\{M(x_{n+1}, x_{n+1}, x_n, t)\} \\ &\quad \Rightarrow \quad \text{Either } \psi\{M(x_{n+1}, x_{n+1}, x_n, t)\} = 1 \text{ or } \psi\{M(x_{n+1}, x_{n+1}, x_n, t)\} = 1 \\ &\quad \Rightarrow \quad M(x_{n+1}, x_{n+1}, x_n, t) = 1. \end{aligned}$$

On taking lim

$$\lim_{x \to \infty} M(x_{n+1}, x_{n+1}, x_n, t) = 1. \tag{19.2}$$

Now we prove $\{x_n\}$ is a Cauchy sequence, i. e., for each $0 < \epsilon < 1$ and $t > 0$ there exist $n_0 \in N$ such that

$$M(x_n, x_n, x_m, t) > 1 - \epsilon \quad \forall\, n, m \geq n_0. \tag{19.3}$$

Suppose that $\{x_n\}$ is not a Cauchy sequence. Then, for every integer $k > 0$, let $n(k)$ be the least positive integer exceeding $m(k)$ with

$$n(k) > m(k) > k$$

and for these integer there are two subsequence $\{x_{n(k)}\}$ and $\{x_{m(k)}\}$ of $\{x_n\}$ are such that

$$M(x_{m(k)}, x_{m(k)}, x_{n(k)}, t) \leq 1 - \epsilon. \tag{19.4}$$

Since $n(k)$ is least integer exceeding $m(k)$. Then

$$M(x_{m(k)}, x_{m(k)}, x_{n(k)-1}, t) > 1 - \epsilon. \tag{19.5}$$

Now

$$\begin{aligned}(1-\epsilon) > M(x_{m(k)}, x_{m(k)}, x_{n(k)}, t) &= M(x_{n(k)}, x_{n(k)}, x_{m(k)}, t)\\ &\geq M\left(x_{n(k)}, x_{n(k)}, x_{n(k)-1}, \frac{t}{2}\right) * M\left(x_{n(k)-1}, x_{m(k)}, x_{m(k)}, \frac{t}{2}\right)\\ &\geq M\left(x_{n(k)}, x_{n(k)}, x_{n(k)-1}, \frac{t}{2}\right) * M\left(x_{m(k)}, x_{m(k)}, x_{n(k)-1}, \frac{t}{2}\right).\end{aligned}$$

On taking lim

$$(1-\epsilon) > \lim_{k\to\infty} M(x_{m(k)}, x_{m(k)}, x_{n(k)}, t) > 1 * (1-\epsilon).$$

So

$$\lim_{k\to\infty} M(x_{m(k)}, x_{m(k)}, x_{n(k)}, t) = (1-\epsilon). \tag{19.6}$$

Thus, for limit value

$$\begin{aligned}\psi\{1-\epsilon\} = \psi\{M(x_{m(k)}, x_{m(k)}, x_{n(k)}, t)\} &= \psi\{M(Tx_{m(k)+1}, Tx_{m(k)+1}, Tx_{n(k)+1}, t)\}\\ &\geq \mathcal{F}\{\psi\{M_g(x_{m(k)+1}, x_{m(k)+1}, x_{n(k)+1}, t)\}, \phi\{M_g(x_{m(k)+1}, x_{m(k)+1}, x_{n(k)+1}, t)\}\},\end{aligned} \tag{19.7}$$

where

$$\begin{aligned}&M_g(x_{m(k)+1}, x_{m(k)+1}, x_{n(k)+1}, t)\\ &\geq \min\left\{\begin{array}{c} M(x_{n(k)+1}, Tx_{m(k)+1}, Tx_{n(k)+1}, t),\\ M(x_{n(k)+1}, x_{n(k)+1}, Tx_{m(k)+1}, t),\\ M(x_{n(k)+1}, Tx_{m(k)+1}, Tx_{m(k)+1}, t),\\ M(x_{n(k)+1}, x_{n(k)+1}, Tx_{n(k)+1}, t),\\ \frac{M(x_{n(k)+1}, Tx_{m(k)+1}, Tx_{n(k)+1}, t) + M(x_{n(k)+1}, Tx_{m(k)+1}, Tx_{n(k)+1}, t)}{2M(x_{n(k)+1}, Tx_{m(k)+1}, Tx_{m(k)+1}, t)} \end{array}\right\}\\ &\geq \min\left\{\begin{array}{c} M(x_{n(k)+1}, x_{m(k)}, x_{n(k)}, t),\\ M(x_{n(k)+1}, x_{n(k)+1}, x_{m(k)}, t),\\ M(x_{n(k)+1}, x_{m(k)}, x_{m(k)}, t),\\ M(x_{n(k)+1}, x_{n(k)+1}, x_{n(k)}, t),\\ \frac{M(x_{n(k)+1}, x_{m(k)}, x_{n(k)}, t) + M(x_{n(k)+1}, x_{m(k)}, x_{n(k)}, t)}{2M(x_{n(k)+1}, x_{m(k)}, x_{m(k)}, t)} \end{array}\right\}\end{aligned}$$

$$\geq \min\left\{\begin{array}{c} M(x_{m(k)}, x_{n(k)}, x_{n(k)+1}, t), \\ M(x_{n(k)+1}, x_{n(k)+1}, x_{m(k)}, t), \\ M(x_{n(k)+1}, x_{m(k)}, x_{m(k)}, t), \\ M(x_{n(k)+1}, x_{n(k)+1}, x_{n(k)}, t), \\ \frac{M(x_{m(k)}, x_{n(k)}, x_{n(k)+1}, t)}{M(x_{m(k)}, x_{m(k)}, x_{n(k)+1}, t)} \end{array}\right\}$$

$$> \min\left\{\begin{array}{c} M(x_{m(k)}, x_{m(k)}, x_{n(k)}, \frac{t}{2}) * M(x_{n(k)}, x_{n(k)+1}, x_{n(k)+1}, \frac{t}{2}), \\ M(x_{n(k)+1}, x_{n(k)+1}, x_{n(k)}, \frac{t}{2}) * M(x_{n(k)}, x_{m(k)}, x_{m(k)}, \frac{t}{2}), \\ M(x_{m(k)}, x_{m(k)}, x_{n(k)}, \frac{t}{2}) * M(x_{n(k)}, x_{n(k)+1}, x_{n(k)+1}, \frac{t}{2}), \\ M(x_{n(k)+1}, x_{n(k)+1}, x_{n(k)}, t), \\ \frac{M(x_{m(k)}, x_{n(k)}, x_{n(k)}, \frac{t}{2}) * M(x_{n(k)}, x_{n(k)+1}, x_{n(k)+1}, \frac{t}{2})}{M(x_{m(k)}, x_{m(k)}, x_{n(k)}, \frac{t}{2}) * M(x_{n(k)}, x_{n(k)+1}, x_{n(k)+1}, \frac{t}{2})} \end{array}\right\}$$

Now taking $\lim_{k\to\infty}$ and using (19.2), (19.6), and Lemma 2.13(3)

$$\lim_{k\to\infty} M_g(x_{m(k)+1}, x_{m(k)+1}, x_{n(k)+1}, t) > (1 - \epsilon). \tag{19.8}$$

Again, taking $\lim_{k\to\infty}$ in (19.7) and using (19.8).

$$\begin{aligned} \psi\{1-\epsilon\} &\geq \mathcal{F}\Big\{\psi\Big\{\lim_{k\to\infty} M_g(x_{m(k)+1}, x_{m(k)+1}, x_{n(k)+1}, t)\Big\}, \\ &\qquad \phi\Big\{\lim_{k\to\infty} M_g(x_{m(k)+1}, x_{m(k)+1}, x_{n(k)+1}, t)\Big\}\Big\} \\ &> \mathcal{F}\{\psi\{(1-\epsilon)\}, \phi\{(1-\epsilon)\}\} \geq \psi\{1-\epsilon\}. \end{aligned}$$

Thus

$$\begin{aligned} &\mathcal{F}\{\psi\{(1-\epsilon)\}, \phi\{(1-\epsilon)\}\} = \psi\{1-\epsilon\} \\ &\Rightarrow \quad \text{Either } \psi\{(1-\epsilon)\} = 1 \text{ or } \psi\{(1-\epsilon)\} = 1. \end{aligned}$$

A contradiction, because $0 < \epsilon < 1$, hence $\{x_n\}$ is a Cauchy sequence in complete M-fuzzy metric space. Then there exist a point $z \in X$ such that

$$\lim_{n\to\infty} x_n = z.$$

Now we show that z is a fixed point of T.

We start with

$$\begin{aligned} \psi\{M(x_n, x_n, Tz, t)\} &= \psi\{M(Tx_{n+1}, Tx_{n+1}, Tz, t)\} \\ &\geq \mathcal{F}\{\psi\{M_g(x_{n+1}, x_{n+1}, z, t)\}, \phi\{M_g(x_{n+1}, x_{n+1}, z, t)\}\}, \end{aligned} \tag{19.9}$$

where

$$M_g(x_{n+1}, x_{n+1}, z, t) \geq \min \begin{Bmatrix} M(z, Tx_{n+1}, Tz, t), \\ M(z, z, Tx_{n+1}, t), \\ M(z, Tx_{n+1}, Tx_{n+1}, t), \\ M(z, z, Tz, t), \\ \frac{M(z,Tx_{n+1},Tz,t)+M(z,Tx_{n+1},Tz,t)}{2M(z,Tx_{n+1},Tx_{n+1},t)} \end{Bmatrix}$$

$$\geq \min \begin{Bmatrix} M(z, x_n, Tz, t), \\ M(z, z, x_n, t), \\ M(z, x_n, x_n, t), \\ M(z, z, Tz, t), \\ \frac{M(z,x_n,Tz,t)+M(z,x_n,Tz,t)}{2M(z,x_n,x_n,t)} \end{Bmatrix}$$

$$\lim_{n\to\infty} M_g(x_{n+1}, x_{n+1}, z, t) \geq \min \begin{Bmatrix} M(z, z, Tz, t), \\ M(z, z, z, t), \\ M(z, z, z, t), \\ M(z, z, Tz, t), \\ \frac{M(z,z,Tz,t)+M(z,z,Tz,t)}{2M(z,z,z,t)} \end{Bmatrix}$$

$$\geq \min \begin{Bmatrix} M(z, z, Tz, t), \\ 1, \\ 1, \\ M(z, z, Tz, t), \\ M(z, z, Tz, t), \end{Bmatrix}$$

$$> \{M(z, z, Tz, t)\}.$$

Again, taking lim in (19.9)

$$\begin{aligned} &\lim_{k\to\infty} \psi\{M(x_n, x_n, Tz, t)\} \\ &\quad \geq \mathcal{F}\Big\{\psi\Big\{\lim_{k\to\infty} M_g(x_{n+1}, x_{n+1}, z, t)\Big\}, \phi\Big\{\lim_{k\to\infty} M_g(x_{n+1}, x_{n+1}, z, t)\Big\}\Big\}, \\ &\quad \psi\{M(z, z, Tz, t)\} \geq \mathcal{F}\{\psi\{M(z, z, Tz, t)\}, \phi\{M(z, z, Tz, t)\}\} \\ &\quad\quad \geq \psi\{M(z, z, Tz, t)\}. \end{aligned}$$

Thus

$$\begin{aligned} &\mathcal{F}\{\psi\{M(z, z, Tz, t)\}, \phi\{M(z, z, Tz, t)\}\} = \psi\{M(z, z, Tz, t)\} \\ &\quad \Rightarrow \quad \text{either } \psi\{M(z, z, Tz, t)\} = 1 \text{ or } \psi\{M(z, z, Tz, t)\} = 1 \\ &\quad \Rightarrow \quad M(z, z, Tz, t) = 1. \end{aligned}$$

Hence

$$Tz = z.$$

To prove uniqueness of fixed point, suppose w is another fixed point of T, i. e.,

$$\begin{aligned} Tz = z \quad & \quad Tw = w \\ \psi\{M(z,z,w,t)\} &= \psi\{M(Tz,Tz,Tw,t)\} \\ &\geq \mathcal{F}\{\psi\{M_g(z,z,w,t)\},\phi\{M_g(z,z,w,t)\}\} \end{aligned}$$

where

$$\begin{aligned} M_g(z,z,w,t) &\geq \min\left\{\begin{matrix} M(w,Tz,Tw,t), \\ M(w,w,Tz,t), \\ M(w,Tz,Tz,t), \\ M(w,w,Tw,t), \\ \frac{M(w,Tz,Tw,t)+M(w,Tz,Tw,t)}{2M(w,Tz,Tz,t)} \end{matrix}\right\} \\ &\geq \min\left\{\begin{matrix} M(w,z,w,t), \\ M(w,w,z,t), \\ M(w,z,z,t), \\ M(w,w,w,t), \\ \frac{M(w,z,w,t)+M(w,z,w,t)}{2M(w,z,z,t)} \end{matrix}\right\} \\ &\geq \min\left\{\begin{matrix} M(z,z,w,t), \\ M(z,z,w,t), \\ M(w,z,z,t),1, \\ 1, \end{matrix}\right\} \\ &> \{M(z,z,w,t)\}. \end{aligned}$$

By using of A-class contraction, we get $M(z,z,w) = 1$. Hence fixed point is unique. □

Corollary 3.3. *Let $(X,M,*)$ be a complete M-fuzzy metric space and $T : X \to X$, be self-mapping with A-class contraction.*

$$\psi\{M(Tx,Ty,Tz,t)\} \geq \mathcal{F}\{\psi\{M_g(x,y,z,t)\},\phi\{M_g(x,y,z,t)\}\},$$

where

$$M_g(x,y,z,t) \geq \min\left\{\begin{matrix} M(z,Tx,Tz,2t), \\ M(z,Ty,Tz,2t), \\ M(z,Tx,Ty,2t). \end{matrix}\right\}$$

Then T has a unique fixed point.

Following example is in backing of our results.

Example 3.4. Let $\mathcal{F}(x,y)$ be any arbitrary function of class A. Suppose $X = [01]$ is the set of real number, and $* : [0,1] \times [0,1] \to [0,1]$ is continuous t-norm for all $t \in (0,\infty)$ and for each $x, y, z \in X$, define

$$M(x,y,z,t) = \frac{t}{t + D^*(x,y,z)},$$

where

$$D^*(x,y,z) = \begin{cases} 0 & \text{if } x = y = z, \\ \max\{x,y,z\} & \text{otherwise.} \end{cases}$$

Clearly $(X, M, *)$ is a M-fuzzy metric space.

Let $\psi : [0,1] \to [0,1]$ and $\phi : (0,1] \to (0,1]$ be two functions are defined as follows

$$\psi(v) = \sqrt{v} \quad \text{for all } 0 \le v \le 1$$

and

$$\phi(\tau) = \frac{\sqrt{\tau} + 1}{2} \quad \text{for all } 0 < \tau \le 1.$$

Suppose that $T : X \to X$ is a self-mapping defined as $T(x) = \frac{x^2}{9}$. Now we establish contractive condition of theorem.

Case I: When $x = y = z$, A-class contractive condition is satisfied.

Case II: When $p\{x > y > z\}$, where p is permutation function.

$$\begin{aligned} \psi\{M(Tx, Ty, Tz, t)\} &= \psi\left\{M\left(\frac{x^2}{9}, \frac{y^2}{9}, \frac{z^2}{9}, t\right)\right\} \\ &= \psi\left\{\frac{t}{t + \max\{\frac{x^2}{9}, \frac{y^2}{9}, \frac{z^2}{9},\}}\right\} \\ &= \psi\left\{\frac{t}{t + \frac{x^2}{9}}\right\} > \frac{t}{t + \frac{x^2}{9}} \qquad \text{(I)} \end{aligned}$$

and

$$\begin{aligned} M_g(x,y,z,t) &> \min\left\{\tfrac{t}{t+z}, \tfrac{t}{t+z}, \tfrac{t}{t+z}, \tfrac{t}{t+z}, 1\right\} \\ &> \frac{t}{t+z}. \qquad \text{(II)} \end{aligned}$$

Thus for all value of $t \in (0, \infty)$, and from (I) and (II)

$$\frac{t}{t + \frac{x^2}{9}} > \frac{t}{t + z} \qquad \forall x, y, z \in X.$$

Therefore A-class contractive condition is satisfied.

Case III: Contractive condition is also satisfied in exception condition, when

$$\{z > p\{x > y\}\}.$$

Hence $x_0 = 0 \in X$ is a fixed point of T.

4 Application in dynamic programming

Dynamic programming is defined as a computer programming technique, which is mathematical optimization method and algorithms paradigms. In 1950, Richard Bellman [1] introduced the basic form of the functional equations arising in dynamic programming as follows

$$f(p) = \max_q \{H(p, q, f(T(p, q)))\},$$

where p and q are state and decision vectors, respectively, $f(p)$ is optimal return function with initial state p, and T is transformation of the process. Many authors [2, 10, 11, 13] worked in dynamic programming to find the common and bounded solution of the functional equations arising in above field.

In the same way, we use our results to find the bounded solution of following system of functional equations arising in dynamic programming.

Let A, B, be two Banach space, and $S \subset A$ & $D \subset B$ are state and decision space, respectively.

$$\left.\begin{aligned} P(x) &= \sup_{y \in D} \{u(x,y) + \Phi(x,y,P(\rho(x,y)))\} \quad \text{for } x \in S, \\ Q(x) &= \sup_{y \in D} \{u(x,y) + \Psi(x,y,Q(\rho(x,y)))\} \quad \text{for } x \in S. \end{aligned}\right\} \tag{19.10}$$

Where

$$\rho : S \times D \to S, \quad u : S \times D \to R,$$
$$\Phi, \Psi : S \times D \times R \to R \quad \text{and} \quad P, Q : S \to R.$$

Now we investigate the existence and uniqueness of the bounded solution of the system (19.10) of functional equations.

For this, we start further process.

Let $B(S)$ be the set of all bounded real valued functions defined on S.

For $f, g, h \in B(S)$, define

$$D^* : B(S)^3 \to R^+ \quad \text{by}$$
$$D^*(f, g, h) = d(f, g) + d(g, h) + d(h, f).$$

Then $(B(S), D^*)$ is generalized metric space.

and $a * b = ab\ \forall a, b \in [0,1]$ and for each $t > 0$, for all $f, g, h \in B(S)$ we define

$$M(f,g,h,t) = e^{-\dfrac{D^*(f,g,h)}{t}},$$

where

$$d(f,g) = \sup_{x \in S}\{|f(x) - g(x)| : \forall f,g \in B(S)\}$$

$\Rightarrow$

$$D^*(f,g,h) = \sup_{x \in S}\{|f(x) - g(x)| + |g(x) - h(x)| + |h(x) - f(x)|\}, \quad \forall f,g,h \in B(S)$$

or

$$D^*(f,g,h) = \|f - g\| + \|g - h\| + \|h - f\|.$$

Then $(B(S), M, *)$ is a complete M-fuzzy metric space, as shown in Example 2.12.

We take functions ϕ and ψ as defined in Example 3.4 and $\mathcal{F} \in \mathcal{A}$ is any function of class A.

Now for every $(x,y) \in S \times D, f, g \in B(S)$ define

$$\left.\begin{aligned} Tf(x) &= \sup_{y \in D}\{u(x,y) + \Phi(x,y,f(\rho(x,y)))\} \quad \text{for } x \in S,\\ Ig(x) &= \sup_{y \in D}\{u(x,y) + \Psi(x,y,g(\rho(x,y)))\} \quad \text{for } x \in S. \end{aligned}\right\} \tag{19.11}$$

Theorem 4.1. *Let T be a mapping defined on $B(S)$ and suppose that the following conditions are satisfied.*

(I) *Functions u, Φ, and Ψ are bounded.*

(II) *For $x, y \in S$, and $f, g \in B(S)$ there exist non-decreasing functions ϕ & ψ and a A-class function $\mathcal{F} \in \mathcal{A}$ such that*

$$e^{\frac{-2|\Phi(x,v_1,f(\rho_1))-\Phi(x,v_2,g(\rho_2))|}{t}} \geq \mathcal{F}(\psi\{e^{\frac{-2d(If(x),Jg(x))}{t}}\}, \phi\{e^{\frac{-2d(If(x),Jg(x))}{t}}\}),$$

where

$$e^{\frac{-2d(If(x),Jg(x))}{t}} \geq \min\left\{\begin{matrix} M(g,Tg,Tg,t),\\ M(g,g,Tf,t),\\ M(g,Tf,Tg,t),\\ M(g,g,Tg,t),\\ \frac{M(g,Tf,Tg,t)+M(g,Tg,Tg,t)}{2M(g,Tf,Tg,t)} \end{matrix}\right\}$$

$\rho_1 = \rho(x, v_1)$ and $\rho_2 = \rho(x, v_2)$. Then the system of functional equations (19.10) *has a unique bounded solution in $B(S)$.*

Proof. According to condition (I), $T : B(S) \to B(S)$ is a self-map, and $(B(S), M, *)$ is a complete M-fuzzy metric space.

Now for $f, g \in B(S)$, $x \in S$, and $\beta > 0$ there exist $v_1, v_2 \in D$ such that

$$T(f(x)) < u(x, v_1) + \Phi(x, v_1, f(\rho(x, v_1))) + \beta, \tag{19.12}$$

$$T(g(x)) < u(x, v_2) + \Phi(x, v_2, g(\rho(x, v_2))) + \beta. \tag{19.13}$$

But

$$T(f(x)) \geq u(x, v_2) + \Phi(x, v_1, f(\rho(x, v_1))), \tag{19.14}$$

$$T(g(x)) \geq u(x, v_1) + \Phi(x, v_2, g(\rho(x, v_2))). \tag{19.15}$$

Thus from eqns. (19.12) and (19.15)

$$\begin{aligned} T(f(x)) - T(g(x)) &< \Phi(x, v_1, f(\rho(x, v_1))) - \Phi(x, v_2, g(\rho(x, v_2))) + \beta \\ &\leq \left|\Phi(x, v_1, f(\rho(x, v_1))) - \Phi(x, v_2, g(\rho(x, v_2)))\right| + \beta. \end{aligned}$$

Again, from eqns. (19.13) and (19.14)

$$\begin{aligned} T(g(x)) - T(f(x)) &< \Phi(x, v_2, g(\rho(x, v_2))) - \Phi(x, v_1, f(\rho(x, v_1))) + \beta \\ -\left(T(f(x)) - T(g(x))\right) &\leq \left|\Phi(x, v_1, f(\rho(x, v_1))) - \Phi(x, v_2, g(\rho(x, v_2)))\right| + \beta \end{aligned}$$

$\Rightarrow$

$$\left|T(f(x)) - T(g(x))\right| \leq \left|\Phi(x, v_1, f(\rho(x, v_1))) - \Phi(x, v_2, g(\rho(x, v_2)))\right| + \beta.$$

Since $\beta > 0$ is arbitrary, then

$$\begin{aligned} D^*(Tf, Tg, Tg) &= 2 \sup\left|T(f(x)) - T(g(x))\right| \\ &\leq 2\left|\Phi(x, v_1, f(\rho(x, v_1))) - \Phi(x, v_2, g(\rho(x, v_2)))\right|. \end{aligned}$$

Therefore,

$$\begin{aligned} e^{\frac{-D^*(Tf,Tg,Tg)}{t}} &\geq e^{\frac{-2|\Phi(x,v_1,f(\rho(x,v_1)))-\Phi(x,v_2,g(\rho(x,v_2)))|}{t}} \\ &\geq \mathcal{F}(\psi\{e^{\frac{-2d(If(x),Jg(x))}{t}}\}, \phi\{e^{\frac{-2d(If(x),Jg(x))}{t}}\}). \end{aligned}$$

Hence it provides

$$\psi\{M(Tf, Tg, Tg, t)\} \geq \mathcal{F}\{\psi\{M_g(f, g, g, t)\}, \phi\{M_g(f, g, g, t)\}\},$$

where $M_g(f, g, g, t)$ is defined according to condition (II).

Therefore, all condition of Theorem 3.2 are satisfied and mapping T has a unique fixed point in $B(S)$. Hence system of functional equations (19.10) has a unique bounded solution. □

5 Conclusion

In this paper, we have established fixed point theorem without using non-decreasing condition of the degree of nearness with the help of A-class contraction in the framework of M-fuzzy metric space. And we used our outcome as an application to obtain a bounded solution of functional equations arising in dynamic programming.

Funding: No grant has been received for this research work.

Bibliography

[1] R. Bellman and E. Stanley Lee, Functional equations in dynamic programming, *Aequ. Math.* **17** (1978), 1–8.

[2] S. Beloul, Common fixed point theorems for weakly subsequentially continuous generalized contractions with applications, *Appl. Math. E-Notes* **15** (2015), 173–186.

[3] B. C. Dhage, Generalized metric spaces and mappings with fixed point, *Bull. Calcutta Math. Soc.* **84**(4) (1992), 329–336.

[4] S. Gahler, 2-Metrische Raume and there topologische Struktur, *Math. Nachr.* **26** (1963), 115–148.

[5] A. George and P. Veeramani, On some results in fuzzy metric spaces, *Fuzzy Sets Syst.* **64** (1994), 395–399.

[6] M. Grabiee, Fixed points in fuzzy metric space, *Fuzzy Sets Syst.* **27** (1988), 385–389.

[7] G. Jungck, Compatible mappings and common fixed point, *Int. J. Math. Math. Sci.* **9**(4) (1986), 771–779.

[8] G. Jungck and B. E. Rhoades, Fixed points for set valued functions without continuity, *Indian J. Pure Appl. Math.* **29**(3) (1998), 227–238.

[9] I. Kromosil and J. Michalek, Fuzzy metric and statistical metric space, *Kybernetika* **11** (1975), 326–334.

[10] J. Li, M. Fu, Z. Liu and S. M. Kang, A common fixed point theorem and its application in dynamic programming, *Appl. Math. Sci.* **2**(17) (2008), 829–842.

[11] Z. Liu, L. Wang et al., Common fixed point theorems for contractive type mappings and their applications in dynamic programming, *Bull. Korean Math. Soc.* **45**(3) (2008), 573–585.

[12] Z. Mustafa and B. Sims, A new approach to generalized metric spaces, *J. Nonlinear Convex Anal.* **7**(2) (2006), 289–297.

[13] N. Saleem, B. Ali et al., Fixed points of Suzuki type generalized multivalued mappings in fuzzy metric spaces with applications, *Fixed Point Theory Appl.* **2015** (2015), 36. https://doi.org/10.1186/s13663-015-0284-7.

[14] S. Sedghi and N. Shobe, A common fixed point theorem in *M*-fuzzy metric space with property (E), *Adv. Fuzzy Math.* **1**(1) (2006), 55–65.

[15] S. Sedghi and N. Shobe, A common fixed point theorem in two *M*-fuzzy metric space, *Commun. Korean Math. Soc.* **22**(4) (2007), 513–526.

[16] S. Sedghi, N. Shobe and A. Aliouche, A common fixed point theorem for weakly compatible mappings in fuzzy metric spaces, *Gen. Math.* **18**(3) (2010), 3–12.

[17] D. Singh, M. Sharma and R. Sharma, Fixed point theorems in *M*-fuzzy metric spaces, *Ann. Fuzzy Math. Inform.* **5**(1) (2013), 147–155.

[18] W. Sintunavarat and P. Kumam, Common fixed point theorems for a pair of weakly compatible mappings in fuzzy metric spaces, *J. Appl. Math.* (2011), 1–14, Hindwani publishing corporation.

[19] T. Veerapandi et al., Some fixed point and coincident point theorem in generalized *M*-fuzzy metric space, *Int. J. Math. Anal.* **3**(13) (2009), 627–635.

[20] K. Wadhwa and A. K. Gupta, Common fixed point theorems for four self-mappings satisfying (CLRst) property via A-class functions in FMS, *Glob. J. Pure Appl. Math.* **17**(2) (2021), 319–332.

[21] I. A. Zadeh, Fuzzy sets, *Inf. Control* **89** (1965), 338–353.

Aronthung S Odyuo, P. Mercy, and Manoj Kumar Patel

FR-labeling on hybrid partition shell graphs

Abstract: A shell $S^1_{n,n-3}$ is a cycle of size n with $n-3$ chords on n vertices. A hybrid non-uniform shell is an extension of doublet shell with order m and n where $m \neq n$. And by hybrid partition shell $S^1_{n,n-3}, S^2_{n,n-3}, \ldots, S^t_{n,r}$ is a graph formed by a set of t-copies of shell of width n partition by a union of λK_2. In this paper, we consider an FR-labeling function $f^*(\alpha\beta)$, where α and β as its set of vertices and prove that it holds for all hybrid non-uniform shell and hybrid partition shell graphs.

Keywords: FR-labeling, non-uniform shell, partition shell

MSC 2010: Primary 05C78

1 Introduction

An FR-labeling, known as Fibonacci range labeling, is a graph $G = (V, E)$ on vertices $x \in V$ with distinct labels $f(x) \to \{F_2, F_3, F_4, \ldots, F_{p+1}\}$ such that, when the edge $e = (\alpha, \beta)$ is labeled with $f^*(e = \alpha\beta) = \lceil \frac{f(\alpha)^2 + f(\beta)^2}{f(\alpha)+f(\beta)} \rfloor$, then the resulting edge labels are distinct and the ratio of each edge to the subsequent edge is in the form of the golden ratio given by $R_t(E_{i,i+1}) = \frac{E_{i+1}}{E_i} \approx \psi$ (for larger i), where $R_t(E_{i,i+1})$ is the ratio of the resulting induced edges $(e_1, e_2), (e_2, e_3), \ldots, (e_{n-1}, e_n)$ and $\psi = 1.618$, known as the golden ratio. Deb and Limaye [1] originated the concept of shell graph and is defined as a shell $S^1_{n,n-3}$ is a cycle of size n with $n-3$ chords on n vertices, known as chords or central vertex or apex vertex. They prove that it is harmonious for uniform shell and open a conjecture for all multiple shells to be harmonious. It has been proven to be harmonious for double shell [3], balanced triple shell [3] and balanced quadruple shell [8]. And for more survey on labeling of shell see [4, 5]. In this paper, we consider FR-labeling, see [6, 7] to generalize the conjecture and prove it for all hybrid non-uniform shell graphs and hybrid partition shell graphs.

Acknowledgement: The authors would like to express their graditude to the anonymous referee for their valuable comments, which greatly contributed to the improvement of this paper.

Aronthung S Odyuo, Department of Mathematics, Kohima Science College, Jotsoma-797002, Nagaland, India
P. Mercy, Department of Mathematics, St Joseph University, Ikishe Model Village, Dimapur-797115, Nagaland, India
Manoj Kumar Patel, Department of Mathematics, National Institute of Technology, Chumukedima-797103, Nagaland, India

https://doi.org/10.1515/9783111313634-020

2 On hybrid graph

We consider only finite and undirected graph with no multiple edges, for basic concepts and notation refer [2]. Throughout the paper, we indicate the Fibonacci number for labeling function where it is given by $\{0, 1, 1, 2, 3, 5, 8, 13, 21, 34, \ldots, \}$ here $F_0 = 0$, $F_1 = 1$, $F_2 = 1$, etc., but we consider the vertices label from F_2, since every corresponding vertex label need to distinct. And by hybrid non-uniform shell is an extension of doublet shell with order m and n, where $m \neq n$ (Figure 20.1). And by hybrid partition shell or subdivided shell $S^1_{n,n-3}, S^2_{n,n-3}, \ldots, S^t_{n,r}$ is formed by a set of t-copies of a shells of width n partition by a union of λK_2 where $\lambda \geq 1$ (Figure 20.2).

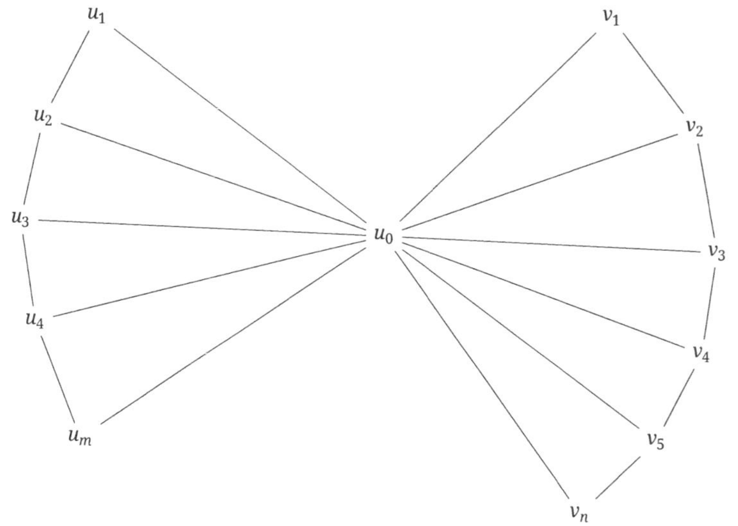

Figure 20.1: Hybrid non-uniform shell with order m and n.

Lemma 2.1. *The generating function for FR-labeling with αβ are, respectively:*

$$\phi(x) = 1 + x\phi + x^2\phi$$

and

$$s(x) = \frac{\phi(x)}{x}.$$

Proof. The recurrence relation for Fibonacci sequence is given by

$$F_n = F_{n-1} + F_{n-2} \quad \text{for } n \geq 2$$

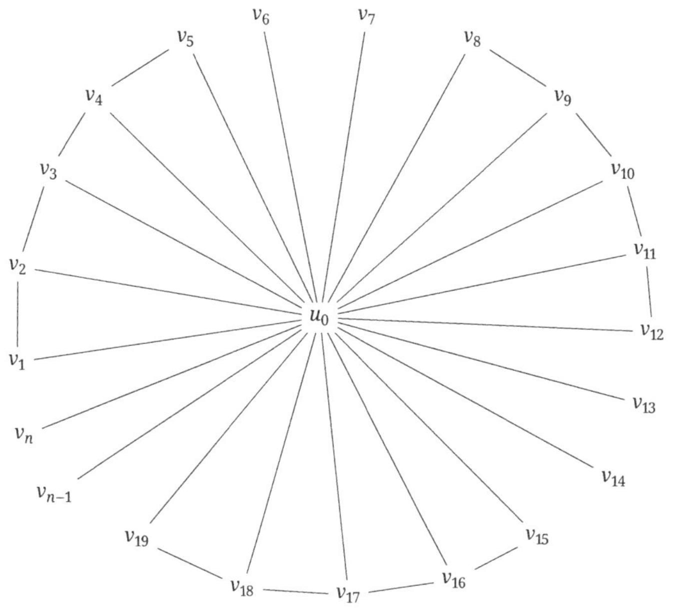

Figure 20.2: A hybrid partition shell with union of λK_2.

and the generating function for Fibonacci sequence is

$$F(x) = \sum_{n=0}^{\infty} F_n x^n = \sum_{n=1}^{\infty} F_n x^n \quad \text{since } F_0 = 0$$

or

$$\phi(x) = \sum_{n=0}^{\infty} \phi(x) = \sum_{n=1}^{\infty} \phi(x).$$

For instance, we ignore the first two initial values F_0, F_1, as we consider distinct labels. We define the generating function from F_2 as $F_1 = F_2 = 1$, so the second sum is in terms of generating function $\sum_{n=2}^{\infty}(\alpha_{n-1}\beta_{n-1})x = x\sum_{n=2}^{\infty}(\alpha_{n-1}\beta_{n-1})x^{n-1} = x\sum_{n=1}^{\infty}(\alpha_{n-1}\beta_{n-1})x^n$

$$\begin{aligned}
\text{Let } \phi(x) &= 1 + 2x + 3x^2 + \cdots = \sum_{n=0}^{\infty} F_n x^n \qquad (20.1)\\
&= 1 + \sum_{n=2}^{\infty} (\alpha\beta)x^n\\
&= 1 + \sum_{n=2}^{\infty} (\alpha_{n-1}\beta_{n-1} + \alpha_{n-2}\beta_{n-2})x^n
\end{aligned}$$

$$
\begin{aligned}
&= 1 + \sum_{n=2}^{\infty}(\alpha_{n-1}\beta_{n-1})x^n + \sum_{n=2}^{\infty}(\alpha_{n-2}\beta_{n-2})x^n \\
&= 1 + x\sum_{n=2}^{\infty}(\alpha_{n-1}\beta_{n-1})x^{n-1} + x^2\sum_{n=2}^{\infty}(\alpha_{n-2}\beta_{n-2})x^{n-2} \\
&= 1 + x\sum_{n=1}^{\infty}(\alpha_n\beta_n)x^n + x^2\sum_{n=0}^{\infty}(\alpha_n\beta_n)x^n \\
\phi(x) &= 1 + x\phi(x) + x^2\phi(x) \\
\phi(x) &= \frac{1}{1-x-x^2}, \qquad (20.2)
\end{aligned}
$$

which is the generating function for FR-labeling and we know that the *generating function* for Fibonacci series is given by $s(x) = \frac{x}{1-x-x^2}$ and by (20.2), we have

$$\phi(x) = \frac{s(x)}{x}. \qquad (20.3)$$

By equation (20.2) and equation (20.3), we have the required result. □

Remark 2.1. For any graph G with corresponding vertices label as a series of Fibonacci numbers $\sum_{n=2}^{\infty} F_n$ then the edge gets labeled between the values $L_2, L_3, L_4, \ldots, L_n$, where L_n is the Lucas number and the relation between the vertices and edges gives

$$
v_i, v_{i+1} \to e_j = \begin{cases}
1,2 \to 1 & \text{for } i = 1,\ j = 1 \\
2,3 \to 3 & \text{for } i = 2,\ j = 2 \\
3,5 \to 4 & \text{for } i = 3,\ j = 3 \\
5,8 \to 7 & \text{for } i = 4,\ j = 4 \\
8,13 \to 11 & \text{for } i = 5,\ j = 5 \\
\ldots & \\
F_{n-1}, F_n \to E_n & \text{for } i = n,\ j = n.
\end{cases} \qquad (20.4)
$$

The above generate the relation $E_n \approx [\psi^n + \frac{1}{2}]$, which is similar to the Lucas number identities.

Theorem 2.1. *A hybrid non-uniform doublet shell with order m and n (order exclude the apex), where m ≠ n admits an FR-labeling.*

Proof. Let G be the non-uniform double shell graph (see Figure 20.1). The order of the graph G is $p = m + n$ and the size is $q = \sum_{m,n=1}^{\infty} 2m + 2n - 2$. Consider u_0 as the central or apex vertex and $v_1, v_2, v_3, \ldots, v_n$ as the vertices of the first shell and $u_1, u_2, u_3, \ldots, u_m$ as the vertices of the second shell adjacent to u_0. For any given $n, m \geq 1$, define a function $f : V(G) \to \{F_2, F_3, F_4, \ldots, F_{p+1}\}$ defined by $f^*(e = \alpha\beta)$. Label the apex vertex $f(u_0) = 1$ and the consecutive vertex by $f(v_i) = F_{i+2};\ 1 \leq i \leq m$ and $f(u_j) = F_{j+m+2}; 1 \leq j \leq n < m$ this simplifies the edge label to be distinct by the operation

Case 1.

$$f(v_iv_{i+1})/f(u_ju_{j+1}) = \begin{cases} i+1 & \text{for odd} \\ i+2 & \text{for even} \\ f(v_{m-1}) + F_{m-1}; & 3 \le i \le m-1 \\ f(u_{n-1}) + F_{m+n-1}; & j = n-1. \end{cases} \tag{20.5}$$

Case 2.

$$f(u_0v_i)/f(u_0u_j) = \begin{cases} f(v_1) - 1; & i = 1 \\ f(v_i); & 2 \le i \le m \\ f(u_i); & 1 \le j \le n \end{cases} \tag{20.6}$$

and the composition table is given below:

Table 20.1: Composition table for hybrid non-uniform graph.

v_i	v_1	v_2	v_3	v_4	v_5	v_6	v_7	v_n
u_0	F_2	F_3	F_4	F_5	F_6	F_7	F_8	F_n
u_j	u_1	u_2	u_3	u_4	u_5	u_6	u_n	
u_0	F_{n+j}	F_{10}	F_{11}	F_{12}	F_{13}	F_{14}	F_m	
v_i	v_1	v_2	v_3	v_4	v_5	v_6		
v_{i+1}	F_3	F_4+1	F_5+2	F_6+3	F_7+5	F_8+8	$F_{k-1}+F_{k-3}$	
u_i	u_1	u_2	u_3	u_4	u_5	u_6		
u_{j+1}	$F_{k+1}+F_{k-1}$	–	–	–	–	–	F_p+F_{p-2}	

Observe that $k > p$, otherwise if $p = k$, then its a contradiction with regard to non-uniform shell graph. Then, by Lemma 2.1 and from the above computation the ratio $R_t(e_i^m e_{i+1}^n) \implies \lceil \frac{\max(e_i^m e_{i+1}^n)}{\min(e_i^m e_{i+1}^n)} \rfloor$ converges to ψ and can be determined as an FR-labeling. □

Theorem 2.2. *A hybrid partition shell graph $\sum_{n=4}^{\infty} S_{n,n-3}^t$ of width n is an FR-labeling iff there exists a union of λK_2, where $\lambda \ge 1$.*

Proof. Without loss of generality, we consider the shell graph $S_{n,n-3}$ with λK_2 partition by a series of $S_{n,n-1}^1, S_{n,n-2}^2, \ldots, S_{n,r}^t$ as a set of t-copies. Consider p and q as its vertex set and edge set, then we have the vertex operation given by:

$$S_{n,n-3}^t = \begin{cases} f(u_0) = 1 \\ f(v_i) = F_{i+2}; & 1 \le i \le n \\ f(u_j) = F_{j+m+2}; & 1 \le j \le m. \end{cases} \tag{20.7}$$

Case 3. If $n = 4$, then $p = \sum_{n=4}^{\infty} p = t(n-1) + t$ and dimension $\sum_{n=4}^{\infty} q = t(n+2)$ and by the labeling function $f^*(\alpha\beta)$, we get the required iteration for the resultant edge label

$$f(u_0v_i) = \begin{cases} f(v_i) - 1; & i = 1 \\ f(v_i); & \text{otherwise} \end{cases} \tag{20.8}$$

which is followed by the partition shell and its iteration is given by

$$f(v_{m-2}v_{m-1}) = \begin{cases} 2, 4; & t = 1 \\ f(v_i) + F_i; & t = 2 \\ f(v_{m-2}) + F_{m-2} & t = n. \end{cases} \tag{20.9}$$

Fixed $\lambda = 1$ and for $t = 1$, the value of the edge label is fixed at 2, 4 and from $t = 2$, we get the required recurrence relation up to the nth term. Observe that:

$$\begin{aligned} t &= 2; \quad i = 5, 6, \\ t &= 3; \quad i = 9, 10, \\ &\dots \\ t &= n; \quad m = (n-2), (n-1). \end{aligned}$$

Case 4. If $n = 6$, then $\sum_{n=6}^{\infty} p = tn + t$ and dimension $\sum_{n=6}^{\infty} q = t(2n - 2) + t$ and the required iteration for edge label follows the same as in equation (20.8) and

$$f(v_{m-3}v_{m-2}) = \begin{cases} 2, 4, 7, 11; & \lambda = 1 \\ f(v_i) + F_i; & \lambda = 2 \\ f(v_{m-3}) + F_{m-3} & \lambda = n \end{cases} \tag{20.10}$$

similarly fixed $\lambda = 2$ and for $t = 1$, the edge label is fixed at 2, 4, 7, 11 and for

$$\begin{aligned} t &= 2; \quad i = 8, 9, 10, 11, \\ t &= 3; \quad i = 15, 16, 17, 18, \\ &\dots \\ t &= n; \quad m = (n-6), (n-5), (n-4), (n-3). \end{aligned}$$

Case 5. If $n = 8$, then $\sum_{n=8}^{\infty} p = tn + 2t$ and dimension $\sum_{n=8}^{\infty} q = t(2n - 2) + 2t$ and from Case (3), Case (4) it is sufficient enough to show

$$f(v_{m-4}v_{m-3}) = \begin{cases} 2, 4, 7, 11, 18, 29; & \lambda = 1 \\ f(v_i) + F_i; & \lambda = 2 \\ f(v_{m-4}) + F_{m-4} & \lambda = n \end{cases} \tag{20.11}$$

and the edge label is fixed at 2, 4, 7, 11, 18, 29 for $t = 1$ and $\lambda = 3$, then for

$$t = 2; \quad i = 11, 12, 13, 14, 15, 16,$$

$$
\begin{aligned}
t &= 3; \quad i = 21, 22, 23, 24, 25, 26,\\
&\ldots\\
t &= n; \quad m = (n-9), (n-8), \ldots, (n-4),
\end{aligned}
$$

then we form the composition table

Table 20.2: Composition table for hybrid partition shell graph.

v_i	v_1	v_2	v_3	v_4	v_5	v_6	v_7	v_n
u_0	F_2	F_3	F_4	F_5	F_6	F_7	F_8	F_n
v_i	v_1	v_2	v_3	v_4	v_5	v_6		
v_{i+1}	F_3	F_4+1	F_5+2	F_6+3	F_7+5	F_8+8	$F_{k-1}+F_{k-3}$	

The above computation will give the resultant edge to be distinct and $\lceil \frac{\max(e_i^m e_{i+1}^n)}{\min(e_i^m e_{i+1}^n)} \rfloor$ will satisfy the condition of the golden ratio. Hence, all hybrid partition shell graph with union of λK_2 will satisfy the condition of FR-labeling. □

3 Conclusion

In this paper, we have proved FR-labeling for hybrid non-uniform and hybrid partition shell graph with union of λK_2. And for the set of t-copies $\sum_{n=4}^{\infty} S_{n,n-3}^t$ the values of λ can be further extended for higher values of n.

Bibliography

[1] P. K. Deb and N. B. Limaye, On harmonious labeling of some cycles related graph, *Ars Comb.* **65** (1980), 177–197.

[2] J. A. Gallian, A survey: A dynamic survey of graph labeling, *Electron. J. Comb.* **1** (2018), DS6.

[3] R. L. Graham and N. J. A. Sloane, On additive bases and harmonious graphs, *SIAM J. Algorithm Discrete Methods* **1** (1980), 382–404.

[4] K. E. Hilda and J. J. Jesintha, Subdivided shell flower graph, *South East Asian J. Math. Math. Sci.* **14** (2018), 79–88.

[5] J. J. Jesintha and K. E. Hilda, All uniform bow graphs are graceful, *Math. Comput. Sci.* **9** (2015), 185–191.

[6] A. S. Odyuo, P. Mercy and M. K. Patel, On Fibonacci range labeling based on vertex switching of a graph, *Bull. Calcutta Math. Soc.* **114**(6) (2022), 829–840.

[7] A. S. Odyuo, P. Mercy and M. K. Patel, Fibonacci range labeling for shell-flower related graphs, in: *Advances in Pure and Applied Algebra*, De Gruyter Proceeding, Berlin, Boston, 2023, pp. 71–80.

[8] Y. Yang, X. Xu and Y. Xi, On harmonious labeling of balance quadruple shells, *Ars Comb.* **75** (2005), 289–296.

R. K. Singh, R. G. Ghumde, and M. K. Patel

On essential *M*-Artinian modules

Abstract: The idea of an e-M-Artinian module is a generalization of Artinian and e-Artinian modules, which is defined as every descending chain of essential M-cyclic submodules gets terminated. In this paper, we proved that, if N is an e-N-Artinian R-module and M is a simple R-module, then the direct sum of N and M is an e-$(N \oplus M)$-Artinian. Also, we prove that for $M_1, M_2, M_3, \ldots, M_r$ are an R-modules, then $\sum_{i=1}^{r} \oplus M_i$ is an e-$\sum_{i=1}^{r} \oplus M_i$-Artinian if and only if each $M_i's$ are an e-M_i-Artinian.

Keywords: Essential submodules, essential M-cyclic submodules, e-M-Artinian modules

MSC 2010: 16P20, 16P60, 16P70

1 Introduction

In this paper, all modules are unital left R-modules and all rings are associative with identity. Recently, the concept of iso-Noetherian (iso-Artinian) modules was introduced by Facchini et al. in [2]. They provided several characterizations of these notions and studied some properties in [3] and [4]. Suppose M and N are left R-modules, then N is known as M-cyclic if $N \cong M/K$ for some submodule K of M. In [5], authors have introduced the concept of chain condition for M-cyclic submodules and proved several important results. With this motivation, here we discuss the descending chain condition of essential M-cyclic submodules in details. A submodule K of a module M is called an essential if for every submodule L of M which is nonzero, has non-trivial intersection with K. A module M is known as Artinian module if every descending chain of submodules of M gets terminated, for example any finite abelian group over integers. In [6], introduced the idea of essential submodues and [1], introduced the concept of essential artinian module and defined as A module M is an e-Artinian if for any chain of $L_1 \supseteq_e L_2 \supseteq_e L_3 \supseteq_e \cdots \supseteq_e L_n \supseteq_e \cdots$ of essential submodules of M gets stabilized, i. e., there exists some positive integer r, such that $L_r = L_{r+1}$. For example, if we consider $\mathbb{Z}_{32}$, then the descending chain $\mathbb{Z}_{32} \supseteq_e 2\mathbb{Z}_{32} \supseteq_e 4\mathbb{Z}_{32} \supseteq_e 8\mathbb{Z}_{32} \supseteq_e 16\mathbb{Z}_{32}$ is stationary. We recall that a module L is called uniform if every non-zero submodule is essential in L.

In this paper, we introduce the notion of an e-M-Artinian modules and studied about the characterizations and properties of this modules. The paper is split into

R. K. Singh, M. K. Patel, Department of Science and Humanities (Mathematics), National Institute of Technology Nagaland, Dimapur-797103, Nagaland, India, e-mails: nitranjitbhu@gmail.com, mkpitb@gmail.com
R. G. Ghumde, Department of Mathematics, Shri Ramdeobaba College of Engineering and Management, Nagpur, Maharastra 440013, India, e-mail: ghumderg@rknec.edu

https://doi.org/10.1515/9783111313634-021

two sections, where first section is devoted to introduction, including motivation and exploratory ideas related to the article. In the second section, we discuss several fundamental properties of an e-M-Artinian modules, where we extend few results of Artinian and e-Artinian modules and also established some new statements. We proved that an e-Artinian module is always an e-M-Artinian and indeed converse is true only if $\mathrm{End}(M)$ is a division ring (Theorem 3). We then proved that, if N is an e-N-Artinian R-module and M is a simple R-module, then the direct sum of N and M is an e-$(N \oplus M)$-Artinian (Theorem 6). We also showed that each $M_i's$ are e-M_i-Artinian if and only if $\sum_{i=1}^{r} \oplus M_i$ is an e-$\sum_{i=1}^{r} \oplus M_i$-Artinian for finite index i (Proposition 2).

Remark 1. Every essential L-cyclic submodules of an essential M-cyclic submodule L is again an essential M-cyclic.

2 Essential *M*-Artinian modules

A module M is said to be an essential-M-Artinian (in short, e-M-Artinian), if any descending chain $h_1(M) \supseteq_e h_2(M) \supseteq_e h_3(M) \supseteq_e \cdots \supseteq_e h_n(M) \supseteq_e \cdots$ of essential M-cyclic submodules of M is stationary, so there exist $r \in \mathbb{N}$, such that $h_r(M) = h_{r+1}(M)$ where $h_i \in \mathrm{End}(M)$ for all i. Clearly, every Artinian, simple, semisimple, uniform modules are an e-M-Artinian. An example of an e-M-Artinian module will be $\mathbb{Z}_{27}$. Consider the terminating chain $\mathbb{Z}_{27} \supseteq_e 3\mathbb{Z}_{27} \supseteq_e 9\mathbb{Z}_{27}$, here every submodule in this descending chain are essential and hence $\mathbb{Z}_{27}$ is an e-$\mathbb{Z}_{27}$-Artinian. Clearly, if a module is an uniform and Artinian then it is an e-M-Artinian.

Theorem 1. *Let M be an R-module. Then the following statements holds.*
1. *Descending chain conditions holds for essential M-cyclic submodules of M.*
2. *Minimum condition holds for essential M-cyclic submodules of M.*

Proof. Proof is to straight forward. □

Theorem 2. *An R-module M is an e-M-Artinian if and only if every family of non-empty essential M-cyclic submodules of M has a minimal element.*

Proof. Assume that M is an e-M-Artinian module and suppose Ω be a non-empty family of essential M-cyclic submodules of M. If there is no minimal element of Ω, we can get non terminating chain of essential M-cyclic submodules $g_0(M) \supseteq_e g_1(M) \supseteq_e \cdots$ from Ω, which arrive at contradiction to our assumption that M is an e-M-Artinian. Hence, we get family of essential M-cyclic submodules of M has a minimal element. Conversely, consider a chain $g_0(M) \supseteq_e g_1(M) \supseteq_e \cdots$ of essential M-cyclic submodules of M. By assumption, the set of essential M-cyclic submodules of M as $\{g_n(M) : n \in \mathbb{N}\}$ contains a minimal element $g_n(M)$. Nonetheless, after that we have $g_k(M) = g_n(M)$ for all $k \geq n$, hence M is an e-M-Artinian. □

Corollary 1. *Every submodule N of an e-M-Artinian module M is also an e-N-Artinian.*

Theorem 3. *Every e-Artinian module is an e-M-Artinian, however, converse is true when End(M) forms a group under multiplication.*

Proof. Proof is straightforward. □

Recall from [7], An R-module M is called epi-retractable if every submodule of M is a homomorphic image of M.

Proposition 1. *An epi-retractable module M is an e-M-Artinian if and only if it is an e-Artinian.*

Proof. Suppose M is an epi-retractable module which is an e-M-Artinian. As every submodule in epi-retractable modules are M-cyclic, hence any descending chain of essential M-cyclic submodules turns into descending chain of essential submodules of M that terminates, since M is an e-M-Artinian. Converse part is obvious by Theorem 3. □

Recall from [8], A sequence of the type $\cdots \rightarrow A_{n-1} \xrightarrow{\pi_{n-1}} A_n \xrightarrow{\pi_n} A_{n+1} \rightarrow \cdots$ of R-homomorphisms is called an exact sequence, if $img(\pi_{n-1}) = \ker(\pi_n)$ for all n. An exact sequence of the type $0 \rightarrow P \xrightarrow{\phi} Q \xrightarrow{\psi} R \rightarrow 0$ is known as short exact sequence, where ϕ and ψ is one-one and onto homomorphisms respectively.

Theorem 4. *Suppose $0 \rightarrow M_1 \xrightarrow{\phi} M \xrightarrow{\psi} M_2 \rightarrow 0$ be a short exact sequence of R-modules and R-homomorphisms. Then M is an e-M-Artinian if and only if both M_1 and M_2 are e-M_1-Artinian and e-M_2-Artinian respectively.*

Proof. We suppose that M_1 is a submodule of M and that $M/M_1 \cong M_2$.

Assume that M is an e-M-Artinian. Since M_1-cyclic submodule of M_1 are also M-cyclic submodule of M by Remark 1, so we have M_1 is an e-M_1-Artinian by Corollary 1. For next part, consider a chain of essential M_2-cyclic submodules of M_2 is $g_1(M_2) \supseteq_e g_2(M_2) \supseteq_e g_3(M_2) \cdots \supseteq_e g_n(M_2) \supseteq_e \cdots$ of M/M_1 where $g_i \in \mathrm{End}(M_2)$ and corresponding to this chain by [8, Proposition 1.9], we get $h_1(M) \supseteq_e h_2(M) \supseteq_e h_3(M) \supseteq_e \cdots \supseteq_e h_n(M) \supseteq_e \cdots$ of essential M-cyclic submodules of M, which contains M_1 and there exists natural number n such that $h_r(M) = h_n(M)$ for all $r \geq n$, as M is an e-M-Artinian. Returning to M/M_1, we see that $g_r(M_2) = g_n(M_2)$ for all $r \geq n$. Hence, $M/M_1 \cong M_2$ is an e-M_2-Artinian module.

Conversely, now suppose that M_1 and $M/M_1 \cong M_2$ are e-M_1-Artinian and e-M_2-Artinian respectively. Now consider the descending chain of essential M-cyclic submodules is $g_1(M) \supseteq_e g_2(M) \supseteq_e g_3(M) \supseteq_e \cdots \supseteq_e g_n(M) \supseteq_e \cdots$ of M. Claim that $g_n(M) = g_m(M)$ for all $m \geq n$. This gives descending chains $g_1(M) \cap M_1 \supseteq_e g_2(M) \cap M_1 \supseteq_e g_3(M) \cap M_1 \supseteq_e \cdots \supseteq_e g_r(M) \cap M_1 \supseteq_e \cdots$ and $(g_1(M) + M_1)/M_1 \supseteq_e (g_2(M) + M_1)/M_1 \supseteq_e (g_3(M) + M_1)/M_1 \supseteq_e \cdots \supseteq_e (g_r(M) + M_1)/M_1 \supseteq_e \cdots$. of M_1 and $M/M_1 \cong M_2$ respectively. Clearly both of these chains terminates, so there exists a natural number $m \in \mathbb{N}$ such that $(g_n(M) \cap M_1) = (g_m(M) \cap M_1)$ and $(g_n(M) + M_1)/M_1 = (g_m(M) + M_1)/M_1$ for all $m \geq n$. Now we proceed to the final step. Since

$$g_n(M) = g_n(M) \cap (g_n(M) + M_1)$$

$$\begin{aligned}
&= g_n(M) \cap (g_m(M) + M_1) \\
&= g_m(M) + (g_n(M) \cap M_1) \quad \text{(by modular property)} \\
&= g_m(M) + (g_m(M) \cap M_1) \quad (\text{since } g_n(M) \cap M_1 = g_m(M) \cap M_1) \\
&= g_m(M) \cap (g_m(M) + M_1) \quad \text{(by modular property)} \\
&= g_m(M).
\end{aligned}$$

Hence $g_n(M) = g_m(M)$ for all $m \geq n$. Hence M is an e-M-Artinian module. □

Corollary 2. *A module M is an e-M-Artinian if and only if its submodule K and quotient module M/K are e-K-Artinian and e-(M/K)-Artinian respectively.*

Proof. If we assume an exact sequence $0 \to K \xrightarrow{\phi} M \xrightarrow{\psi} M/K \to 0$, where ϕ is one-one homomorphism and ψ is an onto homomorphism. Clearly the proof holds from above result. □

Corollary 3. *The homomorphic image of an e-M-Artinian module is an e-M-Artinian.*

Proof. Suppose M is an e-M-Artinian and K be an R-module. If $g : M \to K$ be R-module homomorphism, then we have $M/\ker(g) \cong g(M)$ by first isomorphism theorem. Hence $g(M)$ is an e-M-Artinian by Corollary 2. □

Corollary 4.
1. *Every submodules and quotient modules of an uniform e-Artinian module M are an e-M-Artinian.*
2. *If M be an uniform module such that it contains a submodule K with K as e-K-Artinian and M/K as e-(M/K)-Artinian, then M is an e-M-Artinian.*

Proof. The above theorem leads directly to proof. □

Proposition 2. *Let $M_1, M_2, M_3, \ldots, M_r$ be R-modules. Then $\sum_{i=1}^{r} \oplus M_i$ is an e-$\sum_{i=1}^{r} \oplus M_i$-Artinian if and only if each $M_i^{\prime}s$ are e-M_i-Artinian modules.*

Proof. Let $\sum_{i=1}^{r} \oplus M_i$ be an e-$\sum_{i=1}^{r} \oplus M_i$-Artinian. It is obvious that the result is true for $r = 1$. Consider for $r \geq 2$, an exact sequence $0 \to \sum_{i=1}^{r-1} \oplus M_i \xrightarrow{\phi} \sum_{i=1}^{r} \oplus M_i \xrightarrow{\pi} M_r \to 0$, where the mappings ϕ and π are injection and projection respectively. Using Theorem 4, we have $\sum_{i=1}^{r-1} \oplus M_i$ and M_r are an e-$\sum_{i=1}^{r-1} \oplus M_i$-Artinian and e-$M_r$-Artinian respectively. Taking $i = r$, then clearly we get M_i is an e-M_i-Artinian.

Conversely, assume that each $M_i^{\prime}s$ are e-M_i-Artinian. By using the mathematical induction, suppose $\sum_{i=1}^{r-1} \oplus M_i$ is an e-$\sum_{i=1}^{r-1} \oplus M_i$-Artinian and M_r is an e-M_r-Artinian respectively. Taking the exact sequence $0 \to \sum_{i=1}^{r-1} \oplus M_i \xrightarrow{\phi} \sum_{i=1}^{r} \oplus M_i \xrightarrow{\pi} M_r \to 0$, where ϕ and π are the injection and projection mapping, respectively, and so by Theorem 4, we have $\sum_{i=1}^{r} \oplus M_i$ is an e-$\sum_{i=1}^{r} \oplus M_i$-Artinian. □

Corollary 5. *Finite sum of e-M-Artinian modules are an e-M-Artinian.*

Proof. Proof is similar to above Proposition 2. □

Corollary 6. *Let $M_1, M_2, M_3, \dots, M_r$ be simple R-Modules. Then $\sum_{i=1}^{r} \oplus M_i$ is an e-$\sum_{i=1}^{r} \oplus M_i$-Artinian.*

Proposition 3. *A non-zero e-M-Artinian module possesses a simple M-cyclic submodule.*

Proof. Let M be a non-zero e-M-Artinian module. By Theorom 2, the collection of all non-zero essential M-cyclic submodules of M has a minimal member, and this is a simple M-cyclic submodule of M. □

Theorem 5. *Every finite semisimple module M is an e-M-Artinian.*

Proof. Assume that M is a finite semisimple module, so M is a direct sum of finite number of simple submodules of M, as every direct summand of M is an M-cyclic submodule of M. Now with the help of this we get terminating descending chain of essential M-cyclic submodule of M. Thus M is an e-M-Artinian. □

Theorem 6. *Let N be an e-N-Artinian module and M be a simple R-module, then $(N \oplus M)$ is an e-$(N \oplus M)$-Artinian.*

Proof. Let $f(M_1) \supseteq_e f(M_2) \supseteq_e \cdots$ be any chain of non-zero essential $(N \oplus M)$-cyclic submodules of $(N \oplus M)$. We discuss two cases.

Case 1. If $f(M_i) \cap M = 0$ for all i, then all $f(M_i) \cong K$, where K is any submodule of N. Given that N is an e-N-Artinian, this implies that $f(M_n) = f(M_n)$ for all n, proof is done.

Case 2. If $M \cap f(M_i) \neq 0$ for $f(M_i)$ is nonzero. As M is simple so this implies that $f(M_i) \cap M = M$ and thus $M \subseteq f(M_i)$ for all $i \geq p$. Applying modular property, we get $f(M_i) = f(M_i) \cap (N \oplus M) = (f(M_i) \cap N) \oplus M$, similarly $f(M_{i+1}) = (f(M_{i+1}) \cap N) \oplus M$. Suppose that $f(M_k) \cap N \supseteq_e f(M_{k+1}) \cap N \supseteq_e \cdots$ is a descending chain of essential N-cyclic submodules of N. Because N is an e-N-Artinian, it follows that there exist $s \geq k, f(M_i) \cap N = f(M_{i+1}) \cap N$ for all $i \geq s$. Thus $(f(M_i) \cap N) \oplus M = (f(M_{i+1}) \cap N) \oplus M$ for all $i \geq s$ and hence $f(M_i) = f(M_{i+1})$ for all $i \geq s$. Thus $(N \oplus M)$ is an e-$(N \oplus M)$-Artinian. □

Theorem 7. *Let A and B be any two modules, where $A \subseteq B$ and B is an e-B-Artinian module. Suppose M is an injective module, then M is an e-M-Artinian.*

Proof. Consider the row exact diagram

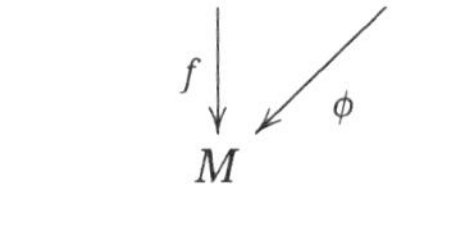

Since M is an injective, so we have $\phi \circ i = f$. We know by Corollary 3, homomorphic image of an e-M-Artinian module is an e-M-Artinian. Since B is an e-B-Artinian and $\phi(B) \subseteq M$, so $\phi(B)$ is also an e-M-Artinian. Claim that M is an e-M-Artinian. Consider the descending chain of essential M-cyclic submodules of M, i. e., $h(M_1) \supseteq_e h(M_2) \supseteq_e h(M_3) \supseteq_e \cdots \supseteq_e h(M_n) \supseteq_e \cdots \longrightarrow (1)$.

Which arises two cases. Case 1: If $\phi(B)$ is in (1) and as $\phi(B)$ is an e-M-Artinian, hence M is an e-M-Artinian. Case 2: If $\phi(B)$ is not in (1), so the descending chain of $\phi(B)$ as $\phi(B) \cap h(M_1) \supseteq_e \phi(B) \cap h(M_2) \supseteq_e \cdots \supseteq_e \phi(B) \cap h(M_1) \supseteq_e \cdots \longrightarrow (2)$. Since $\phi(B)$ is an e-M-Artinian, so (2) terminates, i. e., we have $\phi(B) \cap h(M_r) = \phi(B) \cap h(M_{r+1})$ for all $r \in \mathbb{N}$. Now if $h(M_r) \subseteq \phi(B)$, then also $h(M_{r+1}) \subseteq \phi(B)$, it follows that $h(M_r) = h(M_{r+1})$ and hence M is an e-M-Artinian. If $\phi(B) \in h(M_r)$ then from case 1, we have $h(M_r) = h(M_{r+1})$. Thus M is an e-M-Artinian. □

Bibliography

[1] H. Chakraborty, P. Nath and R. K. Singh, Essential Artinian modules and rings, in: *Advances in Pure and Applied Algebra*, Vol. 730, De Gruyter, 2023, pp. 89–94, https://doi.org/10.1515/9783110785807-011, ISBN 9783110785807.

[2] A. Facchini and A. Ghorbani, Modules with chain conditions upto isomorphism, *J. Algebra* **453** (2016), 578–601. https://doi.org/10.1016/j.jalgebra.2016.01.025.

[3] A. Facchini and Z. Nazemian, Artinian dimension and isoradical of modules, *J. Algebra* **484** (2017), 66–87. https://doi.org/10.1016/j.jalgebra.2017.03.039.

[4] A. Facchini and Z. Nazemian, On isonoetherian and isoartinian modules, *Contemp. Math.* **730** (2019), 1–22. https://doi.org/10.1090/conm/730/14707.

[5] T. Gera, S. C. Gannamaneni and M. K. Patel, Chain conditions on M-cyclic submodules, in: *Advances in Pure and Applied Algebra*, Vol. 730, De Gruyter, 2023, pp. 81–88, https://doi.org/10.1515/9783110785807-010, ISBN 9783110785807.

[6] B. L. Osofsky, Chain conditions on essential submodules, *Proc. Am. Math. Soc.* **114**(1) (1994), 11–19.

[7] B. M. Pandeya, A. K. Chaturdevi and A. J. Gupta Applications of epi-rectractable modules, *Bull. Iranian Math. Soc.* **38**(2) (2012), 469–477.

[8] D. W. Sharpe and P. Vamos, *Injective Modules*, Cambridge University Press, 2008.

Priti Singh and Sharvan Kumar

Properties of reductions of multiplication modules

Abstract: In this paper, we introduce the notion of reduction of submodules and examine the properties of reduction of submodules of multiplication modules over a commutative ring R. For a finitely generated faithful R-module M, K is reduction of N iff $\phi(K)$ is a reduction of $\phi(N)$, where $\phi : M \longrightarrow M$ is a module homomorphism, K and N are submodules of M s. t. $K \subseteq N$.

Keywords: Multiplication modules, cancellation modules, Reduction number

MSC 2010: 13C03, 13D02, 13D05

1 Introduction

The concept of multiplication modules was introduced by A. Barnard [4] in 1981. He showed that every distributive modules are characterized as modules for which every finitely generated submodule is a multiplication module and is cyclic over quasi-local ring. Later, P. F. Smith [14] also examined local properties of multiplication modules and such modules are cyclic or finitely generated. There are many researchers to introduce the concept of multiplication module ([1, 7–12]).

The notion of R-closure and R-dependence of an ideal with respect to M was considered by Toroghy and Dorostkar [2] as generalization of reduction and integral closure of ideals of local rings [13]. In [15], Tiras gave the relation between the R-closure of an ideal I in a commutative Noetherian ring R and the integral closure of the ideal I relative to a Noetherian R-module. Foster and Johnson [6] concerned with the integral and asymptotic closure operations on a multiplicative lattice relative to the greatest element of a lattice module having the ascending chain condition. They have shown that a cancelation law holds for the asymptotic closure of elements of the multiplicative lattice. A'zami and Khajepour [3] defined these concepts for submodules of multiplication modules and presented similar results as for ideals. In this paper, first we define the notion of prod-

Acknowledgement: The authors thank the anonymous referee for helpful comments and the second author gratefully acknowledges the financial support in the form of fellowship, given by the Council of Scientific and Industrial Research (CSIR), India.

Priti Singh, Department of Mathematics, Patna Science College, Patna University, Patna, Bihar, India, e-mail: pritisingh.mnnit@gmail.com
Sharvan Kumar, Department of Mathematics, Patna University, Patna, Bihar, India, e-mail: skgu2810@gmail.com

https://doi.org/10.1515/9783111313634-022

uct of two submodules of a multiplication module with examples. Also, we introduce the reduction of submodule of a localized multiplication module and give some results on reduction of submodule of a localized multiplication submodule. Throughout the discussion, R is a commutative ring with unity and define the set $(N : M) = \{x \in R \mid xM \subseteq N\}$.

2 Reduction over multiplication module

Definition 1 (Multiplication module). A module M is said to be multiplication R-module, if for every submodule N of M, there exists an ideal I of R such that $N = IM$.

Definition 2 (Reduction of modules). Let N and K be two submodules of M. The product of N and K is defined as $NK = (N : M)(K : M)M$. A submodule K is said to be a reduction of N, if $K \subseteq N$ and there exists $s \in \mathbb{N}$ such that $KN^s = N^{s+1}$. The least positive integer s such that $KN^s = N^{s+1}$ is called the reduction number of N with respect to K and is denoted by $r_K(N)$.

Remark 1.
1. If M is a multiplication R-module, then for every submodule N of M such that $N = (N : M)M$.
2. Homormorphic image of a multiplication module is a multiplication module.
3. If $N = IM$, $K = JM$ for some ideals I, J of R, then $NK = (IM)(JM) = (IM : M)(JM : M)M = IJM$.

Example 1. Let $N = 2\mathbb{Z}$ and $K = 3\mathbb{Z}$ be two submodules of $\mathbb{Z}$ over ring $\mathbb{Z}$. Then $(2\mathbb{Z})(3\mathbb{Z}) = (2\mathbb{Z} : \mathbb{Z})(3\mathbb{Z} : \mathbb{Z})\mathbb{Z}$ and $(2\mathbb{Z})(3\mathbb{Z}) = 6\mathbb{Z}$.

Example 2. Every cyclic R-module is a multiplication module but converse is not true.

Proof. Let $M = Rx$ be cyclic module and N be the submodule of M, where $x \in M$. We have to show that $N = (N : M)M$ for every submodule N of M. Let $y \in N$ such that $y = rx$ for some $r \in R$. Therefore, $rM \subseteq N$ and $r \in (N : M)$. So that $y \in (N : M)M$ and $N \subseteq (N : M)M$. Conversely, let $y \in (N : M)M$ such that $y = am$ for some $a \in (N : M)$ and $m \in M$. Since $a \in (N : M)$, $aM \subseteq N$ for all $m \in M$. Thus $am \in N$ and $y \in N$. Therefore, $N = (N : M)M$ and M multiplication module.

A module $\mathbb{Z}[x]$-$\mathbb{Z}$-module. Then $\mathbb{Z}[x]$ is multiplication $\mathbb{Z}$-module and is not a cyclic over $\mathbb{Z}$. □

Example 3. Let $R = k[X, Y]$ be the ring and $M = k[X, Y]$ be an R-module, where k is a field. Suppose $N = (X^3, Y^3)$ and $K = (X^3, X^2Y, Y^3)$ are two submodules of M. Then N is a reduction of K and $r_K(N) = 2$.

3 Properties of multiplication modules

Lemma 1. *Let M be a multiplication R-module and N, K be two submodules of M. Then:*

1. *For each ideal I of R such that $I(NK) = (IN)K = N(IK)$.*
2. *The $W^{-1}M$ is multiplication $W^{-1}R$-module, where W is a multiplicative closed subset of R.*

Proof. 1. Since M is a multiplication R-module, $N = I_1M$ and $K = I_2M$ for some ideals I_1 and I_2 of R. Then

$$I(NK) = I(I_1MI_2M) = I(I_1I_2M) = (II_1I_2)M = (II_1M)I_2M = (IN)K,$$
$$(IN)K = (II_1M)I_2M = (II_1I_2)M = (I_1II_2)M, = (I_1M)(II_2M) = N(IK),$$

for R is a commutative ring.

2. Suppose M is a multiplication R-module. Then there exists an ideal I of R such that $N = IM$ for every submodule N of M. Note that $W^{-1}M$ is a $W^{-1}R$-module. By the property of localization,

$$W^{-1}(N) = W^{-1}(IM),$$
$$W^{-1}(N) = W^{-1}(I)W^{-1}(M).$$

This implies that $W^{-1}M$ is a multiplication $W^{-1}R$-module. □

Theorem 1. *Let M be a multiplication R-module and N, K be two submodules of M. Suppose W is a multiplicative closed subset of R. Then:*

1. *If K is a reduction of N, then $W^{-1}K$ is a reduction of $W^{-1}N$.*
2. *If $W^{-1}K$ is a reduction of $W^{-1}N$, then $(W^{-1}I)(W^{-1}K)$ is a reduction of $(W^{-1}I)(W^{-1}N)$ for each ideal I of R.*
3. *If $W^{-1}I$ is a reduction of $W^{-1}J$, then $(W^{-1}I)(W^{-1}N)$ is a reduction of $(W^{-1}J)(W^{-1}N)$ for each ideal I and J of R.*
4. *If $W^{-1}I$ is a reduction of $W^{-1}J$ and $W^{-1}K$ is a reduction of $W^{-1}N$ for every $I \subseteq J$ and $K \subseteq N$, then $(W^{-1}I)(W^{-1}K)$ is a reduction of $(W^{-1}J)(W^{-1}N)$.*
5. *If $W^{-1}K \subseteq W^{-1}N$ and $(W^{-1}K : W^{-1}M)$ is a reduction of $(W^{-1}N : W^{-1}M)$, then $W^{-1}K$ is a reduction of $W^{-1}N$.*
6. *If $W^{-1}M$ is a finitely generated faithful module and $W^{-1}K$ is a reduction of $W^{-1}N$, then $(W^{-1}K : W^{-1}M)$ is a reduction of $(W^{-1}N : W^{-1}M)$.*

Proof. 1. Let K be a reduction of N. Then $KN^s = N^{s+1}$ for some $s > 0$. By the property of localization,

$$W^{-1}(KN^s) = W^{-1}(N^{s+1})$$
$$W^{-1}\{K(\underbrace{N \cdots N}_{s\text{-times}})\} = W^{-1}(\underbrace{N \cdots N}_{s+1\text{-times}})$$

$$(W^{-1}K)\{\underbrace{(W^{-1}N)\cdots(W^{-1}N)}_{s\text{-times}}\} = \underbrace{(W^{-1}N)\cdots(W^{-1}N)}_{s+1\text{-times}}$$
$$(W^{-1}K)(W^{-1}N)^s = (W^{-1}N)^{s+1}.$$

2. Suppose $W^{-1}K$ is a reduction of $W^{-1}N$. Then $(W^{-1}K)(W^{-1}N)^s = (W^{-1}N)^{s+1}$ for some $s \in \mathbb{N}$. Since $W^{-1}K \subseteq W^{-1}N$, $(W^{-1}I)(W^{-1}K) \subseteq (W^{-1}I)(W^{-1}N)$. Therefore,

$$\begin{aligned}
&((W^{-1}I)(W^{-1}K))((W^{-1}I)(W^{-1}N))^s \\
&= ((W^{-1}I)(W^{-1}K))\{\underbrace{((W^{-1}I)(W^{-1}N))\cdots((W^{-1}I)(W^{-1}N))}_{s\text{-times}}\} \\
&= ((W^{-1}I)(W^{-1}K))\{\underbrace{(W^{-1}I)\cdots(W^{-1}I)}_{s\text{-times}}\}\{\underbrace{(W^{-1}N)...(W^{-1}N)}_{s\text{-times}}\} \\
&= ((W^{-1}I)(W^{-1}K))((W^{-1}I)^s(W^{-1}N)^s) \\
&= (W^{-1}I)(W^{-1}I)^s((W^{-1}K)(W^{-1}N)^s) \\
&= (W^{-1}I)^{s+1}(W^{-1}N)^{s+1} \\
&= (W^{-1}I)^s((W^{-1}I)(W^{-1}N))(W^{-1}N)^s \\
&= \underbrace{((W^{-1}I)(W^{-1}N))\cdots((W^{-1}I)(W^{-1}N))}_{s+1\text{-times}} \\
&= ((W^{-1}I)(W^{-1}N))^{s+1}.
\end{aligned}$$

3. Suppose $W^{-1}I$ is a reduction of $W^{-1}J$. Then $(W^{-1}I)(W^{-1}J)^s = (W^{-1}J)^{s+1}$ for some $s \in \mathbb{N}$. Since $W^{-1}I \subseteq W^{-1}J$, $(W^{-1}I)(W^{-1}N) \subseteq (W^{-1}J)(W^{-1}N)$. Therefore,

$$\begin{aligned}
&((W^{-1}I)(W^{-1}N))((W^{-1}J)(W^{-1}N))^s \\
&= ((W^{-1}I)(W^{-1}N))\{\underbrace{((W^{-1}J)(W^{-1}N))\cdots((W^{-1}J)(W^{-1}N))}_{s\text{-times}}\} \\
&= ((W^{-1}I)(W^{-1}N))\{\underbrace{(W^{-1}J)\cdots(W^{-1}J)}_{s\text{-times}}\}\{\underbrace{(W^{-1}N)\cdots(W^{-1}N)}_{s\text{-times}}\} \\
&= (W^{-1}I)(W^{-1}J)^s(W^{-1}N)(W^{-1}N)^s \\
&= ((W^{-1}I)(W^{-1}J)^s(W^{-1}N)^{s+1} \\
&= (W^{-1}J)^{s+1}(W^{-1}N)^{s+1} \\
&= ((W^{-1}J)(W^{-1}N))^{s+1}.
\end{aligned}$$

4. Let $W^{-1}I$ be a reduction of $W^{-1}J$. Then $(W^{-1}I)(W^{-1}J)^s = (W^{-1}J)^{s+1}$ for some $s \in \mathbb{N}$ and $W^{-1}K$ is a reduction of $W^{-1}N$ such that $(W^{-1}K)(W^{-1}N)^t = (W^{-1}N)^{t+1}$ for some $t \in \mathbb{N}$. Since $W^{-1}I \subseteq W^{-1}J$ and $W^{-1}K \subseteq W^{-1}N$, $(W^{-1}I)(W^{-1}K) \subseteq (W^{-1}J)(W^{-1}N)$. Hence,

$$((W^{-1}I)(W^{-1}K))((W^{-1}J)(W^{-1}N))^{s+t}$$

$$= (W^{-1}I)(W^{-1}J)^s\{\underbrace{(W^{-1}J)\cdots(W^{-1}J)}_{t\text{-times}}\}(W^{-1}K)(W^{-1}N)^t\{\underbrace{(W^{-1}N)\cdots(W^{-1}J)}_{s\text{-times}}\}$$
$$= (W^{-1}J)^{s+1}\{\underbrace{(W^{-1}J)\cdots(W^{-1}J)}_{t\text{-times}}\}(W^{-1}N)^{t+1}\{\underbrace{(W^{-1}N)\cdots(W^{-1}N)}_{s\text{-times}}\}$$
$$= (W^{-1}J)^{s+t+1}(W^{-1}N)^{s+t+1}$$
$$= ((W^{-1}J)(W^{-1}N))^{s+t+1}.$$

5. Let $(W^{-1}K : W^{-1}M)$ be a reduction of $(W^{-1}N : W^{-1}M)$. Then $(W^{-1}K : W^{-1}M)(W^{-1}N : W^{-1}M)^s = (W^{-1}N : W^{-1}M)^{s+1}$ for some $s \in \mathbb{N}$. Now,

$$\begin{aligned}
&(W^{-1}K)(W^{-1}N)^s\\
&\quad= (W^{-1}K)\,\underbrace{(W^{-1}N)\cdots(W^{-1}N)}_{s\text{-times}}\\
&\quad= (W^{-1}K : W^{-1}M)\{\underbrace{(W^{-1}N : W^{-1}M)\cdots(W^{-1}N : W^{-1}M)}_{s\text{-times}}\}(W^{-1}M)\\
&\quad= (W^{-1}K : W^{-1}M)(W^{-1}N : W^{-1}M)^s W^{-1}M\\
&\quad= (W^{-1}N : W^{-1}M)^{s+1} W^{-1}M\\
&\quad= (W^{-1}N)^{s+1}.
\end{aligned}$$

6. Suppose $W^{-1}K$ is a reduction of $W^{-1}N$. Then $(W^{-1}K)(W^{-1}N)^s = (W^{-1}N)^{s+1}$ for some $s \in \mathbb{N}$. And

$$\begin{aligned}
&(W^{-1}K)(W^{-1}N)^s\\
&\quad= (W^{-1}K)\,\underbrace{(W^{-1}N)\cdots(W^{-1}N)}_{s\text{-times}}\\
&\quad= (W^{-1}K : W^{-1}M)\{\underbrace{(W^{-1}N : W^{-1}M)\cdots(W^{-1}N : W^{-1}M)}_{s\text{-times}}\}(W^{-1}M)\\
&\quad= (W^{-1}K : W^{-1}M)(W^{-1}N : W^{-1}M)^s W^{-1}M.
\end{aligned}$$

Also,

$$\begin{aligned}
(W^{-1}N))^{s+1} &= \underbrace{(W^{-1}N)\cdots(W^{-1}N)}_{s+1\text{-times}}\\
&= \{\underbrace{(W^{-1}N : W^{-1}M)\cdots(W^{-1}N : W^{-1}M)}_{s+1\text{-times}}\}(W^{-1}M)\\
&= (W^{-1}N : W^{-1}M)^{s+1} W^{-1}M.
\end{aligned}$$

Note that $W^{-1}M$ is a cancelation $W^{-1}R$-module [5]. Therefore, $(W^{-1}K : W^{-1}M)$ is a reduction of $(W^{-1}N : W^{-1}M)$. □

Proposition 1. *Suppose M is an R-module and K_1, K_2, N_1 and N_2 are submodules of M such that K_1, K_2 are reduction of N_1 and N_2 respectively. Then:*
1. $K_1 + K_2$ *is a reduction of* $N_1 + N_2$.
2. K_1K_2 *is a reduction of* N_1N_2.

Proof. 1. Let $I_1 = (K_1 : M)$, $I_2 = (K_2 : M)$, $J_1 = (N_1 : M)$ and $J_2 = (N_2 : M)$. Assume that $K_1{N_1}^{s_1} = {N_1}^{s_1+1}$, $K_2{N_1}^{s_2} = {N_1}^{s_2+1}$, $K_2{N_2}^{t_1} = {N_2}^{t_1+1}$ and $K_2{N_2}^{t_2} = {N_2}^{t_2+1}$ for some s_1, s_2, t_1, $t_2 \in \mathbb{N}$. Consequently we have the following relations:

$$
\begin{aligned}
&(K_1 + K_2)(N_1 + N_2)^{t_1+t_2+s_1+s_2} \\
&\quad = ((K_1 + K_2) : M)\{\underbrace{((N_1 + N_2) : M)\cdots((N_1 + N_2) : M)}_{t_1+t_2+s_1+s_2\text{-times}}\}M \\
&\quad = ((K_1 : M) + (K_2 : M))\{\underbrace{((N_1 : M) + (N_2 : M))\cdots((N_1 : M) + (N_2 : M))}_{t_1+t_2+s_1+s_2\text{-times}}\}M \\
&\quad = ((K_1 : M) + (K_2 : M))((N_1 : M) + (N_2 : M))^{t_1+t_2+s_1+s_2}M \\
&\quad = (I_1 + I_2)(J_1 + J_2)^{t_1+t_2+s_1+s_2}M \\
&\quad = I_1(J_1 + J_2)^{t_1+t_2+s_1+s_2}M + I_2(J_1 + J_2)^{t_1+t_2+s_1+s_2}M \\
&\quad = I_1 \sum_{i=0}^{t_1+t_2+s_1+s_2} {J_1}^i{J_2}^{t_1+t_2+s_1+s_2-i}M \\
&\qquad + I_2 \sum_{i=0}^{t_1+t_2+s_1+s_2} {J_2}^i{J_1}^{t_1+t_2+s_1+s_2-i}M \\
&\quad = \sum_{i=0}^{s_1} {J_1}^i{J_2}^{t_1+t_2+s_1+s_2-i}I_1M \\
&\qquad + \sum_{i=s_1}^{t_1+t_2+s_1+s_2} {J_1}^i{J_2}^{t_1+t_2+s_1+s_2-i}I_1M \\
&\qquad + \sum_{i=0}^{s_2} {J_2}^i{J_1}^{t_1+t_2+s_1+s_2-i}I_2M \\
&\qquad + \sum_{i=s_2}^{t_1+t_2+s_1+s_2} {J_2}^i{J_1}^{t_1+t_2+s_1+s_2-i}I_2M \\
&\quad = \sum_{i=0}^{s_1} {J_1}^i{J_2}^{t_1+t_2+s_1+s_2-i+1}M \\
&\qquad + \sum_{i=s_1}^{t_1+t_2+s_1+s_2} {J_1}^{i+1}{J_2}^{t_1+t_2+s_1+s_2-i}M \\
&\qquad + \sum_{i=0}^{s_2} {J_2}^i{J_1}^{t_1+t_2+s_1+s_2-i+1}M
\end{aligned}
$$

$$
\begin{aligned}
&+ \sum_{i=s_2}^{t_1+t_2+s_1+s_2} J_2^{\,i+1} J_1^{\,t_1+t_2+s_1+s_2-i} M \\
&= \sum_{i=0}^{t_1+t_2+s_1+s_2+1} J_1^{\,i} J_2^{\,t_1+t_2+s_1+s_2-i} M \\
&+ \sum_{i=0}^{t_1+t_2+s_1+s_2+1} J_2^{\,i} J_1^{\,t_1+t_2+s_1+s_2-i} M \\
&= \sum_{i=0}^{t_1+t_2+s_1+s_2+1} J_1^{\,i} J_2^{\,t_1+t_2+s_1+s_2-i} M \\
&= (J_1 + J_2)^{t_1+t_2+s_1+s_2+1} M = (N_1 + N_2)^{t_1+t_2+s_1+s_2+1}.
\end{aligned}
$$

2. Assume that $K_1 N_1^{\,s_1} = N_1^{\,s_1+1}$ and $K_2 N_2^{\,t_1} = N_2^{\,t_1+1}$, for some $s_1, t_1 \in \mathbb{N}$ also $K_1 \subseteq N_1$ and $K_2 \subseteq N_2$. Let $(K_1 : M) = I_1$, $(K_2 : M) = I_2$, $(N_1 : M) = J_1$ and $(N_2 : M) = J_2$. Therefore,

$$
\begin{aligned}
&(K_1K_2)(N_1N_2)^{s_1+t_1} \\
&= (K_1K_2)\underbrace{(N_1N_2)\cdots(N_1N_2)}_{s_1+t_1\text{-times}} \\
&= ((K_1 : M)(K_2 : M)M)\underbrace{((N_1 : M)(N_2 : M)M)\cdots((N_1 : M)(N_2 : M)M)}_{s_1+t_1\text{-times}} \\
&= ((K_1 : M)(K_2 : M))\underbrace{((N_1 : M)(N_2 : M))\cdots((N_1 : M)(N_2 : M))}_{s_1+t_1\text{-times}} M \\
&= ((K_1 : M)(K_2 : M))((N_1 : M)(N_2 : M))^{s_1+t_1} M \\
&= (I_1I_2)(J_1J_2)^{s_1+t_1} M \\
&= I_1(J_1)^{s_1+t_1} I_2(J_2)^{s_1+t_1} M \\
&= (I_1(J_1)^{s_1+t_1} M)(I_2(J_2)^{s_1+t_1} M) \\
&= (I_1M)(J_1^{s_1+t_1}M)(I_2M)(J_2^{s_1+t_1}M) \\
&= K_1(N_1)^{s_1+t_1} K_2(N_2)^{s_1+t_1} \\
&= N_1^{\,s_1+t_1+1} N_2^{\,s_1+t_1+1} \\
&= J_1^{\,s_1+t_1+1} M J_2^{\,s_1+t_1+1} M \\
&= J_1^{\,s_1+t_1+1} J_2^{\,s_1+t_1+1} M \\
&= (J_1J_2)^{s_1+t_1+1} M \\
&= (N_1N_2)^{s_1+t_1+1}.
\end{aligned}
$$

□

Proposition 2. *For a finitely generated faithful R-module M, K is reduction of N iff $\phi(K)$ is a reduction of $\phi(N)$, where $\phi : M \longrightarrow M$ is a module homomorphism, K and N are submodules of M s. t. $K \subseteq N$.*

Proof. Let $N = IM, K = JM$ for some ideals I, J of R. Assume that $KN^s = N^{s+1}$, for some $s \in \mathbb{N}$. Then

$$\begin{aligned}\phi(N^{s+1}) &= \phi(KN^s)\\ &= \phi((JM)(IM)^s)\\ &= \phi(JI^sM)\\ &= JI^s\phi(M).\\ JI^s\phi(M) &= (I\phi(M))(J^s\phi(M))\\ &= \phi(IM)(\phi(JM))^s\\ &= \phi(K)(\phi(N))^s.\end{aligned}$$

Also,

$$\begin{aligned}\phi(N^{s+1}) &= \phi(J^{s+1}M)\\ &= J^{s+1}\phi(M)\\ &= (\phi(JM))^{s+1}\\ &= (\phi(N))^{s+1}.\end{aligned}$$

Therefore, $(\phi(N))^{s+1} = \phi(K)(\phi(N))^s$.

Conversely, assume that $\phi(K)$ is a reduction of $\phi(N)$. Then $\phi(K)\phi(N)^s = (\phi(N))^{s+1}$ for some $s \in \mathbb{N}$. This implies that $I\phi(M)(J^s\phi(M)) = (J^{s+1}\phi(M))$. Therefore, $IJ^s\phi(M) = J^{s+1}\phi(M)$. Since M is a finitely generated faithful R-module, $\phi(M)$ is a finitely generated faithful R-module. Therefore by (Theorem 6.1, [12]) $IJ^s = J^{s+1}$. For every M, $IM(JM)^s = IJ^sM = J^{s+1}M = (JM)^{s+1}$. Thus K is a reduction of N. □

Bibliography

[1] D. Anderson, Some remarks on multiplication ideals, *Math. Jpn.* **25** (1980), 463–469.

[2] H. Ansari-Toroghy and F. Farshadifar, The dual notion of multiplication modules, *Taiwan. J. Math.* (2007), 1189–1201.

[3] J. Azami and M. Khajepour, Generalization of reduction and closure of ideals, *Algebraic Struct. Appl.* **8**(1) (2021), 147–161.

[4] A. Barnard, Multiplication modules, *J. Algebra* **71**(1) (1981), 174–178.

[5] Z. A. El-Bast and P. P. Smith, Multiplication modules, *Commun. Algebra* **16**(4) (1988), 755–779.

[6] S. M. Foster and J. A. Johnson, The asymptotic and integral closure operations in multiplicative lattice modules, *Tamkang J. Math.* **36**(4) (2005), 345–358.

[7] G. M. Low and P. F. Smith, Multiplication modules and ideals, *Commun. Algebra* **18**(12) (1990), 4353–4375.

[8] A. Naoum, B. Al-Hashimi and J. Kider, On the module of homomorphisms on finitely generated multiplication modules ii, *Period. Math. Hung.* **22**(3) (1991), 153–160.

[9] A. Naoum, B. Al-Hashimi and K. Sharaf, The dual of a finitely generated multiplication module, *Period. Math. Hung.* **20** (1989), 65–74.

[10] A. Naoum, H. Mustafa and M. Abdullah, On finitely generated projective modules and multiplication modules, *Period. Math. Hung.* **20**(1) (1989), 57–63.

[11] A. G. Naoum and M. Hasan, The residual of finitely generated multiplication modules, *Arch. Math.* **46** (1986), 225–230.

[12] A. G. Naoum and A. S. Mijbass, Weak cancellation modules, *Kyungpook Math. J.* **37**(1) (1997), 73.

[13] D. G. Northcott and D. Rees, Reductions of ideals in local rings, *Math. Proc. Camb. Philos. Soc.* **50**(2) (1954), 145–158.

[14] P. F. Smith, Some remarks on multiplication modules, *Arch. Math.* **50** (1988), 223–235.

[15] Y. Tiraş, Integral closure of an ideal relative to a module and Delta-closure, *Turk. J. Math.* **21**(3) (1997), 381–386.

S Priyadharsini

Computational analysis of fractional vibrational equations

Abstract: Fractional vibrational equations have emerged as a powerful tool for modeling complex systems with non-local and memory-dependent vibrational behaviors. This computational analysis explores methods and techniques employed to solve fractional vibrational equations, enabling a deeper understanding of dynamic systems. In this work, a brief overview of the recent results of fractional differential equations and the analytical methods used is given. We delve into analytical method, Laplace transform technique, numerical approaches, generalized Euler's method, and specialized algorithms designed for fractional calculus. The solution of non-linear fractional system and its convergence are discussed. Some insightful illustrations are proposed to validate of the given theory. Recent updates highlight advancements in machine learning applications and interdisciplinary research, expanding the scope of fractional vibrational equations in fields such as structural engineering, materials science, and biophysics. The abstract encapsulates the ongoing efforts to develop robust computational tools, contributing to a more comprehensive exploration of fractional vibrational phenomena and their practical implications across diverse scientific disciplines.

Keywords: Vibrational equation, fractional oscillatory equations, analytic solution, Laplace transform

MSC 2010: 11A25, 11A67

1 Introduction

Fractional differential equation (FDE) is crucial for modeling of real-life systems for several years, as the addition of delay enriches real-world occurrences. The key domains that use the idea of fractional include epidemiology, neurobiology, traffic models, population dynamics, rheology, viscoelastic mechanics, control theory, data processing, stochastic processes, anomalous diffusion, relaxation, vibration, and epidemiology. In recent decades, there has been a notable surge in interest surrounding fractional calculus, driven by its capacity to elucidate intermediate processes, memory phenomena, hereditary properties, and intricate dynamics. As early as the mid-20th century, fractional calculus found utility in describing the stress–strain relationship of viscoelastic materials. Subsequent experiments corroborated these applications [2].

S Priyadharsini, Department of Mathematics, Sri Krishna Arts and Science College, 9/31 Parvathy Puram, Kovai Pudur Post, 64142 Coimbatore, Tamil Nadu, India, e-mail: priyamaths24@gmail.com

https://doi.org/10.1515/9783111313634-023

Various fractional constitutive equations emerged to capture the stress–strain relationship of viscoelastic bodies [19]. The fractional constitutive relationship itself constitutes a fractional relaxation-oscillation system, with stress and strain as input and output (or vice versa), and response characteristics were scrutinized in [11, 32]. Enormous records can be found on this matter of solution such as Laplace transform method, fixed-point method and matrix measure, etc. See [3–9, 15–17, 20, 21, 23–31]. A fractional oscillator, comprising a mass connected to a series of parallel springs and viscoelastic materials governed by fractional stress–strain relationships, leads to a fractional ordinary differential equation. This type of equation has been the subject of investigation in [12]. It is worth mentioning that various alternative forms and methodologies exist for the fractional oscillator equation, incorporating different fractional definitions. In [14], a more general fractional integral operator and its boundedness were detailed. In [18] introduced and analyzed different classes of fractional oscillators within the framework of Weyl fractional calculus. [10] explored distributed-order cases.

Additionally, it is pertinent to note that the method of fractional power series proves practical for numerous fractional ordinary differential equations [1] and particularly effective for addressing nonlinear problems, exemplified in [13], where a convergence analysis was also conducted. Fundamental problems in dynamics can be solved by the theory of fractional differential equations. However, fractional vibrational system is not yet established properly. It is always possible to get a consistent results, which meet the realities. This scenario happens only when the fractional term is deeply analyzed. This study aims to investigate the characteristics of fractional force-free and forced oscillation equations when subjected to fractional operators. Various scenarios of the proposed model are scrutinized using specified parameters, providing a novel insight into the precision and efficacy of the outlined approach. The dynamics and outcomes are meticulously showcased through tables and graphs, ensuring the credibility of the proposed scheme.

Additionally, the exhibited solutions' behavior is thoroughly illustrated. The obtained results reveal that the proposed scheme is remarkably straightforward and effective for scrutinizing the dynamics of nonlinear fractional models, prevalent in engineering and scientific domains. As a result, fractional vibrational equations with oscillatory systems were analyzed. The main advantage of this proposed theory is that one may explicitly guarantee the convergence solution of given system. Recently, it has been found out that vibrational equation is practically applied in the following areas: the inverted pendulum, stability of a floating body, vibrations in an elliptic drum and frequency modulation.

This paper is organized as follows. Section 2 includes the prevailing facts, basic definitions, theorems and lemmas. Section 3 summarizes the concept of Method of Laplace transform in order to solve fractional dynamical system analytically. In Section 4, validity of the proposed procedure is given through vibrational and oscillatory equations with delay. The numerical method in which fractional Euler's method is used as a prediction, and the modified trapezoidal rule is used to make correction to obtain the finite value. For further details, see [22]. In addition, a conclusion is given in Section 5.

2 Preliminaries

This section describes the basic definition of stability and the prevailing research work on the stability of a nonlinear dynamical system, which is related to this work.

Definition 1 (Riemann–Liouville fractional integral [20]). The Riemann–Liouville fractional integral operator of order $\alpha > 0$ of a function $f \in L^1(\mathcal{R}^+)$ is defined by

$$I^\alpha f(t) = \frac{1}{\Gamma(\alpha)} \int_0^t (t-s)^{\alpha-1} f(s) ds, \tag{23.1}$$

where $\Gamma(.)$ is the Euler gamma function.

Definition 2 (Riemann–Liouville fractional derivative [20]). The Riemann–Liouville fractional derivative of order $\alpha > 0, n-1 < \alpha < n, n \in N$, is defined as

$$D^\alpha f(t) = \frac{1}{\Gamma(n-\alpha)} \left(\frac{d}{dt}\right)^n \int_0^t (t-s)^{n-\alpha-1} f(s) ds, \tag{23.2}$$

where D^n is the ordinary differential operator and the function $f(t)$ has absolutely continuous derivative upto order $(n-1)$.

Definition 3 (Caputo fractional derivative [20]). The Caputo fractional derivative of order $\alpha > 0, n-1 < \alpha < n, n \in N$, is defined as

$${}^C D^\alpha f(t) = \frac{1}{\Gamma(n-\alpha)} \int_0^t (t-s)^{n-\alpha-1} f^n(s) ds, \tag{23.3}$$

where the function $f(t)$ has absolutely continuous derivative upto order $(n-1)$.

Mittag–Leffler function and its asymptotic approximation

The exponential function plays a big role in the theory of ordinary differential equations. Another function, which plays a very important role in the fractional calculus, which is the generalization of the exponential function, it is proved that they are the solutions of superior differential equations called the fractional differential equations. Since a few decades the special transcendental function known as ML function has attracted an increasing attention of researchers because of its key role in treating problems related to integral and differential equations of fractional order. Due to its attractive feature it is referred as *the Queen Function of Fractional Calculus.*

Definition 4 (Mittag–Leffler function [20]). The one-parameter Mittag–Leffler function is defined as

$$E_\alpha(z) = \sum_{k=0}^{\infty} \frac{z^k}{\Gamma(\alpha k + 1)}, \quad (\alpha > 0, z \in \mathbb{C}). \tag{23.4}$$

The two-parameter Mittag–Leffler function is defined as

$$E_{\alpha,\beta}(z) = \sum_{k=0}^{\infty} \frac{z^k}{\Gamma(\alpha k + \beta)}, \quad (\alpha, \beta > 0, z \in \mathbb{C}). \tag{23.5}$$

The Mittag–Leffler function of a matrix A is defined by

$$E_{\alpha,\beta}(At) = \sum_{k=0}^{\infty} \frac{(At)^k}{\Gamma(\alpha k + \beta)}, \quad (\alpha, \beta > 0, A \in \mathbb{R}^{n\times n}). \tag{23.6}$$

3 Laplace Transforms

3.1 Laplace transforms of the Mittag–Leffler function

The Laplace transform of Mittag–Leffler functions are given by

$$\mathcal{L}\{t^{\beta-1}E_{\alpha,\beta}(-\lambda t^\alpha)\} = \frac{s^{\alpha-\beta}}{s^\alpha + \lambda}, \quad (\mathcal{R}(s) > |\lambda|^{\frac{1}{\alpha}}), \tag{23.7}$$

where $t \geq 0, \lambda \in \mathbb{R}$.

Integration of Mittag–Leffler function

By the fractional-order term-by-term integration

$$\int_0^z E_{\alpha,\beta}(\lambda t^\alpha) t^{\beta-1} \mathrm{d}t = z^\beta E_{\alpha,\beta+1}(\lambda z^\alpha), \quad (\beta > 0), \tag{23.8}$$

$$\frac{1}{\Gamma(\nu)} \int_0^z E_{\alpha,\beta}(\lambda t^\alpha) t^{\beta-1} \mathrm{d}t = z^{\beta+\nu-1} E_{\alpha,\beta+\nu}(\lambda z^\alpha), \quad (\beta > 0,\ \nu > 0). \tag{23.9}$$

3.2 Laplace transforms of the Caputo derivative

Caputo derivative can be written in the form,

$$ {}^{c}D^{\alpha}f(t) = D^{-(n-\alpha)}g(t), \quad g(t) = f^{(n)}(t), \quad (n-1 < \alpha \leq n). \tag{23.10} $$

By using the Laplace transforms of the Riemann–Liouville fractional integral:

$$ \mathcal{L}\{D^{-\alpha}f(t); s\} = s^{-\alpha}F(s). \tag{23.11} $$

From this, we have

$$ \mathcal{L}\{{}^{c}D^{\alpha}f(t); s\} = s^{-(n-\alpha)}G(s). \tag{23.12} $$

By using the usual property for the Laplace transforms of the derivative of an integer order n of the function $f(t)$ is given by

$$ \mathcal{L}\{f^{n}(t)\} = s^{n}F(s) - \sum_{k=0}^{n-1} s^{n-k-1}f^{(k)}(0) = s^{n}F(s) - \sum_{k=0}^{n-1} s^{n}f^{(n-k-1)}(0), \tag{23.13} $$

we have

$$ L\{{}^{c}D^{\alpha}f(t)\} = s^{\alpha}F(s) - \sum_{k=0}^{n-1} s^{\alpha-k-1}f^{(k)}(0), \quad (n-1 < \alpha \leq n). \tag{23.14} $$

3.3 Laplace Transforms of a simple function

$$ \mathcal{L}[t^{\beta-1}E_{\alpha,\beta}(\lambda t^{\alpha})](s) = \frac{s^{\alpha-\beta}}{s^{\alpha}-\lambda}. \tag{23.15} $$

Special values

- $\mathcal{L}[E_{1/2,1}(\lambda t^{1/2})](s) = \frac{s^{-1/2}}{s^{1/2}-\lambda}$,
- $\mathcal{L}[E_{3/2,1}(\lambda t^{3/2})](s) = \frac{s^{1/2}}{s^{3/2}-\lambda}$,
- $\mathcal{L}[t^{-1/2}E_{3/2,2}(\lambda t^{3/2})](s) = \frac{s^{-1/2}}{s^{3/2}-\lambda}$.

4 Analytical solution of oscillatory equation

Theorem 1. *Let $\alpha > 0$ and $n = -[-\alpha]$. Let G be an open set in $\mathbb{C}$ and for any $x \in G$, $f[t,x] \in C_{\nu}[a,b]$ with $0 \leq \nu < 1$ and $\nu \leq \alpha$. Let $r = n$ for $\alpha \in \mathbb{N}$ and $r = n-1$ for $\alpha \notin \mathbb{N}$. If $x(t) \in C^{r}[a,b]$, then $x(t)$ satisfies the relation*

$$({}^cD^\alpha x)(t) = f[t, x(t)] \quad (\alpha > 0;\ a \le t \le b), \tag{23.16}$$

with the initial conditions

$$x^{(k)}(a) = b_k, b_k \in \mathbb{R} \quad (k = 0, 1, \ldots, n-1; n = -[-\alpha]),$$

iff $x(t)$ satisfies the Volterra integral

$$x(t) = \sum_{j=0}^{n-1} \frac{b_j}{j!}(t-a)^j + \frac{1}{\Gamma(\alpha)} \int_a^t \frac{f[s, x(s)]}{(t-s)^{1-\alpha}} ds \quad (a \le t \le b). \tag{23.17}$$

Let $\alpha > 0$ and $n = -[-\alpha]$, and $0 \le \nu < 1$ and $\nu \ge \alpha$. Let G be ab open set in $\mathbb{C}$ and let $f : (a, b] \times G \to \mathbb{C}$ be a function such that for any $x \in G, f[t, x] \in C_\nu[a, b]$ Lipchitz condition holds. If $n - 1 < \alpha < n$ $(n \in \mathbb{N})$, then there exist a unique solution $x(t)$ to the Cauchy problem in the space $C_\nu^{\alpha,n-1}[a, b]$.

Linear differential system in Cartesian form is considered

$${}^CD^\alpha x(t) = w^2 x(t) + f(t, x(t)), \tag{23.18}$$

with initial value $x(0) = x_0 = (x_{10}, x_{20}, \ldots, x_{n0})^T$, where $x = (x_1, x_2, \ldots, x_n)^T$, and $f(t, x(t)) \in \mathbb{R}^{n \times n}, 0 < \alpha < 1$. The given equation (23.18) can be written as

$${}^CD^\alpha x(t) - w^2 x(t) = f(t, x(t)). \tag{23.19}$$

The solution of the system can be written as,

$$x(t) = E_\alpha(w^2(t)^\alpha)x_0 + \int_0^t (t-s)^{\alpha-1} E_{\alpha,\alpha}(w^2(t-s)^\alpha)\{f(s, x(s))\} ds. \tag{23.20}$$

4.1 Homogeneous equation with constant coefficients

Consider the homogeneous equation,

$$\sum_{k=1}^{m} A_k({}^cD^{\alpha_k} x)(t) + A_0 x(t) = 0, \quad (t > 0; m \in \mathbb{N}, 0 < \alpha_1 < \cdots < \alpha_m), \tag{23.21}$$

involving Caputo derivative ${}^cD^{\alpha_k} x (k = 1, 2, \ldots, m)$ given by

$$({}^cD^\alpha x)(t) = \left(D^\alpha \left[x(s) - \sum_{k=0}^{n-1} \frac{x^{(k)}(a)}{k!}(s-a)^k \right] \right)(t), \tag{23.22}$$

with real constants $A_k \in R(k = 0, 1, \ldots, m-1)$ and $A_m = 1$.

We give the conditions when the solutions $x_1(t), x_2(t), \dots, x_l(t)$ of with $l-1 < \alpha := \alpha_m < l$ will be linearly independent, and when these solutions form the fundamental, system of solutions

$$x_j^{(k)}(0) = 0 \quad (k, j = 0, 1, \dots, l-1, k \neq j), \tag{23.23}$$

$$x_k^{(k)}(0) = 1 \quad (k = 0, 1, \dots, l-1, k \neq j). \tag{23.24}$$

Theorem 2. *Let $l-1 < \alpha \leq l$, $(l \in \mathbb{N})$ and $\lambda \in R$.*

$$({}^cD^\alpha x)(t) - \lambda x(t) = 0, \quad (t > 0; l-1 < \alpha < l; l \in \mathbb{N}, \lambda \in \mathbb{R}),\ (t > 0). \tag{23.25}$$

Then the fundamental system of solutions of the above equation given in terms of the following functions

$$x_j(t) = t^j E_{\alpha, j+1}(\lambda t^\alpha) \quad (j = 0, 1, \dots, l-1), \tag{23.26}$$

Proof. Applying Laplace transforms to the equation (23.25), we have

$$s^\alpha (\mathcal{L}x)(s) - \sum_{j=0}^{l-1} d_j s^{\alpha-j-1} - \lambda(\mathcal{L}x)(s) = 0,$$

$$(\mathcal{L}x)(s) = \sum_{j=0}^{l-1} \frac{d_j s^{\alpha-j-1}}{s^\alpha - \lambda},$$

where d_j are given by $d_j = x^{(j)}(0) \qquad (j = 0, 1, \dots, l-1)$.

Applying inverse Laplace transforms,

$$x(t) = \sum_{j=0}^{l-1} d_j \mathcal{L}^{-1}\left[\frac{s^{\alpha-j-1}}{s^\alpha - \lambda}\right].$$

We know that

$$\mathcal{L}[t^{\beta-1} E_{\alpha,\beta}(\lambda t^\alpha)](s) = \frac{s^{\alpha-\beta}}{s^\alpha - \lambda},$$

take $\beta = j + 1$

$$x(t) = \sum_{j=0}^{l-1} d_j x_j(t) \quad \text{where } x_j(t) = t^j E_{\alpha, j+1}(\lambda t^\alpha),$$

which is the required solution.

In particular, the equation

$$({}^cD^\alpha x)(t) - \lambda x(t) = 0, \quad (t > 0; 0 < \alpha < 1; \lambda \in \mathbb{R}),$$

has its solution given by,

$$x(t) = E_\alpha(\lambda t^\alpha),$$

while the equation

$$({}^cD^\alpha x)(t) - \lambda x(t) = 0, \quad (t > 0; 1 < \alpha < 2; \lambda \in \mathbb{R}),$$

has its fundamental system of solutions given by

$$x_1(t) = E_\alpha(\lambda t^\alpha), \quad x_2(t) = tE_{\alpha,2}(\lambda t^\alpha).$$ □

4.2 Non-homogeneous equation with constant coefficients

Consider the non-homogeneous equation

$$\sum_{k=1}^{m} A_k({}^cD^{\alpha_k} x)(t) + A_0 x(t) = f(t), \quad (t > 0; m \in \mathbb{N}, 0 < \alpha_1 < \cdots < \alpha_m), \tag{23.27}$$

with real coefficients $A_k \in \mathbb{R}$ and a given function $f(t)$ defined on $\mathbb{R}_+$.

The general solution of the equation (23.27) is the sum of its particular solution and the general solution to the corresponding homogeneous equation.

If $l_k - 1 < \alpha_k < l_k$ and $0 = l_0 \le l_1 \ldots \le l_m$, then applying the Laplace transforms, we derive the solution to the equation (23.27) in the form

$$x(t) = \sum_{j=0}^{l_m-1} c_j x_j(t) + \int_0^t G_{\alpha_1,\ldots,\alpha_m}(t-s)f(s)ds,$$

where $x_j(t)$ $(j = 1, \ldots, l_m)$ are solutions to the corresponding homogeneous equation.

$$x_j(t) = \sum_{k=q}^{l_m-1} A_k\left(\mathcal{L}^{-1}\left[\frac{s^{\alpha_k-j-1}}{P_{\alpha_1,\ldots,\alpha_m}(s)}\right]\right)(t),$$

where $l_q \le j \le l_{q+1} - 1$ for $q = 0, \ldots, m-1$, $P_{\alpha_1,\ldots,\alpha_m}(s)$ is given by

$$A_0 + \sum_{k=1}^{m} A_k s^{\alpha_k},$$

and c_j are arbitrary real constants.

Theorem 3. *Let $l - 1 < \alpha \le l$ $(\lambda \in \mathbb{R})$ and let $f(t)$ be a given real function defined on $\mathbb{R}_+$, then the equation*

$$({}^cD^\alpha x)(t) - \lambda x(t) = f(t), \quad (x > 0), \tag{23.28}$$

is solvable and its general solution is given by

$$x(t) = \sum_{j=0}^{l-1} c_j t^j E_{\alpha,j+1}(\lambda t^\alpha) + \int_0^t (t-s)^{\alpha-1} E_{\alpha,\alpha}[\lambda(t-s)^\alpha] f(s)\mathrm{d}s, \tag{23.29}$$

where c_j $(j = 0, \dots, l-1)$ *are arbitrary real constants.*

In particular, the general solutions to the equation (23.28) with $0 < \alpha \le 1$ and $1 < \alpha \le 2$ have the forms

$$x(t) = \int_0^t (t-s)^{\alpha-1} E_{\alpha,\alpha}[\lambda(t-s)^\alpha] f(s)\mathrm{d}s + c_1 E_\alpha(\lambda t^\alpha), \tag{23.30}$$

$$x(t) = \int_0^t (t-s)^{\alpha-1} E_{\alpha,\alpha}[\lambda(t-s)^\alpha] f(s)\mathrm{d}s + c_1 E_\alpha(\lambda t^\alpha) + c_2 E_{\alpha,2}(\lambda t^\alpha), \tag{23.31}$$

respectively, where c_1, c_2 are arbitrary real constants.

4.3 Method of reduction to Volterra integral equations

Theorem 4. *Let $n-1 < \alpha < n$ $(n \in \mathbb{N})$ and let $0 \le \nu < 1$ be such that $\nu \le \alpha$. Also let $\lambda \in \mathbb{R}$. If $f(t) \in C_\nu[a,b]$ the Cauchy problem has a unique solution $x(t) \in C_\nu^{\alpha,n-1}[a,b]$ and the solution is given by equation*

$$x(t) = \sum_{j=0}^{n-1} b_j (t-a)^j E_{\alpha,j+1}(\lambda(t-a)^\alpha) + \int_a^t (t-s)^{\alpha-1} E_{\alpha,\alpha}[\lambda(t-s)^\alpha] f(s)\mathrm{d}s.$$

We construct the explicit solutions of the linear fractional differential equations with the Caputo derivative $({}^cDx)(t)$ of order $\alpha > 0$ $(\alpha \notin \mathbb{N})$ in the space $C_\nu^{\alpha,n-1}[a,b]$ $(n = [\alpha] + 1)$ defined by

$$C_\nu^{\alpha,r}[a,b] = x(t) \in C^r[a,b] :{}^c D^\alpha x \in c_\nu[a,b].$$

We consider the Cauchy problem for the fractional differential equation of order α with the initial conditions

$$\begin{aligned} ({}^cD^\alpha x)(t) - \lambda x(t) &= f(t) \quad (a \le t \le b; n-1 < \alpha < n; n \in \mathbb{N}; \lambda \in R), \\ x^{(k)}(a) &= b_k \quad (b_k \in \mathbb{R}; k = 0, 1, \dots, n-1). \end{aligned}$$

Suppose $f(t) \in C_\nu[a,b]$ with $0 \le \nu < 1$ and $\nu \le \alpha$. Then the given problem is equivalent to the Volterra integral equation of second kind in the space $C^{n-1}[a,b]$

$$x(t) = \sum_{j=0}^{n-1} \frac{b_j}{j!}(t-a)^j + \frac{\lambda}{\Gamma(\alpha)} \int_a^t \frac{x(s)}{(t-s)^{1-\alpha}} \mathrm{d}s + \frac{1}{\Gamma(\alpha)} \int_a^t \frac{f(s)}{(t-s)^{1-\alpha}} \mathrm{d}s. \tag{23.32}$$

To solve the equation (23.32), we apply the method of successive approximation by setting

$$x_0(t) = \sum_{j=0}^{n-1} \frac{b_j}{j!}(t-a)^j,$$

and

$$x_m(t) = x_0(t) + \frac{\lambda}{\Gamma(\alpha)} \int_a^t \frac{x_{m-1}(s)\mathrm{d}s}{(t-s)^{1-\alpha}} + \frac{1}{\Gamma(\alpha)} \int_a^t \frac{f(s)\mathrm{d}s}{(t-s)^{1-\alpha}}.$$

By using this method, we have the following expression for $x_m(t)$

$$x_m(t) = \sum_{j=0}^{n-1} b_j \sum_{k=0}^{m} \frac{\lambda^k (t-a)^{\alpha k+j}}{\Gamma(\alpha k + j + 1)} + \int_a^t \left[\sum_{k=1}^{m} \frac{\lambda^{k-1}}{\Gamma(\alpha k)} (t-s)^{\alpha k - 1} \right] f(s)\mathrm{d}s.$$

Taking the limit as $m \to \infty$ and by using the definition of Mittag–Leffler function, we obtain

$$x(t) = \sum_{j=0}^{n-1} b_j (t-a)^j E_{\alpha,j+1}(\lambda(t-a)^\alpha) + \int_a^t (t-s)^{\alpha-1} E_{\alpha,\alpha}[\lambda(t-s)^\alpha] f(s)\mathrm{d}s.$$

This gives an explicit solution of Volterra integral equation and hence of the cauchy problem.

Since $f[x,t] = \lambda x + f(t)$ satisfies the Lipchitz condition, by the theorem, there exist a unique solution $x(t)$ to the Cauchy problem in the space $C_\nu^{\alpha,n-1}[a,b]$ from this we have the required result.

5 Numerical method

In this chapter, we use the fractional Euler's method, which is generalization of the classical Euler's method. Most nonlinear fractional differential equations do not have analytic solutions, so approximations and numerical techniques must be used. This approach is used to provide an analytical approximate solution to linear and nonlinear problems.

Because they provide immediate and visible symbolic terms of analytic solutions, as well as numerical approximate solutions to both linear and nonlinear differential equations.

For the concept of fractional derivative we will adopt Caputo's definition which is a modification of the Riemann–Liouville definition and has the advantage of dealing properly with initial value problems in which the initial conditions, which are the case in most physical processes.

Consider the IVP

$$\begin{aligned} {}^{C}D^{\alpha}x(t) &= f(t,x(t)), \quad 0<\alpha<1, t>0 \\ x(0) &= x_0. \end{aligned} \tag{23.33}$$

Let $[0,a]$ be the interval over which we want to find the solution of the problem (23.33). We subdivide the interval $[0,a]$ into k subintervals $[t_j,t_{j+1}]$ of equal width $h=a/k$ by using nodes $t_j=jh$ for $j=0,1,\dots,k$.

The general formula for fractional Euler's method is

$$\begin{aligned} t_{j+1} &= t_j + h, \\ x(t_{j+1}) &= x(t_j) + \frac{h^{\alpha}}{\Gamma(\alpha+1)}f(t_j,x(t_j)), \end{aligned} \tag{23.34}$$

for $j=0,1,\dots,k-1$. It is clear that if $\alpha=1$, then the fractional Euler's method (23.34) reduces to the classical Euler's method. The modified trapezoidal rule is introduced.

The new algorithm is based on the modified trapezoidal rule and the fractional Euler's method. It depends on the analytical property that the IVP (23.33) is equivalent to the integral equation

$$x(t) = J^{\alpha}f(t,x(t)) + x(0). \tag{23.35}$$

To obtain the solution point $(t_1,x(t_1))$, we substitute $t=t_1$ into (23.35) and get

$$x(t_1) = (J^{\alpha}f(t,x(t)))(t_1) + x(0). \tag{23.36}$$

By using the modified trapezoidal rule with step size $h=t_1-t_0$, then the result is

$$x(t_1) = \alpha\frac{h^{\alpha}f(t_0,x(t_0))}{\Gamma(\alpha+2)} + \frac{h^{\alpha}f(t_1,x(t_1))}{\Gamma(\alpha+2)} + x(0). \tag{23.37}$$

We use an Euler's method estimate for $x(t_1)$. Substituting (23.34) into (23.37), yields

$$x(t_1) = \alpha\frac{h^{\alpha}f(t_0,x(t_0))}{\Gamma(\alpha+2)} + \frac{h^{\alpha}f(t_1,x(t_1)) + \frac{h^{\alpha}}{\Gamma(\alpha+1)}f(t_0,x(t_0))}{\Gamma(\alpha+2)} + x(0). \tag{23.38}$$

This process is repeated and at each step, the fractional Euler's method is used as a prediction, and the modified trapezoidal rule is used for correction to obtain the finite value. The general formula for our algorithm is given by

$$\begin{aligned}x(t_j) = {} & \frac{h^\alpha}{\Gamma(\alpha+2)}((j-1)^{\alpha+1} - (j-\alpha-1)j^\alpha)f(t_0, x(t_0)) \\ & + \frac{h^\alpha}{\Gamma(\alpha+2)}\sum_{i=1}^{j-1}((j-i+1)^{\alpha+1} - 2(j-i)^{\alpha+1} + (j-i-1)^{\alpha+1})f(t_j, y(t_j)) \\ & + \frac{h^\alpha}{\Gamma(\alpha+2)}f\left(t_j, x(t_{j-1}) + \frac{h^\alpha}{\Gamma(\alpha+1)}f(t_{j-1}, x(t_{j-1}))\right) + x(0).\end{aligned} \tag{23.39}$$

The new algorithm is simple for computational performance for all values of α and h. It is clear that the behavior of the method is independent of the parameter α and the accuracy of the approximation depends on the step size h. By increasing the step size one can reach the desired accuracy. For linear systems of equations, the algorithm gives exact solution; for nonlinear FDE, it provides an approximate solution with good accuracy.

Moreover, this method does not change the problem into a linear one. The algorithm was used in a direct way without using linearization, perturbation or restrictive assumptions. It, therefore, provides more realistic solutions that generally converge very rapidly in real physical problems.

The algorithm is a reliable and very useful one for the numerical evaluation of the Mittag–Leffler function and the other functions arising in fractional calculus. In the present work, the proposed method was successfully applied to system of nonlinear FDE's.

6 Examples

Example 1. The equation

$$({}^cD^{1/2}x)(t) - x(t) = t, \tag{23.40}$$

with the boundary condition $x(0) = 1$.

Solution. This equation is of the form of equation (23.28) with $\lambda = 1, f(t) = t$ and $\alpha = 1/2$, $0 < \alpha \leq 1$.

Consider the corresponding homogeneous equation

$$({}^cD^{1/2}x)(t) - x(t) = 0.$$

Applying Laplace transforms, we get

$$\begin{gathered}\mathcal{L}[({}^cD^{1/2}x)(t)] - \mathcal{L}[x(t)] = 0, \\ s^{1/2}(\mathcal{L}x)(s) - d_0 s^{1/2} - (\mathcal{L}x)(s) = 0,\end{gathered}$$

using the initial condition $d_0 = x(0) = 1$

$$s^{1/2}(\mathcal{L}x)(s) - s^{-1/2} - (\mathcal{L}x)(s) = 0,$$
$$(\mathcal{L}x)(s) = \frac{s^{1/2}}{s^{\frac{1}{2}} - 1}.$$

Applying inverse Laplace transforms, we get

$$x(t) = \mathcal{L}^{-1}\frac{s^{1/2}}{s^{1/2} - 1},$$
$$x(t) = E_{1/2,1}(\sqrt{t}).$$

Particular solution is given by

$$x(t) = \int_0^t (t-s)^{\alpha-1}E_{\alpha,\alpha}[\lambda(t-s)^{\alpha}]f(s)\mathrm{d}s,$$
$$= \int_0^t (t-s)^{-1/2}E_{1/2,1/2}[(t-s)^{1/2}]s\mathrm{d}s.$$

Put $z = t - s$

$$= \int_0^x (z)^{-1/2}E_{1/2,1/2}(\sqrt{z})(t-z)\mathrm{d}t.$$

By using the formula (23.9) with $\alpha = -1/2$, $\beta = -1/2$, $\nu = 2$,

$$x(t) = t^{3/2}E_{1/2,5/2}(\sqrt{t}).$$

The general solution of the given equation is

$$x(t) = E_{1/2,1}(\sqrt{t}) + t^{3/2}E_{1/2,5/2}(\sqrt{t}). \tag{23.41}$$

Example 2. The equation

$$({}^{c}D^{1/2}x)(t) - 2x(t) = t, \tag{23.42}$$

with the boundary condition $x(0) = 1$.

Solution. This equation is of the form of equation (23.28) with $\lambda = 2, f(t) = t$ and $\alpha = 1/2$, $0 < \alpha \leq 1$.

Consider the corresponding homogeneous equation

$$({}^{c}D^{1/2}x(t)) - 2x(t) = 0.$$

Applying Laplace transforms, we get

$$\mathcal{L}[({}^{c}D^{1/2}x)(t)] - 2\mathcal{L}[x(t)] = 0,$$
$$s^{1/2}(\mathcal{L}x)(s) - d_0 s^{1/2} - 2(\mathcal{L}x)(s) = 0,$$

using the initial condition $d_0 = y(0) = 1$,

$$s^{1/2}(\mathcal{L}x)(s) - s^{-1/2} - 2(\mathcal{L}x)(s) = 0,$$
$$(\mathcal{L}x)(s) = \frac{s^{1/2}}{s^{1/2} - 2}.$$

Applying inverse Laplace transforms, we get

$$x(t) = \mathcal{L}^{-1}\frac{s^{1/2}}{s^{1/2} - 2},$$
$$x(t) = E_{1/2,1}(2\sqrt{(t)}).$$

Particular solution is given by,

$$x(t) = \int_0^t (t-s)^{\alpha-1} E_{\alpha,\alpha}[\lambda(t-s)^{\alpha}]f(s)ds,$$
$$= \int_0^t (t-s)^{-1/2} E_{1/2,1/2}[(t-s)^{1/2}]sds,$$

Put $z = t - s$

$$= \int_0^t (z)^{-1/2} E_{1/2,1/2}(\sqrt{z})(t-z)ds.$$

By using the formula (23.9) with $\alpha = -1/2$, $\beta = -1/2$, $\nu = 2$

$$x(t) = x^{3/2} E_{1/2,5/2}(2\sqrt{t}).$$

The general solution of the given equation is

$$x(t) = E_{1/2,1}(2\sqrt{t}) + t^{3/2} E_{1/2,5/2}(2\sqrt{t}). \tag{23.43}$$

Example 3. The equation

$$({}^{c}D^{3/2}x)(t) - x(t) = t, \tag{23.44}$$

with the boundary condition $x(0) = 1$ and $x'(0) = 1$.

Solution. This equation is of the form of equation (23.28) with $\lambda = 1, f(t) = t$ and $\alpha = 3/2$, $1 < \alpha \leq 2$.

Consider the corresponding homogeneous equation

$$({}^{c}D^{3/2}x)(t) - x(t) = 0.$$

Applying Laplace transforms, we get

$$\mathcal{L}[({}^{c}D^{3/2}x)(t)] - \mathcal{L}[x(t) = t] = 0,$$
$$s^{3/2}(\mathcal{L}x)(s) - \sum_{j=0}^{l-1} d_j s^{3/2-j} - (\mathcal{L}x)(s) = 0,$$

using the initial condition $d_0 = x(0) = 1$ and $d_1 = x'(0) = 1$,

$$s^{3/2}(\mathcal{L}x)(s) - s^{1/2} + s^{-1/2} - (\mathcal{L}x)(s) = 0,$$
$$(\mathcal{L}x)(s) = \frac{s^{1/2} + s^{-1/2}}{s^{3/2} - 1}.$$

Applying inverse Laplace transforms, we get

$$x(t) = \mathcal{L}^{-1}\frac{s^{1/2}}{s^{3/2} - 1} + \mathcal{L}^{-1}\frac{s^{-1/2}}{s^{3/2} - 1},$$
$$x(t) = E_{3/2,1}(t^{3/2}) + t^{-1/2}E_{3/2,2}(t^{3/2}).$$

Particular solution is given by

$$x(t) = \int_0^t (t-s)^{\alpha-1}E_{\alpha,\alpha}[\lambda(t-s)^{\alpha}]f(s)\mathrm{d}s,$$
$$= \int_0^s (t-s)^{1/2}E_{3/2,3/2}[(t-s)^{3/2}]s\mathrm{d}s.$$

Put $z = t - s$

$$= \int_0^t (z)^{1/2}E_{3/2,3/2}(z^{3/2})(t-z)\mathrm{d}s.$$

By using the formula (23.9) with $\alpha = \beta = 3/2$, $\nu = 2$,

$$x(t) = t^{5/2}E_{3/2,7/2}(t^{3/2}).$$

The general solution of the given equation is

$$x(t) = E_{3/2,1}(t^{3/2}) + t^{-1/2}E_{3/2,2}(t^{3/2}) + t^{5/2}E_{3/2,7/2}(t^{3/2}). \tag{23.45}$$

Example 4 (Duffing equation). We shall discuss the stability of the NFDE,

$$^{C}D^{\alpha}x(t) = -x(t) - x(t)^3, \quad 0 < \alpha < 2. \tag{23.46}$$

When $\alpha = 1/2$ the nonlinear first-order Duffing equation is given by

$$\begin{aligned} ^{C}D^{\frac{1}{2}}x(t) &= -x(t) - x(t)^3, \\ x(0) &= 1. \end{aligned} \tag{23.47}$$

The homogeneous system $^{C}D^{\frac{1}{2}}x(t) = -x(t)$, which satisfies $|\arg(-1)| > \frac{\pi}{4}$. Here $f(t,x) = -x^3$, which satisfies $f(t,0) = 0$. Hence the given system (23.47) is asymptotically stable.

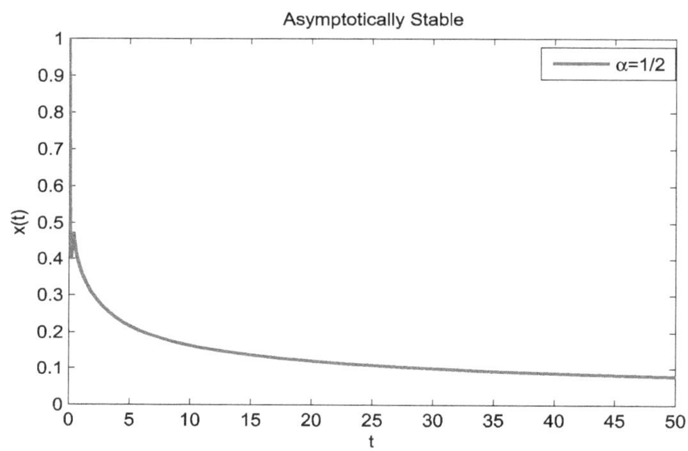

When $\alpha = 3/2$, the nonlinear second-order Duffing equation is given by

$$\begin{aligned} ^{C}D^{\frac{3}{2}}x(t) &= -x(t) - x(t)^3, \\ x(0) &= 0, \ x'(0) = 1. \end{aligned} \tag{23.48}$$

The given equation can be converted into a system of fractional differential equation, by using the substitution

$$\begin{aligned} ^{C}D^{\frac{1}{2}}x_1(t) &= x_2(t), \\ ^{C}D^{\frac{1}{2}}x_2(t) &= x_3(t), \\ ^{C}D^{\frac{1}{2}}x_3(t) &= -x_1(t) - x_1^3(t), \end{aligned} \tag{23.49}$$

where $x_1(t) = x(t)$, with initial condition $x_1(0) = 0$, $x_2(0) = 1$, $x_3(0) = 0$. This can be written in the standard form, $^{C}D^{1/2}x(t) = Ax(t) + f(t,x)$, and $f(t,x) = (0, 0, x_1^3(t))^T$, where

$$A = \begin{bmatrix} 0 & 1 & 0 \\ 0 & 0 & 1 \\ -1 & 0 & 0 \end{bmatrix}.$$

This can be solved numerically.

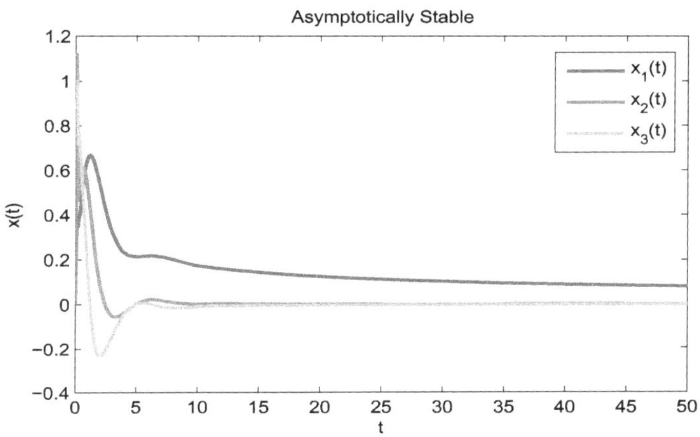

Example 5. Consider the fractional non-homogeneous system

$$^{C}D^{\alpha}x(t) + x(t) = f_0(t), \tag{23.50}$$

where $\alpha = 3/2$, and $f_0(t) = e^{-t}\sin(0.2t)$ with $x(0) = x'(0) = 0$.

The equation (23.50) can be converted into a system of fractional differential equation by using the substitution

$$\begin{aligned} {}^{C}D^{\frac{1}{2}}x_1(t) &= x_2(t), \\ {}^{C}D^{\frac{1}{2}}x_2(t) &= x_3(t), \\ {}^{C}D^{\frac{1}{2}}x_3(t) &= -x_1(t) + f_0(t), \quad \text{where } x_1(t) = x(t). \end{aligned} \tag{23.51}$$

The given system can be written in the standard form ${}^{C}D^{1/2}X(t) = AX(t) + F(t)$, and $F(t) = (0, 0, e^{-t}\sin(0.2t))^T$ with initial condition $x_1(t) = x_2(t) = x_3(t) = 0$, here $X(t) = (x_1(t), x_2(t), x_3(t))$, where

$$A = \begin{bmatrix} 0 & 1 & 0 \\ 0 & 0 & 1 \\ -1 & 0 & 0 \end{bmatrix}.$$

This can be solved numerically.

Example 6 (Bagley–Torvik equation). The effect that makes the vibration of the system to stop after time t, is known as the Damping effect. It is induced by air and internal friction. Because the corresponding damping force is related to the air resistance to the movement of the mass, and the resistance is proportional to the velocity of the moving mass. Consequently, the damping force also has the fractional form arises in the modeling of a rigid plate immersed in a Newtonian fluid. This type of system discussed in [24].

The fractional derivative is used to describe a real physical system, which was formulated in a conventional manner. For further details, see [22]. Now we shall discuss the stability of the system

$$A\,{}^{C}D^{2}x(t) + B\,{}^{C}D^{\alpha}x(t) + Cx(t) = f_1(t), \tag{23.52}$$

where $A \neq 0$ and $B, C \in \mathbb{R}$, and $\alpha = 3/2$, with the initial condition $x(0) = x'(0) = 0$,

$$f_1(t) = \begin{cases} 15, & 0 < t \leq 1, \\ 0 & \text{otherwise.} \end{cases}$$

This equation can be written as

$${}^{C}D^{2}x(t) + \frac{B}{A}{}^{C}D^{\alpha}x(t) + \frac{C}{A}x(t) = \frac{f_1(t)}{A}. \tag{23.53}$$

The equation (23.53) can be converted into a system of fractional differential equation by using the substitution

$$\begin{aligned} {}^{C}D^{\frac{1}{2}}x_1(t) &= x_2(t), \\ {}^{C}D^{\frac{1}{2}}x_2(t) &= x_3(t), \\ {}^{C}D^{\frac{1}{2}}x_3(t) &= x_4(t), \\ {}^{C}D^{\frac{1}{2}}x_4(t) &= -\frac{C}{A}x_1(t) - \frac{B}{A}x_4(t) + \frac{f_1(t)}{A}, \end{aligned} \tag{23.54}$$

where $x_1(t) = x(t)$, with the initial condition $x_1(0) = x_2(0) = x_3(0) = x_4(0) = 0$. The stability of the Bagley–Torvik equation depends on the values of A, B and C. It can be seen from the following.

If $A = B = C = 1$, it can be written in the standard form ${}^{C}D^{1/2}X(t) = AX(t) + F(t)$ and $F(t) = (0, 0, 0, 15)^T$ $0 < t \leq 1$, where

$$A = \begin{bmatrix} 0 & 1 & 0 & 0 \\ 0 & 0 & 1 & 0 \\ 0 & 0 & 0 & 1 \\ -1 & 0 & 0 & -1 \end{bmatrix}.$$

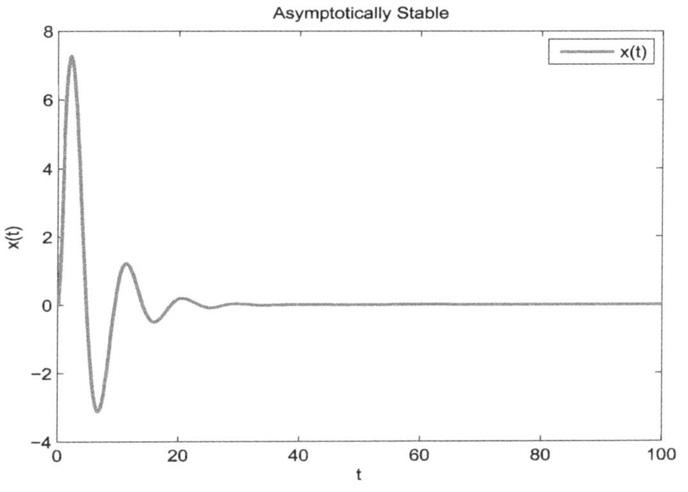

This can be solved numerically.

If $A = 1, B = -1, C = 1$, it can be written in the standard form ${}^{C}D^{1/2}X(t) = AX(t) + F(t)$ and $F(t) = (0, 0, 0, 15)^T \ \ 0 < t \leq 1$, where

$$A = \begin{bmatrix} 0 & 1 & 0 & 0 \\ 0 & 0 & 1 & 0 \\ 0 & 0 & 0 & 1 \\ -1 & 0 & 0 & 1 \end{bmatrix}.$$

This can be solved numerically.

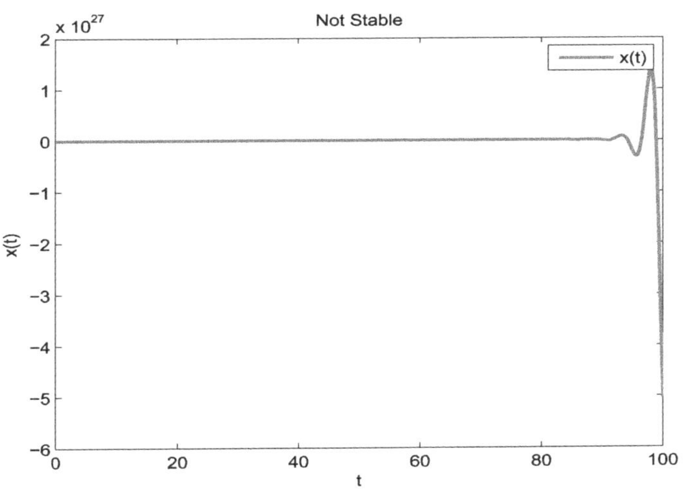

Example 7 (Vibrational differential equation). Consider the system

$$ {}^{C}D''x(t) + 2\mu\omega^{2-\alpha}\,{}^{C}D^{\alpha}x(t) + \omega^2 x(t) = f_0(t), \tag{23.55}$$

where $\alpha = 1/2$, with the initial condition $x(0) = 1/4$, $x'(0) = 0$, and

$$ f_0(t) = \begin{cases} 8, & 0 < t \le 1, \\ 0 & \text{otherwise} \end{cases} $$

Solution. The given equation can be converted into a system of fractional differential equation by using the substitution

$$ \begin{aligned} {}^{C}D^{\alpha}x_1(t) &= x_2(t), \\ {}^{C}D^{\alpha}x_2(t) &= x_3(t), \\ {}^{C}D^{\alpha}x_3(t) &= x_4(t), \\ {}^{C}D^{\alpha}x_4(t) &= -\omega^2 x_1(t) - 2\mu\omega^{3/2}x_2(t) + f_0(t), \end{aligned} \tag{23.56}$$

where $x_1(t) = x(t)$, with the initial condition $x_1(0) = 1/4$, $x_2(0) = x_3(0) = x_4(0) = 0$. This can be written in the standard form ${}^{C}D^{1/2}X(t) = AX(t) + F(t)$ and $F(t) = (0, 0, 0, 8)^T$, where

$$ A = \begin{bmatrix} 0 & 1 & 0 & 0 \\ 0 & 0 & 1 & 0 \\ 0 & 0 & 0 & 1 \\ \omega^2 & -mu\omega^{3/2} & 0 & 0 \end{bmatrix}. \tag{23.57}$$

This can be solved numerically by taking $\omega = m = 1$, $\mu = 1/2$.

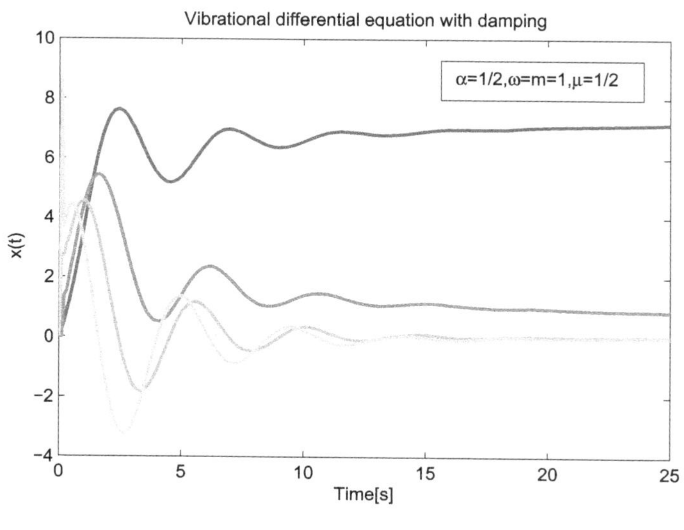

7 Conclusion

In nature, every system occurs in the fractional order. From this it can be concluded that the fractional order system has attractive feature comparing with integer order system. In this work, stability of nonlinear system having fractional order between 0 and 2 is studied. In this study, some simple sufficient conditions on the matrix and nonlinear term guarantees the stability of FDE's, were obtained. Also stability of the fractional neutral and fractional integrodifferential systems was discussed. To show the applicability of the obtained conditions, some examples were presented.

Future works include the stability of nonlinear fractional neutral and integrodifferential equations with delay by analyzing the eigenvalues and applying conditions on the nonlinear term.

Bibliography

[1] I. Area and J. J. Nieto, Power series solution of the fractional logistic equation, *Physica A* **573** (2021), 125947.

[2] R. L. Bagley and P. J. Torvik, A theoretical basis for the application of fractional calculus to viscoelasticity, *J. Rheol.* **27** (1983), 201–210.

[3] K. Bao, P. gong and L. Wang, Exponential synchronization of a class of fuzzy recurrent neural networks with time varying delays and impulses on time scales, *Int. J. Appl. Math. Stat.* **48**(18) (2013), 325–338.

[4] M. Bardi and A. Schiano, Asymptotic behavior of positive solutions of periodic delay logistic equations, *J. Math. Biol.* **14** (1982), 95–100.

[5] M. E. Bauer, Existence and stability of patterns arising from square wave forcing of the damped Mathieu equation, Senior honors Thesis, Duke University, 2008.

[6] E. Bicer, On the stability of functional differential equations with advanced argument, *Int. J. Appl. Math. Stat.* **57**(5) (2018), 84–90.

[7] F. Bozkurt, Stability and density analysis of a monoclonal population growth with piecewise constant arguments, *Int. J. Appl. Math. Stat.* **57** (2018), 70–79.

[8] F. Bozkurt and A. Yousef, Oscillation and global stability of an impulsive delay logistic equation, *Int. J. Appl. Math. Stat.* **57**(6) (2018), 84–91.

[9] J. P. Diaz and D. Bhatta, Stability analysis and simulation results of an SIR mathematical model for the Dengue fever, *Int. J. Appl. Math. Stat.* **57**(2) (2018), 142–157.

[10] J. S. Duan and D. C. Hu, Vibration systems with fractional-order and distributed-order derivatives characterizing viscoinertia, *Fractal Fract.* **5** (2021), 67.

[11] J. S. Duan, D. C. Hu and Y. Q. Chen, Simultaneous characterization of relaxation, creep, dissipation, and hysteresis by fractional-order constitutive models, *Fractal Fract.* **5** (2021), 36.

[12] J. S. Duan, D. C. Hu and M. Li, Comparison of two different analytical forms of response for fractional oscillation equation, *Fractal Fract.* **5** (2021), 188.

[13] J. S. Duan, J. Y. Zhang and X. Qiu, Exact solutions of fractional order oscillation equation with two fractional derivative terms, *J. Nonlinear Math. Phys.* **30** (2023), 531–552.

[14] Q. Fan, G. C. Wu and H. Fu, A note on function space and boundedness of the general fractional integral in continuous time random walk, *J. Nonlinear Math. Phys.* **29** (2022), 95–102.

[15] J. Gao and Y. Ding, A generalized Gronwall inequality and its application to a fractional differential equation, *J. Math. Anal. Appl.* **328** (2007), 1075–1081.

[16] K. Gopalsamy and P. Liu, Persistence and global stability in a population model, *J. Math. Anal. Appl.* **224** (1998), 5980.

[17] Han Liang Huang, The delayed dependent design method of state time delay feedback controller for uncertain fuzzytime delay systems, *Int. J. Appl. Math. Stat.* **48**(18) (2013), 440–448.

[18] M. Li, Three classes of fractional oscillators, *Symmetry* **10** (2018), 40.

[19] F. Mainardi and G. Spada, Creep, relaxation and viscosity properties for basic fractional models in rheology, *Eur. Phys. J. Spec. Top.* **193** (2011), 133–160.

[20] S. Priyadharsini, Stability of fractional neutral and integrodifferential systems, *J. Fract. Calc. Appl.* **7** (2016), 87–102.

[21] S. Priyadharsini, Stability analysis of fractional differential systems with constant delay, *J. Indian Math. Soc.* **83** (2016), 337–350.

[22] S. Priyadharsini, Stability analysis of visco-elastically damped structure through Bagley Torvik equation, *Mapana J. Sci.* **22**(2) (2023), 37–53.

[23] S. Priyadharsini and V. Govindaraj, Asymptotic stability of Caputo fractional singular differential systems with multiple delays, *J. Discontinuity, Nonlinearity, Complexity* **7**(3) (2018), 243–251.

[24] S. Priyadharsini and R. Spandana, Controllability of second order linear and nonlinear delay dynamical systems, *J. Appl. Sci. Comput.* **6**(2) (2019), 475–483.

[25] H. Singh, Chebyshev spectral method for solving a class of local and nonlocal elliptic boundary value problems, *Int. J. Nonlinear Sci. Numer. Simul.* **24**(3) (2023), 899–915.

[26] H. Singh, D. Kumar and D. Baleanu, *Methods of Mathematical Modelling: Fractional Differential Equations*, CRC Press, 2019.

[27] H. Singh, A. Kumar Singh, R. K. Pandey, D. Kumar and J. Singh, An efficient computational approach for fractional Bratu's equation arising in electrospinning process, *Math. Methods Appl. Sci.* **44**(13) (2021), 10225–10238.

[28] H. Singh, H. M. Srivastava and D. Baleanu, *Methods of Mathematical Modelling Infectious Diseases*, 1st edn, June 10, (2022).

[29] H. Singh, H. M. Srivastava and D. Kumar, A reliable numerical algorithm for the fractional vibration equation, *Chaos Solitons Fractals* **103** (2017), 131–138.

[30] H. Singh, H. M. Srivastava and J. J. Nieto, *Handbook of Fractional Calculus for Engineering and Science*, Chapman and Hall/CRC, 2022.

[31] H. Singh and A. M. Wazwaz, Computational method for reaction diffusion-model arising in a spherical catalyst, *Int. J. Appl. Comput. Math.* **7** (2021), 65.

[32] S. M. Yang and J. S. Duan, Response analysis of six-parameter fractional constitutive model, *Phys. Scr.* **96** (2021), 025220.

Preety Kalra and Nisha Malhotra

Analysis of modified fractional order Lotka–Volterra model

Abstract: This study analyzes the fractional-order Lotka–Volterra system with logistic growth and additive Allee effect in the prey population, by making use of the Caputo fractional order derivative. The analysis is done by considering the weak Allee effect as well as the strong Allee effect. The effect of the capture rate of prey by predators has been studied numerically. Hopf bifurcation is studied at the coexistence point when populations are subjected to weak as well as strong Allee effects, taking fractional order and capture rate as the bifurcation parameter. It is seen that fractional-order models provide better stability regions, as this study finds the equilibrium points that are unstable in the ordinary integer order model but stable in the fractional-order model.

Keywords: Lotka–Volterra system, Fractional-order derivative, Caputo derivative, Prey–predator system, Additive Allee effect

MSC 2010: 37N25, 37N30

1 Introduction

All populations in an ecosystem have interspecific as well as intraspecific relationships, known as prey-predatory, competitive, mutualism, and parasitism [57]. The prey-predator relationship is the most common in the natural world and has thus been studied extensively by ecologists [27]. Mathematical biology is the branch of mathematics that studies such biological interactions. Mathematical biology is concerned with the modeling, study, analysis, and interpretation of biological phenomena such as evolution, coexistence, existence, and species interaction. Mathematical modeling is a more effective tool because it conserves time, energy, and resources. Mathematical models are created to accurately represent biological interactions by carefully designing the variables, identifying the governing laws, and making some assumptions to make the model tractable. The famous Lotka–Volterra model was developed for the first time by biophysicist A. J. Lotka [37] and mathematician V. Volterra [50]. The conventional Lotka–Volterra model structure is oversimplified because it makes numerous assumptions that are not precisely true. This include an uninterrupted pattern of prey species' food availability, the absence of adapting of all interacting species, and the initial relationship between the size of every species and its rate of growth, leading to

Preety Kalra, Nisha Malhotra, Department of Mathematics, School of Chemical Engineering and Physical Sciences, Lovely Professional University, 144411 Phagwara, Punjab, India, e-mails: preety.19056@lpu.co.in, nishamz1974@gmail.com

https://doi.org/10.1515/9783111313634-024

exponential expansion if the species remains unaffected by others. Regardless of these assumptions, the Lotka–Volterra model has successfully explained the dynamics of a wide variety of biological interaction systems, including the interactions among numerous species of fish, bacterial lake communities, and freshwater ciliates. In the past few years, Lotka–Volterra systems have proven to be an excellent analytical tool, and indeed the standard for quantifying interactions within massive microbial communities. Lotka–Volterra models have also been presented in non-ecological contexts, such as to address pollution issues. Although Lotka–Volterra models are frequently regarded as archaic and insufficient for much more modern systems biology techniques, it has been discovered that the Lotka–Volterra model structure is remarkably rich and capable of capturing any differentiable nonlinearities if the required number of supplementary variables are introduced into the model. Despite its simplicity and intuitive nature, the Lotka–Volterra model has proven to be an excellent baseline model for comparison with modern complex models. Lotka–Volterra models have been used in nearly 500 PubMed-listed studies in the last decade alone [17].

It has been witnessed that individuals of many species benefit from the presence of conspecifics. Many population's per capita growth rates have been observed to decline when population density is critically low. This could result in a zero-growth rate or even a negative growth rate. The primary reason for this could be a lack of a mate for reproduction when the population is quite low. This eventually results in low reproductivity, which may lead to population extinction. So, for a population to grow, there must be a minimum density of that population, which is known as the critical population density or threshold level, below which the population cannot exist. This phenomenon is known as the Allee effect [5]. The Allee effect is not only caused by a lack of ability to find mates at low population density, but due to social dysfunction, inbreeding depression, swamping of enemies, allelic diversity, and food exploitation [31, 45]. The Allee effect can be weak or strong. The strong Allee effect is attributed to a negative growth rate, whereas continuously decreasing but positive growth rates give rise to a weak Allee effect. For the growth rate to remain positive, a population density is required, known as threshold density. The population is driven to extinction if the population density falls below this threshold [14, 51]. The possibility of positive density dependence in small populations, increasing the risk of extinction, has incited a lot of interest in potential applications, including small population conservation [9], fishing industry management [29], and predatory insect control [33, 48]. Several studies published examined the Allee effect in various population models and conclude that it can have a significant influence on system dynamics. The introduction of the Allee effect can change the dynamic behavior of the system by stabilizing or destabilizing it, or by changing the time to reach the equilibrium point [39].

During the last few decades, fractional derivatives and integrals have attracted the attention of scientists in various fields due to their broader scope and applications. Fractional calculus is a generalization of ordinary differentiation and integration to arbitrary order, real or complex. Due to notable link of fractional calculus with memory and fractal nature, which are abundant in physical processes, its usage is growing remarkably quickly in the development of new models. The incorporation of a fractional-order derivative in a model minimizes inaccuracy caused by a lack of knowledge of significant real parameters. When compared to the corresponding classical systems, the fractional derivative operators allow for a greater degree of freedom in models of real-life situations [20]. Fractional-order differential equations are provided with powerful methods for characterizing hereditary and memory properties that classical systems simply ignore. Furthermore, modeling with a fractional-order derivative accurately expresses the behavior of real-world systems and is useful in the study of dynamical systems. The most important and useful property of fractional-order differential equations is their non-local property, and memory preservance which focuses on the fact that the future state is dependent on the current state as well as all previous states. Additionally, fractional order systems converge to classical-order systems. As a result, the concept of fractional calculus provides many extensions in terms of non-local natures, increased degrees of freedom, extended stability region, and maximum information usage. Nowadays, the applications of fractional calculus are growing rapidly. In recent years, the theory of fractional calculus has been widely applied in a variety of fields, especially in physics, viscoelasticity, engineering, signal processing, epidemiology, etc. [2, 10, 11, 23, 25, 26, 28, 44, 49]. The fractional calculus is useful in image processing techniques such as edge detection, texture segmentation, revealing faint objects, and image restoration [42]. Fractional calculus-based dynamical systems are proving to be helpful in biosciences. There are studies published on fractional-order diffusive dynamical models to study the concentration of cancer tumor cells and the killing rate, interactions of the immune effectors with cancer cells [30], and arterial viscoelasticity [13]. In robotics, it has been proved that fractional calculus helps in improving the control methods and generalization [41]. Fractional calculus has been widely used for epidemic models too. It has been observed that the usage of fractional calculus to handle the modeling process is better than that of the integral order approach because results obtained using fractional order modeling in epidemiology has good fitting with real data [34]. Therfore, Fractional calculus can be categorized as applicable mathematics and can help to solve many real-world problems.

There are several fractional derivatives, including the Riemann–Liouville fractional derivative, the Caputo fractional derivative, the Caputo–Fabrizio derivative, the Atangana derivative, and the Hadamard derivative. In contrast to other types of fractional-order derivatives, the Caputo fractional derivative allows the initial boundary conditions to be included in the formulation of differential equation similar to those of the integer-order differential equation, allowing for a better understanding of the physical significance and hence a wider range of practical applications. Therefore, the Caputo fractional derivative is easily capable of resolving the initial value problem of nonlinear fractional dynamical systems, which is widely used in the modeling of real-world practical systems [6, 40]. Studies on population dynamics using fractional-order models were first published by Ahmed et al. (2007), in which the author studied the stability and numerical analysis of the fractional-order prey-predator model and rabies model [1]. Since then, various studies have been published on fractional-order models of prey-predator interactions incorporating harvesting [32, 38, 43, 46, 55], Monod–Haldane functional response and time delay [12], Holling type III functional response [15], ratio-dependent functional response [53], prey refuge [56], constant prey refuge and feedback control [36], group defense [3, 16, 47], and Allee effect [4, 8, 52]. Various studies have been reported modeling the Lotka–Volterra model using fractional differential equations too. A fractional-order Lotka–Volterra system with logistic growth of prey population and discontinuous predator harvesting with Holling type III functional response is studied by the authors of [54]. Kumar et al. [35] used the Haar wavelet and Adams–Bashforth–Moulton methods to compare a fractional Lotka–Volterra model in the Caputo sense. A discrete fractional order Lotka–Volterra model with logistic growth in prey species was introduced in [22]. Mesady et al. [21] investigated an extended Lotka–Volterra model with immegration using Caputo fractional order derivative. The purposed study aims to analyze the traditional Lotka–Volterra model with logistic growth and additive Allee effect in the prey population using Caputo fractional order derivative. In this chapter, the authors studied the model behavior under the weak Allee effect as well as the strong Allee effect by changing the rate of predation and studied Hopf bifurcation at the coexistence point under weak as well as strong Allee effects by taking fractional derivative and rate of predation as the bifurcation parameter. To the best of our knowledge, no study is published on the Lotka–Volterra model with additive Allee effect using Caputo fractional derivative. The organization of the rest of the chapter is as follows: In section 2, the formation of the mathematical model to be analyzed is explained. Section 3, contributes the mathematical definitions, lemmas, and theorems required for the mathematical analysis of the proposed study. Section 4, presents the mathematical analysis of the model by proving the existence and uniqueness of the solutions, positivity, and boundedness of the solutions. The section 5, a dynamical analysis of the model is presented by evaluating the equilibrium points and by testing the local and global stability of equilibrium points. Section 6, is dedicated to numerical simulations and discussions, and in Section 7 concluding remarks are given.

2 Mathematical model

In the prey–predator model, the Allee effect can be studied in two ways, additively and multiplicatively. Bazykin [7] introduced the multiplicative Allee effect for a single species logistic growth model as

$$\frac{dx}{dt} = rx\left(1 - \frac{x}{K}\right)(x - m), \tag{24.1}$$

where x is the population at any time t. r, K, m are all positive constants. r is the per capita intrinsic growth rate of the population, K is the carrying capacity of the environment, and m is Allee's threshold. The growth rate of the population will remain positive if $0 < m < x < k$ and will be negative outside this interval. Another single species logistic growth model was introduced by Brian Dennis [18] in which he introduced the additive Allee effect, the model equation of which is given by

$$\frac{dx}{dt} = \left[r\left(1 - \frac{x}{k}\right) - \frac{m}{x + a}\right]x, \tag{24.2}$$

where x is population density at any time t, r, K, a, m are all positive constants defined r as the per capita intrinsic growth rate of the population, K is the carrying capacity of the environment, and a is population density at which fitness is half of its maximum value. Here m and a depict the degree of the Allee effect. According to this model if $0 < m < ar$ then the model has a weak Allee effect and if $m > ar$, then the model has a strong Allee effect. We have considered the fractional order Lotka–Volterra model with logistic growth and additive Allee effect in prey population as,

$$\begin{cases} D^{\alpha}x_1 = x_1[r(1 - \frac{x_1}{k}) - \frac{m}{x_1+a}] - fx_1x_2, \\ D^{\alpha}x_2 = fcx_1x_2 - ex_2, \end{cases} \tag{24.3}$$

with initial conditions $x_1(0) \geq 0$, $x_2(0) \geq 0$. Here $x_1 > 0$ is the Prey population at time t, $x_2 > 0$ is the predator population at any time t, all parameters r, K, m, a, f, c, e are all positive depicting the intrinsic growth rate of prey population, carrying capacity of the environment for prey, level of the Allee effect, prey population having fitness half of its maximum value, the capture rate of prey by the predator, the conversion rate of prey to predator, and intrinsic death rate of predators respectively. D^{α} is the time-fractional Caputo derivative, where $\alpha \in (0, 1]$.

3 Mathematical preliminaries

Definition 1 ([12]). Caputo fractional derivative of a function f is defined as, $D^{\alpha}f(x) = \frac{1}{\Gamma(1-\alpha)} \int_0^x (x - t)^{-\alpha} f'(t)\, dt$, where $0 < \alpha < 1$.

Lemma 1 ([16]). *Let us assume that $\alpha \in (0, 1]$ and consider that the function $f(t)$ and $D^{\alpha}f(t) \in \mathrm{C}[\mathrm{a}, \mathrm{b}]$ for all $t \in [\mathrm{a}, \mathrm{b}]$, where $\mathrm{C}[\mathrm{a}, \mathrm{b}]$ is the class of continuous functions on $[\mathrm{a}, \mathrm{b}]$ and D^{α} stands for fractional-order Caputo derivative. The function $f(t)$ is said to be non-decreasing on $[\mathrm{a}, \mathrm{b}]$ if $D^{\alpha}f(t) \geq 0$, and the function $f(t)$ is said to be non-increasing on $[\mathrm{a}, \mathrm{b}]$ and if $D^{\alpha}f(t) \leq 0$.*

Lemma 2 ([16]). *If $x(t) \in \mathbb{R}^+$ be a continuous and differentiable function, then any time for*

$$t > t_0, \quad D^a\left(x(t) - x_* - x_* \ln \frac{x(t)}{x_*}\right) \leq \left(\frac{x - x_*}{x}\right) D^a x(t).$$

Lemma 3 ([16]). *Let $J(X)$ be the Jacobian matrix of a fractional-order system at the equilibrium point X and λ_i $(i = 1, 2, 3 \ldots n)$ be the respective eigenvalues of Jacobian matrix $J(X)$. Then the equilibrium point X is locally asymptotically stable $\Leftrightarrow$ $|\arg\lambda_i| > \frac{a\pi}{2}$ for all λ_i $(i = 1, 2, 3 \ldots n)$, where a is the fractional order.*

Theorem 1 ([16]). *Let $a > 0$, $n - 1 < a < n$, where $n \in \mathbb{N}$. Assume $U(t)$ is n times continuously differentiable function and $D^a\, U(t)$ is piecewise continuous on $[t_0, \infty)$, then $\mathcal{L}\{D^a U(t)\} = s^a \mathcal{F}(s) - \sum_{j=0}^{n-1} s^{a-j-1} U^j(t_0)$, where $\mathcal{F}(s) = \mathcal{L}\{U(t)\}$, i. e., Laplace transform of $U(t)$, D^a is the Caputo fractional order derivative.*

Theorem 2 ([16]). *Let C be the complex plane. For each $m > 0$, $p > 0$, $k \in C^{n\times n}$. The Laplace transform of $t^{p-1}E_{m,p}(kt^m)$ is defined as, $\mathcal{L}\{t^{p-1}E_{m,p}(kt^m)\} = \frac{s^{m-p}}{s^m - K}$, $\Re(s) > \|K\|^{1\backslash m}$, where $\Re(s)$ is the real part of the complex number s, and $E_{m,p}$ is the Mittag–Leffler function defined as $E_{m,p}(z) = \sum_{n=0}^{\infty} \frac{z^m}{\Gamma(mn+p)}$, and Γ is the Gamma function.*

Theorem 3 ([16] Global stability theorem). *Let $D^a x(t) = f(t, x)$, where $x(t_0) > 0$ be the non-autonomous system and $x_* \in \Lambda \subset \mathbb{R}^n$ be the equilibrium point of the system. If $F(t, x) : [0, \infty) \times \Lambda \to \mathbb{R}$ is a continuous and differentiable function such that*

$$D^a f(t, x) \leq -\Omega_1(x),$$
$$\Omega_2(x) \leq f(t, x) \leq \Omega_3(x), \quad \forall\, a \in (0, 1) \text{ and } \forall\, x(t) \in \Lambda,$$

where Ω_i, $(i = 1, 2, 3)$ are continuous positive definite functions on Λ, then the equilibrium point x_* is globally asymptotically stable.

4 Mathematical analysis of the model

4.1 Existence and uniqueness of the solution

Theorem 4. *Consider the interval $I = [t_0, T]$, $T < \infty$ and the region $D = \{(x_1, x_2) \in \mathbb{R}^2 : |x_1|, |x_2| \leq M\}$. Let $C(I)$ be the class of all real-valued functions defined on I, which have continuous first-order derivatives on $C(I)$. Then the initial value problem $D^a x_1(t) = h(x_1 x_2, t)$, $D^a x_2(t) = g(x_1 x_2, t)$, where*

$$h(x_1 x_2, t) = x_1\left[r\left(1 - \frac{x_1}{k}\right) - \frac{m}{x_1 + a}\right] - f x_1 x_2,$$
$$g(x_1 x_2, t) = f c x_1 x_2 - e x_2,$$

with initial condition $x_1(t = t_0) = x_{10}$, $x_2(t = t_0) = x_{20}$ *and* $f : I \times D \to \mathbb{R}^2; \alpha \in (0,1]$ *have unique solution* $X(x_1(t), x_2(t)) \in D$, *with initial condition* $X(x_1(t_0), x_2(t_0)) = X_{t0}$.

Proof. Consider a mapping $F : D \to \mathbb{R}^2$ by $F(X) = (F_1(X),\ F_2(X))$, where $X = (x_1, x_2)$ and $Y = (y_1, y_2)$:

$$F_1(X) = x_1\left[r\left(1 - \frac{x_1}{k}\right) - \frac{m}{x_1 + a}\right] - fx_1x_2, \quad F_2(X) = fcx_1x_2 - ex_2.$$

For any $X, Y \in D$;

$$\begin{aligned}
\|F(X) - F(Y)\| &= |F_1(X) - F_1(Y)| + |F_2(X) - F_2(Y)| \\
&= \left|x_1\left[r\left(1 - \frac{x_1}{k}\right) - \frac{m}{x_1 + a}\right] - fx_1x_2 - y_1\left[r\left(1 - \frac{y_1}{k}\right) - \frac{m}{y_1 + a}\right] + fy_1y_2\right| \\
&\quad + |fcx_1x_2 - ex_2 - fcy_1y_2 + ey_2| \\
&= \left|rx_1 - \frac{r}{K}x_1^2 - \frac{mx_1}{x_1 + a} - fx_1x_2 - ry_1 + \frac{r}{K}y_1^2 + \frac{my_1}{y_1 + a} + fy_1y_2\right| \\
&\quad + |fcx_1x_2 - ex_2 - fcy_1y_2 + ey_2| \\
&\le |r(x_1 - y_1)| + \left|\frac{r}{K}(x_1^2 - y_1^2)\right| + \left|m\left(\frac{x_1}{x_1 + a} - \frac{y_1}{y_1 + a}\right)\right| + |f(x_1x_2 - y_1y_2)| \\
&\quad + |e(x_2 - y_2)| \\
&\le |x_1 - y_1|\left|r + \frac{r}{K}(x_1 + y_1) + \frac{ma}{(x_1 + a) + (y_1 + a)}\right| \\
&\quad + f|x_1(x_2 - y_2) + y_2(x_1 - y_1)| + fc|x_1(x_2 - y_2) + y_2(x_1 - y_1)| + e|x_2 - y_2| \\
&\le |x_1 - y_1|\left(r + \frac{2rM}{K} + ma\right) + fM|(x_2 - y_2)| + fM|(x_1 - y_1)| \\
&\quad + fcM|(x_2 - y_2)| + fcM|(x_1 - y_1)| + e|(x_2 - y_2)| \\
&= |x_1 - y_1|\left(r + \frac{2rM}{K} + ma + fM + fcM\right) + |x_2 - y_2|(fM + fcM + e) \\
&= K_1|x_1 - y_1| + K_2|x_2 - y_2| \le \mathcal{K}|x - y|,
\end{aligned}$$

where

$$\begin{aligned}
K_1 &= r + \frac{2rM}{K} + ma + fM + fcM, \\
K_2 &= (fM + fcM + e) \quad \text{and} \quad \mathcal{K} = \max\{K_1, K_2\}.
\end{aligned}$$

As F satisfies the Lipschitz condition. Therefore, the fractional-order prey–predator system given by (24.3) has a unique solution. □

4.2 Uniform boundedness of the solutions

Theorem 5. *The solution of the system* (24.3) *starts in* $\mathbb{R}_+^2$ *are uniformly bounded.*

Proof. Let us consider a function $F(t) = x_1(t) + \frac{1}{c}x_2(t)$.

Now

$$\begin{aligned}
D^\alpha F(t) &= D^\alpha\left[x_1(t) + \frac{1}{c}x_2(t)\right] = D^\alpha x_1(t) + \frac{1}{c}D^\alpha x_2(t) \\
&= x_1\left[r\left(1 - \frac{x_1}{k}\right) - \frac{m}{x_1 + a}\right] - fx_1x_2 + fx_1x_2 - \frac{e}{c}x_2 \\
&= rx_1 - \frac{r}{K}x_1^2 - \frac{m}{x_1 + a}x_1 - \frac{e}{c}x_2 \\
&< rx_1 - \frac{r}{K}x_1^2 - \frac{m}{K + a}x_1 - \frac{e}{c}x_2 \\
&= \left(r - \frac{m}{K + a}\right)x_1 - \frac{r}{K}x_1^2 - \frac{e}{c}x_2 \\
&= \left(r - \frac{m}{K + a}\right)x_1 - \frac{r}{K}x_1^2 - eF(t) + ex_1,
\end{aligned}$$

$$\begin{aligned}
D^\alpha F(t) + eF(t) &< \left(r + e - \frac{m}{K + a}\right)x_1 - \frac{r}{K}x_1^2 \\
&= M_1x_1 - M_2x_1^2,
\end{aligned}$$

where $M_1 = r + e - \frac{m}{K+a}$ and $M_2 = \frac{r}{K}$. Therefore,

$$\begin{aligned}
D^\alpha F(t) + eF(t) &< -M_2\left(x_1^2 - \frac{M_1}{M_2}x_1 + \frac{M_1^2}{4M_2^2}\right) + \frac{M_1^2}{4M_2} \\
&= \frac{M_1^2}{4M_2} - M_2\left(x_1 - \frac{M_1}{2M_2}\right)^2 \le \frac{M_1^2}{4M_2} = M_3.
\end{aligned}$$

Therefore, we have

$$D^\alpha F(t) + eF(t) < M_3.$$

Applying Laplace transform on both sides and using Theorem 3

$$s^\alpha G(s) - s^{\alpha-1}F(t_0) + eG(s) \le \frac{M_3}{s},$$

where $G(s) = \mathcal{L}\{F(t)\}$ gives

$$(s^\alpha + e)G(s) \le \frac{M_3}{s} + s^{\alpha-1}F(t_0),$$

which gives

$$G(s) \leq \frac{M_3}{s(s^\alpha + e)} + \frac{s^{\alpha-1}F(t_0)}{(s^\alpha + e)} = \frac{s^{\alpha-1}F(t_0)}{(s^\alpha + e)} + \frac{s^{\alpha-(\alpha+1)}}{(s^\alpha + e)}M_3.$$

By taking inverse Laplace transform, we get

$$\begin{aligned} F(t) &\leq F(t_0)\mathcal{L}^{-1}\left\{\frac{s^{\alpha-1}}{(s^\alpha + e)}\right\} + M_3\mathcal{L}^{-1}\left\{\frac{s^{\alpha-(\alpha+1)}}{(s^\alpha + e)}\right\} \\ &= F(t_0)E_{\alpha,1}\{-et^\alpha\} + M_3 t^\alpha E_{\alpha,\alpha+1}\{-et^\alpha\}. \end{aligned}$$

Using $E_{\alpha,\beta}\{z\} = zE_{\alpha,\alpha+\beta}\{z\} + \frac{1}{\Gamma_\beta}$.

$$\begin{aligned} F(t) &\leq F(t_0)E_{\alpha,1}\{-et^\alpha\} - \frac{M_3}{e}[E_{\alpha,1}\{-et^\alpha\} - 1] \\ &= \left(F(t_0) - \frac{M_3}{e}\right)E_{\alpha,1}\{-et^\alpha\} + \frac{M_3}{e}. \end{aligned}$$

Now as $t \longrightarrow \infty$, $E_{\alpha,1}\{-et^\alpha\} \longrightarrow 0$.

Therefore, all solutions to the given system start in $\mathbb{R}^2_+$ lie in the region

$$\left\{X \in \mathbb{R}^2_+ : X \leq \frac{M_3}{e} + v,\ v > 0\right\}.$$

□

5 Dynamical analysis of the model

5.1 Equilibrium points

The given fractional-order system (24.3) has,

1. $(0, 0)$ as the equilibrium point without any condition,
2. $(x_1', 0)$ as the equilibrium point if $m \leq \min\{ar, \frac{Kr}{4}(1 + \frac{a}{K})^2\}$,
3. $(x_1', 0), (x_1'', 0)$ as equilibrium points, if $ar \leq m \leq \frac{Kr}{4}(1 + \frac{a}{K})^2$,
4. (x_{1s}, x_{2s}) as equilibrium point if $m < \frac{r}{fc}(1 - \frac{e}{fcK})(e + afc)$ and $\frac{e}{fcK} < 1$.

Where,

$$x_1' = \frac{K}{2}\left[\left(1 - \frac{a}{K}\right) + \sqrt{\left(1 + \frac{a}{K}\right)^2 - \frac{4m}{Kr}}\right],$$

$$x_1'' = \frac{K}{2}\left[\left(1 - \frac{a}{K}\right) - \sqrt{\left(1 + \frac{a}{K}\right)^2 - \frac{4m}{Kr}}\right],$$

$$x_{1s} = \frac{e}{cf},$$
$$x_{2s} = \frac{r}{f} - \frac{re}{cf^2K} - \frac{mc}{e + afc}.$$

5.2 Local asymptotical stability analysis of the equilibrium points

Given system of equations is

$$D^\alpha x_1 = x_1\left[r\left(1 - \frac{x_1}{k}\right) - \frac{m}{x_1 + a}\right] - fx_1x_2 = h(x_1, x_2),$$
$$D^\alpha x_2 = fcx_1x_2 - ex_2 = g(x_1, x_2),$$
$$\frac{\partial h}{\partial x_1} = r - \frac{2rx_1}{K} - \frac{ma}{(x_1 + a)^2} - fx_2, \frac{\partial h}{\partial x_2} = -fx_1, \frac{\partial g}{\partial x_1} = fcx_2 \frac{\partial g}{\partial x_2} = fcx_1 - e,$$
$$\mathbb{J}(h, g) = \begin{bmatrix} \frac{\partial h}{\partial x_1} & \frac{\partial h}{\partial x_2} \\ \frac{\partial g}{\partial x_1} & \frac{\partial g}{\partial x_2} \end{bmatrix} = \begin{bmatrix} r - \frac{2rx_1}{K} - \frac{ma}{(x_1+a)^2} - fx_2 & -fx_1 \\ fcx_2 & fcx_1 - e \end{bmatrix}.$$

Theorem 6. *If $m > \max\{ar, \frac{Kr}{4}(1 + \frac{a}{K})^2\}$, the extinction point $(0, 0)$ is always locally asymptotically stable.*

Proof. The value of the Jacobian matrix at $(0, 0)$ is given by,

$$\mathbb{J}(0, 0) = \begin{bmatrix} r - \frac{m}{a} & 0 \\ 0 & -e \end{bmatrix}.$$

Being a diagonal matrix, its eigenvalues are given by $\lambda_1 = r - \frac{m}{a}, \lambda_2 = -e < 0$. So $(0, 0)$ will be stable if $r - \frac{m}{a} < 0$, which gives $r < \frac{m}{a}$ or $m > ar$. Also, if $m > \frac{Kr}{4}(1 + \frac{a}{K})^2$, then $(0, 0)$ is the only equilibrium point of the system, therefore, $(0, 0)$ is locally asymptotically stable if $m > \max\{ar, \frac{Kr}{4}(1 + \frac{a}{K})^2\}$. □

Theorem 7. *If $ar - \frac{re}{fc}(\frac{e}{Kfc} + \frac{a}{K} - 1) < m < ar$, where $m > 0$, the equilibrium point $(x_1', 0)$ will be asymptotically stable.*

Proof. The value of the Jacobian matrix at $(x_1', 0)$ is given by,

$$\mathbb{J}(x_1', 0) = \begin{bmatrix} r - \frac{2rx_1'}{K} - \frac{ma}{(x_1'+a)^2} & -fx_1' \\ 0 & fcx_1' - e \end{bmatrix}.$$

Eigenvalues of the matrix are given by $\lambda_1 = r - \frac{2rx_1'}{K} - \frac{ma}{(x_1'+a)^2}, \lambda_2 = fcx_1' - e.$

The point $(x_1', 0)$ will be stable if $\lambda_1 < 0$ and $\lambda_2 < 0$. This gives $r - \frac{2rx_1'}{K} - \frac{ma}{(x_1'+a)^2} < 0$ and $fcx_1' - e < 0$.

On solving the inequation $r - \frac{2rx_1'}{K} - \frac{ma}{(x_1'+a)^2} < 0$, one can see $0 < m < ar$, and on solving the inequation $fcx_1' - e < 0$, it is found that $m > ar - \frac{re}{fc}(\frac{e}{Kfc} + \frac{a}{K} - 1)$. Hence the predator free equilibrium point $(x_1', 0)$ will be locally asymptotically stable if $0 < m < ar$ and $m > ar - \frac{re}{fc}(\frac{e}{Kfc} + \frac{a}{K} - 1)$. Combining the both obtained results and considering the positivity of m, it is concluded that $(x_1', 0)$ is locally asymptotically stable if $ar - \frac{re}{fc}(\frac{e}{Kfc} + \frac{a}{K} - 1) < m < ar$, where $m > 0$. □

Theorem 8. *If $ar \leq m \leq \frac{Kr}{4}(1 + \frac{a}{K})^2$, the predator-free equilibrium point $(x_1', 0)$ is locally stable provided the initial prey population is more than x_1'' or when the prey population is subjected to a strong Allee effect, the threshold level for the survival of the prey population is given by x_1'' in the absence of predators.*

Theorem 9. *The predator-free equilibrium point $(x_1'', 0)$ is always unstable.*

Proof. The Jacobian matrix at $(x_1'', 0)$ is given as,

$$\mathbb{J}(x_1'', 0) = \begin{bmatrix} r - \frac{2rx_1''}{K} - \frac{ma}{(x_1''+a)^2} & -fx_1'' \\ 0 & fcx_1'' - e \end{bmatrix}.$$

Eigenvalues of the matrix are given by $\lambda_1 = r - \frac{2rx_1''}{K} - \frac{ma}{(x_1''+a)^2}$, $\lambda_2 = fcx_1'' - e$.

The point $(x_1'', 0)$ will be stable if $\lambda_1 < 0$ and $\lambda_2 < 0$. This gives $r - \frac{2rx_1''}{K} - \frac{ma}{(x_1'+a)^2} < 0$ and $fcx_1'' - e < 0$. By solving the inequality $r - \frac{2rx_1''}{K} - \frac{ma}{(x_1''+a)^2} < 0$, it is found that $0 < m < ar$. This is contrary to the existence condition of $(x_1'', 0)$, which is given by $m \geq ar$. Therefore, the point $(x_1'', 0)$ is always unstable. □

Theorem 10. *If any one of the following conditions is satisfied,*
1. $T^2 - 4\Delta \geq 0$ *and* $T < 0$,
2. $T^2 - 4\Delta < 0$, *and* $T < 0$,
3. $T^2 - 4\Delta < 0$, $T > 0$, *and* $0 < \alpha < \frac{2}{\pi} \tan^{-1} |\frac{\sqrt{4\Delta - T^2}}{T}|$,
4. $T = 0$, *and* $\Delta > 0$,

the coexistence point (x_{1s}, x_{2s}) is locally asymptotically stable, where

$$T = \frac{mecf}{(e + afc)^2} - \frac{re}{Kcf} \quad \text{and} \quad \Delta = e\left[r - \frac{re}{cfK} - \frac{mcf}{e + afc}\right].$$

Proof. The Jacobian matrix at (x_{1s}, x_{2s}) is given as,

$$\mathbb{J}(x_{1s}, x_{2s}) = \begin{bmatrix} r - \frac{2rx_{1s}}{K} - \frac{ma}{(x_{1s}+a)^2} - fx_{2s} & -fx_{1s} \\ fcx_{2s} & fcx_{1s} - e \end{bmatrix} = \begin{bmatrix} a_{11} & a_{12} \\ a_{21} & a_{22} \end{bmatrix},$$

$$a_{11} = r - \frac{2rx_{1s}}{K} - \frac{ma}{(x_{1s}+a)^2} - fx_{2s}, \quad a_{12} = -fx_{1s},$$

$$a_{21} = fcx_{2s}, \quad a_{22} = fcx_{1s} - e = 0.$$

Now the characteristic equation of the matrix is given by $\lambda^2 - T\lambda + \Delta = 0$,

Here

$$T = a_{11} + a_{22} = a_{11} = r - \frac{2rx_{1s}}{K} - \frac{ma}{(x_{1s}+a)^2} - fx_{2s}$$

$$= r - \frac{2re}{Kcf} - \frac{maf^2c^2}{(e+afc)^2} - r + \frac{re}{cfK} + \frac{mfc}{e+afc} = \frac{mecf}{(e+afc)^2} - \frac{re}{Kcf},$$

$$\Delta = a_{11}a_{22} - a_{12}a_{21} = -a_{12}a_{21} = f^2cx_{1s}x_{2s} = e\left[r - \frac{re}{cfK} - \frac{mcf}{e+afc}\right].$$

Eigenvalues of the Jacobian matrix are given by $\lambda_i = \frac{T \pm \sqrt{T^2-4\Delta}}{2}$, $i = 1, 2$.

1. If $T^2 - 4\Delta \geq 0$ and $T < 0$, both eigenvalues are negative, but real, which gives $|\arg(\lambda_1)| = |\arg(\lambda_1)| = \pi > \alpha\frac{\pi}{2}$. Therefore, (x_{1s}, x_{2s}) will be locally stable.
2. If $T^2 - 4\Delta < 0$ and $T < 0$, both eigenvalues will be complex conjugates of each other with negative real parts. Therefore, $|\arg(\lambda_{1,2})| = |-\pi + \tan^{-1}(\frac{\sqrt{4\Delta-T^2}}{T})| = |\pi - \tan^{-1}(\frac{\sqrt{4\Delta-T^2}}{T})| > \alpha\frac{\pi}{2}$, and hence the coexistence point will be locally asymptotically stable.
3. If $T^2 - 4\Delta < 0$, $T > 0$, then λ_1 and λ_2 are complex conjugate of each other with their real parts positive. Therefore, $|\arg(\lambda_{1,2})| = |\tan^{-1}(\frac{\sqrt{4\Delta-T^2}}{T})| = \tan^{-1}|\frac{\sqrt{4\Delta-T^2}}{T}|$. The coexistence point (x_{1s}, x_{2s}) will be stable only if $\tan^{-1}|\frac{\sqrt{4\Delta-T^2}}{T}| > \alpha\frac{\pi}{2}$. In other words, if $0 < \alpha < \frac{2}{\pi}\tan^{-1}|\frac{\sqrt{4\Delta-T^2}}{T}|$, the equilibrium point will be stable.
4. If $T = 0$, and $\Delta > 0$. Here $(\lambda_{1,2}) = \frac{\pm\sqrt{-4\Delta}}{2} = \pm\sqrt{-\Delta}$.
   Now if $\Delta > 0$, then λ_1 and λ_2 will be the complex conjugate of each other and $|\arg(\lambda_{1,2})| = \frac{\pi}{2} > \alpha\frac{\pi}{2}$, the equilibrium point will be stable. □

Theorem 11. *If any one of the following conditions is satisfied,*

1. $T^2 - 4\Delta \geq 0$, *and* $T > 0$,
2. $T^2 - 4\Delta < 0$, $T > 0$, *and* $\tan^{-1}|\frac{\sqrt{4\Delta-T^2}}{T}| < \alpha\frac{\pi}{2}$,
3. $T = 0$, *and* $\Delta < 0$,

the coexistence point (x_{1s}, x_{2s}) *is locally asymptotically unstable.*

Proof. The Jacobian matrix at (x_{1s}, x_{2s}) is given as,

$$\mathbb{J}(x_{1s}, x_{2s}) = \begin{bmatrix} r - \frac{2rx_{1s}}{K} - \frac{ma}{(x_{1s}+a)^2} - fx_{2s} & -fx_{1s} \\ fcx_{2s} & fcx_{1s} - e \end{bmatrix} = \begin{bmatrix} a_{11} & a_{12} \\ a_{21} & a_{22} \end{bmatrix},$$

$$a_{11} = r - \frac{2rx_{1s}}{K} - \frac{ma}{(x_{1s}+a)^2} - fx_{2s}, \quad a_{12} = -fx_{1s},$$

$$a_{21} = fcx_{2s}, \quad a_{22} = fcx_{1s} - e = 0.$$

Now the characteristic equation of the matrix is given by $\lambda^2 - T\lambda + \Delta = 0$.

Here

$$T = a_{11} + a_{22} = a_{11} = r - \frac{2rx_{1s}}{K} - \frac{ma}{(x_{1s}+a)^2} - fx_{2s}$$

$$= r - \frac{2re}{Kcf} - \frac{maf^2c^2}{(e+afc)^2} - r + \frac{re}{cfK} + \frac{mfc}{e+afc} = \frac{mecf}{(e+afc)^2} - \frac{re}{Kcf},$$

$$\Delta = a_{11}a_{22} - a_{12}a_{21} = -a_{12}a_{21} = f^2cx_{1s}x_{2s} = e\left[r - \frac{re}{cfK} - \frac{mcf}{e+afc}\right].$$

Eigenvalues of the Jacobian matrix are given by $\lambda_i = \frac{T \pm \sqrt{T^2-4\Delta}}{2}$, $i = 1, 2$.

1. If $T^2 - 4\Delta \geq 0$, and $T > 0$, both eigenvalues will be positive reals, and hence $|\arg(\lambda_1)| = |\arg(\lambda_2)| = 0 < \alpha\frac{\pi}{2}$. In this situation, (x_{1s}, x_{2s}) will be unstable.
2. If $T^2 - 4\Delta < 0$, $T > 0$, then λ_1 and λ_2 are complex conjugates of each other with their real parts positive. Therefore, $|\arg(\lambda_{1,2})| = |\tan^{-1}(\frac{\sqrt{4\Delta - T^2}}{T})| = \tan^{-1}|\frac{\sqrt{4\Delta-T^2}}{T}|$. The coexistence point (x_{1s}, x_{2s}) will be stable only if $\tan^{-1}|\frac{\sqrt{4\Delta-T^2}}{T}| > \alpha\frac{\pi}{2}$. Therefore, if $\tan^{-1}|\frac{\sqrt{4\Delta-T^2}}{T}| < \alpha\frac{\pi}{2}$, the coexistence point will be unstable.
3. If $T = 0$, and $\Delta < 0$, then the eigenvalues λ_1 and λ_2 will be real numbers with one of the eigenvalues positive, say $\lambda_1 > 0$. Therefore, $|\arg(\lambda_1)| = 0 < \alpha\frac{\pi}{2}$. So the equilibrium point will be unstable. □

5.3 Bifurcation analysis

Theorem 12. *If $T^2 - 4\Delta < 0$, $T > 0$, and when fractional-order α crosses the critical value $\alpha_0 = \frac{2}{\pi}\tan^{-1}|\frac{\sqrt{4\Delta-T^2}}{T}|$, the given fractional-order system shows Hopf bifurcation around the coexistence point.*

Proof. The eigenvalues of the system (24.3) at coexistence point are given by $\lambda_i = \frac{T \pm i\sqrt{4\Delta-T^2}}{2} = \frac{T}{2} \pm i\frac{\sqrt{4\Delta-T^2}}{2} = u \pm iv$. Here $u = \frac{T}{2} > 0$, and $v = \frac{\sqrt{4\Delta-T^2}}{2}$.

Therefore, we have two eigenvalues which are complex conjugates of each other with positive real parts. Now min $|\arg(\lambda_{1,2})| = \tan^{-1}(\frac{v}{u})$. Now consider a function

$$F(\alpha) = \frac{\pi\alpha}{2} - \min|\arg(\lambda_{1,2})| = \frac{\pi\alpha}{2} - \tan^{-1}\left(\frac{v}{u}\right).$$

Now $F(\alpha) = 0$ gives $\alpha = \frac{2}{\pi}\tan^{-1}(\frac{v}{u})$, and $F'(\frac{2}{\pi}\tan^{-1}(\frac{v}{u})) = \frac{\pi}{2} \neq 0$. Therefore, when α crosses the critical value $\frac{2}{\pi}\tan^{-1}(\frac{v}{u})$, the given fractional-order system will show Hopf bifurcation near the coexistence point. □

6 Numerical analysis and discussion

In this chapter, we have analyzed the fractional-order Lotka–Volterra model with logistic growth and additive Allee effect in prey population. For numerical simulations, we have used the Adam–Bashforth–Moulton predictor-corrector method [19] and PECE scheme for fractional differential equation developed by Roberto Garappa [24] on MATLAB, to validate the theoretical results obtained and to compare the integer order model with fractional-order model under the weak Allee effect and strong Allee effect.

For numerical simulations, we have taken the set of parameters as,

$$K = 1, \quad r = 1, \quad a = 0.2, \quad c = 0.5, \quad e = 0.1.$$

From this, we get values of $ar = 0.2$ and $\frac{Kr}{4}(1 + \frac{a}{K})^2 = 0.36$.

For first simulation, by selecting $f = 2$, $m = 0.1$ the prey population is subjected to the weak Allee effect ($0 < m < ar$). The coexistence point obtained is $(0.1, 0.3)$. Values obtained of $T = \frac{1}{90} > 0$, $\Delta = \frac{17}{300}$, $T^2 - 4\Delta = -\frac{1835}{8100} < 0$, $\frac{2}{\pi}\tan^{-1}|\frac{\sqrt{4\Delta - T^2}}{T}| = 0.985$.

Figures 24.1, 24.2, and 24.3 are plotted by taking $\alpha = 0.98, 0.99$, and integral value 1, respectively, with initial populations (0.2, 0.5). It is clear that when value of α is less than its critical value of 0.985, the populations stabilize. As value of α crosses 0.985, populations start fluctuating and hopf bifurcation starts, and in the integer order model size of fluctuations increases.

For next simulation, the level of predation is decreased to $f = 1$. The value of the coexistence point was found to be (0.2, 0.55). Values obtained of $T = -\frac{3}{40} > 0$, $\Delta = 0.055$, $T^2 - 4\Delta = -0.214375 < 0$. Therefore, all stability conditions are satisfied. Figure 24.4 is showing that populations are stable at coexistence points in integer order as well as in fractional-order models. In lower-order models, populations stabilize faster.

In Fig. 24.5, we have considered $f = 0.5$, the value of the coexistence point found to be (0.4, 0.87). Values obtained of $T = -\frac{13}{45} < 0$, $\Delta = \frac{13}{300}$, $T^2 - 4\Delta = -0.0898 < 0$. Therefore, all stability conditions are satisfied. Figures are showing that populations are stable at coexistence points in integer order as well as in fractional-order models.

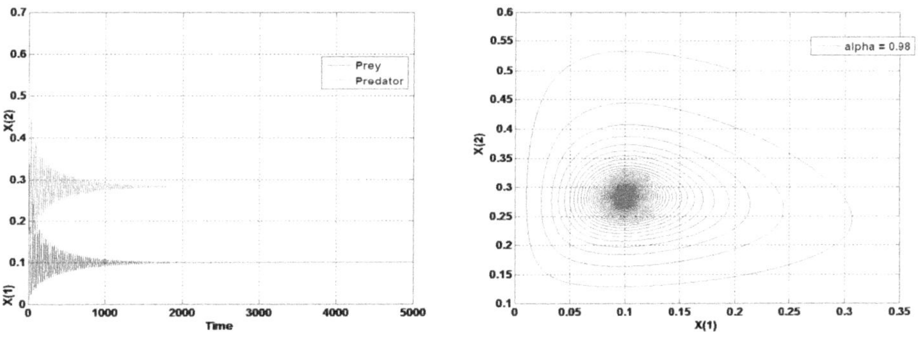

Figure 24.1: Stability analysis of the coexistence point (0.1, 0.3) in fractional model with fractional-order alpha = 0.98, when prey population is sublected to weak Allee effect ($m = 0.1$), and rate of predation as $f = 2$.

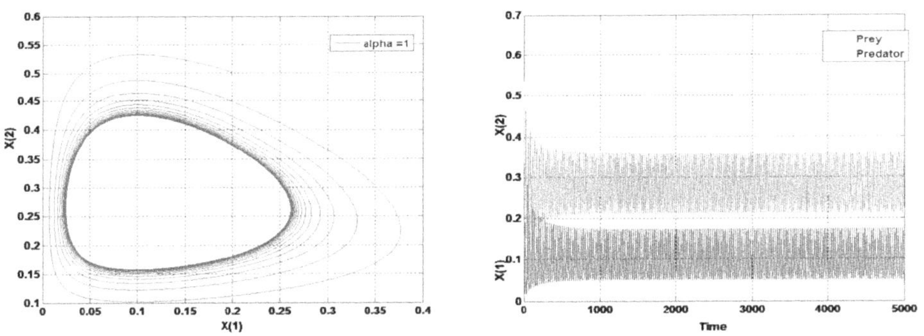

Figure 24.2: Stability analysis of the coexistence point (0.1, 0.3) in fractional model with fractional-order alpha = 0.99, when prey population is sublected to weak Allee effect ($m = 0.1$), and rate of predation as $f = 2$.

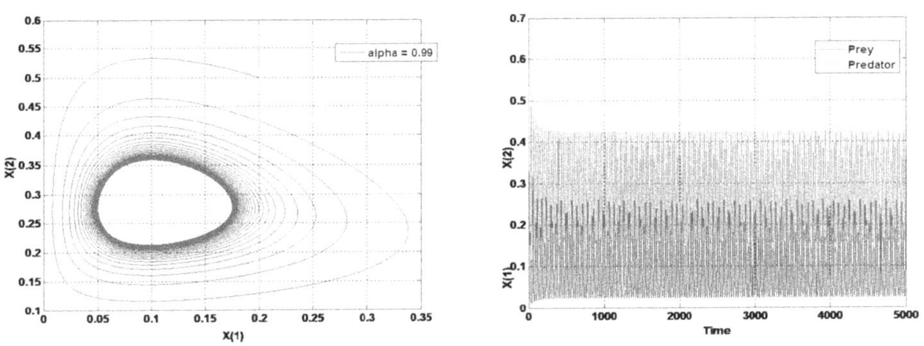

Figure 24.3: Stability analysis of the coexistence point (0.1, 0.3) in the integer order model, when prey population is sublected to weak Allee effect ($m = 0.1$) and rate of predation as $f = 2$.

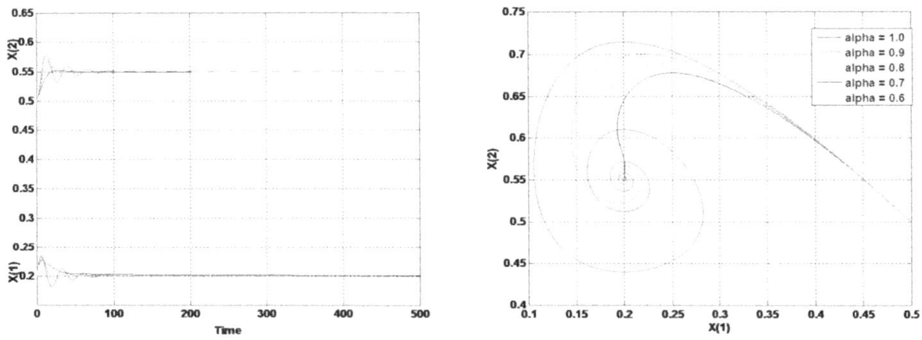

Figure 24.4: Comparison of integer order model with fractional-order model with weak Allee effect (m = 0.1), and rate of predation f = 1.

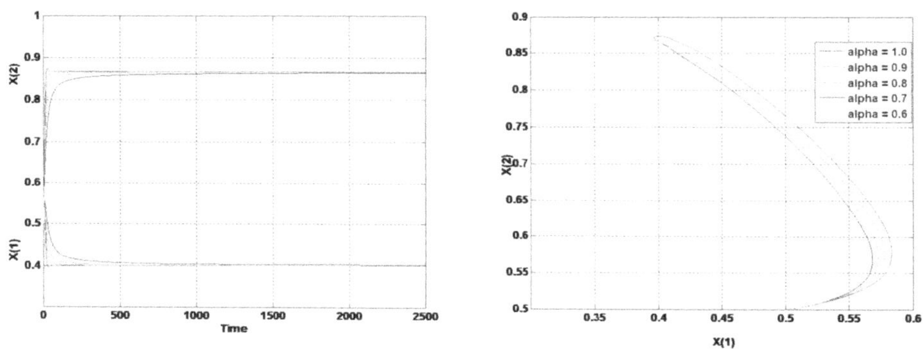

Figure 24.5: Comparison of integer-order model with fractional-order model with weak Allee effect (m = 0.1), and rate of predation f = 0.5.

For Fig. 24.6, considering $f = 0.1$, we get $ar - \frac{re}{fc}(\frac{e}{Kfc} + \frac{a}{K} - 1) = -271\backslash 45$. As $m = 0.1$, $(ar - \frac{re}{fc}(\frac{e}{Kfc} + \frac{a}{K} - 1) < m < ar)$, therefore, all conditions of Theorem 7 are satisfied and we have

$$x_1' = \frac{K}{2}\left[\left(1 - \frac{a}{K}\right) + \sqrt{\left(1 + \frac{a}{K}\right)^2 - \frac{4m}{Kr}}\right] = 0.91.$$

Figure 24.6 is plotted by taking the initial population (0, 5, 0.5) and different fractional order and integer order models, showing that the point (0.91, 0) is locally asymptotically stable and Fig. 24.7 is drawn by taking alpha = 0.9 and different initial populations establishing the global stability of point (0.91, 0) in fractional model.

In Fig. 24.8, $K = 1, r = 1, a = 0.2, c = 0.5, e = 0.1, f = 2, m = 0.25$ the prey population is subjected to the strong Allee effect ($ar < m < \frac{Kr}{4}(1 + \frac{a}{K})^2$). The coexistence point found to be (0.1, 0.03). Values obtained of $T = -\frac{13}{180} < 0$, $\Delta = \frac{1}{150}$, $T^2 - 4\Delta = -0.0214 < 0$.

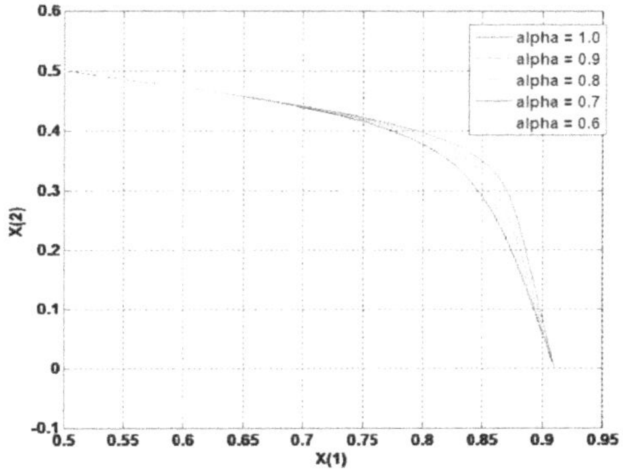

Figure 24.6: Stability analysis of the predator-free point (0.91, 0) in fractional model and integer-order model.

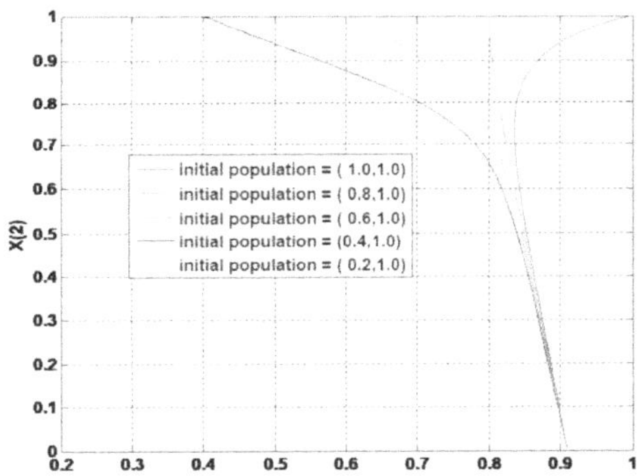

Figure 24.7: Stability analysis of the predator-free point (0.91, 0) in fractional model with fractional order (alpha) = 0.9 and different initial populations.

Though this point satisfies all conditions of stability, due to the high predation rate and the strong Allee effect populations fail to coexist. This shows that populations subjected to the strong Allee effect need a threshold level to survive.

In next simulation, the level of predation is decreased to $f = 1$. The value of the coexistence point found to be $(0.2, 0.175)$. Values obtained of $T = 0.1125 > 0$, $\Delta = 0.0175$, $T^2 - 4\Delta = -0.05734375 < 0$, $\frac{2}{\pi}\tan^{-1}|\frac{\sqrt{4\Delta-T^2}}{T}| = 0.7204$. Figure 24.9 is showing hopf bifurcation around the coexistence point when the initial population is $(0.5, 0.2)$, as alpha takes a value of 0.73 crossing the critical level of 0.7204. However, if we take the initial population of $(0.2, 0.2)$, as shown in Fig. 24.10, the system is collapsed, which shows the requirement of the threshold level of population for survival when subjected to the strong Allee effect. In Fig. 24.11, by taking alpha $= 0.7$ and initial population $(0.5, 0.2)$ system is stabilizing.

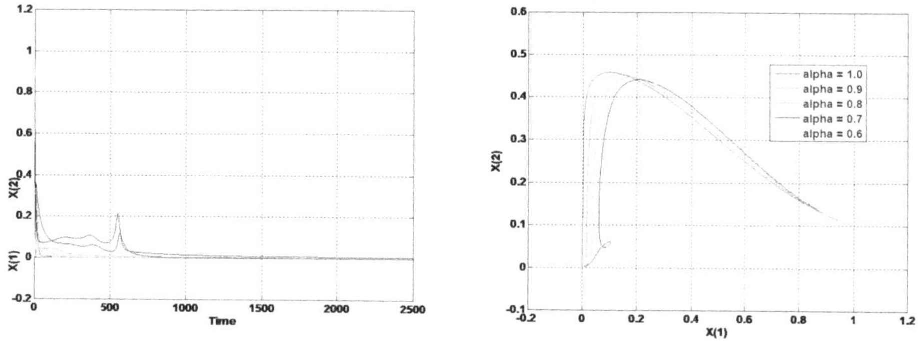

Figure 24.8: Analysis of the fractional order as well as integer-order model, when prey population is subjected to a strong Allee effect $m = 0.25$, with rate of predation $f = 2$.

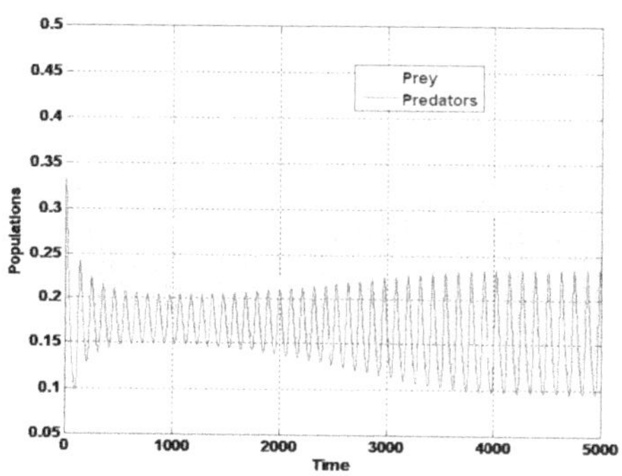

Figure 24.9: Stability analysis of the coexistence point (0.2, 0.175) in fractional-order model by taking alpha $= 0.73$ greater than critical value of 0.7204, when prey population is subjected to strong Allee effect ($m = 0.25$), rate of predation as $f = 1$, and initial population $= (0.5, 0.2)$.

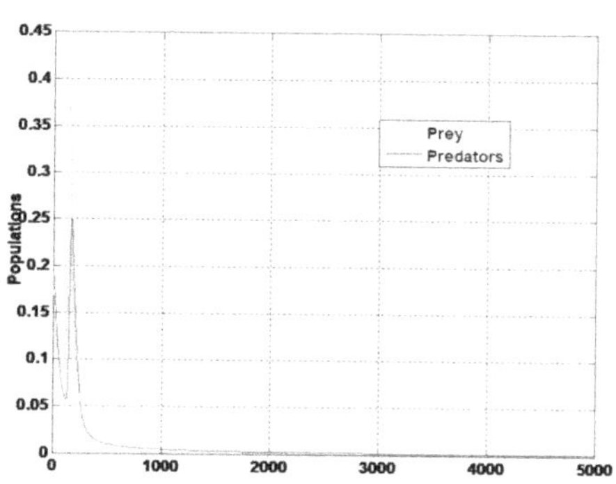

Figure 24.10: Stability analysis of the coexistence point (0.2, 0.175) in fractional-order model by taking alpha $=$ 0.73 greater than critical value of 0.7204, when prey population is subjected to strong Allee effect ($m = 0.25$), rate of predation as $f = 1$, and initial population $= (0.5, 0.2)$.

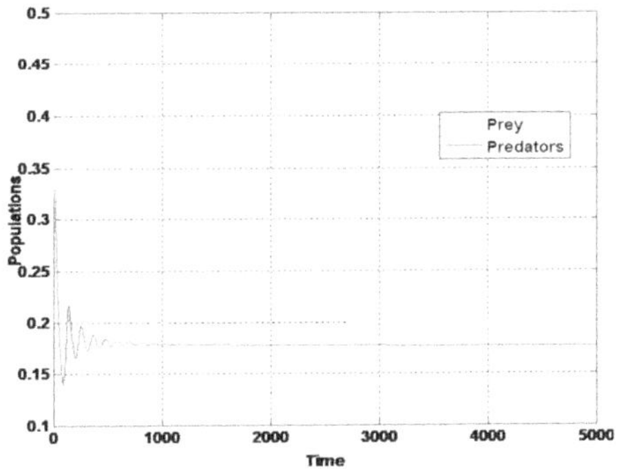

Figure 24.11: Stability analysis of the coexistence point (0.2, 0.175) in fractional-order model by taking alpha = 0.7 less than critical value of 0.7204, when prey population is subjected to strong Allee effect (m = 0.25), rate of predation as f = 1, and initial population = (0.5, 0.2).

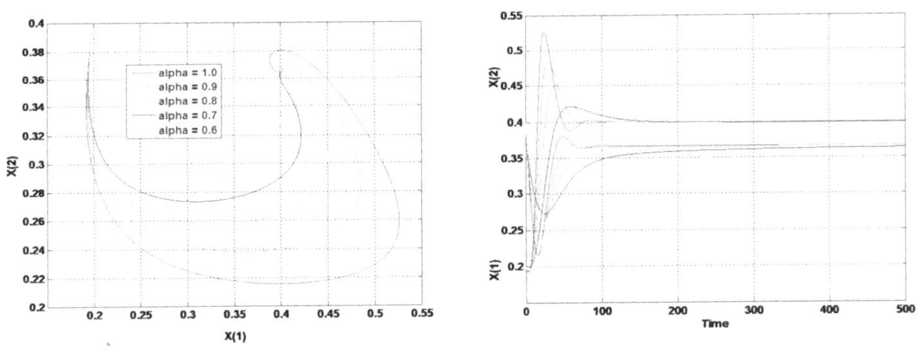

Figure 24.12: Stability analysis of the coexistence point (0.4, 0.37) in fractional order as well as integer-order model, when prey population is subjected to a strong Allee effect m = 0.25, with rate of predation f = 0.5.

In Fig. 24.12, considering $f = 0.5$, the value of the coexistence point was found to be (0.4, 0.37). Values obtained of $T = -\frac{1}{15} < 0$, $\Delta = \frac{11}{600}$, $T^2 - 4\Delta = -61\backslash 900 < 0$. Therefore, all stability conditions are satisfied. Figure 24.12 is showing that populations are stable at coexistence points in integer order as well as in fractional-order models.

In next simulations, analysis is done in fractional-order model by varying the rate of predation, when prey population is subjected to weak and strong Allee effect.

For numerical simulations, we have taken the set of parameters as

$$K = 1, \quad r = 1, \quad a = 0.2, \quad c = 0.5, \quad e = 0.1, \quad \alpha = 0.9.$$

From this data, we get values of $ar = 0.2$, and $\frac{Kr}{4}(1 + \frac{a}{K})^2 = 0.36$.

By selecting $m = 0.15$, the prey population is subjected to a weak Allee effect. Figure 24.13 is plotted by taking the value of f as 0.19. It has been seen that the predator-free

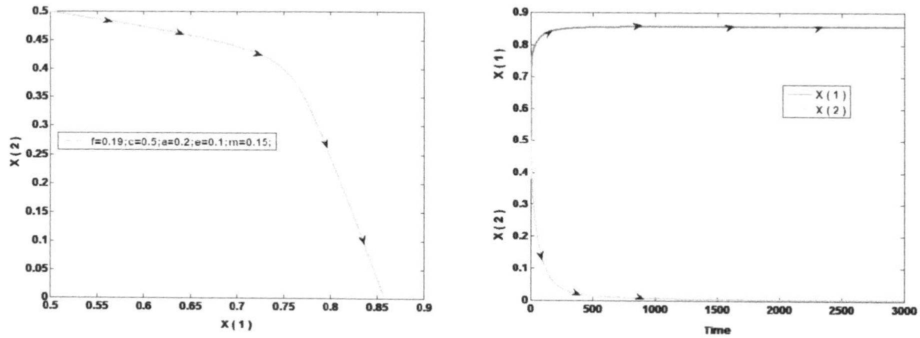

Figure 24.13: Populations converging to predator-free point, when prey population is subjected to weak Allee effect.

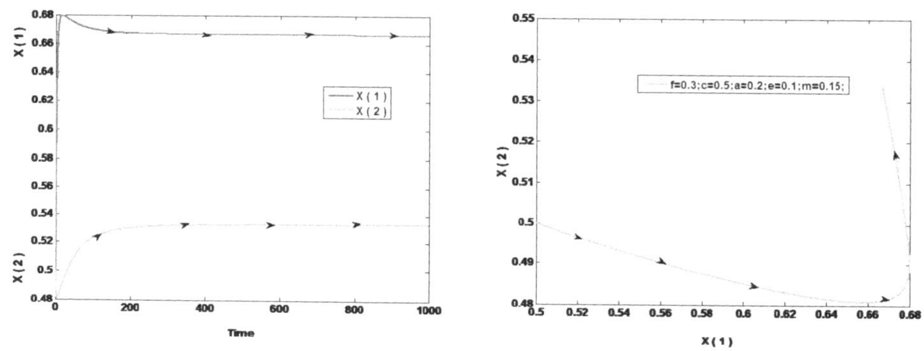

Figure 24.14: Populations are converging to coexistence point without fluctuating, when subjected to the weak Allee effect.

point is asymptotically stable. Also it has been observed numerically that predator-free point remains asymptotically stable whenever f remains less than or equal to 0.19.

In the next simulation, the value of f is considered to be 0.3 and it has been found that the system (24.3) exists as coexistence point, as shown in Fig. 24.14. For Fig. 24.15, the value of f is considered 1.5, system (24.3) is found to exists as coexistence point, but it has started fluctuating. Figure 24.16 is plotted by taking $f = 1.9$, the system (24.3) undergoes Hopf bifurcation.

In fact, it is found numerically that when the value f varies from 0.2 to 0.5 the system (24.3) persists as coexistence point which is asymptotically stable, but as f varies from 0.6 to 1.7, the system (24.3) starts oscillating, and stable limit cycles appear around the coexistence point, which are increasing on increasing the value of f, but as f crosses the critical value 1.8, system starts oscillating infinitely and hopf bifurcation appears.

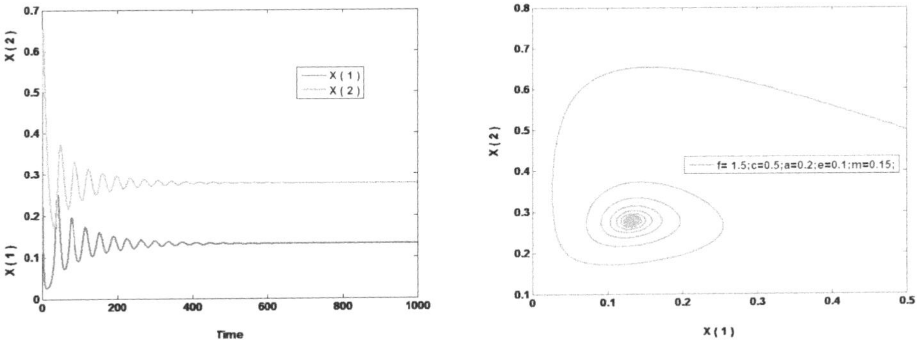

Figure 24.15: Populations are converging to coexistence point, but after fluctuating, when subjected to weak Allee effect.

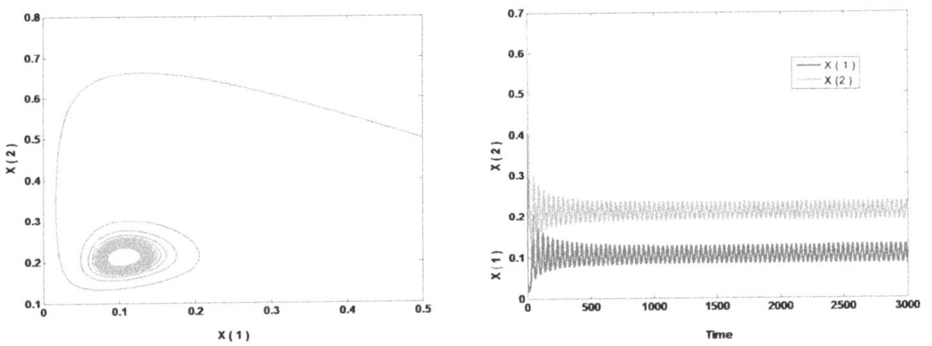

Figure 24.16: Populations undergoes Hopf bifurcation, when $f = 1.9$ and subjected to weak Allee effect.

For the next simulation, by selecting $m = 0.25$, the prey population is subjected to the strong Allee effect. Figure 24.17 is plotted by taking $f = 0.2$, and the system (24.3) exists as predator free asymptotically stable. Figure 24.18 is plotted by taking $f = 0.3$, the system (24.3) exists as coexistence point that is asymptotically stable.

Figure 24.19 is plotted by taking $f = 0.59$, and it is observed that system (24.3) move towards the coexistence equilibrium point after fluctuating. In Fig. 24.20, value of f is taken 0.6 and it is observed that if initial predator population is considered less than or equal to 0.4 with the prey's initial population as 0.5 the system persists as a stable coexistence point, but if the predator's initial population is considered more than or equal to 0.5, whenever the prey initial population is 0.5, the extinction point becomes stable. It has been seen numerically that when the value of f remains less than or equal to 0.25, the system persists with a predator-free point as asymptotically stable. As f takes values from 0.26 to 0.59, the system persists as a coexistence point that is asymptotically stable, however, very small stable limit cycles appear when f crosses 0.4, but as f takes

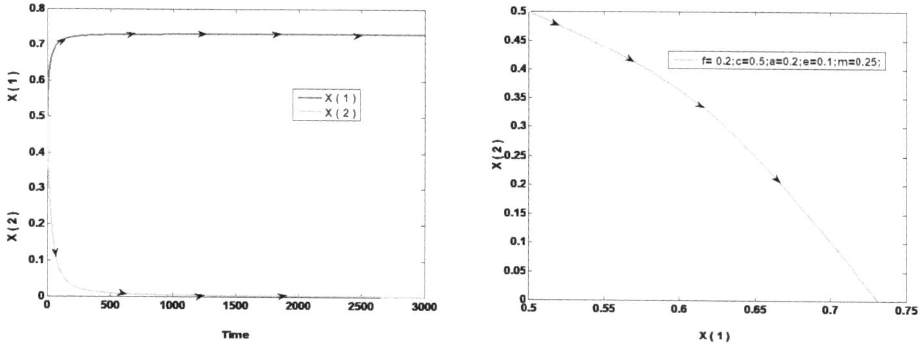

Figure 24.17: Populations are converging to predator-free point, when subjected to strong Allee effect.

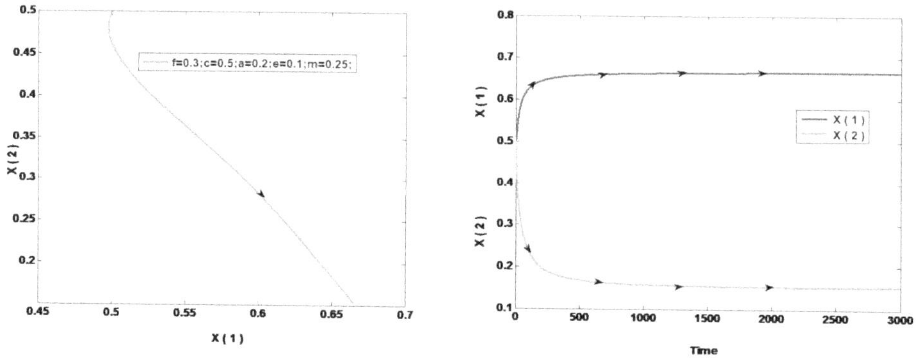

Figure 24.18: Populations are converging to coexistence point, when subjected to the strong Allee effect.

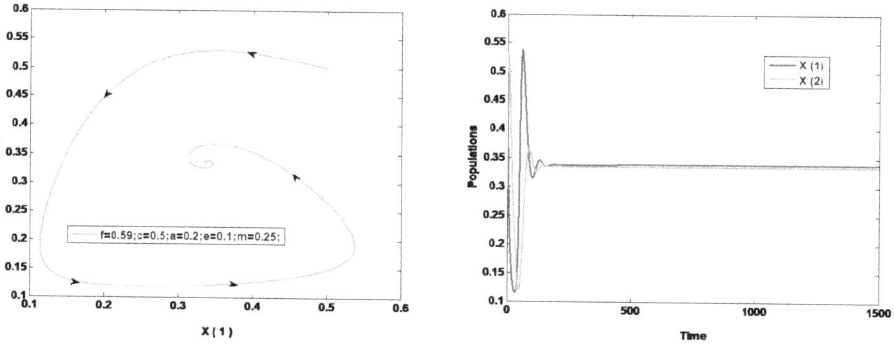

Figure 24.19: Populations are converging to the coexistence point after fluctuating, when subjected to the strong Allee effect.

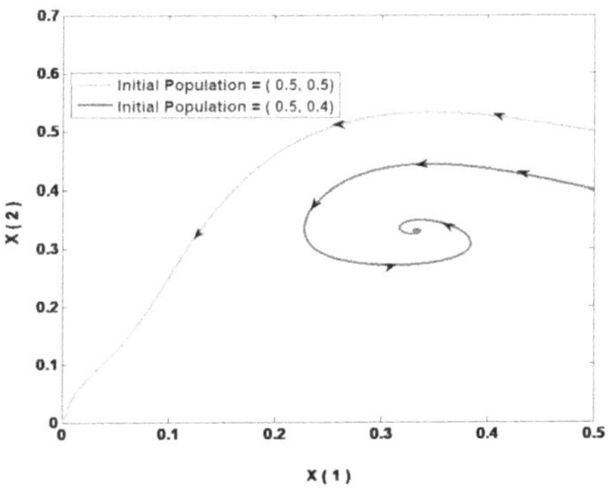

Figure 24.20: Populations moving towards the two different equilibrium points under the influence of strong Alee effect.

values slightly more than or equal to 0.6, a threshold level of prey populations for given initial predator population is required for the existence of both populations.

7 Conclusion

The present study is based on the Lotka–Volterra prey–predator model with logistic growth and additive Allee effect on the prey population. In this study, fractional derivative in the Caputo sense is used. Theorems like existence and uniqueness, and boundedness of the solutions are derived. Local asymptotical stability of the predator free and interior coexistence equilibrium points have been discussed. It is found that when prey populations subjected to weak Allee effect $0 \leq m < ar$, with different rates of predation $f = 0.1, 0.5, 1, 2$, the system (24.3) exists either as stable predator free or stable coexistence or fluctuating about coexistence point in fractional order as well as in integer-order model. However, when prey population is subjected to the strong Allee effect, $ar < m < \frac{Kr}{4}(1 + \frac{a}{K})^2$ and the value of $f = 2$ is taken system shows requirement of minimum initial prey population to exist. For $f = 1$ under strong Allee effect, system exists as coexistence point whenever α is less than 0.7204, but as α crosses the value 0.7204, system start fluctuating about coexistence point and fails to exist in integer order model. For $f = 0.5$ system shows coexistence in fractional as well as in integer order model. But if Allee effect crosses the level $m > \frac{Kr}{4}(1 + \frac{a}{K})^2$, no initial prey population can help in survival of the both populations. The combination of both the Allee effect and rate of predation gives rise to complex dynamics of populations. The computational effectiveness of the fractional-order model over the integer-order model is also demonstrated. In the integer order model with the introduction of weak Allee effect ($m = 0.1$) with a higher rate of predation ($f = 2$), Hopf bifurcation is shown, whereas in the fractional-order model, a critical value of the fractional order ($\alpha = 0.999$) is obtained, below which populations stabilize always. Under the influence of the strong Allee effect ($m = 0.25$) with different rates of predation ($f = 2, 1$) populations always move towards extinction in the ordinary integer order model, but in the case of the fractional order model, it has been observed that with a suitably chosen rate of predation ($f = 1$), it is possible to find the value fractional order ($\alpha = 0.72$) below which populations can coexist stably whenever the initial prey population is above a threshold level. Hence it is observed that memory plays an important role in stabilizing the populations, as lower fractional order models stabilize faster. Therefore fractional order Lotka,–Volterra model with additive Allee effect shows rich dynamics by providing many stable coexistence points.

Bibliography

[1] E. Ahmed, A. M. A. El-Sayed and H. A. A. El-Saka, Equilibrium points, stability and numerical solutions of fractional-order predator–prey and rabies models, *J. Math. Anal. Appl.* **325** (2007), 542–553. https://doi.org/10.1016/j.jmaa.2006.01.087, ISSN:0022-247X.

[2] H. Al-Sulami, M. El-Shahed, J. J. Nieto and W. Shammakh, On fractional order dengue epidemic model, *Math. Probl. Eng.* **2014**, 456537 (2014). https://doi.org/10.1155/2014/456537, ISSN:1563-5147.

[3] J. Alidousti and E. Ghafari, Dynamic behavior of a fractional order prey-predator model with group defense, *Chaos Solitons Fractals* **134**, 109688 (2020). https://doi.org/10.1016/j.chaos.2020.109688, ISSN:0960-0779.

[4] J. Alidousti and E. Ghafari, A fractional order predator–prey model with strong allee effect and Michaelis–Menten type of predator harvesting, *AIP Conf. Proc.* **2498**, 020018 (2022). https://doi.org/10.1063/5.0082684, ISSN:1551-7616.

[5] W. C. Allee, *Animal Aggregations: A Study in General Sociology*, University of Chicago Press, Chicago, 1931.

[6] A. Atangana, A. Secer, et al., A note on fractional order derivatives and table of fractional derivatives of some special functions, *Abstr. Appl. Anal.* **2013**, 1 (2013). https://doi.org/10.1155/2013/279681, ISSN:0245-2322.

[7] A. D. Bazykin, *Nonlinear Dynamics of Interacting Populations*, World Scientific, 1998.

[8] A. Ben Saad and O. Boubaker, A New fractional-order predator-prey system with Allee effect, *Stud. Comput. Intell.* **688** (2017), 857–877. https://doi.org/10.1007/978-3-319-50249-6, ISBN: 978-3-319-50249-6.

[9] L. Berec, E. Angulo and F. Courchamp, Multiple Allee effects and population management, *Trends Ecol. Evol.* **22** (2007), 185–191. https://doi.org/10.1016/j.tree.2006.12.002, ISSN:0169-5347.

[10] M. A. Bhat and A. Khan, Multi-switching combination synchronization of different fractional-order non-linear dynamical systems, *Int. J. Model. Simul.* **38** (2018), 254–261. https://doi.org/10.1080/02286203.2018.1442988, ISSN:0228-6203.

[11] Y. Chatibi, E. H. El Kinani and A. Ouhadan, Lie symmetry analysis and conservation laws for the time fractional Black–Scholes equation, *Int. J. Geom. Methods Mod. Phys.* **17**, 2050010 (2020). https://doi.org/10.1142/S0219887820500103, ISSN:0219-8878.

[12] R. Chinnathambi and F. A. Rihan, Stability of fractional-order prey–predator system with time-delay and Monod–Haldane functional response, *Nonlinear Dyn.* **92** (2018), 1637–1648. https://doi.org/10.1007/s11071-018-4151-z, ISSN:0924-090X.

[13] A. Coccarelli, J. M. Carson, A. Aggarwal and S. Pant, A framework for incorporating 3D hyperelastic vascular wall models in 1D blood flow simulations, *Biomech. Model. Mechanobiol.* **20** (2021), 1231–1249. https://doi.org/10.1007/s10237-021-01437-5, ISSN:1617-7959.

[14] F. Courchamp, T. Clutton-Brock and B. Grenfell, Inverse density dependence and the Allee effect, *Trends Ecol. Evol.* **14** (1999), 405–410. https://doi.org/10.1016/S0169-5347(99)01683-3, ISSN:0169-5347.

[15] L. Dai, J. Wang, Y. Ni and B. Xu, Dynamical analysis of a new fractional-order predator–prey system with Holling type-III functional, *Adv. Differ. Equ.* **2021**, 1 (2021). https://doi.org/10.1186/s13662-020-03169-9, ISSN:1687-1839.

[16] M. Das and G. P. Samanta, A prey-predator fractional order model with fear effect and group defense, *Int. J. Dyn. Control* **9** (2021), 334–349. https://doi.org/10.1007/s40435-020-00626-x, ISSN:1687-1839.

[17] J. D. Davis, D. V. Olivença, S.. P. Brown and E. O. Voit, Methods of quantifying interactions among populations using Lotka-Volterra models, *Bull. Aust. Math. Soc.* **2**, 1021897 (2022). https://doi.org/10.3389/fsysb.2022.1021897, ISSN:2674-0702.

[18] B. Dennis, Allee effects: population growth, critical density, and the chance of extinction, *Nat. Resour. Model.* **3** (1989), 481–538. https://doi.org/10.1111/j.1939-7445.1989.tb00119.x, ISSN:0890-8575.

[19] K. Diethelm, Efficient solution of multi-term fractional differential equations using P (EC) m E methods, *Computing* **71** (2003), 305–319. https://doi.org/10.1007/s00607-003-0033-3, ISSN:0890-8575.

[20] M. Du, Z. Wang and H. Hu, Measuring memory with the order of fractional derivative, *Sci. Rep.* **3** (2013), 3431. https://doi.org/10.1038/srep03431, ISSN:0245-2322.

[21] A. El-Mesady, O. Bazighifan, S. Araci, et al., Insight into Caputo fractional-order extension of Lotka–Volterra model with emphasis on immigration effect, *J. Math.* **2023**, 8795660 (2023). https://doi.org/10.1155/2023/8795660, ISSN:2147-5520.

[22] A. Elsonbaty and A. A. Elsadany, On discrete fractional-order Lotka-Volterra model based on the Caputo difference discrete operator, *Math. Sci.* **17** (2023), 67–79. https://doi.org/10.1007/s40096-021-00442-0, ISSN:2147-5520.

[23] Y. Fan, X. Huang, Z. Wang and Y. Li, Nonlinear dynamics and chaos in a simplified memristor-based fractional-order neural network with discontinuous memductance function, *Nonlinear Dyn.* **93** (2018), 611–627. https://doi.org/10.1007/s11071-018-4213-2, ISSN:0924-090X.

[24] R. Garrappa, Predictor-corrector PECE method for fractional differential equations. MATLAB Central File Exchange, 2011, Available at https://in.mathworks.com/matlabcentral/fileexchange/32918-predictor-corrector-pece-method-for-fractional-differential.

[25] G. González-Parra, A. J. Arenas and B. M. Chen-Charpentier, A fractional order epidemic model for the simulation of outbreaks of influenza A (H1N1), *Math. Methods Appl. Sci.* **37** (2014), 2218–2226. https://doi.org/10.1002/mma.2968, ISSN:0170-4214.

[26] D. Guha, P. K. Roy and S. Banerjee, Grasshopper optimization algorithm scaled fractional order PI-D controller applied to reduced order model of load frequency control system, *Int. J. Model. Simul.* **40** (2020), 217–242. https://doi.org/10.1080/02286203.2019.1596727, ISSN:0228-6203.

[27] F. C. Hoppensteadt, *Mathematical Methods of Population Biology*, Cambridge University Press, 1982.

[28] X. Huang, Y. Fan, J. Jia, Z. Wang and Y. Li, Quasi-synchronisation of fractional-order memristor-based neural networks with parameter mismatches, *IET Control Theory Appl.* **11** (2017), 2317–2327. https://doi.org/10.1049/iet-cta.2017.0196, ISSN:1751-8644.

[29] J. A. Hutchings, Thresholds for impaired species recovery, *Proc. R. Soc. B.* **282**, 20150654 (2015). https://doi.org/10.1098/rspb.2015.0654, ISSN:1471-2954.

[30] O. S. Iyiola and F. D. Zaman, A fractional diffusion equation model for cancer tumor, *AIP Adv.* **4**, 107121 (2014). https://doi.org/10.1063/1.4898331, ISSN:2158-3226.

[31] J. Jacobs, Cooperation, optimal density and low density thresholds: yet another modification of the logistic model, *Oecologia* **64** (1984), 389–395. https://doi.org/10.1007/BF00379138, ISSN:0029-8549.

[32] M. Javidi and N. Nyamoradi, Dynamic analysis of a fractional order prey–predator interaction with harvesting, *Appl. Math. Model.* **37** (2018), 8946–8956. https://doi.org/10.1016/j.apm.2013.04.024, ISSN:0307-904X.

[33] D. M. Johnson, A. M. Liebhold, P. C. Tobin and O. N. Bjørnstad, Allee effects and pulsed invasion by the gypsy moth, *Nature* **444** (2006), 361–363. https://doi.org/10.1038/nature05242, ISSN:2193-1801.

[34] M. A. Khan, A. Khan, A. Elsonbaty and A. A. Elsadany, Modeling and simulation results of a fractional dengue model, *Eur. Phys. J. Plus* **44** (2019), 379. https://doi.org/10.1140/epjp/i2019-12765-0, ISSN:2190-5444.

[35] S. Kumar, R. Kumar, R. P.Agarwal and B. Samet, A study of fractional Lotka-Volterra population model using Haar wavelet and Adams-Bashforth-Moulton methods, *Math. Methods Appl. Sci.* **43** (2020), 5564–5578. https://doi.org/10.1002/mma.6297, ISSN:0170-4214.

[36] H.-L. Li, A. Muhammadhaji, L. Zhang and Z. Teng, Stability analysis of a fractional-order predator–prey model incorporating a constant prey refuge and feedback control, *Adv. Differ. Equ.* **2018**, 1 (2018). https://doi.org/10.1186/s13662-018-1776-7, ISSN:1687-1839.

[37] A. J.Lotka, *Elements of Physical Biology*, Williams and Wilkins, 1925.

[38] M. Mandal, S. Jana, S. K. Nandi and T. K. Kar, Modeling and analysis of a fractional-order prey-predator system incorporating harvesting, *Model. Earth Syst. Environ.* **7** (2021), 1159–1176. https://doi.org/10.1007/s40808-020-00970-z, ISSN:2363-6211.

[39] H. Merdan, Stability analysis of a Lotka–Volterra type predator–prey system involving Allee effects, *ANZIAM J.* **52** (2010), 139–145. https://doi.org/10.1017/S1446181111000630, ISSN:1446-1811.

[40] I. Petráš, *Fractional-Order Nonlinear Systems: Modeling, Analysis and Simulation*, Springer, 2011.

[41] I. Podlubny, Fractional-order systems and PI/sup/spl lambda//D/sup/spl mu//-controllers, *IEEE Trans. Autom.* **44** (1999), 208–214. https://doi.org/10.1109/9.739144, ISSN:1558-2523.

[42] Y. Pu, W. Wang, J. Zhou, Y. Wang and H. Jia, Fractional differential approach to detecting textural features of digital image and its fractional differential filter implementation, *Sci. China Inf. Sci.* **51** (2008), 1319–1339. https://doi.org/10.1007/s11432-008-0098-x, ISSN:1009-2757.

[43] P. Song, H. Zhao and X. Zhang, Dynamic analysis of a fractional order delayed predator–prey system with harvesting, *Theory Biosci.* **135** (2016), 59–72. https://doi.org/10.1007/s12064-016-0223-0, ISSN:1611-7530.

[44] Q. Song, X. Dong, Z. Bai, B. Chen, et al., Existence for fractional Dirichlet boundary value problem under barrier strip conditions, *J. Nonlinear Sci. Appl.* **10** (2017), 3592–3598. https://doi.org/10.22436/jnsa.010.07.19, ISSN:2008-1898.

[45] P. A. Stephens and W. J. Sutherland, Consequences of the Allee effect for behaviour, ecology and conservation, *Trends Ecol. Evol.* **14** (1999), 401–405. https://doi.org/10.1016/S0169-5347(99)01684-5, ISSN:0169-5347.

[46] A. Suryanto, I. Darti, S. Panigoro Hasan and A. Kilicman, A fractional-order predator–prey model with ratio-dependent functional response and linear harvesting, *Mathematics* **7** (2019), 1100. https://doi.org/10.3390/math7111100, ISSN:2227-7390.

[47] B. Tang, Dynamics for a fractional-order predator-prey model with group defense, *Sci. Rep.* **10** (2020), 4906. https://doi.org/10.1038/s41598-020-61468-3, ISSN:2045-2322.

[48] P. C. Tobin, L. Berec and A. M. Liebhold, Exploiting Allee effects for managing biological invasions, *Ecol. Lett.* **14** (2011), 615–624. https://doi.org/10.1111/j.1461-0248.2011.01614.x, ISSN:1461-0248.

[49] S. Ullah, M. A. Khan, M. Farooq, A fractional model for the dynamics of TB virus, *Chaos Solitons Fractals* **116** (2018), 63–71. https://doi.org/10.1016/j.chaos.2018.09.001, ISSN:0960-0779.

[50] V. Volterra, *Variazioni e fluttuazioni del numero d'individui in specie animali conviventi*, Societá anonima tipografica "Leonardo da Vinci" 1927.

[51] M.-H. Wang and M. Kot, Speeds of invasion in a model with strong or weak Allee effects, *Math. Biosci.* **171** (2001), 83–97. https://doi.org/10.1016/S0025-5564(01)00048-7, ISSN:0025-5564.

[52] B. Xie and Z. Zhang, Impact of Allee and fear effects in a fractional order prey–predator system incorporating prey refuge, *Chaos* **33**, 013131 (2023). https://doi.org/10.1063/5.0130809, ISSN:1551-7616.

[53] Y. Xie, J. Lu and Z. Wang, Stability analysis of a fractional-order diffused prey–predator model with prey refuges, *Phys. A: Stat. Mech.* **526**, 120773 (2019). https://doi.org/10.1016/j.physa.2019.04.009, ISSN:0378-4371.

[54] Y. Xie, Z. Wang, B. Meng and X. Huang, Dynamical analysis for a fractional-order prey–predator model with Holling III type functional response and discontinuous harvest, *Appl. Math. Lett.* **106**, 106342 (2020). https://doi.org/10.1016/j.aml.2020.106342, ISSN:0893-9659.

[55] M. Yavuz and N. Sene, Stability analysis and numerical computation of the fractional predator–prey model with the harvesting rate, *Chaos, Fractal Fract.* **4** (2020), 35. https://doi.org/10.3390/fractalfract4030035, ISSN:2504-3110.

[56] H. Zhang, Y. Cai, S. Fu and W. Wang, Impact of the fear effect in a prey-predator model incorporating a prey refuge, *Appl. Math. Comput.* **356** (2019), 328–337. https://doi.org/10.1016/j.amc.2019.03.034, ISSN:0096-3003.

[57] X.-Q. Zhao, *Dynamical Systems in Population Biology*, Springer, 2003.

Gautham B. Shenoy, A. Ragavaendra, S. B. Chandrakala, Badekara Sooryanarayana, and Muhammad Kamran Siddiqui

Topological indices of two chemical compounds used for drugs in the treatment of H1N1

Abstract: In this paper, we find several topological polynomials and indices of the molecular graph of the chemicals zanamivir and oseltamivir. These drugs are used for the treatment of diseases caused by H1N1. The topological indices depend on the degree of vertices. They will be indicators for the stability of the structures. H1N1 is a deadly virus. It requires a serious attempt to cure or control, the diseases, caused by such viruses. Our work is going to help the researchers to find the properties of the chemical required in all aspects. To predict the properties, we need statistically and numerically developed models.

Keywords: Topological polynomials, topological indices, zanamivir , oseltamivir , H1N1

MSC 2010: 05C12

1 Introduction

In recent years, scientific study has shifted its focus to the causes and treatments of dreaded diseases. In the past few years, infectious diseases, their causes, and treatments have become the focus of research. In May–June 2009, scientists discovered a new virus — influenza A(H1N1). This virus was first found in the United States, then spread to other parts of the world. This novel H1N1 virus had a rare combination of influenza genes that had never been seen before in a creature or a human. In the first year alone, the virus has caused nearly three lakh deaths, directly or by associated causes. The incidence of nar-

Gautham B. Shenoy, Department of Mathematics, Poornaprajna College, 1-102, Shantheri, Sudhindra Nagar road, Belman Post, Karkala Taluku, 576111 Udupi, Karnataka, India, e-mail: gautham.shenoy1996@gmail.com
A. Ragavaendra, Department of Mathematics, Poornaprajna College, Volakkadu road, 576101 Udupi, Karnataka, India, e-mail: raghavendra.bhat.a@gmail.com
S. B. Chandrakala, Department of Mathematics, Nitte Meenakshi Institute of Technology, C1, Rishab Enclave, 2nd cross, 1st Main, Patelappa Lyt, Nagashettyhalli, 2nd stage RMV, 560094 Bengaluru, Karnataka, India, e-mail: chandrakalasb14@gmail.com
Badekara Sooryanarayana, Department of Mathematics, Dr. Ambedkar Institute of Technology, #83, Sumkukha, 17th Cross, 4th Block, 4th Stage, Behind Rehjija Park, Basaveshwara Nagara, 560079 Bengaluru, Karnataka State, India, e-mail: dr_bsnrao@yahoo.co.in
Muhammad Kamran Siddiqui, Department of Mathematics, COMSATS University Islamabad, Lahore Campus, Lahore, Pakistan, e-mail: kamransiddiqui@cuilahore.edu.pk

https://doi.org/10.1515/9783111313634-025

colepsy increased dramatically after vaccination against the influenza A(H1N1)pdm09 virus. The influenza A(H1N1)pdm09 virus was treated with a variety of vaccinations. Adamantanes (amantadine and rimantadine) and neuraminidase inhibitors (zanamivir, peramivir and oseltamivir) and adamantanes (rimantadine and amantadine) are the two antiviral drugs, which are widely used to treat influenza in humans. Because of the extensive use of these antiviral medications as monotherapies, influenza viruses resistant to both treatment classes have emerged [17–19, 24].

Chemical structure of zanamivir	**Chemical structure of oseltamivir**
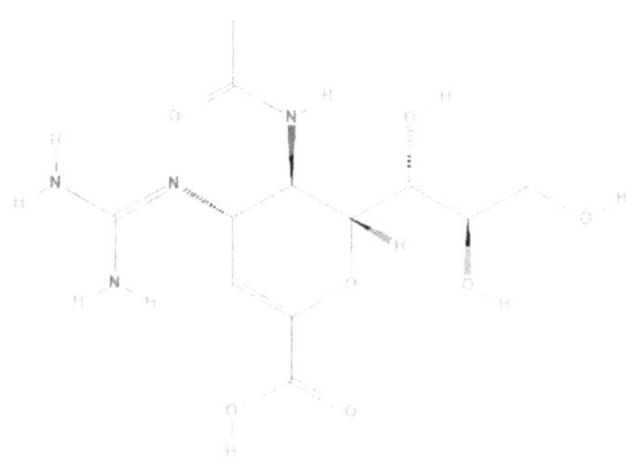	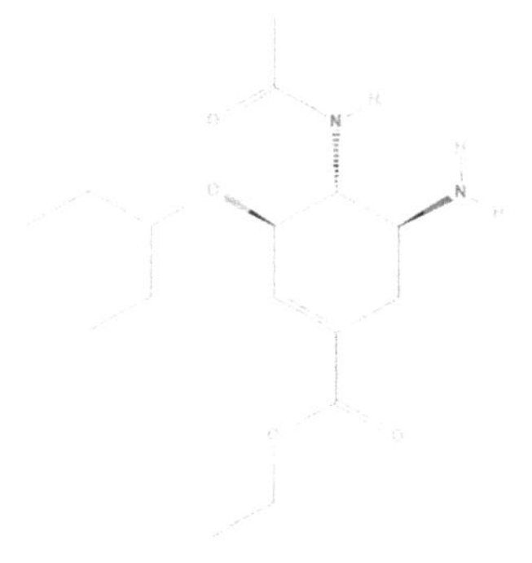

Without using quantum physics, mathematical tools such as polynomials and numbers are used to predict the properties of chemical compounds in mathematical chemistry. Degree-based topological indices are the most well-known invariants of this type, which correlate the structure with different physical properties, chemical reactivity, and biological activities of the chemical compounds, such as strain energy, fracture toughness, boiling point and rigidity are strongly linked to its graphical structure, which plays a vital role in chemical graph theory. Some of the degree-based topological descriptors are defined in Table 25.1 and were computed for molecular structure of oseltamivir and zanamivir.

For a graph G, let $V(G)$ and $E(G)$ denotes vertex and edge set respectively. For a vertex $v \in V(G)$, d_v denotes the degree of vertex v.

Gautham B. Shenoy, A. Ragavaendra, S. B. Chandrakala, Badekara Sooryanarayana, and Muhammad Kamran Siddiqui

Topological indices of two chemical compounds used for drugs in the treatment of H1N1

Abstract: In this paper, we find several topological polynomials and indices of the molecular graph of the chemicals zanamivir and oseltamivir. These drugs are used for the treatment of diseases caused by H1N1. The topological indices depend on the degree of vertices. They will be indicators for the stability of the structures. H1N1 is a deadly virus. It requires a serious attempt to cure or control, the diseases, caused by such viruses. Our work is going to help the researchers to find the properties of the chemical required in all aspects. To predict the properties, we need statistically and numerically developed models.

Keywords: Topological polynomials, topological indices, zanamivir , oseltamivir , H1N1

MSC 2010: 05C12

1 Introduction

In recent years, scientific study has shifted its focus to the causes and treatments of dreaded diseases. In the past few years, infectious diseases, their causes, and treatments have become the focus of research. In May–June 2009, scientists discovered a new virus — influenza A(H1N1). This virus was first found in the United States, then spread to other parts of the world. This novel H1N1 virus had a rare combination of influenza genes that had never been seen before in a creature or a human. In the first year alone, the virus has caused nearly three lakh deaths, directly or by associated causes. The incidence of nar-

Gautham B. Shenoy, Department of Mathematics, Poornaprajna College, 1-102, Shantheri, Sudhindra Nagar road, Belman Post, Karkala Taluku, 576111 Udupi, Karnataka, India, e-mail: gautham.shenoy1996@gmail.com
A. Ragavaendra, Department of Mathematics, Poornaprajna College, Volakkadu road, 576101 Udupi, Karnataka, India, e-mail: raghavendra.bhat.a@gmail.com
S. B. Chandrakala, Department of Mathematics, Nitte Meenakshi Institute of Technology, C1, Rishab Enclave, 2nd cross, 1st Main, Patelappa Lyt, Nagashettyhalli, 2nd stage RMV, 560094 Bengaluru, Karnataka, India, e-mail: chandrakalasb14@gmail.com
Badekara Sooryanarayana, Department of Mathematics, Dr. Ambedkar Institute of Technology, #83, Sumkukha, 17th Cross, 4th Block, 4th Stage, Behind Rehjija Park, Basaveshwara Nagara, 560079 Bengaluru, Karnataka State, India, e-mail: dr_bsnrao@yahoo.co.in
Muhammad Kamran Siddiqui, Department of Mathematics, COMSATS University Islamabad, Lahore Campus, Lahore, Pakistan, e-mail: kamransiddiqui@cuilahore.edu.pk

https://doi.org/10.1515/9783111313634-025

colepsy increased dramatically after vaccination against the influenza A(H1N1)pdm09 virus. The influenza A(H1N1)pdm09 virus was treated with a variety of vaccinations. Adamantanes (amantadine and rimantadine) and neuraminidase inhibitors (zanamivir, peramivir and oseltamivir) and adamantanes (rimantadine and amantadine) are the two antiviral drugs, which are widely used to treat influenza in humans. Because of the extensive use of these antiviral medications as monotherapies, influenza viruses resistant to both treatment classes have emerged [17–19, 24].

Chemical structure of zanamivir	Chemical structure of oseltamivir
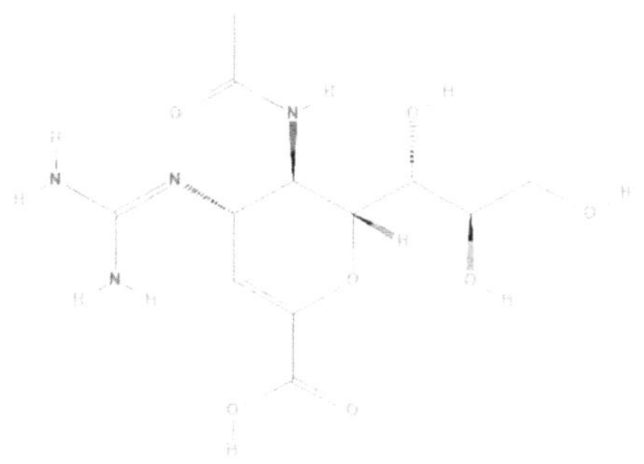	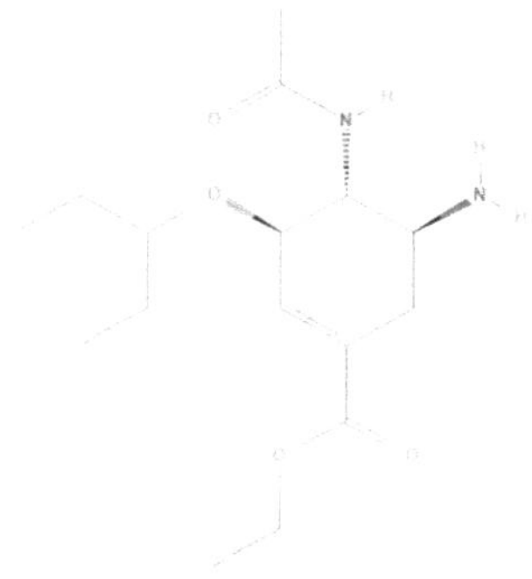

Without using quantum physics, mathematical tools such as polynomials and numbers are used to predict the properties of chemical compounds in mathematical chemistry. Degree-based topological indices are the most well-known invariants of this type, which correlate the structure with different physical properties, chemical reactivity, and biological activities of the chemical compounds, such as strain energy, fracture toughness, boiling point and rigidity are strongly linked to its graphical structure, which plays a vital role in chemical graph theory. Some of the degree-based topological descriptors are defined in Table 25.1 and were computed for molecular structure of oseltamivir and zanamivir.

For a graph G, let $V(G)$ and $E(G)$ denotes vertex and edge set respectively. For a vertex $v \in V(G)$, d_v denotes the degree of vertex v.

Gautham B. Shenoy, A. Ragavaendra, S. B. Chandrakala,
Badekara Sooryanarayana, and Muhammad Kamran Siddiqui

Topological indices of two chemical compounds used for drugs in the treatment of H1N1

Abstract: In this paper, we find several topological polynomials and indices of the molecular graph of the chemicals zanamivir and oseltamivir. These drugs are used for the treatment of diseases caused by H1N1. The topological indices depend on the degree of vertices. They will be indicators for the stability of the structures. H1N1 is a deadly virus. It requires a serious attempt to cure or control, the diseases, caused by such viruses. Our work is going to help the researchers to find the properties of the chemical required in all aspects. To predict the properties, we need statistically and numerically developed models.

Keywords: Topological polynomials, topological indices, zanamivir , oseltamivir , H1N1

MSC 2010: 05C12

1 Introduction

In recent years, scientific study has shifted its focus to the causes and treatments of dreaded diseases. In the past few years, infectious diseases, their causes, and treatments have become the focus of research. In May–June 2009, scientists discovered a new virus — influenza A(H1N1). This virus was first found in the United States, then spread to other parts of the world. This novel H1N1 virus had a rare combination of influenza genes that had never been seen before in a creature or a human. In the first year alone, the virus has caused nearly three lakh deaths, directly or by associated causes. The incidence of nar-

Gautham B. Shenoy, Department of Mathematics, Poornaprajna College, 1-102, Shantheri, Sudhindra Nagar road, Belman Post, Karkala Taluku, 576111 Udupi, Karnataka, India, e-mail: gautham.shenoy1996@gmail.com
A. Ragavaendra, Department of Mathematics, Poornaprajna College, Volakkadu road, 576101 Udupi, Karnataka, India, e-mail: raghavendra.bhat.a@gmail.com
S. B. Chandrakala, Department of Mathematics, Nitte Meenakshi Institute of Technology, C1, Rishab Enclave, 2nd cross, 1st Main, Patelappa Lyt, Nagashettyhalli, 2nd stage RMV, 560094 Bengaluru, Karnataka, India, e-mail: chandrakalasb14@gmail.com
Badekara Sooryanarayana, Department of Mathematics, Dr. Ambedkar Institute of Technology, #83, Sumkukha, 17th Cross, 4th Block, 4th Stage, Behind Rehjija Park, Basaveshwara Nagara, 560079 Bengaluru, Karnataka State, India, e-mail: dr_bsnrao@yahoo.co.in
Muhammad Kamran Siddiqui, Department of Mathematics, COMSATS University Islamabad, Lahore Campus, Lahore, Pakistan, e-mail: kamransiddiqui@cuilahore.edu.pk

https://doi.org/10.1515/9783111313634-025

colepsy increased dramatically after vaccination against the influenza A(H1N1)pdm09 virus. The influenza A(H1N1)pdm09 virus was treated with a variety of vaccinations. Adamantanes (amantadine and rimantadine) and neuraminidase inhibitors (zanamivir, peramivir and oseltamivir) and adamantanes (rimantadine and amantadine) are the two antiviral drugs, which are widely used to treat influenza in humans. Because of the extensive use of these antiviral medications as monotherapies, influenza viruses resistant to both treatment classes have emerged [17–19, 24].

Chemical structure of zanamivir	Chemical structure of oseltamivir
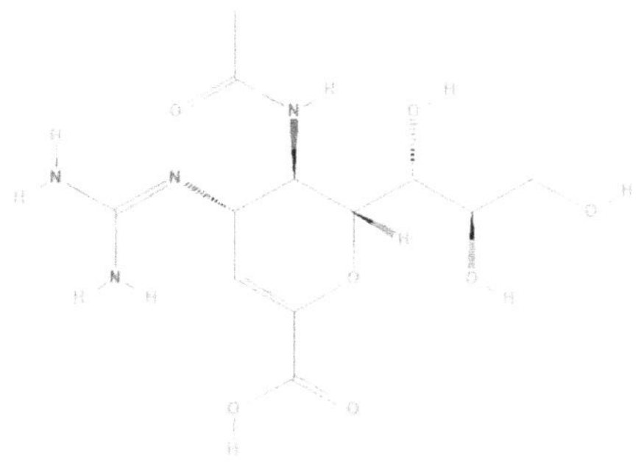	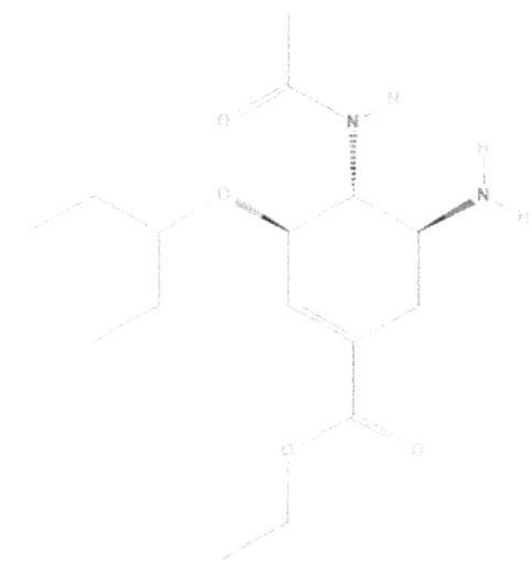

Without using quantum physics, mathematical tools such as polynomials and numbers are used to predict the properties of chemical compounds in mathematical chemistry. Degree-based topological indices are the most well-known invariants of this type, which correlate the structure with different physical properties, chemical reactivity, and biological activities of the chemical compounds, such as strain energy, fracture toughness, boiling point and rigidity are strongly linked to its graphical structure, which plays a vital role in chemical graph theory. Some of the degree-based topological descriptors are defined in Table 25.1 and were computed for molecular structure of oseltamivir and zanamivir.

For a graph G, let $V(G)$ and $E(G)$ denotes vertex and edge set respectively. For a vertex $v \in V(G)$, d_v denotes the degree of vertex v.

Table 25.1: Notations and formulas of topological descriptors.

Topological indices and polynomial	Notation	Expression
First Zagreb index [12]	$M_1(G)$	$\sum_{s\in V(G)} d_s^2$
Second Zagreb index [12]	$M_2(G)$	$\sum_{st\in E(G)} d_s d_t$
Third Zagreb index [8]	$ZG_3(G)$	$\sum_{st\in E(G)} \lvert d_s - d_t\rvert$
First modified Zagreb index [13]	${}^mM_1(G)$	$\sum_{s\in V(G)} \frac{1}{d_s^2}$
Second modified Zagreb index [13]	${}^mM_2(G)$	$\sum_{st\in E(G)} \frac{1}{d_s d_t}$
Hyper Zagreb index [31]	$HM(G)$	$\sum_{st\in E(G)} (d_s + d_t)^2$
Forgotten topological index [10]	$F(G)$	$\sum_{st\in E(G)} d_s^2 + d_t^2$
Randic connectivity index [4]	$R_\alpha(G)$	$\sum_{st\in E(G)} (d_s d_t)^\alpha$
Reciprocal Randic index [11]	$RR(G)$	$\sum_{st\in E(G)} \sqrt{(d_s d_t)}$
Reduced second Zagreb index [11]	$RM_2(G)$	$\sum_{st\in E(G)} (d_s - 1)(d_t - 1)$
Reduced reciprocal Randic index [11]	$RRR(G)$	$\sum_{st\in E(G)} \sqrt{(d_s - 1)(d_t - 1)}$
Atomic Bond Connectivity index [5]	$ABC(G)$	$\sum_{st\in E(G)} \sqrt{\frac{d_s + d_t - 2}{d_s d_t}}$
Geometric Arithmatic index [34]	$GA(G)$	$\sum_{st\in E(G)} \frac{2\sqrt{d_s d_t}}{d_s + d_t}$
Harmonic index [6]	$H(G)$	$\sum_{st\in E(G)} \frac{2}{d_s + d_t}$
Symmetric Degree Division index	$S_{DD}(G)$	$\sum_{st\in E(G)} \frac{d_s^2 + d_t^2}{d_s d_t}$
First Gourava index [20]	$GO_1(G)$	$\sum_{ts\in E(G)} d_t + d_s + d_t d_s$
Second Gourava index [20]	$GO_2(G)$	$\sum_{ts\in E(G)} (d_t + d_s)(d_t d_s)$
Product connectivity Gourava index [21]	$PGO(G)$	$\sum_{ts\in E(G)} \frac{1}{\sqrt{(d_t + d_s)(d_t d_s)}}$
Sum connectivity Gourava index [22]	$SGO(G)$	$\sum_{ts\in E(G)} \frac{1}{\sqrt{(d_t + d_s) + (d_t d_s)}}$
First hyper Gourava index [23]	$HGO_1(G)$	$\sum_{ts\in E(G)} (d_t + d_s + d_t d_s)^2$
Second hyper connectivity Gourava index [23]	$HGO_2(G)$	$\sum_{ts\in E(G)} [(d_t + d_s)(d_t d_s)]^2$
Inverse sum indeg index [33]	$ISI(G)$	$\sum_{ts\in E(G)} \frac{d_t d_s}{d_t + d_s}$
Augmented Zagreb index [9]	$A(G)$	$\sum_{ts\in E(G)} [\frac{d_t d_s}{d_t + d_s - 2}]^3$
Sum connectivity index [35]	$X_{1/2}(G)$	$\sum_{st\in E(G)} (d_t + d_s)^{-1/2}$
Redefined first Zagreb index [30]	$ReZG_1(G)$	$\sum_{ts\in E(G)} \frac{d_t + d_s}{d_t d_s}$
Redefined second Zagreb index [30]	$ReZG_2(G)$	$\sum_{ts\in E(G)} \frac{d_t d_s}{d_t + d_s}$
First Zagreb polynomial [7]	$ZG_1(G, x)$	$\sum_{ts\in E(G)} x^{d_t + d_s}$
Second Zagreb polynomial [7]	$ZG_2(G, x)$	$\sum_{ts\in E(G)} x^{d_t d_s}$
Third Zagreb polynomial [7]	$ZG_3(G, x)$	$\sum_{ts\in E(G)} x^{\lvert d_t + d_s\rvert}$
Forgotten polynomial [1]	$F(G, x)$	$\sum_{ts\in E(G)} x^{d_t^2 + d_s^2}$

Although all of the indices described above have their own significance, degree-based indices are well known and play a significant role in chemical graph theory and biology [2, 3, 14–16, 25, 27–29, 32].

2 Topological indices of molecular graph of zanamivir compound ($C_{12}H_{20}N_4O_7$)

Based on degree of the vertices, vertex set of molecular graph of $C_{12}H_{20}N_4O_7$ compound can be partioned into 3 partions. The first vertex set partition contains of 9 vertices s, where $d_s = 1$. The second vertex set contains of 5 vertices s, where $d_s = 2$. The third partition contains of 9 vertices s, where $d_s = 3$. The edge set of molecular graph of $C_{12}H_{20}N_4O_7$ compound can be partitioned 4 subsets, depending on the degree of end vertices. The first edge set partition contains of 1 edge st, where $d_s = 1$ and $d_t = 2$. The second edge set partition contains of 9 edges st, where $d_s = 2$ and $d_t = 3$. The third edge set partition contains of 8 edges st, where $d_s = 3$ and $d_t = 1$. The fourth edge set partition contains of 5 edges st, where $d_s = 3$ and $d_t = 3$. The cardinality of vertices and edges in $C_{12}H_{20}N_4O_7$ molecular graph are 23 each. The tabular representation of degree of the vertices, edge set partition and their frequencies are listed in Tables 25.2 and 25.3.

Table 25.2: Degrees of the vertices and their frequencies of partitions of vertex set of $C_{12}H_{20}N_4O_7$.

d_s	1	2	3
Frequency	9	5	9

Table 25.3: Frequencies of the edge set partition of $C_{12}H_{20}N_4O_7$ based on (d_s, d_t).

(d_s, d_t)	(1, 2)	(2, 3)	(3, 1)	(3, 3)
Frequency	1	9	8	5

Theorem 1. *Let G be a molecular graph of $C_{12}H_{20}N_4O_7$ compound. Then, the first, second and the third Zagreb indices of G respectively are given by,*

$$M_1(G) = 110; \quad M_2(G) = 125; \quad ZG_3(G) = 26.$$

Proof. From Table 25.2, we compute $M_1(G)$ as below,

$$\begin{aligned} M_1(G) &= \sum_{s \in V(G)} d_s^2, \\ &= 9(1^2) + 5(2^2) + 9(3^2), \\ &= 110. \end{aligned}$$

From Table 25.3, we compute $M_2(G)$ as below,

$$M_2(G) = \sum_{st \in E(G)} d_s d_t,$$

$$= 1 \cdot 2 + 9(2 \cdot 3) + 8(3 \cdot 1) + 5(3 \cdot 3),$$
$$= 125.$$

From Table 25.3, we compute $ZG_3(G)$ as below,

$$\begin{aligned} ZG_3(G) &= \sum_{st \in E(G)} |d_s - d_t|, \\ &= |1-2| + 9|2-3| + 8|3-1| + 5|3-3|, \\ &= 26. \end{aligned}$$
□

Theorem 2. *Let G be the molecular graph of* $C_{12}H_{20}N_4O_7$ *compound. Then, the first, second and third Zagreb polynomial of G respectively are given by,*

$$\begin{aligned} ZG_1(G,x) &= 5x^6 + 9x^5 + 8x^4 + x^3, \\ ZG_2(G,x) &= 5x^9 + 9x^6 + 8x^3 + x^2, \\ ZG_3(G,x) &= 8x^2 + 10x + 5. \end{aligned}$$

Proof. From Table 25.3, we compute $ZG_1(G,x)$ as below,

$$\begin{aligned} ZG_1(G,x) &= \sum_{st \in E(G)} x^{d_s+d_t}, \\ &= x^{1+2} + 9x^{2+3} + 8x^{3+1} + 5x^{3+3}, \\ &= 5x^6 + 9x^5 + 8x^4 + x^3. \end{aligned}$$

From Table 25.3, we compute $ZG_2(G,x)$ as below,

$$\begin{aligned} ZG_2(G,x) &= \sum_{st \in E(G)} x^{d_s d_t}, \\ &= x^{1 \cdot 2} + 9x^{2 \cdot 3} + 8x^{3 \cdot 1} + 5x^{3 \cdot 3}, \\ &= 5x^9 + 9x^6 + 8x^3 + x^2. \end{aligned}$$

From Table 25.3, we compute $ZG_3(G,x)$ as below,

$$\begin{aligned} ZG_3(G,x) &= \sum_{st \in E(G)} x^{|d_s+d_t|}, \\ &= x^{|1-2|} + 9x^{|2-3|} + 8x^{|3-1|} + 5x^{|3-3|}, \\ &= 8x^2 + 10x + 5. \end{aligned}$$
□

Theorem 3. *Let G be the molecular graph of* $C_{12}H_{20}N_4O_7$ *compound. Then the first modified and the second modified Zagreb indices, hyper Zagreb index, forgotten topological index of the molecular graph of zanamivir compound are given by,*

$$^{m}M_1(G) = 11.25 \quad ^{m}M_2(G) = 5.222 \quad HM(G) = 542 \quad F(G) = 292.$$

Proof. From Table 25.2, we compute ${}^mM_1(G)$ as below,

$$\begin{aligned} {}^mM_1(G) &= \sum_{s\in V(G)} \frac{1}{d_s^2}, \\ &= 9\frac{1}{1^2} + 5\frac{1}{2^2} + 9\frac{1}{3^2}, \\ &= 11.25. \end{aligned}$$

From Table 25.3, we compute ${}^mM_2(G)$ as below,

$$\begin{aligned} {}^mM_2(G) &= \sum_{st\in E(G)} \frac{1}{d_s d_t}, \\ &= \frac{1}{1\cdot 2} + 9\frac{1}{2\cdot 3} + 8\frac{1}{3\cdot 1} + 5\frac{1}{3\cdot 3}, \\ &= 5.222. \end{aligned}$$

From Table 25.3, we compute $HM(G)$ as below,

$$\begin{aligned} HM(G) &= \sum_{st\in E(G)} (d_s + d_t)^2, \\ &= (1+2)^2 + 9(2+3)^2 + 8(3+1)^2 + 5(3+3)^2, \\ &= 542. \end{aligned}$$

From Table 25.3, we compute the $F(G)$ as below,

$$\begin{aligned} F(G) &= \sum_{st\in E(G)} d_s^2 + d_t^2, \\ &= (1^2+2^2) + 9(2^2+3^2) + 8(3^2+1^2) + 5(3^2+3^2), \\ &= 292. \end{aligned}$$

□

Theorem 4. *Let G be the molecular graph of* $C_{12}H_{20}N_4O_7$ *compound. Then the forgotten polynomial of G is* $F(G,x) = 5x^{18} + 9x^{13} + 8x^{10} + x^5$.

Proof. The forgotten polynomial $F(G,x)$ is obtained using the data in the Table 25.3 as below,

$$\begin{aligned} F(G,x) &= \sum_{st\in E(G)} x^{d_s^2+d_t^2}, \\ &= x^{(1^2+2^2)} + 9x^{(2^2+3^2)} + 8x^{(3^2+1^2)} + 5x^{(3^2+3^2)}, \\ &= 5x^{18} + 9x^{13} + 8x^{10} + x^5. \end{aligned}$$

□

Theorem 5. *Let G be the molecular graph of* $C_{12}H_{20}N_4O_7$ *compound. The Randic connectivity index, reciprocal Randic index, the reduced second Zagreb index and the reduced reciprocal Randic index of G are given by,*

$$R_{1/2}(G) = 10.4597033 \quad RR(G) = 52.3160277079$$
$$RM_2(G) = 38 \quad RRR(G) = 22.7279220614.$$

Proof. The Randic connectivity index $R_{1/2}(G)$ is computed as follows by using the data given in the Table 25.3,

$$\begin{aligned} R_{1/2}(G) &= \sum_{st \in E(G)} (d_s d_t)^{1/2}, \\ &= \frac{1}{\sqrt{1 \cdot 2}} + 9\frac{1}{\sqrt{2 \cdot 3}} + 8\frac{1}{\sqrt{3 \cdot 1}} + 5\frac{1}{\sqrt{3 \cdot 3}}, \\ &= 10.4597033. \end{aligned}$$

The reciprocal Randic connectivity index $RR(G)$ is computed as follows by using the data given in the Table 25.3,

$$\begin{aligned} RR(G) &= \sum_{st \in E(G)} \sqrt{(d_s d_t)}, \\ &= \sqrt{1 \cdot 2} + 9\sqrt{2 \cdot 3} + 8\sqrt{3 \cdot 1} + 5\sqrt{3 \cdot 3}, \\ &= 52.3160277079. \end{aligned}$$

The index $RM_2(G)$ is obtained as follows by using Table 25.3,

$$\begin{aligned} RM_2(G) &= \sum_{st \in E(G)} (d_s - 1)(d_t - 1), \\ &= (1-1) \cdot (2-1) + 9(2-1) \cdot (3-1) + 8(3-1) \cdot (1-1) + 5(3-1) \cdot (3-1), \\ &= 38. \end{aligned}$$

The index $RRR(G)$ is computed as follows from Table 25.3,

$$\begin{aligned} RRR(G) &= \sum_{st \in E(G)} \sqrt{(d_s - 1)(d_t - 1)}, \\ &= \sqrt{(1-1) \cdot (2-1)} + 9\sqrt{(2-1) \cdot (3-1)} \\ &\quad + 8\sqrt{(3-1) \cdot (1-1)} + 5\sqrt{(3-1) \cdot (3-1)}, = 22.7279220614. \end{aligned}$$ □

Theorem 6. *Let G be the molecular graph of* $C_{12}H_{20}N_4O_7$ *compound. Then, the*
- *Atom Bond Connectivity index* $ABC(G) = 20.2023601164$
- *Geometric Arithmatic index* $GA(G) = 21.6891753459$
- *Harmonic index* $H(G) = 9.933334$

- *Symmetric degree division index* $S_{DD}(G) = 58.66667$
- *first and second Gourava indices are* $GO_1(G) = 235$ *and* $GO_2(G) = 642$
- *product connectivity Gourava index* $PGO(G) = 5.04123085717$
- *sum connectivity Gourava index* $SGO(G) = 7.4755259$
- *first and second hyper-Gourava indices* $HGO_1(G) = 2631$ *and* $HGO_2(G) = 23868$
- *Inverse sum indeg index* $ISI(G) = 24.966667$
- *augmented Zagreb* $A(G) = 163.953125$
- *sum-connectivity index* $X_{-\frac{1}{2}}(G) = 10.6435134$
- *redefined first and second index* $ReZG_1(G) = 23$ *and* $ReZG_2(G) = 24.966667$.

Proof. The index $ABC(G)$ is computed as follows by using the data given in the Table 25.3,

$$\begin{aligned}
ABC(G) &= \sum_{st\in E(G)} \sqrt{\frac{d_s + d_t - 2}{d_s d_t}}, \\
&= \sqrt{\frac{1+2-2}{1\cdot 2}} + 9\sqrt{\frac{2+3-2}{2\cdot 3}} + 8\sqrt{\frac{3+1-2}{3\cdot 1}} + 5\sqrt{\frac{3+3-2}{3\cdot 3}}, \\
&= 20.2023601164.
\end{aligned}$$

The index $GA(G)$ is computed as follows by the data given in the Table 25.3,

$$\begin{aligned}
GA(G) &= \sum_{ts\in E(G)} \frac{2\sqrt{d_s d_t}}{d_s + d_t}, \\
&= \frac{2\sqrt{1\cdot 2}}{1+2} + 9\frac{2\sqrt{2\cdot 3}}{2+3} + 8\frac{2\sqrt{3\cdot 1}}{3+1} + 5\frac{2\sqrt{3\cdot 3}}{3+3}, \\
&= 21.6891753459.
\end{aligned}$$

The index $H(G)$ is computed as follows by the data given in the Table 25.3,

$$\begin{aligned}
H(G) &= \sum_{ts\in E(G)} \frac{2}{d_s + d_t}, \\
&= \frac{2}{1+2} + 9\frac{2}{2+3} + 8\frac{2}{3+1} + 5\frac{2}{3+3}, \\
&= 9.933334.
\end{aligned}$$

The index $S_{DD}(G)$ is computed as follows by the data given in the Table 25.3,

$$\begin{aligned}
S_{DD}(G) &= \sum_{ts\in E(G)} \frac{d_s^2 + d_t^2}{d_s d_t}, \\
&= \frac{1^2 + 2^2}{1\cdot 2} + 9\frac{2^2 + 3^2}{2\cdot 3} + 8\frac{3^2 + 1^2}{3\cdot 1} + 5\frac{3^2 + 3^2}{3\cdot 3}, \\
&= 58.66667.
\end{aligned}$$

The index $GO_1(G)$ is computed as follows by using the data given in the Table 25.3,

$$\begin{aligned} GO_1(G) &= \sum_{st\in E(G)} d_s + d_t + d_s d_t, \\ &= (1+2+1\cdot 2) + 9(2+3+2\cdot 3) + 8(3+1+3\cdot 1) + 5(3+3+3\cdot 3), \\ &= 235. \end{aligned}$$

The index $GO_2(G)$ is computed as follows by the data given in the Table 25.3,

$$\begin{aligned} GO_2(G) &= \sum_{ts\in E(G)} (d_s + d_t)(d_s d_t), \\ &= (1+2)(1\cdot 2) + 9(2+3)(2\cdot 3) + 8(3+1)(3\cdot 1) + 5(3+3)(3\cdot 3), \\ &= 642. \end{aligned}$$

The index $PGO(G)$ is computed as follows by the data given in the Table 25.3,

$$\begin{aligned} PGO(G) &= \sum_{ts\in E(G)} \frac{1}{\sqrt{(d_s + d_t)(d_s d_t)}}, \\ &= \frac{1}{\sqrt{(1+2)(1\cdot 2)}} + 9\frac{1}{\sqrt{(2+3)(2\cdot 3)}} \\ &\quad + 8\frac{1}{\sqrt{(3+1)(3\cdot 1)}} + 5\frac{1}{\sqrt{(3+3)(3\cdot 3)}}, = 5.04123085717. \end{aligned}$$

The index $SGO(G)$ is computed as follows by the data given in the Table 25.3,

$$\begin{aligned} SGO(G) &= \sum_{st\in E(G)} \frac{1}{\sqrt{(d_s + d_t) + (d_s d_t)}}, \\ &= \frac{1}{\sqrt{(1+2)+(1\cdot 2)}} + 9\frac{1}{\sqrt{(2+3)+(2\cdot 3)}} \\ &\quad + 8\frac{1}{\sqrt{(3+1)+(3\cdot 1)}} + 5\frac{1}{\sqrt{(3+3)+(3\cdot 3)}}, = 7.4755259. \end{aligned}$$

The index $HGO_1(G)$ is computed as follows by the data given in the Table 25.3,

$$\begin{aligned} HGO_1(G) &= \sum_{ts\in E(G)} (d_s + d_t + d_s d_t)^2, \\ &= (1+2+1\cdot 2)^2 + 9(2+3+2\cdot 3)^2 \\ &\quad + 8(3+1+3\cdot 1)^2 + 5(3+3+3\cdot 3)^2, = 2631. \end{aligned}$$

The index $HGO_2(G)$ is computed as follows by the data given in the Table 25.3,

$$\begin{aligned} HGO_2(G) &= \sum_{ts\in E(G)} [(d_s + d_t)(d_s d_t)]^2, \\ &= [(1+2)(1\cdot 2)]^2 + 9[(2+3)(2\cdot 3)]^2 \\ &\quad + 8[(3+1)(3\cdot 1)]^2 + 5[(3+3)(3\cdot 3)]^2, = 23868. \end{aligned}$$

The index $ISI(G)$ is computed as follows by the data given in the Table 25.3,

$$\begin{aligned} ISI(G) &= \sum_{ts\in E(G)} \frac{d_s d_t}{d_s + d_t}, \\ &= \frac{1\cdot 2}{1+2} + 9\frac{2\cdot 3}{2+3} + 8\frac{3\cdot 1}{3+1} + 5\frac{3\cdot 3}{3+3}, \\ &= 24.966667. \end{aligned}$$

The index $A(G)$ is computed as follows by utilising the data given in the Table 25.3,

$$\begin{aligned} A(G) &= \sum_{st\in E(G)} \left[\frac{d_s d_t}{d_s + d_t - 2}\right]^3, \\ &= \left(\frac{1\cdot 2}{1+2-2}\right)^3 + 9\left(\frac{2\cdot 3}{2+3-2}\right)^3 + 8\left(\frac{3\cdot 1}{3+1-2}\right)^3 + 5\left(\frac{3\cdot 3}{3+3-2}\right)^3, \\ &= 163.953125. \end{aligned}$$

The index $X_{-\frac{1}{2}}(G)$ is computed as follows by utilizing the data given in the Table 25.3,

$$\begin{aligned} X_{-\frac{1}{2}}(G) &= \sum_{st\in E(G)} (d_s + d_t)^{-1/2}, \\ &= (1+2)^{-\frac{1}{2}} + 9(2+3)^{-\frac{1}{2}} + 8(3+1)^{-\frac{1}{2}} + 5(3+3)^{-\frac{1}{2}}, \\ &= 10.6435134. \end{aligned}$$

The index $ReZG_1(G)$ is computed as follows by utilizing the data given in the Table 25.3,

$$\begin{aligned} ReZG_1(G) &= \sum_{st\in E(G)} \frac{d_s + d_t}{d_s d_t}, \\ &= \frac{1+2}{1\cdot 2} + 9\frac{2+3}{2\cdot 3} + 8\frac{3+1}{3\cdot 1} + 5\frac{3+3}{3\cdot 3}, \\ &= 23. \end{aligned}$$

The index $ReZG_2(G)$ is computed as follows by utilizing the data given in the Table 25.3,

$$\begin{aligned} ReZG_2(G) &= \sum_{st\in E(G)} \frac{d_s d_t}{d_s + d_t}, \\ &= \frac{1\cdot 2}{1+2} + 9\frac{2\cdot 3}{2+3} + 8\frac{3\cdot 1}{3+1} + 5\frac{3\cdot 3}{3+3}, \\ &= 24.966667. \end{aligned}$$

□

3 Topological indices of molecular graph of oseltamivir compound ($C_{16}H_{28}N_2O_4$)

The vertex set of molecular graph of the chemical compound $C_{16}H_{28}N_2O_4$ can be partitioned into three partitions depending on the degree of vertices. Degree of the vertices and number of vertices in each of the partitions are given in Table 25.4. The edge set can be partitioned into five subsets, depending on the degree of end vertices. Degree of the end vertices of the edges in each partitions and the number of such edges are given in Table 25.5 Let (d_s, d_t) denotes the degree of the end vertices of an edge uv.

Table 25.4: Degrees of the vertices and their frequencies of partition of vertex set of $C_{16}H_{28}N_2O_4$.

d_s	1	2	3
Frequency	7	8	7

Table 25.5: Frequencies of the edge set partition of $C_{16}H_{28}N_2O_4$ based on (d_s, d_t).

(d_s, d_t)	(1, 2)	(2, 2)	(2, 3)	(3, 1)	(3, 3)
Frequency	3	1	11	4	3

Theorem 7. *Let G be the molecular graph of the chemical compound* $C_{16}H_{28}N_2O_4$. *Then the Zagreb indices are given by,*

$$M_1(G) = 102 \quad M_2(G) = 115 \quad ZG_3(G) = 22.$$

Proof. The first Zagreb index of G is computed utilizing the data given in the Table 25.4,

$$\begin{aligned} M_1(G) &= \sum_{s \in V(G)} d_s^2, \\ &= 7(1^2) + 8(2^2) + 7(3^2), \\ &= 102. \end{aligned}$$

The second and third Zagreb indices of G are computed utilizing the data given in the Table 25.5,

$$\begin{aligned} M_2(G) &= \sum_{st \in E(G)} d_s d_t, \\ &= 3(1 \cdot 2) + 2 \cdot 2 + 11(2 \cdot 3) + 4(3 \cdot 1) + 3(3 \cdot 3), \\ &= 115. \end{aligned}$$

$$\begin{aligned} ZG_3(G) &= \sum_{st\in E(G)} |d_s - d_t|, \\ &= 3|1-2| + |2-2| + 11|2-3| + 4|3-1| + 3|3-3|, \\ &= 22. \end{aligned}$$

□

Theorem 8. *Let G be the molecular graph of chemical compound* $C_{16}H_{28}N_2O_4$. *The Zagreb polynomial of G are given by,*

$$\begin{aligned} ZG_1(G,x) &= 3x^6 + 11x^5 + 5x^4 + 3x^3, \\ ZG_2(G,x) &= 3x^9 + 11x^6 + x^4 + 4x^3 + 3x^2, \\ ZG_3(G,x) &= 4x^2 + 14x + 4. \end{aligned}$$

Proof. Data in the Table 25.5 is used to compute Zagreb polynomials as below,

$$\begin{aligned} ZG_1(G,x) &= \sum_{st\in E(G)} x^{d_s+d_t}, \\ &= 3x^{1+2} + x^{2+2} + 11x^{2+3} + 4x^{3+1} + 3x^{3+3}, \\ &= 3x^6 + 11x^5 + 5x^4 + 3x^3. \\ ZG_2(G,x) &= \sum_{st\in E(G)} x^{d_s d_t}, \\ &= 3x^{1\cdot 2} + x^{2\cdot 2} + 11x^{2\cdot 3} + 4x^{3\cdot 1} + 3x^{3\cdot 3}, \\ &= 3x^9 + 11x^6 + x^4 + 4x^3 + 3x^2. \\ ZG_3(G,x) &= \sum_{st\in E(G)} x^{|d_s-d_t|}, \\ &= 3x^{|1-2|} + x^{|2-2|} + 11x^{|2-3|}| + 4x^{|3-1|} + 3x^{|3-3|}, \\ &= 4x^2 + 14x + 4. \end{aligned}$$

□

Theorem 9. *Let G be the molecular graph of chemical compound* $C_{16}H_{28}N_2O_4$. *Then the modified Zagreb indices, hyper-Zagreb index and the forgotten topological index of G are given by,*

$$\begin{gathered} {}^mM_1(G) = 9.7777778 \quad {}^mM_2(G) = 5.249999 \\ HM(G) = 490 \quad F(G) = 260. \end{gathered}$$

Proof. Using the data given in Table 25.4, ${}^mM_1(G)$ is computed as below,

$$\begin{aligned} {}^mM_1(G) &= \sum_{s\in V(G)} \frac{1}{d_s^2}, \\ &= 7\frac{1}{1^2} + 8\frac{1}{2^2} + 7\frac{1}{3^2}, \\ &= 9.7777778. \end{aligned}$$

Using the data given in Table 25.5, ${}^{m}M_2(G)$ $HM(G)$ and $F(G)$ are computed as below,

$$
\begin{aligned}
{}^{m}M_2(G) &= \sum_{st\in E(G)} \frac{1}{d_s d_t},\\
&= 3\frac{1}{1\cdot 2} + \frac{1}{2\cdot 2} + 11\frac{1}{2\cdot 3} + 4\frac{1}{3\cdot 1} + 3\frac{1}{3\cdot 3},\\
&= 5.249999.\\
HM(G) &= \sum_{st\in E(G)} (d_s + d_t)^2,\\
&= 3(1+2)^2 + (2+2)^2 + 11(2+3)^2 + 4(3+1)^2 + 3(3+3)^2,\\
&= 490.\\
F(G) &= \sum_{st\in E(G)} d_s^2 + d_t^2,\\
&= 3(1^2+2^2) + (2^2+2^2) + 11(2^2+3^2) + 4(3^2+1^2) + 3(3^2+3^2),\\
&= 260.
\end{aligned}
$$

□

Theorem 10. *Let G be the molecular graph of chemical compound* $C_{16}H_{28}N_2O_4$. *The forgotten polynomial of G is* $F(G,x) = 3x^{18} + 11x^{13} + 4x^{10} + x^8 + 3x^5$.

Proof. By the definition of $F(G,x)$ and using the data given in Table 25.5

$$
\begin{aligned}
F(G,x) &= \sum_{st\in E(G)} x^{d_s^2+d_t^2},\\
&= 3x^{(1^2+2^2)} + x^{(2^2+2^2)} + 11x^{(2^2+3^2)} + 4x^{(3^2+1^2)} + 3x^{(3^2+3^2)},\\
&= 3x^{18} + 11x^{13} + 4x^{10} + x^8 + 3x^5.
\end{aligned}
$$

□

Theorem 11. *Let G be the molecular graph of chemical compound* $C_{16}H_{28}N_2O_4$. *Then, the*

- *randic connectivity index* $R_{1/2}(G) = 10.42145261$
- *reciprocal Randic index* $RR(G) = 49.115231088$
- *reduced second Zagreb index* $RM_2(G) = 35$
- *reduced reciprocal Randic index* $RRR(G) = 22.5563491.$

Proof. The randic connectivity index of G is computed utilizing the data in the Table 25.5,

$$
\begin{aligned}
R_{1/2}(G) &= \sum_{st\in E(G)} (d_s d_t)^{1/2},\\
&= 3\frac{1}{\sqrt{1\cdot 2}} + \frac{1}{\sqrt{2\cdot 2}} + 11\frac{1}{\sqrt{2\cdot 3}} + 4\frac{1}{\sqrt{3\cdot 1}} + 3\frac{1}{\sqrt{3\cdot 3}},\\
&= 10.42145261.
\end{aligned}
$$

The reciprocal randic connectivity index of G is computed utilizing the data in the Table 25.5,

$$\begin{aligned}RR(G) &= \sum_{st\in E(G)} \sqrt{(d_s d_t)},\\ &= 3\sqrt{1\cdot 2} + \sqrt{2\cdot 2} + 11\sqrt{2\cdot 3} + 4\sqrt{3\cdot 1} + 3\sqrt{3\cdot 3},\\ &= 49.115231088.\end{aligned}$$

The reduced second Zagreb index of G is computed utilizing the data in the Table 25.5,

$$\begin{aligned}RM_2(G) &= \sum_{st\in E(G)} (d_s-1)(d_t-1),\\ &= 3(1-1)\cdot(2-1) + (2-1)\cdot(2-1) + 11(2-1)\cdot(3-1)\\ &\quad + 4(3-1)\cdot(1-1) + 3(3-1)\cdot(3-1), = 35.\end{aligned}$$

The reduced reciprocal Randic index of G is computed utilizing the data in the Table 25.5,

$$\begin{aligned}RRR(G) &= \sum_{st\in E(G)} \sqrt{(d_s-1)(d_t-1)},\\ &= 3\sqrt{(1-1)\cdot(2-1)} + \sqrt{(2-1)\cdot(2-1)} + 11\sqrt{(2-1)\cdot(3-1)}\\ &\quad + 4\sqrt{(3-1)\cdot(1-1)} + 3\sqrt{(3-1)\cdot(3-1)},\\ &= 22.5563491.\end{aligned}$$

□

Theorem 12. *Let G be the molecular graph of chemical compound* $C_{16}H_{28}N_2O_4$*. Then, the symmetric degree division index, first and second Gourava indices, product connectivity Gourava index, sum connectivity Gourava index, first and second hyper Gourava indices, sum-connectivity index, redefined first and second index of graph G are given by,*

$$\begin{gathered}ABC(G) = 15.8725880411 \quad GA(G) = 19.27399113 \quad H(G) = 9.9\\ S_{DD}(G) = 52.6666667 \quad GO_1(G) = 217 \quad GO_2(G) = 574\\ PGO(G) = 5.04600974442 \quad SGO(G) = 7.29827352 \quad HGO_1(G) = 2341\\ HGO_2(G) = 13756 \quad ISI(G) = 24.7 \quad A(G) = 167.671875\\ X_{-\frac{1}{2}}(G) = 10.364 \quad ReZG_1(G) = 22 \quad ReZG_2(G) = 24.7.\end{gathered}$$

Proof. The index $ABC(G)$ is computed as follows by using the data given in the Table 25.5,

$$\begin{aligned}ABC(G) &= \sum_{st\in E(G)} \sqrt{\frac{d_s+d_t-2}{d_s d_t}},\\ &= 3\sqrt{\frac{1+2-2}{1\cdot 2}} + \sqrt{\frac{2+2-2}{2\cdot 2}} + 11\sqrt{\frac{2+3-2}{2\cdot 3}} + 4\sqrt{\frac{3+1-2}{3\cdot 1}} + 3\sqrt{\frac{3+3-2}{3\cdot 3}},\\ &= 15.8725880411.\end{aligned}$$

The index $GA(G)$ is computed as follows by the data given in the Table 25.5,

$$
\begin{aligned}
GA(G) &= \sum_{ts\in E(G)} \frac{2\sqrt{d_s d_t}}{d_s + d_t},\\
&= 3\frac{2\sqrt{1\cdot 2}}{1+2} + \frac{2\sqrt{2\cdot 2}}{2+2} + 11\frac{2\sqrt{2\cdot 3}}{2+3} + 4\frac{2\sqrt{3\cdot 1}}{3+1} + 3\frac{2\sqrt{3\cdot 3}}{3+3},\\
&= 19.27399113.
\end{aligned}
$$

The index $H(G)$ is computed as follows by the data given in the Table 25.5,

$$
\begin{aligned}
H(G) &= \sum_{ts\in E(G)} \frac{2}{d_s + d_t},\\
&= 3\frac{2}{1+2} + \frac{2}{2+2} + 11\frac{2}{2+3} + 4\frac{2}{3+1} + 3\frac{2}{3+3},\\
&= 9.9.
\end{aligned}
$$

The symmetric degree division index of G is obtained by utilizing the data given in the Table 25.5,

$$
\begin{aligned}
S_{\mathrm{DD}}(G) &= \sum_{st\in E(G)} \frac{d_s^2 + d_t^2}{d_s d_t},\\
&= 3\frac{1^2+2^2}{1\cdot 2} + \frac{2^2+2^2}{2\cdot 2} + 11\frac{2^2+3^2}{2\cdot 3} + 4\frac{3^2+1^2}{3\cdot 1} + 3\frac{3^2+3^2}{3\cdot 3},\\
&= 52.6666667.
\end{aligned}
$$

The first Gourava index of G is obtained by utilizing the data given in the Table 25.5,

$$
\begin{aligned}
GO_1(G) &= \sum_{st\in E(G)} d_s + d_t + d_s d_t = 3(1+2+1\cdot 2) + (2+2+2\cdot 2)\\
&\quad + 11(2+3+2\cdot 3) + 4(3+1+3\cdot 1) + 3(3+3+3\cdot 3), = 217.
\end{aligned}
$$

The second Gourava index of G is obtained by utilizing the data given in the Table 25.5,

$$
\begin{aligned}
GO_2(G) &= \sum_{st\in E(G)} (d_s + d_t)(d_s d_t), = 3(1+2)\cdot(1\cdot 2) + (2+2)\cdot(2\cdot 2)\\
&\quad + 11(2+3)\cdot(2\cdot 3) + 4(3+1)\cdot(3\cdot 1) + 3(3+3)\cdot(3\cdot 3), = 574.
\end{aligned}
$$

The product connectivity Gourava index of G is obtained by utilizing the data given in the Table 25.5,

$$
\begin{aligned}
PGO(G) &= \sum_{st\in E(G)} \frac{1}{\sqrt{(d_s+d_t)(d_s d_t)}}, = 3\frac{1}{\sqrt{(1+2)(1\cdot 2)}} + \frac{1}{\sqrt{(2+2)(2\cdot 2)}}\\
&\quad + 11\frac{1}{\sqrt{(2+3)(2\cdot 3)}} + 4\frac{1}{\sqrt{(3+1)(3\cdot 1)}} + 3\frac{1}{\sqrt{(3+3)(3\cdot 3)}}, = 5.04600974442.
\end{aligned}
$$

The sum connectivity Gourava index of G is obtained by utilizing the data given in the Table 25.5,

$$
\begin{aligned}
SGO(G) &= \sum_{st\in E(G)} \frac{1}{\sqrt{(d_s+d_t)+(d_sd_t)}}, \\
&= 3\frac{1}{\sqrt{(1+2)+(1\cdot 2)}} + \frac{1}{\sqrt{(2+2)+(2\cdot 2)}} + 11\frac{1}{\sqrt{(2+3)+(2\cdot 3)}} \\
&\quad + 4\frac{1}{\sqrt{(3+1)+(3\cdot 1)}} + 3\frac{1}{\sqrt{(3+3)+(3\cdot 3)}}, \\
&= 7.29827352.
\end{aligned}
$$

The first hyper Gourava index of G is obtained by utilizing the data given in the Table 25.5,

$$
\begin{aligned}
HGO_1(G) &= \sum_{st\in E(G)} (d_s+d_t+d_sd_t)^2 = 3(1+2+1.2)^2 + (2+2+2.2)^2 \\
&\quad + 11(2+3+2.3)^2 + 4(3+1+3.1)^2 + 3(3+3+3.3)^2, = 2341.
\end{aligned}
$$

The second hyper Gourava index of G is obtained by utilizing the data given in the Table 25.5,

$$
\begin{aligned}
HGO_2(G) &= \sum_{st\in E(G)} [(d_s+d_t)(d_sd_t)]^2, = 3[(1+2)\cdot(1\cdot 2)]^2 + [(2+2)\cdot(2\cdot 2)]^2 \\
&\quad + 11[(2+3)\cdot(2\cdot 3)]^2 + 4[(3+1)\cdot(3\cdot 1)]^2 + 3[(3+3)\cdot(3\cdot 3)]^2, \\
&= 13756.
\end{aligned}
$$

The index $ISI(G)$ is computed as follows by the data given in the Table 25.5,

$$
\begin{aligned}
ISI(G) &= \sum_{ts\in E(G)} \frac{d_sd_t}{d_s+d_t}, \\
&= 3\frac{1\cdot 2}{1+2} + \frac{2\cdot 2}{2+2} + 11\frac{2\cdot 3}{2+3} + 4\frac{3\cdot 1}{3+1} + 3\frac{3\cdot 3}{3+3}, \\
&= 24.7.
\end{aligned}
$$

The index $A(G)$ is computed as follows by utilising the data given in the Table 25.5,

$$
\begin{aligned}
A(G) &= \sum_{st\in E(G)} \left[\frac{d_sd_t}{d_s+d_t-2}\right]^3, \\
&= 3\left(\frac{1\cdot 2}{1+2-2}\right)^3 + \left(\frac{2\cdot 2}{2+2-2}\right)^3 + 11\left(\frac{2\cdot 3}{2+3-2}\right)^3 + 4\left(\frac{3\cdot 1}{3+1-2}\right)^3 + 3\left(\frac{3\cdot 3}{3+3-2}\right)^3, \\
&= 167.671875.
\end{aligned}
$$

The sum-connectivity index of G is obtained by utilizing the data given in the Table 25.5,

$$
X_{-\frac{1}{2}}(G) = \sum_{st\in E(G)} (d_s+d_t)^{-1/2},
$$

$$= 3(1+2)^{-\frac{1}{2}} + (2+2)^{-\frac{1}{2}} + 11(2+3)^{-\frac{1}{2}} + 4(3+1)^{-\frac{1}{2}} + 3(3+3)^{-\frac{1}{2}},$$
$$= 10.364.$$

The redefined first index of G is obtained by utilizing the data given in the Table 25.5,

$$\begin{aligned} ReZG_1(G) &= \sum_{st\in E(G)} \frac{d_s + d_t}{d_s d_t}, \\ &= 3\frac{1\cdot 2}{1+2} + \frac{2\cdot 2}{2+2} + 11\frac{2\cdot 3}{2+3} + 4\frac{3\cdot 1}{3+1} + 3\frac{3\cdot 3}{3+3}, \\ &= 22. \end{aligned}$$

The redefined second index of G is obtained by utilizing the data given in the Table 25.5,

$$\begin{aligned} ReZG_2(G) &= \sum_{st\in E(G)} \frac{d_s d_t}{d_s + d_t}, \\ &= 3\frac{1\cdot 2}{1+2} + \frac{2\cdot 2}{2+2} + 11\frac{2\cdot 3}{2+3} + 4\frac{3\cdot 1}{3+1} + 3\frac{3\cdot 3}{3+3}, \\ &= 24.7. \end{aligned}$$

□

4 Graphical compression of zanamivir ($C_{12}H_{20}N_4O_7$) and oseltamivir ($C_{16}H_{28}N_2O_4$)

The similarities between statistical and biological behavior, can be analyzed by the results for topological polynomial for the structure of oseltamivir and zanamivir are given in this section. Figure 25.1 shows the graphical behavious of forgotten topological indes for oseltamiv and zanamivir. Graphical nature of first, second and third zagreb indices of Oseltamivir and zanamivir and show in Figure 25.2 and Figure 25.3 respectively.

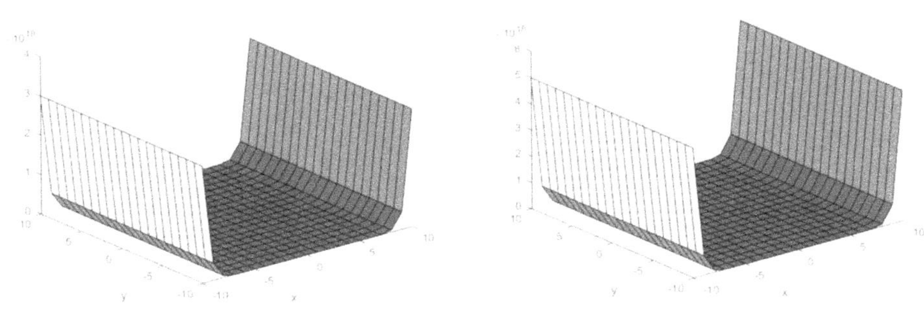

Figure 25.1: Plot of forgotten polynomial of oseltamivir (left) and forgotten polynomial of zanamivir (right).

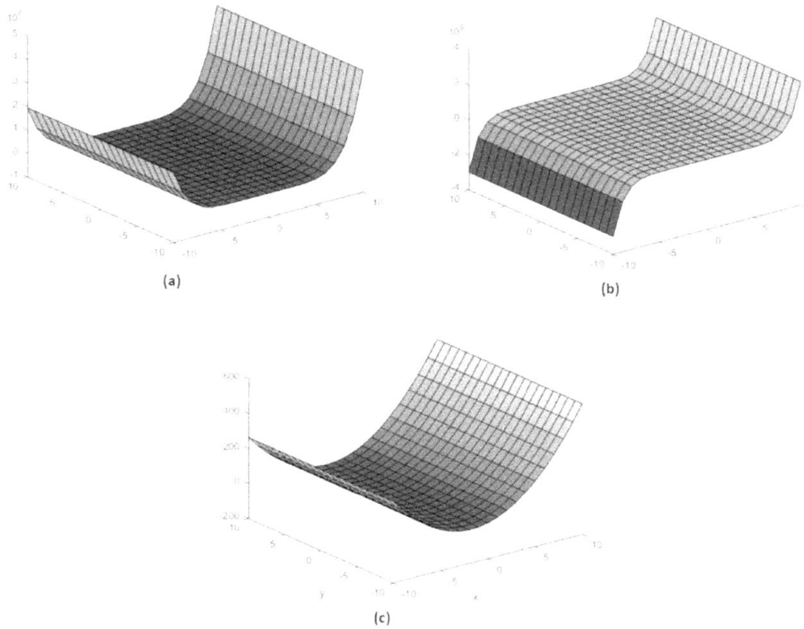

Figure 25.2: Plot of (a) first Zagreb polynomial, (b) second Zagreb polynomial, (b) third Zagreb polynomial of oseltamivir.

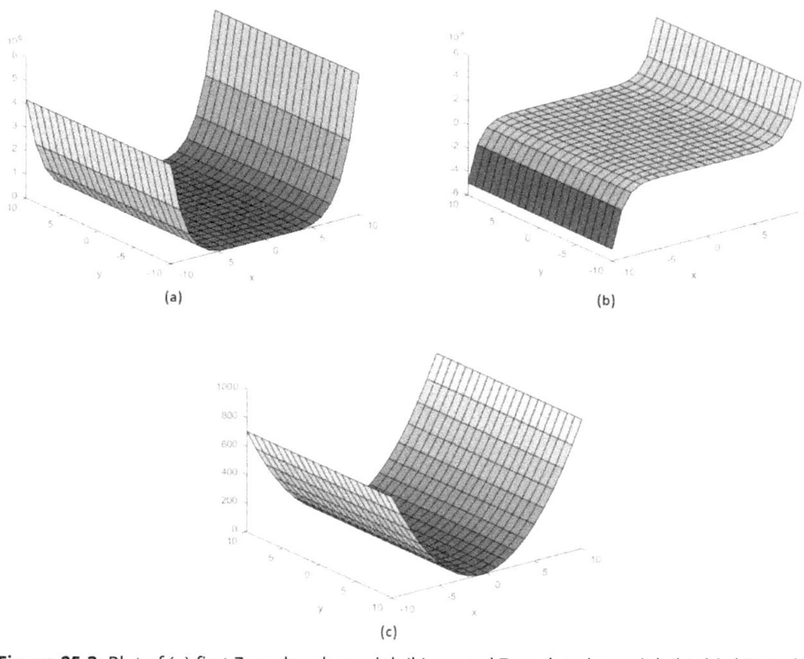

Figure 25.3: Plot of (a) first Zagreb polynomial, (b) second Zagreb polynomial, (b) third Zagreb polynomial of zanamivir.

Bibliography

[1] M. Ajmal, W. Nazeer, W. Khalid and S. M. Kang, Forgotten polynomial and forgotten index for the line graphs of Banana tree graph, firecracker graph and subdivision graphs, *Glob. J. Pure Appl. Math.* **13**(6) (2017), 2673–2682.

[2] S. Akhter, Z. Iqbal, A. Aslam and W. Gao, Computation of Mostar index for some graph operations, *Int. J. Quant. Chem.* (2021), e26674.

[3] A. Asok and J. V. Kureethara, QSPR analysis of certain degree and eccentricity based topological indices and butane derivatives, *Rasayan J. Chem.* **12** (2019), 464–470.

[4] B. Bollobás and P. Erdos, Graphs of extremal weights, *Ars Comb.* **50** (1998), 225–233.

[5] E. Estrada, L. Torres, L. Rodriguez and I. Gutman, An atom-bond connectivity index: modelling the enthalpy of formation of alkanes, *Indian J. Chem., Sect. A: Inorganic, Physical, Theoretical and Analytical* **37**(10) (1998), 849–855.

[6] S. Fajtlowicz, On conjectures of Graffiti—II, *Congr. Numer.* **60** (1987), 187–197.

[7] G. H. Fath-Tabar, Zagreb polynomial and PI indices of some nanostructures, *Digest J. Nanomater. Biostruct.* **4**(1) (2009).

[8] G. H. Fath-Tabar, Old and new Zagreb indices of graphs, *MATCH Commun. Math. Comput. Chem.* **65**(1) (2011), 79–84.

[9] B. Furtula, A. Graovac and D. Vukičević, Augmented zagreb index, *J. Math. Chem.* **48**(2) (2010), 370–380.

[10] B. Furtula and I. Gutman, A forgotten topological index, *J. Math. Chem.* **53**(4) (2015), 1184–1190.

[11] I. Gutman, B. Furtula and C. Elphick, Three new/old vertex-degree based topological indices, *MATCH Commun. Math. Comput. Chem.* **72**(24) (2014), 617–632.

[12] I. Gutman and N. Trinajstic, Graph theory and molecular orbitals. Total ϕ-electron energy of alternant hydrocarbons, *Chem. Phys. Lett.* **17**(4) (1972), 535–538.

[13] J. Hao, Theorems about Zagreb indices and modified Zagreb indices, *MATCH Commun. Math. Comput. Chem.* **65** (2011), 659–670.

[14] Ö.Ç. Havare, Quantitative structure analysis of some molecules in drugs used in the treatment of COVID-19 with topological indices, *Polycycl. Aromat. Compd.* (2021), 1–12.

[15] Ö.Ç. Havare, Topological indices and QSPR modeling of some novel drugs used in the cancer treatment, *Int. J. Quant. Chem.* (2021), 1–23.

[16] M. Ishtiaq, A. Rauf, Q. Rubbab, M. K. Siddiqui, A. U. Rehman and M. Cancan, A degree based topological study of two carbon nanosheets VC_5C_7 and HC_5C_7, *Polycycl. Aromat. Compd.* (2021), 1–21.

[17] M. G. Ison, L. V. Gubareva, R. L. Atmar, J. Treanor and F. G. Hayden, Rcovery of drug-resistant inflenza virus from immunocompromised patients: a case series, *J. Infect. Dis.* **193** (2006), 760–764.

[18] M. G. Ison, V. P. Mishin, T. J. Braciale, F. G. Hayden and L. V. Gubareva, Comparative activities of oseltamivir and A-322278 in immunocompetent and immunocompromised murine models of influenza virus infection, *J. Infect. Dis.* **193** (2006), 765–772.

[19] M. D. Jong, T. T. Tran, H. K. Trong, M. H. Vo, G. J. Smith, V. C. Nguyen, V. C. Bach, V. C. Phan, Q. H. Do, Y. Guan, J. S. Peiris, T. H. Tran and J. Farrar, Oseltamivir resistance during treatment of influenza A(H5N1) infection, *N. Engl. J. Med.* **353** (2005), 2667–2672.

[20] V. R. Kulli, The Gourava indices and indices of graphs, *Ann. Pure Appl. Math.* **14**(1) (2017), 33–38.

[21] V. R. Kulli, The product connectivity Gourava index, *J. Comput. Math. Sci.* **8**(6) (2017), 235–242.

[22] V. R. Kulli, On the sum connectivity Gourava index, *Int. J. Math. Arch.* **8**(6) (2017), 211–217.

[23] V. R. Kulli, On hyper Gourava indices and coindices, *Int. J. Math. Arch.* **8**(12) (2017), 116–120.

[24] T. C. Li, M. C. Chan and N. Lee, Clinical implications of antiviral resistance in influenza, *Viruses* **7** (2015), 4929–4944.

[25] E. D. Molina, J. M. Rodríguez, J. L. Sánchez and J. M. Sigarreta, Some properties of the arithmetic–geometric index, *Symmetry* **13**(5) (2021), 857.

[26] A. Moscona, Oseltamivir resistance-disabling our influenza defenses, *N. Engl. J. Med.* **353** (2005), 2633–2636.

[27] P. Poojary, A. Raghavendra, B. G. Shenoy, M. R. Farahani and B. Sooryanarayana, Certain topological indices and polynomials for the Isaac graphs, *J. Discrete Math. Sci. Cryptogr.* **24**(2) (2021), 511–525.

[28] P. Poojary, G. B. Shenoy, N. N. Swamy, R. Ananthapadmanabha, B. Sooryanarayana and N. Poojary, Reverse topological indices of some molecules in drugs used in the treatment of H1N1, *Biointerface Res. Appl. Chem.* **13**(1) (2023), 71.

[29] J. Rada, J. M. Rodríguez and J. M. Sigarreta, Geometric properties of chemical graphs, *Int. J. Quant. Chem.* (2021).

[30] P. S. Ranjini, V. Lokesha and A. Usha, Relation between phenylene and hexagonal squeeze using harmonic index, *Int. J. Graph Theory* (2013), 116–121.

[31] G. H. Shirdel, H. Rezapour and A. M. Syadi, The hyper-Zagreb index of graph operations, *Iran. J. Math. Chem.* **4**(2) (2013), 213–220.

[32] N. N. Swamy, C. K. Gangappa, P. Poojary, B. Sooryanarayana and N. H. Mudalagiraiah, Topological indices of the subdivision graphs of the nanostructure TUC 4 C 8 (R) using M-polynomials, *J. Discrete Math. Sci. Cryptogr.* **25**(1) (2022), 265–282.

[33] D. Vukicevic and M. Gasperov, Bond aditive mdelling 1. Ariatic indices, *Croat. Chem. Acta* **83** (2010), 243–260.

[34] D. Vuki˘cević and B. Furtula, Topological index based on the ratios of geometrical and arithmetical means of end–vertex degrees of edges, *J. Math. Chem.* **46** (2009), 1369–1376.

[35] B. Zhou and N. Trinajstić, On a novel connectivity index, *J. Math. Chem.* **46**(4) (2009), 1252–1270.

M. Kumari, R. Mohanta, K. Prasad, and H. Mahato

On the Gaussian higher-order Mersenne numbers

Abstract: This paper aims to introduce sequence of Gaussian higher order Mersenne numbers and present their algebraic properties. We present the Binet's formula, Catalan's identity, d'Ocagne's identity, Vajda's identity, Honsberger's identity, generating functions, partial sums, etc. of them. Here, we show results of Gaussian higher order Mersenne numbers in connections with Mersenne and higher order Mersenne numbers.

Keywords: Gaussian higher order Mersenne numbers, partial sum, Catalan's identity, Vajda's identity, Honsberger's identity, generating functions

MSC 2010: 11B37, 11B39, 11B83

1 Introduction

In number theory, a number of the form $2^n - 1$ is known as Mersenne number $\{M_n\}$ and a number of the form $2^n + 1$ is known as Mersenne–Lucas number $\{m_n\}$. For $n \geq 0$, the Mersenne numbers $\{M_n\}$ are given by the recurrence relation [3],

$$M_{n+2} = 3M_{n+1} - 2M_n, \quad n \geq 0, \text{ with } M_0 = 0,\ M_1 = 1. \tag{26.1}$$

The Mersenne–Lucas numbers $\{m_n\}$ satisfy the same recurrence relation (26.1), but with initial condition $m_0 = 2$, $m_1 = 1$. The characteristic equation for the recurrence relation (26.1) is

$$\lambda^2 - 3\lambda + 2 = 0. \tag{26.2}$$

Note that, the roots of (26.2) are $\lambda_1 = 2$ and $\lambda_2 = 1$, thus, the Binet's formulas for the Mersenne and Mersenne–Lucas numbers are given by, respectively,

$$M_n = 2^n - 1 \quad \text{and} \quad m_n = 2^n + 1. \tag{26.3}$$

M. Kumari, K. Prasad, Department of Applied Science & Humanities (Maths), Government Engineering College Bhojpur, Bihar, India, e-mails: muneshnasir94@gmail.com, klkaprsd@gmail.com, https://orcid.org/0000-0002-6541-0284, https://orcid.org/0000-0002-3653-5854
R. Mohanta, Department of Mathematics, Central University of Jharkhand, Ranchi, 835222, India, e-mail: ranjanrabi.511@gmail.com
H. Mahato, Department of Mathematics, Central University of Jharkhand, Cheri Manatu, Kamre, Ranchi, 835222, Jharkhand, India, e-mail: hrishikesh.mahato@cuj.ac.in

https://doi.org/10.1515/9783111313634-026

In recent studies on Mersenne numbers, Soykan [17] studied the generalized Mersenne numbers with arbitrary initial values. Dasdemir [4] studied the Gaussian form and application in quaternion. Kumari et al. [9] and Uysal et al. [20] presented a study on the octonions with Mersenne numbers. Frontczak and Goy [5] showed some connections between Mersenne and Horadam numbers. Some recent contribution on Mersenne numbers are due to Bród and Szynal-Liana [2], Soykan [18], Kumari et al. [10], Tasci and Sevgi [19], etc.

In 1996, Randić et al. [16] proposed and studied the higher order Fibonacci numbers and presented their properties in context of chemistry uses. In some recent studies of higher order numbers, Kizilateş and Kone [8] extended the study for higher order Fibonacci numbers in quaternion algebra and Uysal and Özkan [21] studied some properties of higher order Jacobsthal–Lucas quaternions. Özimamoğlu [12] studied the hyper complex numbers whose components are higher order Pell numbers.

In the same way, for positive integer k the higher-order Mersenne numbers denoted as $\{M_n^{(k)}\}$ are defined as [15]

$$M_n^{(k)} = \frac{M_{kn}}{M_k}, \quad n = 0, 1, 2, \ldots. \tag{26.4}$$

From (26.4), we should note that the following results are verified for higher-order Mersenne numbers:

$$M_0^{(k)} = 0, \quad M_1^{(k)} = 1, \quad \text{and} \quad M_2^{(k)} = 2^k + 1.$$

On going through OEIS, we noted that for $k = 1$, the sequence is indexed as A000225, for $k = 2$ indexed as A002450 and for $k = 3$ indexed as A023001 in OEIS database.

Definition 1 ([15]). For $n \geq 0$, the higher-order Mersenne numbers $M_n^{(k)}$ are given by the recurrence relation

$$M_{n+2}^{(k)} = (2^k + 1)M_{n+1}^{(k)} - 2^k M_n^{(k)}, \quad \text{where } M_0^{(k)} = 0,\ M_1^{(k)} = 1. \tag{26.5}$$

Binet's formula of the higher-order Mersenne numbers $M_n^{(k)}$ is given as

$$M_n^{(k)} = \frac{(2^k)^n - 1}{2^k - 1}. \tag{26.6}$$

Some results on higher-order Mersenne numbers from [15] (eqns. (26.7) and (26.8)) that we need in this article are as follows: The ordinary and exponential generating functions of higher-order Mersenne numbers are, respectively,

$$\sum_{n=0}^{\infty} M_n^{(k)} x^n = \frac{x}{(x-1)(2^k x - 1)} \quad \text{and} \quad \sum_{n=0}^{\infty} M_n^{(k)} \frac{x^n}{n!} = \frac{e^{2^k x} - e^x}{2^k - 1}. \tag{26.7}$$

Let $k, n \in \mathbb{N}$, then summation formulas involving higher-order Mersenne numbers are:

$$\sum_{r=0}^{n} M_r^{(k)} = \frac{2^k M_n^{(k)} - n}{M_k}, \quad \sum_{r=0}^{n} M_{2r}^{(k)} = \frac{2^{2k} M_n^{(2k)} - n}{M_k} \quad \text{and} \quad \sum_{r=0}^{n} M_{2r+1}^{(k)} = \frac{2^k M_{n+1}^{(2k)} - n - 1}{M_k}. \tag{26.8}$$

In 1963, Horadam [7] presented a study on the complex Fibonacci numbers defined as

$$C_n = F_n + iF_{n+1}, \quad \text{where } F_n \text{ is the } n\text{th Fibonacci number.}$$

Later, Berzsenyi [1] in 1977 defined the complex Fibonacci numbers in a different way and referred them as the Gaussian Fibonacci numbers. By using a recurrence relation analogous to the Fibonacci numbers, Horadam [6] also defined the Gaussian form of Fibonacci numbers. Some recent developments on Gaussian form of a number sequence and their properties can be seen in [4, 10, 11, 13, 14, 22].

Motivated by these studies on Gaussian form of a number sequence, here we present a study on Gaussian higher-order Mersenne numbers. We present their some basic properties like the explicit formula, Catalan's identity, d'Ocagne identity, Vajda's identity, partial sums, generating functions, etc.

2 Gaussian higher order Mersenne numbers

Definition 2. Let $k \geq 1$ be any integer, then the Gaussian higher-order Mersenne numbers $\{GM_n^{(k)}\}$ are defined as

$$GM_{n+2}^{(k)} = (2^k + 1)GM_{n+1}^{(k)} - 2^k GM_n^{(k)}, \quad n \geq 0, \tag{26.9}$$

with initial conditions $GM_0^{(k)} = -i/2^k$ and $GM_1^{(k)} = 1$.

For recurrence relation (26.9), the characteristic equation is $x^2 - (2^k + 1)x + 2^k = 0$ and on solving it, we have the characteristic roots as $\lambda_1 = 2^k$ and $\lambda_2 = 1$.

The first few Gaussian higher-order Mersenne numbers are listed in Table 26.1.

Table 26.1: Gaussian higher-order Mersenne numbers for $k = 1, 2, 3, 4$, and 5.

$GM_n^{(k)}$	$k = 1$	$k = 2$	$k = 3$	$k = 4$	$k = 5$
$GM_0^{(k)}$	$-i/2$	$-i/4$	$-i/8$	$-i/16$	$-i/32$
$GM_1^{(k)}$	1	1	1	1	1
$GM_2^{(k)}$	$3 + i$	$5 + i$	$9 + i$	$17 + i$	$33 + i$
$GM_3^{(k)}$	$7 + 3i$	$21 + 5i$	$73 + 9i$	$273 + 17i$	$1057 + 33i$
$GM_4^{(k)}$	$15 + 7i$	$85 + 21i$	$585 + 73i$	$4369 + 273i$	$33825 + 1057i$
$GM_5^{(k)}$	$31 + 15i$	$341 + 85i$	$4681 + 585i$	$69905 + 4369i$	$1082401 + 33825i$

Theorem 1. *For $n \geq 0$, the Binet's formula for Gaussian higher-order Mersenne numbers is*

$$GM_n^{(k)} = \frac{2^{kn} - 1}{2^k - 1} + i\Big(\frac{2^{k(n-1)} - 1}{2^k - 1}\Big).$$

Proof. By the theory of difference equation, the explicit formula for the Gaussian higher-order Mersenne numbers is

$$GM_n^{(k)} = C_1\lambda_1^n + C_2\lambda_2^n = C_1(2^k)^n + C_2.$$

Now using the initial conditions from recurrence relation (26.9), we get $C_1 = \frac{2^k+i}{2^k(2^k-1)}$ and $C_2 = \frac{-1-i}{2^k-1}$. Thus, on substituting these values of C_1 and C_2 in the above expression of explicit formula of $GM_n^{(k)}$, we get the required result. □

Thus from Theorem 1 and Binet's formula (26.6), we conclude that Gaussian higher-order Mersenne numbers $\{GM_n^{(k)}\}$ and higher-order Mersenne numbers $\{M_n^{(k)}\}$ are related as

$$GM_n^{(k)} = M_n^{(k)} + iM_{n-1}^{(k)}, \quad n \geq 1. \tag{26.10}$$

Theorem 2 (Catalan's identity). *For $n > r$, we have*

$$GM_{n-r}^{(k)}GM_{n+r}^{(k)} - (GM_n^{(k)})^2 = (2^k)^{n-r-1}(M_r^{(k)})^2(M_k - im_k).$$

Proof. From relation (26.10), we have

$$\begin{aligned} GM_{n-r}^{(k)}GM_{n+r}^{(k)} - (GM_n^{(k)})^2 &= (M_{n-r}^{(k)} + iM_{n-r-1}^{(k)})(M_{n+r}^{(k)} + iM_{n+r-1}^{(k)}) - (M_n^{(k)} + iM_{n-1}^{(k)})^2 \\ &= (M_{n-r}^{(k)}M_{n+r}^{(k)} - (M_n^{(k)})^2) - (M_{n-1-r}^{(k)}M_{n-1+r}^{(k)} - (M_{n-1}^{(k)})^2) \\ &\quad + i[M_{n+r}^{(k)}M_{n-r-1}^{(k)} - 2M_n^{(k)}M_{n-1}^{(k)} + M_{n+r-1}^{(k)}M_{n-r}^{(k)}]. \end{aligned}$$

For the expression, $M_{n+r}^{(k)}M_{n-r-1}^{(k)} - 2M_n^{(k)}M_{n-1}^{(k)} + M_{n+r-1}^{(k)}M_{n-r}^{(k)}$ using Binet's formula (26.6), we write

$$\begin{aligned} &M_{n+r}^{(k)}M_{n-r-1}^{(k)} - 2M_n^{(k)}M_{n-1}^{(k)} + M_{n+r-1}^{(k)}M_{n-r}^{(k)} \\ &\quad = \frac{1}{M_k^2}[((2^k)^{n+r} - 1)((2^k)^{n-r-1} - 1) - 2((2^k)^n - 1)((2^k)^{n-1} - 1) \\ &\qquad + ((2^k)^{n+r-1} - 1)((2^k)^{n-r} - 1)] \\ &\quad = \frac{-1}{M_k^2}[(2^k)^{n+r} + (2^k)^{n+r-1} - 2((2^k)^n + (2^k)^{n-1}) + (2^k)^{n-r} + (2^k)^{n-r-1}] \\ &\quad = \frac{-1}{M_k^2}[(2^k)^{n+r-1}m_k - 2(2^k)^{n-1}m_k + (2^k)^{n-r-1}m_k] \end{aligned}$$

$$= \frac{-m_k}{M_k^2}[(2^k)^{n+r-1} - 2(2^k)^{n-1} + (2^k)^{n-r-1}]$$
$$= -m_k(2^k)^{n-r-1}(M_r^{(k)})^2.$$

Now, from Binet's formula (26.6) and on substituting the above expression, we have

$$\begin{aligned} GM_{n-r}^{(k)}GM_{n+r}^{(k)} - (GM_n^{(k)})^2 &= -(2^k)^{n-r}(M_r^{(k)})^2 + (2^k)^{n-1-r}(M_r^{(k)})^2 - im_k(2^k)^{n-r-1}(M_r^(k))^2 \\ &= -(2^k)^{n-r-1}M_k(M_r^{(k)})^2 - im_k(2^k)^{n-r-1}(M_r^{(k)})^2 \\ &= -(2^k)^{n-r-1}(M_r^{(k)})^2(M_k + im_k) \end{aligned}$$

as required. □

Corollary 1 (Cassini's identity). *For any positive integer n, we have*

$$GM_{n-1}^{(k)}GM_{n+1}^{(k)} - (GM_n^{(k)})^2 = -(2^k)^{n-2}(M_k + im_k).$$

Proof. Setting $r = 1$ in Theorem 2 gives the above result after simplification. □

Theorem 3 (d'Ocagane identity). *Let $m, n \in \mathbb{N}$ such that $m > n$, we have*

$$GM_{n+1}^{(k)}GM_m^{(k)} - GM_n^{(k)}GM_{m+1}^{(k)} = (2^k)^{n-1}M_kM_{m-n}^{(k)}(M_k + im_k). \tag{26.11}$$

Proof. From relation (26.10), we write

$$\begin{aligned} &GM_{n+1}^{(k)}GM_m^{(k)} - GM_n^{(k)}GM_{m+1}^{(k)} \\ &\quad = (M_{n+1}^{(k)} + iM_n^{(k)})(M_m^{(k)} + iM_{m-1}^{(k)}) - (M_n^{(k)} + iM_{n-1}^{(k)})(M_{m+1}^{(k)} + iM_m^{(k)}) \\ &\quad = (M_{n+1}^{(k)}M_m^{(k)} - M_n^{(k)}M_{m+1}^{(k)}) - (M_n^{(k)}M_{m-1}^{(k)} - M_{n-1}^{(k)}M_m^{(k)}) \\ &\qquad + i(M_{n+1}^{(k)}M_{m-1}^{(k)} - M_{n-1}^{(k)}M_{m+1}^{(k)}). \end{aligned}$$

Since,

$$\begin{aligned} &M_{n+1}^{(k)}M_{m-1}^{(k)} - M_{n-1}^{(k)}M_{m+1}^{(k)} \\ &\quad = \frac{1}{M_k^2}[((2^k)^{n+1} - 1)((2^k)^{m-1} - 1) - ((2^k)^{n-1} - 1)((2^k)^{m+1} - 1)] \\ &\quad = \frac{1}{M_k^2}[(2^k)^{n-1} - (2^k)^{n+1} + (2^k)^{m+1} - (2^k)^{m-1}] \\ &\quad = \frac{1}{M_k^2}[-(2^k)^{n-1}M_{2k} + (2^k)^{m-1}M_{2k}] \\ &\quad = \frac{(2^k)^{n-1}M_{2k}}{M_k^2}[(2^k)^{m-n} - 1] \\ &\quad = (2^k)^{n-1}M_2^{(k)}M_{m-n}^{(k)} \end{aligned}$$

$$= (2^k)^{n-1} m_k M_{m-n}^{(k)}.$$

Thus with the substitution we get

$$\begin{aligned} GM_{n+1}^{(k)} GM_m^{(k)} - GM_n^{(k)} GM_{m+1}^{(k)} &= (2^k)^n M_{m-n}^{(k)} - (2^k)^{n-1} M_{m-n}^{(k)} + i(2^k)^{n-1} m_k M_{m-n}^{(k)} \\ &= (2^k)^{n-1} M_k M_{m-n}^{(k)} (M_k + i m_k). \end{aligned}$$

□

Theorem 4 (Vajda's identity). *For every integer $n, m, r > 0$, we have*

$$GM_{n+m}^{(k)} GM_{n+r}^{(k)} - GM_n^{(k)} GM_{n+m+r}^{(k)} = -(2^k)^{n-1} m_{rk} M_m^{(k)} + i m_k (2^k)^{n-1} M_r^{(k)} M_m^{(k)}.$$

Proof. Using relation (26.10), we write

$$\begin{aligned} & GM_{n+m}^{(k)} GM_{n+r}^{(k)} - GM_n^{(k)} GM_{n+m+r}^{(k)} \\ &= (M_{n+m}^{(k)} + iM_{n+m-1}^{(k)})(M_{n+r}^{(k)} + iM_{n+r-1}^{(k)}) - (M_n^{(k)} + iM_{n-1}^{(k)})(M_{n+m+r}^{(k)} + iM_{n+m+r-1}^{(k)}) \\ &= (M_{n+m}^{(k)} M_{n+r}^{(k)} - M_n^{(k)} M_{n+m+r}^{(k)}) - (M_{n+m-1}^{(k)} M_{n+r-1}^{(k)} - M_{n-1}^{(k)} M_{n+m+r-1}^{(k)}) \\ &\quad + i[M_{n+m}^{(k)} M_{n+r-1}^{(k)} + M_{n+m-1}^{(k)} M_{n+r}^{(k)} - M_n^{(k)} M_{n+m+r-1}^{(k)} - M_{n-1}^{(k)} M_{n+m+r}^{(k)}]. \end{aligned}$$

The imaginary part can be further simplified as

$$\begin{aligned} & M_{n+m}^{(k)} M_{n+r-1}^{(k)} + M_{n+m-1}^{(k)} M_{n+r}^{(k)} - M_n^{(k)} M_{n+m+r-1}^{(k)} - M_{n-1}^{(k)} M_{n+m+r}^{(k)} \\ &= \frac{1}{M_k^2}[((2^k)^{n+m} - 1)((2^k)^{n+r-1} - 1) + ((2^k)^{n+m-1} - 1)((2^k)^{n+r} - 1) \\ &\quad - ((2^k)^n - 1)((2^k)^{n+m+r-1} - 1) - ((2^k)^{n-1} - 1)((2^k)^{n+m+r} - 1)] \\ &= \frac{1}{M_k^2}[-(2^k)^{n+m} - (2^k)^{n+m-1} - (2^k)^{n+r} - (2^k)^{n+r-1} + (2^k)^n + (2^k)^{n-1} \\ &\quad + (2^k)^{n+m+r} + (2^k)^{n+m+r-1}] \\ &= \frac{1}{M_k^2}[-(2^k)^{n+m-1} m_k - (2^k)^{n+r-1} m_k + (2^k)^{n-1} m_k + (2^k)^{n+m+r-1} m_k] \\ &= \frac{m_k}{M_k^2}[-(2^k)^{n+m-1} - (2^k)^{n+r-1} + (2^k)^{n-1} + (2^k)^{n+m+r-1}] \\ &= \frac{-m_k (2^k)^{n-1}}{M_k^2}[(2^k)^m + (2^k)^r - (2^k)^{m+r} - 1] \\ &= \frac{m_k (2^k)^{n-1}}{M_k^2}((2^k)^r - 1)((2^k)^m - 1) \\ &= m_k (2^k)^{n-1} M_r^{(k)} M_m^{(k)}. \end{aligned}$$

Thus on substituting in the Vajda's identity, we get

$$\begin{aligned}
&GM_{n+m}^{(k)}GM_{n+r}^{(k)} - GM_n^{(k)}GM_{n+m+r}^{(k)} \\
&\quad = -(2^k)^n\left(\frac{(2^k)^r+1}{2^k-1}\right)M_m^{(k)} + 2^{k(n-1)}\left(\frac{(2^k)^r+1}{2^k-1}\right)M_m^{(k)} + im_k 2^{k(n-1)}M_r^{(k)}M_m^{(k)} \\
&\quad = -2^{k(n-1)}m_{rk}M_m^{(k)} + im_k(2^k)^{n-1}M_r^{(k)}M_m^{(k)}.
\end{aligned}$$

□

Theorem 5 (Honsberger's identity). *For any integer $p > 1$, we have*

$$\begin{aligned}
GM_{p-1}^{(k)}GM_n^{(k)} + GM_p^{(k)}GM_{n+1}^{(k)} = {}& \frac{m_k}{M_k}[m_{2k}(2^k)^{p+n-3} - ((2^k)^{p-2} + (2^k)^{n-1})] \\
&+ \frac{i}{M_k^2}[2m_{2k}(2^k)^{p+n-2} - m_k^2((2^k)^{p-2} + (2^k)^{n-1}) + 4].
\end{aligned}$$

Proof. Using relation (26.10), we have

$$\begin{aligned}
&GM_{p-1}^{(k)}GM_n^{(k)} + GM_p^{(k)}GM_{n+1}^{(k)} \\
&\quad = (M_{p-1}^{(k)} + iM_{p-2}^{(k)})(M_n^{(k)} + iM_{n-1}^{(k)}) + (M_p^{(k)} + iM_{p-1}^{(k)})(M_{n+1}^{(k)} + iM_n^{(k)}) \\
&\quad = [M_{p-1}^{(k)}M_n^{(k)} + M_p^{(k)}M_{n+1}^{(k)} - M_{p-2}^{(k)}M_{n-1}^{(k)} - M_{p-1}^{(k)}M_n^{(k)}] \\
&\qquad + i[M_{p-2}^{(k)}M_n^{(k)} + M_{p-1}^{(k)}M_{n-1}^{(k)} + M_p^{(k)}M_n^{(k)} + M_{p-1}^{(k)}M_{n+1}^{(k)}] \\
&\quad = [M_p^{(k)}M_{n+1}^{(k)} - M_{p-2}^{(k)}M_{n-1}^{(k)}] \\
&\qquad + i[M_{p-2}^{(k)}M_n^{(k)} + M_{p-1}^{(k)}M_{n-1}^{(k)} + M_p^{(k)}M_n^{(k)} + M_{p-1}^{(k)}M_{n+1}^{(k)}].
\end{aligned} \tag{26.12}$$

The real part of (26.12) can be simplified as

$$\begin{aligned}
&M_p^{(k)}M_{n+1}^{(k)} - M_{p-2}^{(k)}M_{n-1}^{(k)} \\
&\quad = \frac{1}{M_k^2}[((2^k)^p - 1)((2^k)^{n+1} - 1) - ((2^k)^{p-2} - 1)((2^k)^{n-1} - 1)] \\
&\quad = \frac{1}{M_k^2}[(2^k)^{p+n+1} - (2^k)^p - (2^k)^{n+1} - (2^k)^{p+n-3} + (2^k)^{p-2} + (2^k)^{n-1}] \\
&\quad = \frac{1}{M_k^2}[(2^k)^{p+n-3}((2^k)^4 - 1) - (2^k)^{p-2}[(2^k)^2 - 1] - (2^k)^{n-1}[(2^k)^2 - 1]] \\
&\quad = \frac{m_k}{M_k}[(2^k)^{p+n-3}[(2^k)^2 + 1] - (2^k)^{p-2} - (2^k)^{n-1}] \\
&\quad = \frac{m_k}{M_k}[m_{2k}(2^k)^{p+n-3} - ((2^k)^{p-2} + (2^k)^{n-1})],
\end{aligned}$$

and imaginary part of (26.12) can be simplified as

$$\begin{aligned}
&M_{p-2}^{(k)}M_n^{(k)} + M_{p-1}^{(k)}M_{n-1}^{(k)} + M_p^{(k)}M_n^{(k)} + M_{p-1}^{(k)}M_{n+1}^{(k)} \\
&\quad = \frac{1}{M_k^2}[((2^k)^{p-2} - 1)((2^k)^n - 1) + ((2^k)^{p-1} - 1)((2^k)^{n-1} - 1)
\end{aligned}$$

$$
\begin{aligned}
&+ ((2^k)^p - 1)((2^k)^n - 1) + ((2^k)^{p-1} - 1)((2^k)^{n+1} - 1)] \\
&= \frac{1}{M_k^2}[2(2^k)^{p+n} + 2(2^k)^{p+n-2} - (2^k)^{p-2} - 2(2^k)^{p-1} - (2^k)^p - (2^k)^{n-1} \\
&\quad - 2(2^k)^n - (2^k)^{n+1} + 4] \\
&= \frac{1}{M_k^2}[2m_{2k}(2^k)^{p+n-2} - m_k((2^k)^{p-2} + (2^k)^{p-1} + (2^k)^{n-1} + (2^k)^n) + 4] \\
&= \frac{1}{M_k^2}[2m_{2k}(2^k)^{p+n-2} - m_k^2((2^k)^{p-2} + (2^k)^{n-1}) + 4].
\end{aligned}
$$

Now substituting in (26.12), we get

$$
\begin{aligned}
GM_{p-1}^{(k)}GM_n^{(k)} + GM_p^{(k)}GM_{n+1}^{(k)} &= \frac{m_k}{M_k}[m_{2k}(2^k)^{p+n-3} - ((2^k)^{p-2} + (2^k)^{n-1})] \\
&\quad + \frac{i}{M_k^2}[2m_{2k}(2^k)^{p+n-2} - m_k^2((2^k)^{p-2} + (2^k)^{n-1}) + 4]. \qquad \square
\end{aligned}
$$

Theorem 6 (Partial sum). *For any positive integer k, we have*

$$
\sum_{r=0}^{n} GM_r^{(k)} = \left(\frac{2^k M_n^{(k)} - n}{M_k}\right) + i\left(\frac{2^k M_{n-1}^{(k)} - n + 1}{M_k} - \frac{1}{2^k}\right).
$$

Proof. From relation (26.10), we can write

$$
\begin{aligned}
\sum_{r=0}^{n} GM_r^{(k)} &= GM_0^{(k)} + \sum_{r=0}^{n-1}(M_{r+1}^{(k)} + iM_r^{(k)}) \\
&= \frac{-i}{2^k} + \sum_{r=0}^{n-1} M_{r+1}^{(k)} + i\sum_{r=0}^{n-1} M_r^{(k)} \\
&= \frac{-i}{2^k} + \left(\frac{2^k M_n^{(k)} - n}{M_k}\right) + i\left(\frac{2^k M_{n-1}^{(k)} - n + 1}{M_k}\right) \quad \text{(from (26.8))} \\
&= \left(\frac{2^k M_n^{(k)} - n}{M_k}\right) + i\left(\frac{2^k M_{n-1}^{(k)} - n + 1}{M_k} - \frac{1}{2^k}\right). \qquad \square
\end{aligned}
$$

Theorem 7 (Partial sum of even-indexed terms). *For any positive integer k, we have*

$$
\sum_{r=0}^{n} GM_{2r}^{(k)} = \frac{2^{2k} M_n^{(2k)} - n}{M_k} + i\left(\frac{2^k M_n^{(2k)} - n}{M_k} - \frac{1}{2^k}\right).
$$

Proof. Proceeding as the above theorem, we have

$$
\sum_{r=0}^{n} GM_{2r}^{(k)} = GM_0^{(k)} + \sum_{r=0}^{n-1}(M_{2r+2}^{(k)} + iM_{2r+1}^{(k)})
$$

$$= \frac{-i}{2^k} + \sum_{r=0}^{n-1} M_{2r+2}^{(k)} + i\sum_{r=0}^{n-1} M_{2r+1}^{(k)}$$
$$= \frac{2^{2k}M_n^{(2k)} - n}{M_k} + i\left(\frac{2^k M_n^{(2k)} - n}{M_k} - \frac{1}{2^k}\right) \quad \text{(from (26.8))}. \qquad \square$$

Theorem 8 (Partial sum of odd-indexed terms). *For any positive integer k, we have*

$$\sum_{r=0}^{n} GM_{2r+1}^{(k)} = \frac{2^k M_{n+1}^{(2k)} - n - 1}{M_k} + i\left(\frac{2^{2k}M_n^{(2k)} - n}{M_k}\right).$$

Proof. Proceeding as Theorem 6, we have

$$\sum_{r=0}^{n} GM_{2r+1}^{(k)} = \sum_{r=0}^{n}(M_{2r+1}^{(k)} + iM_{2r}^{(k)}) \quad \text{(from relation (26.10))}$$
$$= \left(\frac{2^k M_{n+1}^{(2k)} - n - 1}{M_k}\right) + i\left(\frac{2^{2k}M_n^{(2k)} - n}{M_k}\right) \quad \text{(from (26.8))}. \qquad \square$$

Theorem 9. *The generating function for Gaussian higher-order Mersenne numbers is given by*

$$GM(k,x) = \frac{2^k x + i(2^k x + x - 1)}{2^k(x-1)(2^k x - 1)}.$$

Proof. Let us assume that the ordinary generating function for $\{GM_n^{(k)}\}$ is given by

$$GM(k,x) = \sum_{n=0}^{\infty} GM_n^{(k)} x^n.$$

Then using relation (26.10), we can write

$$GM(k,x) = \sum_{n=0}^{\infty} GM_n^{(k)} x^n = GM_0^{(k)} + \sum_{n=0}^{\infty} GM_{n+1}^{(k)} x^{n+1}$$
$$= GM_0^{(k)} + \sum_{n=0}^{\infty} M_{n+1}^{(k)} x^{n+1} + i\sum_{n=0}^{\infty} M_n^{(k)} x^{n+1}.$$

Since, from (26.7) the generating function of higher-order Mersenne numbers is

$$\sum_{n=0}^{\infty} M_n^{(k)} x^n = \frac{x}{(x-1)(2^k x - 1)}. \quad \text{So,} \quad \sum_{n=0}^{\infty} M_{n+1}^{(k)} x^{n+1} = \frac{x}{(x-1)(2^k x - 1)}.$$

Thus, we get

$$GM(k,x) = \frac{-i}{2^k} + \frac{x}{(x-1)(2^k x - 1)} + \frac{ix^2}{(x-1)(2^k x - 1)}$$

$$= \frac{x}{(x-1)(2^k x-1)} + i\left(\frac{x^2}{(x-1)(2^k x-1)} - \frac{1}{2^k}\right)$$
$$= \frac{x + \frac{i}{2^k}((1+2^k)x-1)}{(x-1)(2^k x-1)}. \qquad \square$$

In particular for $k = 1$, we have the generating function of Gaussian Mersenne numbers, i. e.,

$$GM(1,x) = \frac{x + i(\frac{3}{2}x - \frac{1}{2})}{1 - 3x + 2x^2}.$$

Theorem 10. *The exponential generating function $E(k,x)$ for Gaussian higher-order Mersenne numbers is*

$$E(k,x) = \frac{e^{2^k x} - e^x}{2^k - 1} + i\left(\frac{e^{2^k x} - 2^k e^x}{2^k(2^k-1)}\right).$$

Proof. We have

$$E(k,x) = \sum_{n=0}^{\infty} GM_n^{(k)} \frac{x^n}{n!} = GM_0^{(k)} + \sum_{n=0}^{\infty} GM_{n+1}^{(k)} \frac{x^{n+1}}{(n+1)!}$$
$$= GM_0^{(k)} + \sum_{n=0}^{\infty} M_{n+1}^{(k)} \frac{x^{n+1}}{(n+1)!} + i\sum_{n=0}^{\infty} M_n^{(k)} \frac{x^{n+1}}{(n+1)!} \quad \text{(from (26.10))}$$
$$= \frac{-i}{2^k} + \left(\frac{e^{2^k x} - e^x}{2^k - 1}\right) + i\sum_{n=0}^{\infty}\left(\frac{(2^k)^n - 1}{2^k - 1}\right)\frac{x^{n+1}}{(n+1)!} \quad \text{(from (26.7) and (26.6))}$$
$$= \frac{-i}{2^k} + \left(\frac{e^{2^k x} - e^x}{2^k - 1}\right) + \frac{i}{2^k - 1}\left(\sum_{n=0}^{\infty} \frac{(2^k)^n x^{n+1}}{(n+1)!} - \sum_{n=0}^{\infty} \frac{x^{n+1}}{(n+1)!}\right)$$
$$= \frac{-i}{2^k} + \left(\frac{e^{2^k x} - e^x}{2^k - 1}\right) + \frac{i}{2^k - 1}\left(\frac{(e^{2^k x} - 1)}{2^k} - (e^x - 1)\right)$$
$$= \left(\frac{e^{2^k x} - e^x}{2^k - 1}\right) + \frac{i}{2^k - 1}\left(\frac{(e^{2^k x} - 1)}{2^k} - \frac{2^k(e^x - 1)}{2^k} - \frac{2^k - 1}{2^k}\right).$$

This completes the proof. $\square$

3 Conclusion

In summary, we introduced and studied the Gaussian higher-order Mersenne numbers. Here we discussed their algebraic properties such as Catalan's identity, d'Ocagne's identity, Vajda's identity, Honsberger's identity, generating functions, etc. in closed form. In addition, various series sum formulas are also presented. Here we provided the results in connection with Mersenne and higher order Mersenne numbers.

Study on this sequence can be extended to matrix theory, clifford algebra like quaternions, octonions, etc.

Bibliography

[1] G. Berzsenyi, Gaussian Fibonacci numbers, *Fibonacci Q.* **15** (1977), 233–236.

[2] D. Bród and A. Szynal-Liana, On generalized bihyperbolic Mersenne numbers, *Math. Bohem.* (2023), 1–11.

[3] P. Catarino, H. Campos and P. Vasco, On the Mersenne sequence, *Ann. Math. Inform.* **46** (2016), 37–53.

[4] A. Dasdemir and G. Bilgici, Gaussian Mersenne numbers and Mersenne quaternions, *Notes Number Theory Discrete Math.* **25**(3) (2019), 87–96.

[5] R. Frontczak and T. Goy, Mersenne-Horadam identities using generating functions, *Carpath. Math. Publ.* **12**(1) (2020), 34–45.

[6] C. Harman, Complex Fibonacci numbers, *Fibonacci Q.* **19**(1) (1981), 82–86.

[7] A. F. Horadam, Complex Fibonacci numbers and Fibonacci quaternions, *Am. Math. Mon.* **70**(3) (1963), 289–291.

[8] C. Kizilateş and T. Kone, On higher order Fibonacci quaternions, *J. Anal.* **29**(4) (2021), 1071–1082.

[9] M. Kumari, K. Prasad and H. Mahato, On the k-Mersenne and k-Mersenne-Lucas octonions, *TWMS J. Appl. Eng. Math.* **14**(3) (2024), 1218–1230.

[10] M. Kumari, K. Prasad and J. Tanti, On the generalization of Mersenne and Gaussian Mersenne polynomials, *J. Anal.* **32** (2024), 931–947. https://doi.org/10.1007/s41478-023-00666-4.

[11] M. Kumari, J. Tanti and K. Prasad, On some new families of k-Mersenne and generalized k-Gaussian Mersenne numbers and their polynomials, *Contrib. Discrete Math.* **18**(2) (2023), 244–260.

[12] H. Özimamoğlu, On hyper complex numbers with higher order Pell numbers components, *J. Anal.* **31** (2023), 2443–2457.

[13] E. Özkan and M. Taştan, On gauss Fibonacci polynomials, on Gauss Lucas polynomials and their applications, *Commun. Algebra* **48**(3) (2020), 952–960.

[14] K. Prasad, R. Mohanty, M. Kumari and H. Mahato, Some new families of generalized k-Leonardo and Gaussian Leonardo numbers, *Commun. Comb. Optim.* **9**(3), (2024), 539–553. https://doi.org/10.22049/CCO.2023.28236.1485.

[15] K. Prasad, M. Kumari, R. Mohanta and H. Mahato, The sequence of higher order Mersenne numbers and associated binomial transforms, 2023, arXiv preprint arXiv:2307.08073.

[16] M. Randić, D. A. Morales and O. Araujo, Higher-order Fibonacci numbers, *J. Math. Chem.* **20** (1996), 79–94.

[17] Y. Soykan, A study on generalized Mersenne numbers, *J. Progressive Res. Math.* **18**(3) (2021), 90–108.

[18] Y. Soykan, On generalized p-Mersenne numbers, *Earthline J. Math. Sci.* **8**(1) (2022), 83–120.

[19] D. Taşcı and E. Sevgi, Some properties between Mersenne, Jacobsthal and Jacobsthal-Lucas hybrid numbers, *Chaos Solitons Fractals* **146** (2021), 110862.

[20] M. Uysal, M. Kumari, B. Kuloglu, K. Prasad and E. Özkan, On the hyperbolic k-Mersenne and k-Mersenne-Lucas octonions, *Kragujev. J. Math.* **49**(05) (2025), 765–779.

[21] M. Uysal and E. Özkan, Higher-order Jacobsthal–Lucas quaternions, *Axioms* **11**(12) (2022), 671.

[22] T. Yagmur and N. Karaaslan, Gaussian modified Pell sequence and Gaussian modified Pell polynomial sequence, *Aksaray University J. Sci. Eng.* **2**(1) (2018), 63–72.

Sakshi Gupta and Shelly Garg

Petri net: modeling, properties and analysis techniques

Abstract: Graph theory is a branch of mathematics that is used to study the pairwise relations between the objects of a network. Petri net (PN) is a special type of graph that has emerged as an important mathematical technique to analyze the behavior of the network. It is a mathematical and computational formalism used to model and study the complex networks having concurrent and parallel interactions between their components. Petri net has been widely applied in the application areas, such as computer networks, telecommunication networks, biological networks, social networks, communication networks and many more. The elements of these networks are represented by vertices and the links between these elements is represented using the edges. The study of these networks helps to analyze the complex relation between their elements. Due to the widespread availability of the vast network data sources in the recent years, the study of networks has received a major boost. The current work discuss the basic definitions, properties and analysis methods of Petri nets used to model and analyze any network behavior.

Keywords: Petri net, modeling, structural properties, behavioral properties, invariant analysis

MSC 2010: 05C82, 05C90

1 Introduction

Modeling is an important part of both the design and the analysis of a network. Petri net is among the most successful and oldest modeling formalisms. The Swiss mathematician Leonhard Euler in 1736 laid the foundation of graph theory with the famous mathematical problem Seven Bridges of Königsberg. Since then, it has found applications in the study of wide variety of fields. In the graph representation of a network, the vertices represent the objects and the edges represent the relation between the objects.

Acknowledgement: The authors express their deep gratitude to anonymous reviewers, referees and editors for their valuable guidance and suggestions.

Sakshi Gupta, Department of Applied Sciences and Humanities, Dronacharya College of Engineering, Gurugram, Haryana, India, e-mail: sakshi86.10@gmail.com
Shelly Garg, Department of Electronics and Communication, Dronacharya College of Engineering, Gurugram, Haryana, India

https://doi.org/10.1515/9783111313634-027

Due to the instinctive graphical formalism and significant analysis techniques, PNs are suitable for an extensive field of applications. They have gained broad acceptance as a modeling formalism in areas like communication networks, biological networks, software design, knowledge representation, control engineering and many more [6, 8, 9, 11, 14, 15, 18, 22, 23]. This is because of its understandable representation and ability to model the systems having graph-like structure [5, 18], due to which even non-Petri net experts can easily understand the behavior of the complex networks. Also, PN possesses different analysis techniques that are used to confirm and analyze system representation and behavior [10, 19, 21]. To name a few, we can determine the reachable states, liveness, deadlocks, and cycles. By performing the qualitative analysis, one can easily figure out the fundamental mechanism of different processes that occurs inside the network.

Modeling and analysis of the networks through Petri net had been and will continue to be a viable interest amid the researchers because of the Petri net power to represent the complex situations in an understanding way. It possesses the ability to encompass all of the available modeling formalisms with its different extensions. A system can be studied both qualitatively and quantitatively using Petri net. A qualitative PN model can be easily extended to the quantitative model by considering the kinetic data of the network under consideration. Various extensions of classic Petri net such as Boolean, time, hybrid, stochastic, continuous, fuzzy and colored PNs are available to model and study the networks [1, 3, 4, 13, 17]. These variants have boosted the modeling efficiency of Petri net by considering the concepts like probability and time, allowing to model any real situation.

2 Petri net

It is a special and effective graph structure used to represent and analyze the behavior of the asynchronous systems possessing parallel, distributed and concurrent activities [18]. In 1962, a German mathematician Carl Adam Petri proposed it in his PhD thesis "Communication with Automata"‘ [20]. It is a weighted, directed, bipartite graph [16, 18] and is also called place-transition net.

Petri net has two types of vertices. The first one is place (p), which represent resource and is denoted by circle. The second type is transition (t), which represent events and is represented by rectangular box. These two types of vertices are linked to each other through the directed arcs that show the connection between them. If an edge is directed from the transition towards some place or from the place towards some transition, then that place is called output or input place of that transition respectively. The number of arcs (s) from the place and from the transition are denoted by $I(t,p)$ and $O(t,p)$. This number is shown by putting the weight equal to s on the directed edge. Generally, if arc weight is one, then we do not specify the arc weight. While modeling, some places are assigned tokens that are the non-negative integers. These tokens are modeled by using black dots and they indicate the number of sources present. The system state is represented by the Marking vector μ, whose each entry corresponds to the number of tokens present in the corresponding place.

Mathematically, Petri net is defined as:

Definition 1 (*Petri net*). It is a 5-tuple [13] (P, T, I, O, μ_0), where
1. $P = \{p_1, p_2, \ldots, p_n\}$ and $T = \{t_1, t_2, \ldots, t_m\}$ are the disjoint finite set of places and transitions s. t. their union is non-empty,
2. $I, O : T \times P \rightarrow Z^+$, where I and O is the input and output matrix that specifies the number of arcs joining places to transitions and vice-versa,
3. $\mu_0 : P \rightarrow \{0, 1, 2, \ldots\}$ represents initial marking vector, written as $\mu_0 = (m_1, m_2, \ldots, m_n)$, where each $m_i \in \{0, 1, 2, \ldots\}$ indicates the token number present in the corresponding places.

Definition 2. (i) *Source transition:* A source transition does not have an input place, i. e., if t_j is a source transition, then $I(t_j, p) = 0$. It is always unconditionally enabled and is used to produce the resources.
(ii) *Sink transition:* A sink transition does not have an output place, i. e., if t_j is a sink transition, then $O(t_j, p) = 0$. Its firing will only consume tokens and no new tokens will be produced. It is used to consume the resources.

Along with the simple directed arcs, extended Petri nets also have inhibitor and read arcs. These arcs have increased the modeling power of Petri nets [19]. A read arc is used for modeling the action of such resources which are required for the process execution, but remain unconsumed. It is denoted by a bidirectional arc. An inhibitor arc indicates the inhibitory mechanism of the resource. It is modeled by putting an empty circle at the arc end and always directs towards the transition. This transition is enable only in the absence of a resource. By using inhibitory arcs, the modeling efficiency of Petri net has increased up to the Turing machine level [19]. The components of Petri net are shown in Figure 27.1.

3 Execution of a Petri net

A PN executes by firing of enable transitions. A transition t is called enabled transition if it satisfies the condition $\mu(p) \geq I(t, p)$, p being the input places of t. Any enable transition may fire depending upon the conditions assigned to the model. At some marking μ_i, firing of enable transition changes the marking calculated by:

$$\mu_{(k+1)}(p_i) = \mu_k(p_i) + O(t, p_i) - I(t, p_i), \quad \forall p_i \in P,$$

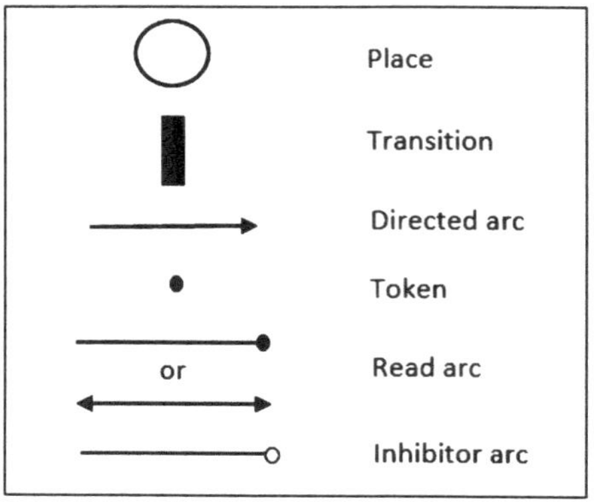

Figure 27.1: Components of a Petri net.

It is denoted by $\mu_k \xrightarrow{t} \mu_{(k+1)}$ and state $\mu_{(k+1)}$ is said to be directly reachable from μ_k via transition t.

Next, we will discuss how a Petri net executes with the help of an example. Consider the PN model shown in Figure 27.2 before firing. It has three transitions and six places. The read arc connects the place p_5 to the transition t_1. Initially, one token is assigned to the places p_0 and p_5 and zero tokens to all the other places, i. e., $\mu_0(p_0) = 1$, $\mu_0(p_1) = 0$, $\mu_0(p_2) = 0$, $\mu_0(p_3) = 0$, $\mu_0(p_4) = 0$ and $\mu_0(p_5) = 1$. Hence, the initial marking in state $S0$ is $\mu_0 = (1, 0, 0, 0, 0, 1)$.

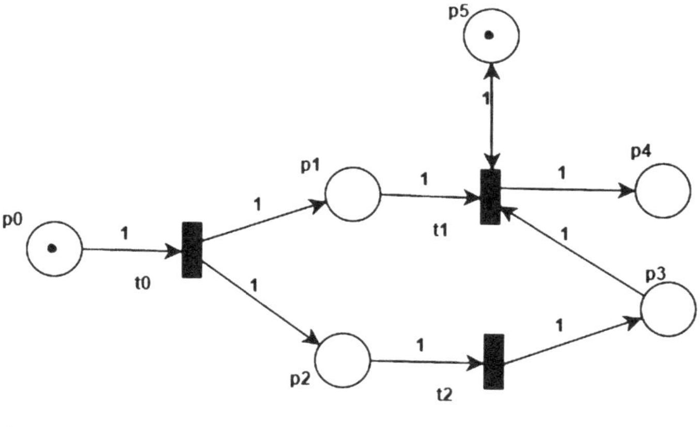

Figure 27.2: PN model before firing.

Now the input place of the transition t_0 is p_0 and $\mu_0(p_0) = 1 \geq 1 = I(t_0, p_0)$. Thus, the transition t_0 satisfies the enabling condition and hence it is enable. It will fire and tokens will move from their input place p_0, towards the places p_1 and p_2. The next state of the PN is calculated below:

$$\mu_1(p_0) = \mu_0(p_0) + O(t_0, p_0) - I(t_0, p_0) = 1 + 0 - 1 = 0,$$
$$\mu_1(p_1) = \mu_0(p_1) + O(t_0, p_1) - I(t_0, p_1) = 0 + 1 - 0 = 1.$$

And

$$\mu_1(p_2) = \mu_0(p_2) + O(t_0, p_2) - I(t_0, p_2) = 0 + 1 - 0 = 1.$$

The marking of other places remains unchanged. The system's new marking in state $S1$ is $\mu_1 = (0, 1, 1, 0, 0, 1)$. Figure 27.3 displays the PN model after firing of transition t_0.

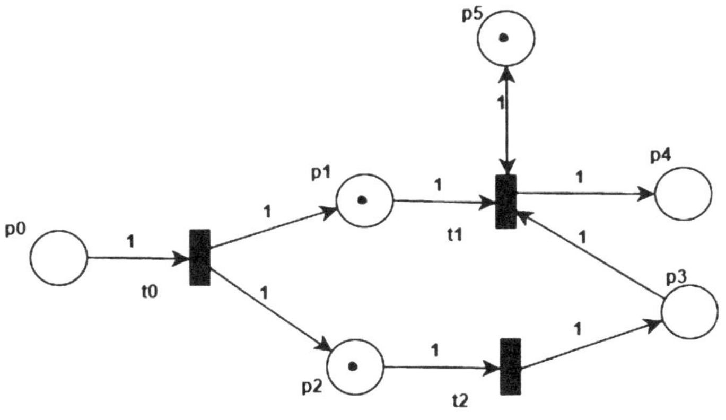

Figure 27.3: Petri net model after firing of the transition t_0.

Next, among the three transitions, the transition t_2 satisfies the enabling condition $\mu_1(p_2) = 1 \geq 1 = I(t_2, p_2)$. Thus, the transition t_2 will fire and this will again change the state of the PN. The next state of the PN is calculated below:

$$\mu_2(p_2) = \mu_1(p_2) + O(t_2, p_2) - I(t_2, p_2) = 1 + 0 - 1 = 0,$$
$$\mu_2(p_3) = \mu_1(p_3) + O(t_2, p_3) - I(t_2, p_3) = 0 + 1 - 0 = 1.$$

Marking of remaining places remain unchanged. The system's new marking in state $S2$ is $\mu_2 = (0, 1, 0, 1, 0, 1)$. Figure 27.4 shows the PN model after firing of the transition t_2.

In the new marking, the transition t_1 satisfies the enabling condition, because $\mu_2(p_1) = 1 \geq 1 = I(t_1, p_1)$, $\mu_2(p_3) = 1 \geq 1 = I(t_1, p_3)$ and $\mu_2(p_5) = 1 \geq 1 = I(t_1, p_5)$. Hence, it will fire and the new marking of the PN can be find as:

$$\mu_3(p_1) = \mu_2(p_1) + O(t_1, p_1) - I(t_1, p_1) = 1 + 0 - 1 = 0,$$
$$\mu_3(p_5) = \mu_2(p_5) + O(t_1, p_5) - I(t_1, p_5) = 1 + 1 - 1 = 1,$$
$$\mu_3(p_3) = \mu_2(p_3) + O(t_1, p_3) - I(t_1, p_3) = 1 + 0 - 1 = 0$$

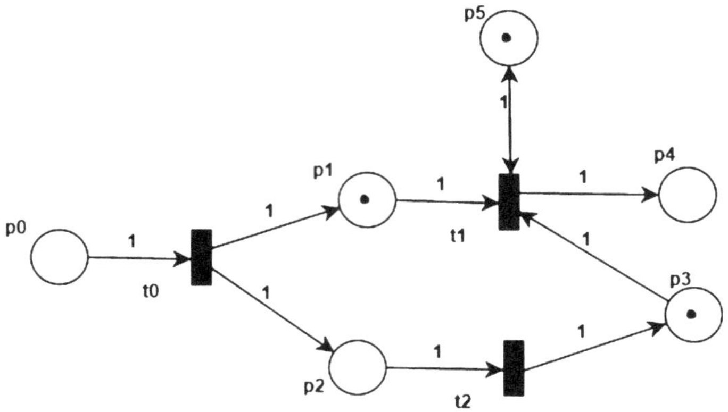

Figure 27.4: PN model after firing of the transition t_2.

and

$$\mu_3(p_4) = \mu_2(p_4) + O(t_1, p_4) - I(t_1, p_4) = 0 + 1 - 0 = 1.$$

Marking of place p_2 will remain unchanged and the system's final marking in state $S3$ is $\mu_3 = (0,0,0,0,1,1)$, because no transition is enabled after firing of t_1. Figure 27.5 shows the PN model after firing of the transition t_1.

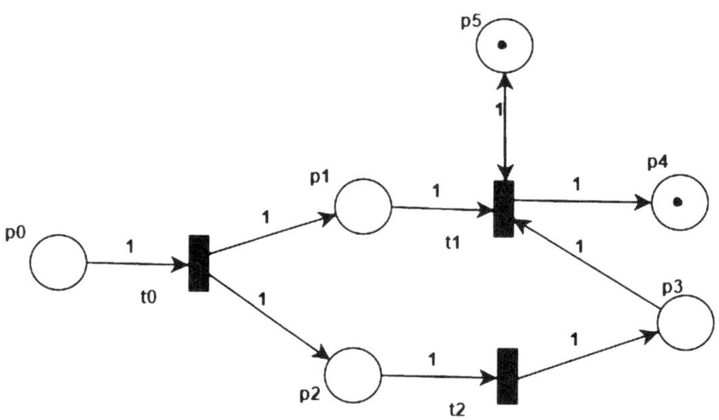

Figure 27.5: PN model after firing of the transition t_1.

4 Properties of Petri net

In addition to being a mathematical and graphical technique, Petri net also possess important properties that can be used to characterize the properties of a network [18, 19,

21]. These properties do not depend on the kinetic parameters and are also independent of the time. So, these are called *qualitative properties* of the Petri net. Each property has some significant meaning and helps to get better insights in the system's behavior.

1. *Ordinary Petri net*: If the weight of each arc in a PN is equal to one, then it is called ordinary Petri net. Biologically, it tells us that the stoichiometric coefficient of every reaction is equal to one.
2. *Boundedness*: If every place of a Petri net has maximum c number of tokens in any marking, i. e., if $\mu_0(p_i) \leq c \; \forall p_i \in P$, then it is called a bounded or c-bounded PN. If $c = 1$, then it is known as a safe Petri net. Boundedness ensures that there is no overflow of the resources. In a Petri net, the input places of the source transition with no restriction on capacity are always unbounded.
3. *Reachability Graph*: In this graph, vertices represent the markings μ' reachable from the initial marking μ_0 and the directed arcs denote the fired transitions. This graph shows distinct states of a network. A state is called directly reachable from another, if these states are connected by a directed arc. Reachability graph has a significant role in the study of the system's behavior.
4. *Liveness*: If a transition can fire at least once in every marking μ' reachable from μ, then it is called a live transition. If each transition in a Petri net is live, then it is called a live PN.
5. *Invariants*: The invariant analysis helps to predict the system behavior, even without the knowledge of kinetic data. The invariants are found from the incidence matrix of a PN. The values of the incidence matrix, C are defined by

$$c_{ij} = O(j,i) - I(j,i),$$

   where $I(j,i)$ and $O(j,i)$ are the input and output arc weight from the transition j. There are two types of invariants — place invariants and transition invariants:
   (a) *Place invariant*: A place invariant, also called a P-invariant, is the solution vector x, which is found by solving the below equation system:

$$C^T \cdot x = 0, \quad \text{where } x = (x_1, x_2, \ldots, x_n)^T,\; x_i \in \{1, 2, \ldots, n\} \text{ for } 1 \leqslant i \leqslant n.$$

   A P-invariant of a PN contains such places for which the sum of tokens does not vary throughout the model processing.
   A PN is called covered by place invariants (CPI), if every place is a member of some minimal P-invariant. If a Petri net possess CPI property, then it is a bounded PN [12]. The Petri net model of a network having places with source transitions only, will not be CPI because it is unbounded.
   (b) *Transition invariant*: A transition invariant, also called a T-invariant, is the solution vector y, which is found by solving the below equation system:

$$C \cdot y = 0, \quad \text{where } y = (y_1, y_2, \ldots, y_m)^T,\; y_i \in \{1, 2, \ldots, m\} \text{ for } 1 \leqslant i \leqslant m.$$

A T-invariant of a Petri net is a multi-set of transitions and has two different perceptions. It either gives the relative rate of transition firing corresponding to the steady state, or the number of times a transition should fire in order to regenerate the given marking of the system [2]. In other words, T-invariants represent the cyclic behavior in a network.
A PN is called covered by transition invariants (CTI), if every transition is a member of some minimal T-invariant [2, 12]. If a bounded PN possess CTI property, then it is live [1]. If a network is CTI, it means each process is executing in the network.

5 Analysis of the Petri net model (Figure 27.2)

In this section, we have discussed how we can use the properties of PN to analyze the behavior of a network by using our running example of Petri net model (Figure 27.2). To do so, we have used the PN tool *PIPE v4.3.0* [7].

The PN is ordinary PN when each arc weight is equal to 1. It is bounded as well as safe, as shown by the state space analysis in Figure 27.6. This result is also supported from the markings obtained in the Subsection 3, because each place receives maximum one token during the transition firing. This makes the PN model 1-bounded and hence safe.

Figure 27.6: Analysis of the PN model presented in Figure 27.2.

The reachability graph of the PN model, representing all reachable markings obtained in the Subsection 3, is displayed in Figure 27.7. The orange state $S3$ is the deadlock state, where no transition is enabled. The transition firing sequence of the deadlock state is given by $t_0 \rightarrow t_2 \rightarrow t_1$, which is shown in Figure 27.6. Since the model attains a deadlock state, thus it is not live.

The incidence matrix of our PN model is a 6×3 matrix and is shown in Figure 27.8. There are 3 P-invariants in the PN and it is covered by place invariants. There is no

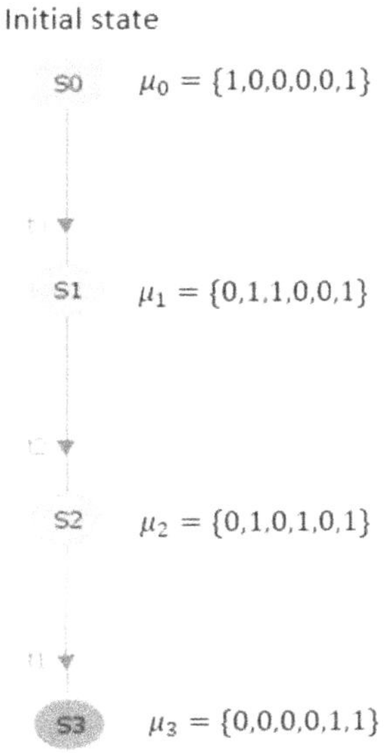

Figure 27.7: Reachability graph of the PN model displayed in Figure 27.2.

Combined incidence matrix

	t0	t1	t2
p0	-1	0	0
p1	1	-1	0
p2	1	0	-1
p3	0	-1	1
p4	0	1	0
p5	0	0	0

Figure 27.8: Incidence matrix of the PN presented in Figure 27.2.

T-invariant in the model, i. e., we can reproduce a state of the system. The sets of minimal P-invariants and T-invariants are presented in Figure 27.9.

6 Conclusion

Petri net is used to model the static properties of a concurrent and discrete event system. The two basic concepts used while modeling are the conditions and events of a system to be studied. It is used to describe how the information or data flows in a network, due to which it is gaining increased interest among the researchers. Its graphical representation makes the modeling more useful for practical purposes. Different variants of PN

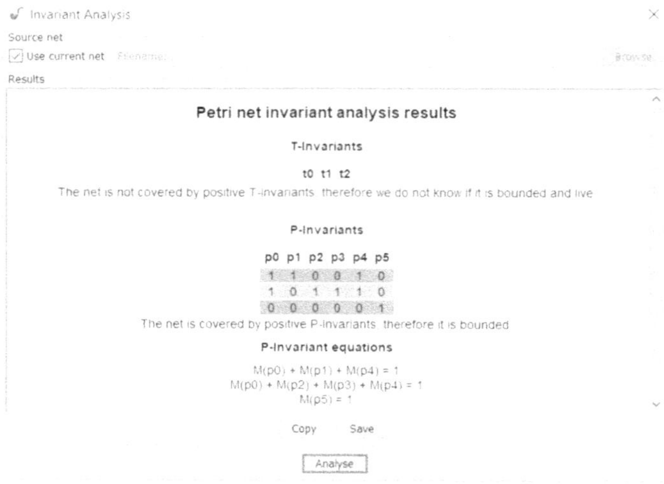

Figure 27.9: Place invariant and transition invariant of the PN model presented in Figure 27.2.

are available to model the concurrent systems and to analyze their behavior through its different properties and analysis techniques. To study the system performance based on the concepts of time, probability and uncertainty, the quantitative extensions of PN such as Continuous, Time, Stochastic, Fuzzy and Hybrid PNs have been introduced. The qualitative PN model can be converted to the quantitative PN by considering the kinetic parameters. Analysis of a PN model of a network will be helpful to get the insights about the complex interactions between its components.

Bibliography

[1] P. Baldan, N. Cocco, A. Marin and M. Simeoni, Petri nets for modelling metabolic pathways: a survey, *Nat. Comput.* **9**(4) (2010), 955–989.

[2] M. A. Blätke, M. Heiner and W. Marwan, Biomodel engineering with Petri nets, in: *Algebraic and Discrete Mathematical Methods for Modern Biology*, Elsevier, 2015, pp. 141–192.

[3] T. Brant-Ribeiro, R. D. Araújo, I. E. Mendonça, M. S. Soares and R. G. Cattelan, Interactive web interfaces modeling, simulation and analysis using colored Petri nets, *Softw. Syst. Model.* **18**(1) (2019), 721–737.

[4] R. V. Carvalho, F. J. Verbeek and C. J. Coelho, Bio-modeling using Petri nets: a computational approach, in: *Theoretical and Applied Aspects of Systems Biology*, Springer, 2018, pp. 3–26.

[5] C. Chaouiya, Petri net modelling of biological networks, *Brief. Bioinform.* **8**(4) (2007), 210–219.

[6] S. Dahiya, V. Prakash, S. Kumawat and P. Singh, Development of Petri net-based design model for energy efficiency in wireless sensor networks, in: *Computational Methods and Data Engineering*, Springer, 2021, pp. 259–272.

[7] N. J. Dingle, W. J. Knottenbelt and T. Suto, Pipe2: a tool for the performance evaluation of generalised stochastic Petri nets, *ACM SIGMETRICS Perform. Eval. Rev.* **36**(4) (2009), 34–39.

[8] S. Gupta, Z. Fatima and S. Kumawat, Study of the bioenergetics to identify the novel pathways as a drug target against mycobacterium tuberculosis using Petri net, *Biosystems* **209** (2021), 104509.

[9] S. Gupta, S. Kumawat and G. P. Singh, Fuzzy Petri net representation of fuzzy production propositions of a rule based system, in: *International Conference on Advances in Computing and Data Sciences*, Springer, 2019, pp. 197–210.

[10] S. Gupta, S. Kumawat and G. P. Singh, Validation and analysis of metabolic pathways using Petri nets, in: *Soft Computing: Theories and Applications*, Springer, 2022, pp. 361–374.

[11] S. Gupta, G. P. Singh and S. Kumawat, Modeling and targeting an essential metabolic pathway of plasmodium falciparum in apicoplast using Petri nets, *Appl. Math.-A Journal of Chinese Universities* **37**(1) (2022), 91–110.

[12] M. Heiner, D. Gilbert and R. Donaldson, Petri nets for systems and synthetic biology, in: *International School on Formal Methods for the Design of Computer, Communication and Software Systems*, Springer, 2008, pp. 215–264.

[13] S. Kansal, M. Acharya and G. P. Singh, Boolean Petri nets, in: *Petri nets-Manufacturing and Computer Science*, P. Pawlewski, ed., 2012, pp. 381–406.

[14] I. Koch, Petri nets in systems biology, *Softw. Syst. Model.* **14**(2) (2015), 703–710.

[15] S. Kumawat, Chinese postman problem: A Petri net based approach, in: *Computational Methods and Data Engineering*, Springer, 2021, pp. 203–222.

[16] S. Kumawat and G. Purohit, Total span of farm work flow using Petri net with resource sharing, *Int. J. Bus. Process Integration Manag.* **8**(3) (2017), 160–171.

[17] F. Liu, M. Heiner and D. Gilbert, Fuzzy Petri nets for modelling of uncertain biological systems, *Brief. Bioinform.* **21**(1) (2020), 198–210.

[18] T. Murata, Petri nets: Properties, analysis and applications, *Proc. IEEE* **77**(4) (1989), 541–580.

[19] J. L. Peterson, *Petri Net Theory and the Modeling of Systems*, Prentice Hall PTR, 1981.

[20] C. A. Petri, Communication with automata, 1966.

[21] W. Reisig, *Petri Nets: An Introduction*, Vol. 4, Springer Science & Business Media, 2012.

[22] G. P. Singh and S. K. Singh, Petri net recommender system for generating of perfect binary tree, *Int. J. Knowl. Syst. Sci. (IJKSS)* **10**(2) (2019), 1–12.

[23] G. P. Singh, S. K. Singh and M. Jha, Existence of forbidden digraphs for crisp boolean Petri nets, *Int. J. Math. Eng. Manag. Sci.* **5**(1) (2020), 83–95.

Ashish Negi and Ompal Singh

An inventory system with time-dependent linear demand for deteriorating items under uncertainty with a completely backlogged

Abstract: The technique used by decision-makers to simulate real-world issues is fuzzy set theory, which focuses primarily on how to statistically cope with uncertainty and imprecision. In this paper, an inventory system for deteriorating items with time-dependent linear demand is proposed under uncertainty. Shortages are permitted and completely backlogged to make the approach more inclusive. Additionally, a constant rate of deterioration is presumed. This model seeks to minimize the total average cost while determining the best ordering strategy. By reducing the overall cost of the inventory system, the solution to this model is found analytically. After that a well-known method that is used in numerous study areas with uncertainty is the signed distance method.

Keywords: Inventory system, deteriorating items, completely backlogging, uncertainty, time dependent demand, signed distance method

1 Introduction

The rate of degradation and beginning point for perishable goods vary depending on the type of product and the storage facilities that are available. Due to issues with the environment, humidity, temperature, and other areas, many products, including food, vegetables, cosmetics, pharmaceuticals, blood banks, etc., fail to preserve their freshness for an extended period of time. Therefore, the most crucial component of inventory management is having the greatest and most intelligent storage facilities for things. When analyzing the inventory models, one of the most crucial elements is degradation, which is significant. Because product storage time raises the overall cost of the inventory, holding costs also play a significant part in the inventory system. Economic Order Quantity (EOQ) inventory model was the first model examined by Harris [5]. Hadley and Whitin [3] studied the theory of inventory systems with deterministic and probabilistic parameters, types of inventory models and inventory costs, etc. Demand has always played a crucial part in the inventory system under the economic order quantity (EOQ) model.

Ashish Negi, Ompal Singh, Department of Mathematics, SRM Institute of Science and Technology, Delhi-NCR Campus, Modinagar, Ghaziabad-201204, Uttar Pradesh, India

https://doi.org/10.1515/9783111313634-028

In recent years, EOQ models were developed for time-varying demand with deterioration [8], fuzzy EOQ models for deteriorating items with price-sensitive and stock-dependent demand function under permissible shortages [17], EOQ models with learning effect, carbon emission under inflationary circumstances, and retailer inspection for naturally deteriorating items [1]. The effect of learning in fuzziness by considering fuzzy demand in the EOQ model for deteriorating items under a finite-time horizon was also introduced [13].

In today's society, a client chooses a product mostly on the basis of its dependability and advertising in the print or electronic media. The frequency of advertising in today's mass media, including TV, newspapers, magazines, websites, and mobile devices, determines the demand for a given product. People are encouraged to purchase more of these products by the glitzy commercials and by the salespeople themselves. Numerous scholars have presented diverse forms of demand rates that are taken into consideration, including price-dependent demand [14, 6, 11], time-dependent demand [20, 9, 17, 18], time-and-price dependent demand [19], fuzzy random demand [16], stock-dependent demand [2], uncertainty demand [16, 4, 7, 10, 15], created a model of deteriorating inventory with underpriced preservation technology and stock-sensitive demand.

It is assumed in several research papers that there is a complete backlog of unmet demand. In reality, particular consumers are prepared to hold off on picking up their orders until the end of the shortage period, while others are not. A fixed proportion of demand during the stock-out period is backordered. The decision variables are the inventory cycle and the ratio between the initial stock and the total quantity demanded throughout the inventory cycle. An inventory system with time-dependent demand and partial backordering under return on inventory investment maximization [12].

2 Assumptions and notations

1. For each corporate organization, creating a suitable inventory system is one of their top concerns. This study looked at a fuzzy inventory model for deteriorating goods in which the objective cost function and inventory cost coefficients are fuzzy in nature and solved using signed distance. In this model, the demand rate follows a time-varying pattern (decreasing function of time). Shortages are not considered in this research work. This work study the fuzzy inventory model with succeeding assumptions and notations {~ sign used for the fuzzy constraints}:
2. The demand of the product is $F(t) = (\beta_1 + \beta_2 t)$.
3. Shortage is not allowed to occur in this research work. Control how often or when the inventory should be replenished.
4. The replenishment rate is instantaneous, i. e., lead time is assumed to be zero.

5. Let $\tilde{P}$ and $\tilde{Q}$ be any two fuzzy numbers:

$$\tilde{P} \oplus \tilde{Q} = (p_1 q_1, p_2 q_2, p_3 q_3, p_4 q_4),$$

   where $\tilde{P} = (p_1, p_2, p_3, p_4)$ and $\tilde{Q} = (q_1, q_2, q_3, q_4)$
6. Triangular fuzzy number $\phi = (\phi_1, \phi_2, \phi_3)$, where $\phi_1 = \phi - \Delta_1$, $\phi_2 = \phi$, $\phi_3 = \phi + \Delta_2$.

The membership function of ϕ is

$$\mu_\phi(\tilde{\phi}) = \begin{cases} \frac{\phi-\phi_1}{\phi_2-\phi_1}, & \phi_1 \le \phi \le \phi_2 \\ \frac{\phi_3-\phi}{\phi_3-\phi_2}, & \phi_2 \le \phi \le \phi_3 \\ 0, & \text{otherwise.} \end{cases}$$

3 Notations

1. A_0: The order cost per unit product.
2. C_h: The holding cost per unit product.
3. $I_1(t)$: The inventory level in the time interval $(0, t_1)$.
4. $I_2(t)$: The inventory level in the time interval (t_1, T).
5. w: The initial inventory level.
6. α: Deterioration rate.
7. C_d: The deteriorating cost per unit product.
8. C_s: The shorting cost per unit product.
9. T_{IC}: The inventory cost function.
10. $\tilde{T}_{IC}$: The inventory cost function in fuzzy model.

4 Mathematical formulation

In this section, we develop mathematical model to optimize total inventory costs under the influence of inflation.

The governing differential equations during the stock period $[0, t_1]$ and shortage period $[t_1, T]$ are:

$$\frac{dI_1(t)}{dt} + \alpha I_1(t) + (\beta_1 + \beta_2 t) = 0, \quad 0 \le t \le t_1, \tag{28.1}$$

$$\frac{dI_2(t)}{dt} + (\beta_1 + \beta_2 t) = 0, \quad t_1 \le t \le T. \tag{28.2}$$

With boundary condition

$$I_1(t) = w, \text{ when } t = 0, \quad \text{and} \quad I_2(t) = 0, \text{ when } t = t_1.$$

Solution (1): Applying boundary condition gives:

$$I_1(t) = \left(-\frac{\beta_1}{\alpha} + \frac{\beta_2}{\alpha^2}\right) - \frac{\beta_2}{\alpha}t + \left(w + \frac{\beta_1}{\alpha} - \frac{\beta_2}{\alpha^2}\right)e^{-\alpha t}. \tag{28.3}$$

Solution (2): Applying boundary condition gives:

$$I_2(t) = \beta_1(t_1 - t) + \frac{\beta_2}{2}(t_1^2 - t^2). \tag{28.4}$$

Ordering costs:

$$O_C = A_0.$$

Holding costs:

$$H_C = C_h\left[\int_0^{t_1} I_1(t)dt\right],$$

$$H_C = C_h\left[\left(-\frac{\beta_1}{\alpha} + \frac{\beta_2}{\alpha^2}\right)t_1 - \frac{\beta_2}{2\alpha}t_1^2 - \frac{1}{\alpha}\left(w + \frac{\beta_1}{\alpha} - \frac{\beta_2}{\alpha^2}\right)e^{-\alpha t_1}\right].$$

Deteriorating costs:

$$D_C = C_d\left[\int_0^{t_1} \alpha I_1(t)dt\right],$$

$$D_C = C_d\left[\left(-\beta_1 + \frac{\beta_2}{\alpha}\right)t_1 - \frac{\beta_2}{2}t_1^2 - \left(w + \frac{\beta_1}{\alpha} - \frac{\beta_2}{\alpha^2}\right)e^{-\alpha t_1}\right].$$

Shorting costs:

$$S_C = C_s\left[\int_{t_1}^{T} S_T dt\right], \quad \text{where } S_T = -I_2(t),$$

$$S_C = C_s\left[-\beta_1\left\{\left(t_1T - \frac{T^2}{2}\right) - \left(t_1^2 - \frac{t_1^2}{2}\right)\right\} - \frac{\beta_2}{2}\left\{\left(t_1^2T - \frac{T^3}{3}\right) - \left(t_1^3 - \frac{t_1^3}{3}\right)\right\}\right].$$

Total inventory costs function:

$$T_{IC}(t_1, T) = \frac{1}{T}(O_C + H_C + D_C + S_C),$$

$$T_{IC}(t_1, T) = \frac{1}{T}\left[A_0 + (C_h + \alpha C_d)\left(-\frac{\beta_1}{\alpha} + \frac{\beta_2}{\alpha^2}\right)t - (C_h + \alpha C_d)\frac{\beta_2}{2\alpha}t_1^2\right.$$
$$- \frac{1}{\alpha}(C_h + \alpha C_d)\left(w + \frac{\beta_1}{\alpha} - \frac{\beta_2}{\alpha^2}\right)e^{-\alpha t}$$
$$\left. + C_s\left[-\beta_1\left\{\left(t_1 T - \frac{T^2}{2}\right) - \left(t_1^2 - \frac{t_1^2}{2}\right)\right\} - \frac{\beta_2}{2}\left\{\left(t_1^2 T - \frac{T^3}{3}\right) - \left(t_1^3 - \frac{t_1^3}{3}\right)\right\}\right]\right].$$

5 Optimum solution method

To obtain the optimum solution, we first get the values of t_1, T„ T_{IC} for this inventory system. To minimize the total inventory costs T_{IC}, we have

$$\frac{\partial T_{IC}}{\partial t_1} = 0, \quad \frac{\partial T_{IC}}{\partial T} = 0,$$
$$\left(\frac{\partial^2 T_{IC}}{\partial t_1^2}\right)\left(\frac{\partial^2 T_{IC}}{\partial T^2}\right) - \left(\frac{\partial^2 T_{IC}}{\partial t_1 \partial T}^2\right) > 0,$$
$$\text{and} \quad \left(\frac{\partial^2 T_{IC}}{\partial t_1^2}\right) > 0.$$

6 Fuzzy model

Most of the researchers proposed their models for forecasting the optimal solution and optimal ordering quantity assumed by constant demand and deterioration rate. Generally, the inventory system with deteriorating products, such as vegetables, fruits, packing food products contains the rough constraints, like inaccurate inventory costs, fuzzy holding costs, ordering costs, etc. In this work, we consider that the inventory ordering cost ($\tilde{A}_0$), inventory holding costs ($\tilde{C}_h$), inventory deteriorating costs ($\tilde{C}_d$) and inventory shorting costs ($\tilde{C}_s$) are in fuzzy form. Therefore, due to our consideration, the objective function ($\tilde{T}_{IC}$) becomes ($\tilde{T}_{IC}$).

7 Signed distance method

A well-known method that is used in numerous study areas with uncertainty is the signed distance method. The suggested inventory model's hazy restrictions are as follows: inventory ordering costs ($\tilde{A}_0$), inventory holding costs ($\tilde{C}_h$), inventory deteriorating costs ($\tilde{C}_d$) and shorting costs ($\tilde{C}_s$) are defined as:

(i) $C_h \in [C_h - \Delta_1, C_h - \Delta_2]$, where $0 < \Delta_1 < C_h$ and $0 < \Delta_1\Delta_2$.
(ii) $A_0 \in [A_0 - \Delta_3, A_0 - \Delta_4]$, where $0 < \Delta_3 < A_0$ and $0 < \Delta_3\Delta_4$.
(iii) $C_d \in [C_d - \Delta_5, C_d - \Delta_6]$, where $0 < \Delta_5 < C_h$ and $0 < \Delta_5\Delta_6$.

Using the signed distance method, the defuzzification of inventory ordering costs ($\tilde{A}_0$), inventory holding costs ($\tilde{C}_h$), inventory deteriorating costs ($\tilde{C}_d$) and inventory shorting costs ($\tilde{C}_s$) are defined as

(i) $d(\tilde{C}_h, 0) = C_h + \frac{1}{4}(\Delta_2 - \Delta_1)$.

(ii) $d(\tilde{A}_0, 0) = A_0 + \frac{1}{4}(\Delta_4 - \Delta_3)$.

(iii) $d(\tilde{C}_d, 0) = C_d + \frac{1}{4}(\Delta_6 - \Delta_5)$.

Fuzziness can be related to objective cost functions with constrained involved. Zimmermann [21] proposed the fuzzy-based both objective function and constraints. Here the fuzzy objective cost function ($\tilde{T}_{IC}$) can be written as in form

$$\tilde{T}_{IC}(t_1, T) = (T_{IC1} + T_{IC2} + T_{IC3})$$
$$T_{IC}(t_1, T) = \frac{1}{T}(O_C + H_C + D_C + S_C)$$

$$\begin{aligned} T_{IC1}(t_1, T) = \frac{1}{T}\bigg[&(A_0 - \Delta_3) + ((C_h - \Delta_1) + \alpha(C_d - \Delta_5))\left(-\frac{\beta_1}{\alpha} + \frac{\beta_2}{\alpha^2}\right)t \\ &- ((C_h - \Delta_1) + \alpha(C_d - \Delta_5))\frac{\beta_2}{2\alpha}t_1^2 - \frac{1}{\alpha}((C_h - \Delta_1) + \alpha(C_d - \Delta_5))\left(w + \frac{\beta_1}{\alpha} - \frac{\beta_2}{\alpha^2}\right)e^{-\alpha t} \\ &+ C_s\left[-\beta_1\left\{\left(t_1 T - \frac{T^2}{2}\right) - \left(t_1^2 - \frac{t_1^2}{2}\right)\right\} - \frac{\beta_2}{2}\left\{\left(t_1^2 T - \frac{T^3}{3}\right) - \left(t_1^3 - \frac{t_1^3}{3}\right)\right\}\right]\bigg]. \end{aligned}$$

Now

$$T_{IC2}(t_1, T) = T_{IC}(t_1, T)$$

and

$$\begin{aligned} T_{IC3}(t_1, T) = \frac{1}{T}\bigg[&(A_0 - \Delta_4) + ((C_h - \Delta_2) + \alpha(C_d - \Delta_6))\left(-\frac{\beta_1}{\alpha} + \frac{\beta_2}{\alpha^2}\right)t \\ &- ((C_h - \Delta_2) + \alpha(C_d - \Delta_6))\frac{\beta_2}{2\alpha}t_1^2 - \frac{1}{\alpha}((C_h - \Delta_2) + \alpha(C_d - \Delta_6))\left(w + \frac{\beta_1}{\alpha} - \frac{\beta_2}{\alpha^2}\right)e^{-\alpha t} \\ &+ C_s\left[-\beta_1\left\{\left(t_1 T - \frac{T^2}{2}\right) - \left(t_1^2 - \frac{t_1^2}{2}\right)\right\} - \frac{\beta_2}{2}\left\{\left(t_1^2 T - \frac{T^3}{3}\right) - \left(t_1^3 - \frac{t_1^3}{3}\right)\right\}\right]\bigg]. \end{aligned}$$

In particular, the inventory system can use the fuzzy method discussed in this work to apply the best inventory regulations. The objective's defuzzification the cost function

$$\begin{aligned} \tilde{F}(T_{IC}) = T_{IC}(t_1, T) + \frac{1}{T}\bigg[&\frac{1}{4}(\Delta_4 - \Delta_3)\left(-\frac{\beta_1}{\alpha} + \frac{\beta_2}{\alpha^2}\right)t \\ &+ \left(\frac{1}{4}(\Delta_2 - \Delta_1)\frac{\beta_2}{2\alpha}t_1^2 + \alpha(\Delta_6 - \Delta_5)\left(w + \frac{\beta_1}{\alpha} - \frac{\beta_2}{\alpha^2}\right)e^{-\alpha t}\right) \\ &+ C_s\left[-\beta_1\left\{\left(t_1 T - \frac{T^2}{2}\right) - \left(t_1^2 - \frac{t_1^2}{2}\right)\right\} - \frac{\beta_2}{2}\left\{\left(t_1^2 T - \frac{T^3}{3}\right) - \left(t_1^3 - \frac{t_1^3}{3}\right)\right\}\right]\bigg]. \end{aligned}$$

8 Numerical analysis

This work examines the significance of a demand- and supply-based inventory system. It focuses on developing a fuzzy-based inventory model for products that decay over time and have demand that varies with time, while considering fuzzy constraints. In order determine the correctness and pertinence of this study, we begin by examining the numerical examples and delving into the sensitivity analysis of the constraints used to analyze this system, alongside the optimal total cost function. The solution methodology for the inventory system, aimed at achieving the most cost-effective outcome and determining the optimal order quantity, is demonstrated through the following examples.

Example 1

Here we assumed the following constraints values for the developed model; demand rate $\beta_1 = 400$ and $\beta_2 = 5$ units per annum; ordering costs $(O_c) = 1500$; deterioration rate $(\alpha) = 0.6$; holding costs $(C_h) = 0.4$; deterioration costs $(C_d) = 0.04$; shorting costs $(C_s) = 0.02$. Using these constraints values, the proposed non-linear optimization techniques with various parameters values and the sensitivity analysis with resultant are presented in Table 28.1. The optimal solution is obtained after some iteration deterioration rate

Table 28.1: Sensitivity analysis for effactive parameters.

Change in		t_1	T	w	T_{IC}
β_1	−10 %	2.42	2.96	3312.23	4567.82
	400	2.92	3.53	3716.52	5102.66
	+10 %	3.36	4.02	4056.36	5832.16
β_2	−10 %	2.56	2.86	3256.25	4483.49
	5	2.92	3.53	3716.52	5102.66
	+10 %	3.45	4.12	4123.78	5784.57
α	−10 %	2.92	2.58	3716.53	5102.56
	0.6	2.92	3.53	3716.52	5102.66
	+10 %	2.92	2.98	3716.53	5102.72
C_d	−10 %	2.92	1.98	3716.51	5102.65
	0.04	2.92	3.53	3716.52	5102.66
	+10 %	2.92	6.28	3716.53	5102.67
O_C	−10 %	2.82	3.46	3716.28	5092.52
	1500	2.92	3.53	3716.52	5102.66
	+10 %	2.98	3.62	3716.76	5116.82
C_h	−10 %	2.79	5.87	3706.39	4583.76
	0.4	2.92	3.53	3716.52	5102.66
	+10 %	2.35	2.24	3725.76	5653.71
C_s	−10 %	1.25	3.92	3616.45	4927.18
	0.02	2.92	3.53	3716.52	5102.66
	+10 %	3.32	2.94	3846.48	5324.79

$t_1 = 2.92$, $T = 3.53$, $w = 3716.52$ and $T_{IC} = 5102.66$. It can be decided that the output of these analyzes (Table 28.1) and the obtained result is acceptable and can be used in business industries for decision making. The sensitivity analysis and graphical results shown below demonstrate the efficiency of our developed system.

9 Conclusion

The effect of constraints involved in this inventory system with total inventory costs and sensitivity analyzes are illustrated to demonstrate the optimal output. In this study, a fuzzy inventory system with time-varying linearly decreasing demand is formulated. This research work has been solved in both cases crisp and fuzzy environment. For crisp system, the optimal inventory costs are minimized with optimization techniques and in fuzzy system the optimal inventory costs are minimized with signed distance method.

This study will help the experts of seasonal items such as fruits, vegetable, packed food, medicine, cosmetic items, etc. Finally, numerical analysis and sensitivity analysis of total inventory costs with various constraints are illustrated to obtain the optimum result. The fuzzy environment is better in comparison of crisp environment for optimizing (minimize) the total inventory cost of the system. This analysis and formulation examined are in general framework and can be protracted to contain some other fuzzy-based method and fuzzy logic relations with various constraints on demand, deterioration, advertisement, price, backlogging, profit, production, trade credit and inflation, etc.

Bibliography

[1] O. A. Alamri et al., An EOQ model with carbon emissions and inflation for deteriorating imperfect quality items under learning effect, *Sustainability* **14**(3) (2022), 1365.

[2] Yang Chih-Te., An inventory model with both stock-dependent demand rate and stock-dependent holding cost rate, *Int. J. Prod. Econ.* (2014), 214–221.

[3] G. Hadley and T. M. Whitin, *Analysis of inventory systems*, Prentice Hall, Englewood Cliffs, 1963.

[4] O. Hansen, S. Transchel and H. Friedrich, Replenishment strategies for lost sales inventory systems of perishables under demand and lead time uncertainty, *Eur. J. Oper. Res.* **308**(2) (2023), 661–675.

[5] F. W. Harris, Operations and cost, factory management, *Prod. Oper. Manag.* (1915), 1–6.

[6] S. Kumar et al., An inventory model with price dependent demand rates as power law form using ant colony optimization, *J. Manag. Inf. Decis. Sci.* (2021), 1–10.

[7] A. K. Malik et al., Effect of uncertainty in optimal inventory policy for manufacturing products, *Int. J. Intell. Syst. Appl. Eng.* **10**(1s) (2022), 102–110.

[8] V. Mohan, M. Ota and A. Kumar, An inventory model for decaying items with Pareto distribution, time-dependent demand and shortages, *Int. J. Math. Oper. Res.* **21**(1) (2022), 83–103.

[9] P. Mondal, A. K. Das and T. K. Roy, An EOQ model for deteriorating item with continuous linear time dependent demand with trade of credit and replenishment time being demand dependent, *Int. J. Math. Oper. Res.* **24**(1) (2023), 104–127.

[10] J. Ray, Deterioration and its uncertainty in inventory systems, *Glob. J. Pure Appl. Math.* **13**(8) (2017), 4003–4014.

[11] S. Saha and N. Sen, An inventory model for deteriorating items with time and price dependent demand and shortages under the effect of inflation, *Int. J. Math. Oper. Res.* **14**(3) (2019), 377–388.

[12] L. A. San-José et al., An inventory system with time-dependent demand and partial backordering under return on inventory investment maximization, *Comput. Oper. Res.* **145** (2022), 105861.

[13] H. N. Soni and S. N. Suthar, EOQ model of deteriorating items for fuzzy demand and learning in fuzziness with finite horizon, *J. Control Decis.* **8**(2) (2021), 89–97.

[14] A. Swami et al., An inventory model with price sensitive demand, variable holding cost and trade-credit under inflation, *Int. J. Curr. Res.* (2015), 17313–17321.

[15] T. Tanthatemee and B. Phruksaphanrat, Fuzzy inventory control system for uncertain demand and supply, in: *Proceedings of the International Multiconference of Engineers and Computer Scientists*, 2012.

[16] E. B. Tirkolaee, A. Goli and G.-W. Weber, Multi-objective aggregate production planning model considering overtime and outsourcing options under fuzzy seasonal demand, in: *Advances in Manufacturing II: Volume 2-Production Engineering and Management*, Springer International Publishing, 2019.

[17] R. Udayakumar, An EOQ model for non-instantaneous deteriorating items with time-dependent demand under partial backlogging, *J. Manag. Anal.* **9**(4) (2022), 514–531.

[18] R. Verma, P. Narang and P. K. De, An EOQ model with time-dependent demand and holding cost under the effect of inspection on deterioration rate, *J. Ind. Integr. Manag.* (2023), 2350015.

[19] A. S. Yadav, A. Kumar, K. K. Yadav et al., Optimization of an inventory model for deteriorating items with both selling price and time-sensitive demand and carbon emission under green technology investment, *Int. J. Interact. Des. Manuf.* (2023), https://doi.org/10.1007/s12008-023-01689-8.

[20] K. K. Yadav, A. S. Yadav and S. Bansal, Interval number approach for two-warehouse inventory management of deteriorating items with preservation technology investment using analytical optimization methods, *Int. J. Interact. Des. Manuf.* (2024), https://doi.org/10.1007/s12008-023-01672-3.

[21] H. J. Zimmermann, Description and optimization of fuzzy systems, *Int. J. Gen. Syst.* (1975), 209–215.

Deepak Kumar, Sukhchain Singh, and Bharat Bhushan

On $*$-reverse derivations of involutive rings

Abstract: An additive map d is called $*$-reverse derivation on a ring R if $d(xy) = d(y)x^* + y^*d(x)$ for all $x, y \in R$. Recently, many algebraists studied $*$-differential identities on a prime ring and semi-prime rings involving a pair of different types of derivations. In this article, we study the structure of an involutive ring R when a pair of $*$-reverse derivation satisfies some $*$-differential identities.

Keywords: Prime-ring, semi-prime ring, involution, reverse derivations, $*$-reverse derivations

MSC 2010: 16W10, 16N60, 16W25

1 Introduction

Throughout this article, R is a ring with center $Z(R)$. A ring R is called a *prime ring* if its (0) ideal is a prime ideal and is called a *semi-prime ring* if its (0) ideal is a semi-prime ideal. A two-ordered anti-endomorphism $*$ of R is called an *involution* on R. A ring with involution is called a $*$*-ring*. An element x of ring R is called *symmetric* (*skew-symmetric*) if $x^* = x$ ($x^* = -x$). $H(R)$ is the collection of all symmetric elements of R and $S(R)$ is the collection of all skew-symmetric elements of R. An involution is called of *first kind* if $Z(R) \subseteq H(R)$ otherwise the involution is called of *second kind*. As usual, $a \circ b$ (resp. $[a, b]$) represents the anti-commutator $ab + ba$ (resp. commutator $ab - ba$).

An additive map d on R is a *derivation* (*reverse derivation*) if $d(xy) = d(x)y + xd(y)$ ($d(xy) = d(y)x + yd(x)$) for every pair of $x, y \in R$. Herstein [7] proved that if there exists a nonzero reverse derivation on a prime ring R, then R is commutative. Moreover Samman [9] proved that if there exists a nonzero reverse derivation on a semi-prime ring R, then it is an ordinary derivation on R. These results indicate that the study of reverse derivation is not appropriate in order to characterize the structure of a ring.

In 1989, Brešar and Vukman [5] introduced the notion of $*$-derivation. This motivates Bhushan et al. [4] to introduce the notion of $*$-reverse derivation as: An additive

Acknowledgement: The research of the second author is supported by University Grant Commission, New Delhi (JRF, grant no. 403382).

Deepak Kumar, Department of Mathematics, Punjabi University, RSF-13, Gol Market, Patiala-147002, Punjab, India, e-mail: deepakmath@pbi.ac.in
Sukhchain Singh, Department of Mathematics, Punjabi University, 67 Urban Estate Phase 3, Patiala-147002, Punjab, India, e-mail: sukhchain_rs@pbi.ac.in
Bharat Bhushan, Department of Mathematics, Akal University, 151302 Talwandi Sabo, Punjab, India, e-mail: bharat_rs18@pbi.ac.in

https://doi.org/10.1515/9783111313634-029

function δ is called a $*$-*reverse derivation* if $\delta(xy) = \delta(y)x^* + y^*\delta(x)$ on R. For any fixed $a \in R$, a mapping $x \mapsto [x^*, a]$ is a natural example of a $*$-reverse derivation. This example ensures the existence of a nonzero $*$-reverse derivation on a prime ring which is different from the ordinary derivation. Hence the study of $*$-reverse derivations on a prime ring is nontrivial and interesting.

On the other hand, it is seen that $*$-differential identities plays a significant role to characterize the ring using different maps. Many algebraists like Aboubakr [1], Susan [6], Bhushan [2, 4], Sandhu [10, 11] and many more obtained crucial results using some $*$-differential identities along with such maps. In particular, Bharat et al. [3] and D. Kumar et al. [8] classified the pair of mappings that satisfies some $*$-differential identities. Continuing in this line of investigation, we study the structure of a prime ring and semi-prime ring on a pair of $*$-reverse derivations satisfying the following assertions for all $x, y \in R$:

(i) $[d_1(x), d_2(y)] = [x^*, y^*]$,
(ii) $[d_1(x), d_2(x^*)] = [x, x^*]$,
(iii) $d_1(x) \circ d_2(y) = x^* \circ y^*$,
(iv) $d_1(x) \circ d_2(x) = x \circ x^*$.

2 Preliminaries

Lemma 2.1 ([2, Theorem 8]). *Let R be a prime ring and d be a $*$-reverse derivation on R. If d is $*$-centralizing on a nonzero one sided ideal of the ring R, then R is commutative.*

Lemma 2.2. *If d is a $*$-reverse derivation on R, then d preserve the center of R.*

Proof. For any $r \in R$ and $z \in Z(R)$, we have

$$d(zr) - d(r)z^* - r^*d(z) = 0 \tag{29.1}$$

and

$$d(rz) - d(z)r^* - z^*d(r) = 0. \tag{29.2}$$

From both equations (29.1) and (29.2), we get $d(z)r^* - r^*d(z) = 0$. Which means $[d(z), r] = 0$ for all $z \in Z(R)$ and for all $r \in R$. Hence $d(z) \in Z(R)$ for all $z \in Z(R)$. □

3 Main results

Theorem 3.1. *Let R be a semi-prime ring and d_1, d_2 be $*$-reverse derivations. If $[d_1(x), d_2(y)] = [x^*, y^*]$ for all $x, y \in R$, then R is commutative.*

Proof. We have

$$[d_1(x), d_2(y)] = [x^*, y^*] \quad \text{for all } x, y \in R. \tag{29.3}$$

Replacing y into zy, where $z \in R$, in the equation (29.3), we get

$$\begin{aligned}&[d_1(x), d_2(y)]z^* + d_2(y)[d_1(x), z^*] + [d_1(x), y^*]d_2(z) + y^*[d_1(x), d_2(z)]\\ &\quad = y^*[x^*, z^*] + [x^*, y^*]z^*.\end{aligned} \tag{29.4}$$

Using (29.3) in the equation (29.4), we find

$$d_2(y)[d_1(x), z^*] + [d_1(x), y^*]d_2(z) = 0. \tag{29.5}$$

Replacing z by $d_1(x)^*$ in (29.5), we get

$$[d_1(x), y^*]d_2(d_1(x)^*) = 0. \tag{29.6}$$

For any $r \in R$, replacing y into yr in the equation (29.6), we get

$$[d_1(x), r^*]y^* d_2(d_1(x)^*) = 0. \tag{29.7}$$

Now changing y into $d_1(r)^* y$ in (29.7), we get

$$[d_1(x), r^*]y^*[d_2(d_1(x)^*), d_1(r)] = 0. \tag{29.8}$$

Using (29.3) to get

$$[d_1(x), r^*]y^*[d_1(x), r^*] = 0. \tag{29.9}$$

In particular, by taking $r = x$ in (29.9), we obtain

$$[d_1(x), x^*] = 0. \tag{29.10}$$

Hence by using the Lemma 2.1, R is commutative. □

Corollary 3.1. *Let R be a prime ring and d be a $*$-reverse derivation. If $[d(x), d(y)] = [x^*, y^*]$ for all $x, y \in R$, then R is commutative.*

Theorem 3.2. *Let R be a 2-torsion free prime ring with involution of second kind. If d_1, d_2 are two $*$-reverse derivations on R such that $[d_1(x), d_2(x^*)] = [x, x^*]$ for all $x \in R$, then R is commutative.*

Proof. We have for all $x \in R$,

$$[d_1(x), d_2(x^*)] = [x, x^*]. \tag{29.11}$$

Polarizing the above relation, we get

$$[d_1(x), d_2(y^*)] + [d_1(y), d_2(x^*)] = [x, y^*] + [y, x^*]. \tag{29.12}$$

Changing x into xh for any $h \in H(R) \cap Z(R)$ and using the equation (29.12) and Lemma 2.1 to find

$$d_1(h)[x^*, d_2(y^*)] + [d_1(y), x]d_2(h) = 0. \tag{29.13}$$

Now replacing x into xk for any $k \in S(R) \cap Z(R)$ and using primeness of R, we obtain

$$-d_1(h)[x^*, d_2(y^*)] + [d_1(y), x]d_2(h) = 0. \tag{29.14}$$

Adding the equation (29.13) and (29.14), we capture

$$2[d_1(y), x]d_2(h) = 0. \tag{29.15}$$

Since R is 2-torsion free prime ring, the above equation can be written as

$$[d_1(y), x]d_2(h) = 0. \tag{29.16}$$

Since R is a prime ring, from the above equation we get either $[d_1(y), x] = 0$ or $d_2(h) = 0$. For the first case, using the Lemma 2.1, we get either d_1 a zero map or R is commutative. For second case, we have $d_2(h) = 0$, taking $h = k^2$ for $k \in S(R) \cap Z(R)$ to obtain $d_2(k) = 0$ for all $k \in S(R) \cap Z(R)$. Similarly, we have $d_1(k) = 0$ for all $k \in S(R) \cap Z(R)$.

Now replacing x into xk in the equation (29.12), we have

$$k(-[d_1(x), d_2(y^*)] + [d_1(y), d_2(x^*)]) = k([x, y^*] - [y, x^*]), \tag{29.17}$$

which implies

$$-[d_1(x), d_2(y^*)] + [d_1(y), d_2(x^*)] = [x, y^*] - [y, x^*]. \tag{29.18}$$

From the equation (29.12) and (29.18), we obtain

$$[d_1(y), d_2(x^*)] = [x, y^*]. \tag{29.19}$$

Taking x^* instead of x and interchanging variables, the above relation can be written as

$$[d_1(x), d_2(y)] = -[x^*, y^*]. \tag{29.20}$$

Now by following the Theorem 3.1, it can be easily obtained that R is commutative. □

Corollary 3.2. *Let R be a 2-torsion free prime ring with involution of second kind. If d is $*$-reverse derivations on R such that $[d(x), d(x^*)] = [x, x^*]$ for all $x \in R$, then R is commutative.*

Corollary 3.3. *Let R be a 2-torsion free prime ring with involution of second kind. If d is $*$-reverse derivations on R such that $[d(x), d(x^*)] = [x, x^*] + x \circ x^*$ for all $x \in R$, then R is commutative.*

Proof. We have for all $x \in R$

$$[d(x), d(x^*)] = [x, x^*] + x \circ x^*.$$

Interchanging x and x^* in the above relation, we find

$$[d(x), d(x^*)] = [x, x^*] - x \circ x^*.$$

From last two equations, we conclude

$$[d(x), d(x^*)] = [x, x^*].$$

Therefore, by using the Theorem 3.2, we get the required result. □

Theorem 3.3. *Let R be a prime ring with involution $*$ and d_1, d_2 be $*$-reverse derivations on R. If $d_1(x) \circ d_2(y) = x^* \circ y^*$ for all $x, y \in R$ and $Z(R) \neq \{0\}$, then R is commutative.*

Proof. We have

$$d_1(x) \circ d_2(y) = x^* \circ y^* \quad \text{for all } x, y \in R. \tag{29.21}$$

Replacing x into xy in the equation (29.21), we obtain

$$d_1(y)x^* \circ d_2(y) + y^* d_1(x) \circ d_2(y) = y^* x^* \circ y^*. \tag{29.22}$$

Simplifying the above equation

$$(d_1(y) \circ d_2(y))x^* - d_1(y)[d_2(y), x^*] + y^*(d_1(x) \circ d_2(y)) + [d_2(y), y^*]d_1(x) = y^*(x^* \circ y^*). \tag{29.23}$$

Using (29.21) in the equation (29.23), we get

$$(d_1(y) \circ d_2(y))x^* - d_1(y)[d_2(y), x^*] + [d_2(y), y^*]d_1(x) = 0. \tag{29.24}$$

For any $r \in R$, changing x into rx in (29.24)

$$-d_1(y)x^*[d_2(y), r^*] + [d_2(y), y^*]x^* d_1(r) = 0. \tag{29.25}$$

Taking $r = z \in Z(R)$ and changing x into x^* in the equation (29.25)

$$[d_2(y), y^*]x d_1(z) = 0 \quad \text{for all } z \in Z(R), y \in R. \tag{29.26}$$

Since R is a prime ring, the above equation implies that either $[d_2(y), y^*] = 0$ for all $y \in R$ or $d_1(z) = 0$ for all $z \in Z(R)$. For the first case, by using [2, Theorem 8], R is commutative.

For second case, in particular taking $y = z$ for any $z \in Z(R)$ in the hypothesis, we obtain $0 = 2zx$ for all $x \in R$, i.e, $z = 0$, which is contradiction as $Z(R) \neq \{0\}$. □

Corollary 3.4. *Let R be a prime ring with involution $*$ and d be a $*$-reverse derivations on R. If $d(x) \circ d(y) = x^* \circ y^*$ and $Z(R) \neq \{0\}$, then R is commutative.*

Theorem 3.4. *Let R be a prime ring with involution of second kind and d_1, d_2 be non-zero $*$-reverse derivations on R. If $d_1(x) \circ d_2(x) = x \circ x^*$ for every $x \in R$, then R is commutative.*

Proof. We have

$$d_1(x) \circ d_2(x) = x \circ x^* \quad \text{for all } x \in R. \tag{29.27}$$

Taking $x + y$ instead of x for any $y \in R$ in the equation (29.27), we get

$$d_1(x) \circ d_2(y^*) + d_1(y) \circ d_2(x^*) = x \circ y^* + y \circ x^*. \tag{29.28}$$

Replacing x into xh, where $h \in H(R) \cap Z(R)$, in the equation (29.28) and using Lemma 2.1, we obtain

$$d_1(h)(x^* \circ d_2(y^*)) + (d_1(y) \circ x)d_2(h) = 0. \tag{29.29}$$

For any $k \in S(R) \cap Z(R)$, changing x into xk in the equation (29.29), we get

$$-d_1(h)(x^* \circ d_2(y^*)) + (d_1(y) \circ x)d_2(h) = 0. \tag{29.30}$$

Adding the equation (29.29) and the equation (29.30) and using 2-torsion freeness, we find

$$(d_1(y) \circ x)d_2(h) = 0. \tag{29.31}$$

Since R is a prime ring, from the above equation we get either $(d_1(y) \circ x) = 0$ or $d_2(h) = 0$. In the first case, taking $x \in Z(R)$, we have $xd_1(y) = 0$ for all $x \in Z(R), y \in R$ which implies $d_1(y) = 0$ for all $y \in R$.

For the second case, we have $d_2(h) = 0$, taking $h = k^2$ for $k \in S(R) \cap Z(R)$ to obtain $d_2(k) = 0$ for all $k \in S(R) \cap Z(R)$. Similarly, we can obtain $d_1(k) = 0$ for all $k \in S(R) \cap Z(R)$.

Now replacing x into xk in the equation (29.28), we have

$$\begin{aligned} &d_1(k)x^* \circ d_2(y^*) - kd_1(x) \circ d_2(y^*) + d_1(y) \circ d_2(x^*)k - d_1(y) \circ d_2(k)x \\ &= kx \circ y^* - ky \circ x^*. \end{aligned} \tag{29.32}$$

On simplifying the above relation

$$k(-d_1(x) \circ d_2(y^*) + d_1(y) \circ d_2(x^*)) = k(x \circ y^* - y \circ x^*). \tag{29.33}$$

Using primeness of R, we get

$$-d_1(x) \circ d_2(y^*) + d_1(y) \circ d_2(x^*) = x \circ y^* - y \circ x^*. \tag{29.34}$$

From the equation (29.28) and (29.34), we obtain

$$d_1(y) \circ d_2(x^*) = x \circ y^*. \tag{29.35}$$

By replacing x with x^* and interchanging the variables, the above equation can be written as

$$d_1(x) \circ d_2(y) = x^* \circ y^*. \tag{29.36}$$

Hence by using the Theorem 3.3, R is commutative. □

Corollary 3.5. *Let R be a prime ring with involution of second kind and d be a non-zero $*$-reverse derivations on R. If $d(x) \circ d(x) = x \circ x^*$ for every $x \in R$, then R is commutative.*

Bibliography

[1] A. Aboubakr and S. González, Generalized reverse derivations on semi-prime rings, *Sib. Math. J.* **56** (2015), 199–205.

[2] B. Bhushan, D. Kumar and G. S. Sandhu, On $*$-reverse derivations in semiprime rings, *AIP Conf. Proc.* **1** (2022), 120006.

[3] B. Bhushan, G. S. Sandhu, S. Ali and D. Kumar, A classification of generalized derivations in rings with involution, *Filomat* **35**(5) (2021), 1439–1452.

[4] B. Bhushan, G. S. Sandhu and D. Kumar, A note on $*$-reverse derivation in rings, *Adv. Math. Sci. J.* **9**(7) (2020), 1–6.

[5] M. Brešar and J. Vukman, On some additive mappings in rings with involution, *Aequ. Math.* **38** (1989), 178–185.

[6] S. F. El-Deken and M. E.-S. Mahmoud, On centrally extended reverse and generalized reverse derivations, *Indian J. Pure Appl. Math.* **51**(3) (2020), 1165–1180.

[7] I. N. Herstein, Jordan derivations of prime rings, *Proc. Am. Math. Soc.* **8** (1957), 1104–1110.

[8] D. Kumar, B. Bhushan and G. S. Sandhu, Some results on $*$-differential identities in prime rings, *Asian-Eur. J. Math.* **15**(08) (2022), 2250159 (1–12).

[9] M. Samman, A note on reverse derivations, *Int. J. Math. Educ. Sci. Technol.* **37** (2006), 98–101.

[10] G. Sandhu, B. Ferreira and D. Kumar, On generalized $*$-reverse derivable maps, *Proyecciones (Antofagasta)* **40**(3) (2021), 767–778.

[11] G. Sandhu and D. Kumar, Annihilator conditions of multiplicative reverse derivation on prime rings, *Int. Electron. J. Algebra* **25** (2019), 87–103.

C. Reena and M. Karthika

More functions via soft pre*-generalized closed sets

Abstract: The aim of this paper is to introduce the contra version of soft pre*-generalized continuous functions. Many other functions associated with soft pre*-generalized closed sets are also discussed. Characterization of these functions are studied together with their interrelation with other existing continuous functions.

Keywords: Contra soft pre*-generalized continuous functions, contra soft pre*-generalized irresolute functions, strongly soft pre*-generalized continuous functions, perfectly soft pre*-generalized continuous functions, totally soft pre*-generalized continuous functions, slightly soft pre*-generalized continuous

MSC 2010: 16P20, 16P60, 16P70

1 Introduction

Many scientists have studied and improved the soft set theory, which is initiated by Molodtsov [5] acts as an effective mathematical tool for dealing with uncertainties. Meanwhile, the concept of soft topological space over an initial universe with a fixed set of parameters have been introduced. In 2010, Athar Kharal and Ahmad [4] introduced the concept of soft mapping. Aras and Sonmez [1] discussed the properties of soft continuous mappings. Later, the contraversions and various other forms of soft continuous functions have been discussed by many others. Recently, the authors [7] of this paper studied on the concept of soft pre*-generalized continuous and irresolute functions. In this paper, we introduce various forms of continuous functions associated with soft pre*-generalized closed sets namely contra soft pre*-generalized continuous functions, contra soft pre*-generalized irresolute functions, strongly soft pre*-generalized continuous functions, perfectly soft pre*-generalized continuous functions, totally soft pre*-generalized continuous functions and slightly soft pre*-generalized continuous.

C. Reena, Department of Mathematics, St. Mary's College (Autonomous), (Affiliated to Manonmaniam Sundaranar University, Abishekapatti, Tirunelveli) 359/10, Mini Sahaya Puram, Lionstown, 628001 Thoothukudi, TamilNadu, India, e-mail: reenastephany@gmail.com
M. Karthika, Department of Mathematics, St. Mary's College (Autonomous), (Affiliated to Manonmaniam Sundaranar University, Abishekapatti, Tirunelveli) 86F, Muthu Krishna Puram, 6th Street, 628001 Thoothukudi, TamilNadu, India, e-mail: karthikamarimuthu97@gmail.com

https://doi.org/10.1515/9783111313634-030

2 Preliminaries

Throughout this paper, $(X,\tilde{\tau},A)$, $(Y,\tilde{\sigma},B)$ and $(Z,\tilde{\mu},C)$ are soft topological spaces. Let (F,A) be a subset of a soft topological space. Then $S_t\mathrm{cl}(F,A)$, $S_t\mathrm{int}(F,A)$, $S_t\mathrm{cl}^*(F,A)$ and $S_t\mathrm{int}^*(F,A)$ denote the soft closure, soft interior, soft generalized closure and soft generalized interior, respectively.

Definition 1 ([6]). A subset (F,A) of a soft topological space $(X,\tilde{\tau},A)$ is said to be *soft* **pre***-generalized closed* if $S_t\mathrm{pcl}(F,A)\tilde{\subseteq}(U,A)$, whenever $(F,A)\tilde{\subseteq}(U,A)$ and (U,A) is soft pre*-open. The complement of soft pre*-generalized closed is called *soft* **pre***-generalized open.*

Definition 2. Let $\tilde{f}:(X,\tilde{\tau},A)\to(Y,\tilde{\sigma},B)$ be a soft function. Then,

i. $\tilde{f}$ is *contra soft continuous* [2] if $\tilde{f}^{-1}(G,B)$ is soft closed in $(X,\tilde{\tau},A)$ for every soft open set (G,B) in $(Y,\tilde{\sigma},B)$.
ii. $\tilde{f}$ is *strongly soft continuous* [2] if $\tilde{f}^{-1}(G,B)$ is both soft open and soft closed in $(X,\tilde{\tau},A)$ for every soft subset (G,B) in $(Y,\tilde{\sigma},B)$.
iii. $\tilde{f}$ is *perfectly soft continuous* [2] if $\tilde{f}^{-1}(G,B)$ is both soft open and soft closed in $(X,\tilde{\tau},A)$ for every soft open set (G,B) in $(Y,\tilde{\sigma},B)$.
iv. $\tilde{f}$ is *slightly soft continuous* [3] if $\tilde{f}^{-1}(G,B)$ is soft open in $(X,\tilde{\tau},A)$ for every soft clopen set (G,B) in $(Y,\tilde{\sigma},B)$.

Definition 3 ([7]). Let $(X,\tilde{\tau},A)$ and $(Y,\tilde{\sigma},B)$ be soft topological spaces. Let $u:X\to Y$ and $p:A\to B$ be a soft mappings. Then the soft function $\tilde{f}:(X,\tilde{\tau},A)\to(Y,\tilde{\sigma},B)$ is said to be *soft* **pre***-generalized continuous* if the inverse image is soft pre*-generalized closed in $(X,\tilde{\tau},A)$ for every soft closed set in $(Y,\tilde{\sigma},B)$.

Definition 4 ([7]). Let $(X,\tilde{\tau},A)$ and $(Y,\tilde{\sigma},B)$ be soft topological spaces. Let $u:X\to Y$ and $p:A\to B$ be a soft mappings. Then the soft function $\tilde{f}:(X,\tilde{\tau},A)\to(Y,\tilde{\sigma},B)$ is said to be *soft* **pre***-generalized irresolute* if the inverse image is soft pre*-generalized closed in $(X,\tilde{\tau},A)$ for every soft pre*-generalized closed set in $(Y,\tilde{\sigma},B)$.

Definition 5. A space $(X,\tilde{\tau},A)$ is said to be *soft locally indiscrete*, if every soft open set in $(X,\tilde{\tau},A)$ is soft closed in $(Y,\tilde{\sigma},B)$.

Remark 1 ([6]). Every soft closed (open) set is soft pre*-generalized closed (open).

3 Contra soft pre*-generalized continuous functions

Definition 6. Let $(X,\tilde{\tau},A)$ and $(Y,\tilde{\sigma},B)$ be soft topological spaces. Let $u:X\to Y$ and $p:A\to B$ be a soft mappings. A soft function $\tilde{f}:(X,\tilde{\tau},A)\to(Y,\tilde{\sigma},B)$ is said to be *contra soft* **pre***-generalized continuous* if the inverse image $\tilde{f}^{-1}(G,B)$ is soft pre*-generalized closed in $(X,\tilde{\tau},A)$ for every soft open set (G,B) in $(Y,\tilde{\sigma},B)$.

Example 1. Let $X = \{a,b\}$, $Y = \{x,y\}$, $A = \{e_1,e_2\}$ and $B = \{k_1,k_2\}$. Define $u : X \to Y$ and $p : A \to B$ as $u(a) = y$, $u(b) = x$, $p(e_1) = k_2$ and $p(e_2) = k_1$. Consider the soft topologies $\widetilde{\tau} = \{F_1,F_5,F_7,F_8,F_{16}\}$, $\widetilde{\tau}^{\widetilde{c}} = \{F_1,F_9,F_{10},F_{12},F_{16}\}$ and $\widetilde{\sigma} = \{G_1,G_6,G_{14},G_{16}\}$, $\widetilde{\sigma}^{\widetilde{c}} = \{G_1,G_3,G_{11},G_{16}\}$. Here, $S_tp^*gc(X,\widetilde{\tau},A) = \{F_1,F_2,F_3,F_4,F_9,F_{10},F_{11},F_{12},F_{16}\}$. Let $\widetilde{f} : (X,\widetilde{\tau},A) \to (Y,\widetilde{\sigma},B)$ be a soft mapping. Since $\widetilde{f}^{-1}(G_6) = F_{11}$ and $\widetilde{f}^{-1}(G_{14}) = F_{12}$. Hence $\widetilde{f}$ is contra soft pre*-generalized continuous.

Theorem 1. *A soft function $\widetilde{f} : (X,\widetilde{\tau},A) \to (Y,\widetilde{\sigma},B)$ is contra soft* pre**-generalized continuous if and only if the inverse image $\widetilde{f}^{-1}(G,B)$ is soft* pre**-generalized open in X for every soft closed set in Y.*

Proof. Let $\widetilde{f} : (X,\widetilde{\tau},A) \to (Y,\widetilde{\sigma},B)$ be contra soft pre*-generalized continuous and (G,B) be a soft closed set in Y. Then $(G,B)^{\widetilde{c}}$ is soft open in Y. Since $\widetilde{f}$ is contra soft pre*-generalized continuous $\widetilde{f}^{-1}((G,B)^{\widetilde{c}})$ is soft pre*-generalized closed in X. That is, $(\widetilde{f}^{-1}(G,B))^{\widetilde{c}}$ is soft pre*-generalized closed in X. Hence $\widetilde{f}^{-1}(G,B)$ is soft pre*-generalized open in X.

Conversely, suppose the inverse image of every soft closed set in Y is soft pre*-generalized open in X. Let (G,B) be soft closed in Y. Then $(G,B)^{\widetilde{c}}$ is soft open in Y. By assumption, $\widetilde{f}^{-1}((G,B)^{\widetilde{c}})$ is soft pre*-generalized open in Y. That is, $(\widetilde{f}^{-1}(G,B))^{\widetilde{c}}$ is soft pre*-generalized open in X. Therefore, $\widetilde{f}^{-1}(G,B)$ is soft pre*-generalized closed in X. Hence, $\widetilde{f}$ is contra soft pre*-generalized continuous. □

Theorem 2. *If $\widetilde{f} : (X,\widetilde{\tau},A) \to (Y,\widetilde{\sigma},B)$ is a contra soft* pre**-generalized continuous function and $\widetilde{g} : (Y,\widetilde{\sigma},B) \to (Z,\widetilde{\mu},C)$ is a soft continuous function, then $\widetilde{g} \circ \widetilde{f} : (X,\widetilde{\tau},A) \to (Z,\widetilde{\mu},C)$ is contra soft* pre**-generalized continuous.*

Proof. Let (H,C) be a soft open set in Z. Since $\widetilde{g}$ is soft continuous function, $\widetilde{g}^{-1}(H,C)$ is soft open in Y. Since $\widetilde{f}$ is contra soft pre*-generalized continuous, $\widetilde{f}^{-1}(\widetilde{g}^{-1}(H,C))$ is soft pre*-generalized closed in X. This implies $\widetilde{f}^{-1}(\widetilde{g}^{-1}(H,C)) = (\widetilde{g} \circ \widetilde{f})^{-1}(H,C)$ is soft pre*-generalized closed in X. Hence $\widetilde{g} \circ \widetilde{f}$ is contra soft pre*-generalized continuous. □

Theorem 3. *If $\widetilde{f} : (X,\widetilde{\tau},A) \to (Y,\widetilde{\sigma},B)$ is soft* pre**-generalized irresolute and $\widetilde{g} : (Y,\widetilde{\sigma},B) \to (Z,\widetilde{\mu},C)$ is contra soft* pre**-generalized continuous, then $\widetilde{g} \circ \widetilde{f} : (X,\widetilde{\tau},A) \longrightarrow (Z,\widetilde{\mu},C)$ is contra soft* pre**-generalized continuous.*

Proof. Let (H,C) be soft open set in Z. Since $\widetilde{g}$ is contra soft pre*-generalized continuous, $\widetilde{g}^{-1}(H,C)$ is soft pre*-generalized closed in Y. Also, since $\widetilde{f}$ is soft pre*-generalized irresolute, $\widetilde{f}^{-1}(\widetilde{g}^{-1}(H,C)) = (\widetilde{g} \circ \widetilde{f})^{-1}(H,C)$ is soft pre*-generalized closed in X. Hence $\widetilde{g} \circ \widetilde{f}$ is contra soft pre*-generalized continuous. □

Theorem 4. *If $\widetilde{f} : (X,\widetilde{\tau},A) \to (Y,\widetilde{\sigma},B)$ is soft* pre**-generalized continuous and $(Y,\widetilde{\sigma},B)$ is soft locally indiscrete, then $\widetilde{f}$ is contra soft* pre**-generalized continuous.*

Proof. Let (G, B) be soft open set in Y. Since Y is soft locally indiscrete, (G, B) is soft closed in Y. Again, since $\widetilde{f}$ is soft pre*-generalized continuous $\widetilde{f}^{-1}(G, B)$ is soft pre*-generalized closed in X. Hence $\widetilde{f}$ is contra soft pre*-generalized continuous. □

Theorem 5. *If a function $\widetilde{f} : (X, \widetilde{\tau}, A) \to (Y, \widetilde{\sigma}, B)$ is soft continuous and $(X, \widetilde{\tau}, A)$ is soft locally indiscrete, then $\widetilde{f}$ is contra soft* pre**-generalized continuous.*

Proof. Let (G, B) be soft open set in Y. Since $\widetilde{f}$ is soft continuous, $\widetilde{f}^{-1}(G, B)$ is soft open in X. Again, since X is soft locally indiscrete, $\widetilde{f}^{-1}(G, B)$ is soft closed in X. By Remark 1, $\widetilde{f}^{-1}(G, B)$ is soft pre*-generalized closed in X. Hence $\widetilde{f}$ is contra soft pre*-generalized continuous. □

4 Contra soft pre*-generalized irresolute functions

Definition 7. A soft function $\widetilde{f} : (X, \widetilde{\tau}, A) \to (Y, \widetilde{\sigma}, B)$ is said to be contra soft pre*-generalized irresolute if $\widetilde{f}^{-1}(G, B)$ is soft pre*-generalized closed in $(X, \widetilde{\tau}, A)$ for every soft pre*-generalized open set (G, B) in $(Y, \widetilde{\sigma}, B)$.

Example 2. Let $X = \{a, b\}$, $Y = \{x, y\}$, $A = \{e_1, e_2\}$ and $B = \{k_1, k_2\}$. Define $u : X \to Y$ and $p : A \to B$ as $u(a) = x$, $u(b) = y$, $p(e_1) = k_1$ and $p(e_2) = k_2$. Consider the soft topologies $\widetilde{\tau} = \{F_1, F_6, F_{14}, F_{16}\}$, $\widetilde{\tau}^{\widetilde{c}} = \{F_1, F_3, F_{11}, F_{16}\}$ and $\widetilde{\sigma} = \{G_1, G_3, G_9, G_{11}, G_{16}\}$, $\widetilde{\sigma}^{\widetilde{c}} = \{G_1, G_6, G_8, G_{14}, G_{16}\}$. Here, $S_t p^* gc(X, \widetilde{\tau}, A) = \{F_1, F_2, F_3, F_4, F_5, F_7, F_9, F_{10}, F_{11}, F_{12}, F_{13}, F_{15}, F_{16}\}$ and $S_t p^* go(Y, \widetilde{\sigma}, B) = \{G_1, G_3, G_9, G_{11}, G_{12}, G_{15}, G_{16}\}$. Let $\widetilde{f} : (X, \widetilde{\tau}, A) \to (Y, \widetilde{\sigma}, B)$ be a soft mapping. Then $\widetilde{f}^{-1}(G_3) = F_3$, $\widetilde{f}^{-1}(G_9) = F_9$, $\widetilde{f}^{-1}(G_{11}) = F_{11}$, $\widetilde{f}^{-1}(G_{12}) = F_{12}$ and $\widetilde{f}^{-1}(G_{15}) = F_{16}$. Hence $\widetilde{f}$ is contra soft pre*-generalized irresolute.

Theorem 6. *A soft function $\widetilde{f} : (X, \widetilde{\tau}, A) \to (Y, \widetilde{\sigma}, B)$ is contra soft* pre**-generalized irresolute if and only if the inverse image $\widetilde{f}^{-1}(G, B)$ is soft* pre**-generalized open in X for every soft* pre**-generalized closed set (G, B) in Y.*

Proof. Let (G, B) be soft pre*-generalized closed in Y. Then $(G, B)^{\widetilde{c}}$ is soft pre*-generalized open in Y. Since $\widetilde{f}$ is contra soft pre*-generalized irresolute, $\widetilde{f}^{-1}((G, B)^{\widetilde{c}})$ is soft pre*-generalized closed in X. That is, $(\widetilde{f}^{-1}(G, B))^{\widetilde{c}}$ is soft pre*-generalized closed in X. Hence $\widetilde{f}^{-1}(G, B)$ is soft pre*-generalized open in X.

Conversely, assume that the inverse image of every soft pre*-generalized closed set in Y is soft pre*-generalized open in X. Let (G, B) be soft pre*-generalized open in Y. Then $(G, B)^{\widetilde{c}}$ is soft pre*-generalized closed in Y. By assumption, $\widetilde{f}^{-1}((G, B)^{\widetilde{c}})$ is soft pre*-generalized open in X. That is $(\widetilde{f}^{-1}(G, B))^{\widetilde{c}}$ is soft pre*-generalized open in X. This implies $\widetilde{f}^{-1}(G, B)$ is soft pre*-generalized closed in X. Hence $\widetilde{f}$ is contra soft pre*-generalized irresolute. □

Theorem 7. *Let $\widetilde{f} : (X, \widetilde{\tau}, A) \to (Y, \widetilde{\sigma}, B)$ be a contra soft* pre**-generalized irresolute function. Then $\widetilde{f}$ is contra soft* pre**-generalized continuous.*

Proof. Let (G,B) be any soft open set in Y. Then (G,B) is soft pre*-generalized open in Y. Since $\widetilde{f}$ is contra soft pre*-generalized irresolute, $\widetilde{f}^{-1}(G,B)$ is soft pre*-generalized closed in X. Hence $\widetilde{f}$ is contra soft pre*-generalized continuous. □

Theorem 8. *If $\widetilde{f} : (X,\widetilde{\tau},A) \to (Y,\widetilde{\sigma},B)$ is contra* pre**-generalized irresolute and $\widetilde{g} : (Y,\widetilde{\sigma},B) \to (Z,\widetilde{\mu},C)$ is soft* pre**-generalized continuous functions, then $\widetilde{g} \circ \widetilde{f} : (X,\widetilde{\tau},A) \to (Z,\widetilde{\mu},C)$ is contra soft* pre**-generalized continuous.*

Proof. Let (H,C) be an soft open set in Z. Since $\widetilde{g}$ is soft pre*-generalized continuous, $\widetilde{g}^{-1}(H,C)$ is soft pre*-generalized open in Y. Since $\widetilde{f}$ is contra soft pre*-generalized irresolute, $\widetilde{f}^{-1}(\widetilde{g}^{-1}(H,C))$ is soft pre*-generalized closed in X. This implies $\widetilde{f}^{-1}(\widetilde{g}^{-1}(H,C)) = (\widetilde{g} \circ \widetilde{f})^{-1}(H,C)$ is soft pre*-generalized closed in X. Hence $\widetilde{g} \circ \widetilde{f}$ is contra soft pre*-generalized continuous. □

Theorem 9. *If $\widetilde{f} : (X,\widetilde{\tau},A) \to (Y,\widetilde{\sigma},B)$ is contra* pre**-generalized irresolute and $\widetilde{g} : (Y,\widetilde{\sigma},B) \to (Z,\widetilde{\mu},C)$ is soft* pre**-generalized irresolute functions, then $\widetilde{g} \circ \widetilde{f} : (X,\widetilde{\tau},A) \to (Z,\widetilde{\mu},C)$ is contra soft* pre**-generalized irresolute.*

Proof. Let (H,C) be a soft pre*-generalized open set in Z. Since $\widetilde{g}$ is soft pre*-generalized irresolute, $\widetilde{g}^{-1}(H,C)$ is soft pre*-generalized open in Y. Since $\widetilde{f}$ is contra soft pre*-generalized irresolute, $\widetilde{f}^{-1}(\widetilde{g}^{-1}(H,C))$ is soft pre*-generalized closed in X. This implies $\widetilde{f}^{-1}(\widetilde{g}^{-1}(H,C)) = (\widetilde{g} \circ \widetilde{f})^{-1}(H,C)$ is soft pre*-generalized closed in X. Hence $\widetilde{g} \circ \widetilde{f}$ is contra soft pre*-generalized irresolute. □

Theorem 10. *If $\widetilde{f} : (X,\widetilde{\tau},A) \to (Y,\widetilde{\sigma},B)$ and $\widetilde{g} : (Y,\widetilde{\sigma},B) \to (Z,\widetilde{\mu},C)$ are contra soft* pre**-generalized irresolute functions, then $\widetilde{g} \circ \widetilde{f} : (X,\widetilde{\tau},A) \longrightarrow (Z,\widetilde{\mu},C)$ is soft* pre**-generalized irresolute.*

Proof. Let (H,C) be soft pre*-generalized open set in Z. Since $\widetilde{g}$ is contra soft pre*-generalized irresolute, $\widetilde{g}^{-1}(H,C)$ is soft pre*-generalized closed in Y. Also, since $\widetilde{f}$ is contra soft pre* generalized-irresolute, by Theorem 6, $\widetilde{f}^{-1}(\widetilde{g}^{-1}(H,C)) = (\widetilde{g} \circ \widetilde{f})^{-1}(H,C)$ is soft pre*-generalized open in X. Hence $\widetilde{g} \circ \widetilde{f}$ is soft pre*-generalized irresolute. □

5 Strongly soft pre*-generalized continuous functions

Definition 8. A soft function $\widetilde{f} : (X,\widetilde{\tau},A) \to (Y,\widetilde{\sigma},B)$ is said to be *strongly soft* **pre***-*generalized continuous* if the inverse image of every soft pre*-generalized closed set (G,B) in $(Y,\widetilde{\sigma},B)$ is soft closed in $(X,\widetilde{\tau},A)$.

Example 3. Let $X = \{a,b\}$, $Y = \{x,y\}$, $A = \{e_1,e_2\}$ and $B = \{k_1,k_2\}$. Define $u : X \to Y$ and $p : A \to B$ as $u(a) = y$, $u(b) = x$, $p(e_1) = k_2$ and $p(e_2) = k_1$. Consider the soft topologies $\widetilde{\tau} = \{F_1,F_2,F_3,F_4,F_5,F_6,F_7,F_8,F_9,F_{10},F_{11},F_{12},F_{13},F_{14},F_{15},F_{16}\}$, $\widetilde{\tau}^{\widetilde{c}} = \{F_1,F_2,$

$F_3, F_4, F_5, F_6, F_7, F_8, F_9, F_{10}, F_{11}, F_{12}, F_{13}, F_{14}, F_{15}, F_{16}\}$ and $\widetilde{\sigma} = \{G_1, G_5, G_7, G_8, G_{16}\}$ $\widetilde{\sigma}^{\widetilde{c}} = \{G_1, G_9, G_{10}, G_{12}, G_{16}\}$. Here, $S_tp^*gc(Y, \widetilde{\sigma}, B) = \{G_1, G_2, G_3, G_4, G_9, G_{10}, G_{11}, G_{12}, G_{16}\}$. Let $\widetilde{f} : (X, \widetilde{\tau}, A) \to (Y, \widetilde{\sigma}, B)$ be a soft mapping. Then $\widetilde{f}^{-1}(G_2) = F_9, \widetilde{f}^{-1}(G_3) = F_5, \widetilde{f}^{-1}(G_4) = F_{13}$, $\widetilde{f}^{-1}(G_9) = F_2, \widetilde{f}^{-1}(G_{10}) = F_{10}, \widetilde{f}^{-1}(G_{11}) = F_6$ and $\widetilde{f}^{-1}(G_{12}) = F_{14}$. Hence $\widetilde{f}$ is strongly soft pre*-generalized continuous.

Theorem 11. *A soft function $\widetilde{f} : (X, \widetilde{\tau}, A) \to (Y, \widetilde{\sigma}, B)$ is strongly soft* pre**-generalized continuous if and only if the inverse image of every soft* pre**-generalized open set in $(Y, \widetilde{\sigma}, B)$ is soft open in X.*

Proof. Let (G, B) be soft pre*-generalized open in Y. Then $(G, B)^{\widetilde{c}}$ is soft pre*-generalized closed in Y. Since $\widetilde{f}$ is strongly soft pre*-generalized continuous, $\widetilde{f}^{-1}((G, B)^{\widetilde{c}})$ is soft closed in X. That is, $(\widetilde{f}^{-1}(G, B))^{\widetilde{c}}$ is soft closed in X. Hence $\widetilde{f}^{-1}(G, B)$ is soft open in X.

Conversely, suppose the inverse image of every soft pre*-generalized open set in Y is open in X. Let (G, B) be soft pre*-generalized closed in Y. Then, $(G, B)^{\widetilde{c}}$ is soft pre*-generalized open in Y. By assumption, $\widetilde{f}^{-1}((G, B)^{\widetilde{c}})$ is soft open in X. That is $(\widetilde{f}^{-1}(G, B))^{\widetilde{c}}$ is soft open in X. Thus $\widetilde{f}^{-1}(G, B)$ is soft closed in X. Hence $\widetilde{f}$ is strongly soft pre*-generalized continuous. □

Theorem 12. *If $\widetilde{f} : (X, \widetilde{\tau}, A) \to (Y, \widetilde{\sigma}, B)$ is soft continuous and $\widetilde{g} : (Y, \widetilde{\sigma}, B) \to (Z, \widetilde{\mu}, C)$ is strongly soft* pre**-generalized continuous, then $\widetilde{g} \circ \widetilde{f} : (X, \widetilde{\tau}, A) \longrightarrow (Z, \widetilde{\mu}, C)$ is strongly soft* pre**-generalized continuous.*

Proof. Let (H, C) be soft pre*-generalized closed set in Z. Since $\widetilde{g}$ is strongly soft pre*-generalized continuous, $\widetilde{g}^{-1}(H, C)$ is soft closed in Y. Also, since $\widetilde{f}$ is soft continuous, $\widetilde{f}^{-1}(\widetilde{g}^{-1}(H, C)) = (\widetilde{g} \circ \widetilde{f})^{-1}(H, C)$ is soft closed in X. Hence $\widetilde{g} \circ \widetilde{f}$ is strongly soft pre*-generalized continuous. □

Theorem 13. *If $\widetilde{f} : (X, \widetilde{\tau}, A) \to (Y, \widetilde{\sigma}, B)$ is soft* pre**-generalized continuous and $\widetilde{g} : (Y, \widetilde{\sigma}, B) \to (Z, \widetilde{\mu}, C)$ is strongly soft* pre**-generalized continuous, then $\widetilde{g} \circ \widetilde{f} : (X, \widetilde{\tau}, A) \longrightarrow (Z, \widetilde{\mu}, C)$ is soft* pre**-generalized irresolute.*

Proof. Let (H, C) be soft pre*-generalized closed set in Z. Since $\widetilde{g}$ is strongly soft pre*-generalized continuous, $\widetilde{g}^{-1}(H, C)$ is soft closed in Y. Also, since $\widetilde{f}$ is soft pre*-generalized continuous, $\widetilde{f}^{-1}(\widetilde{g}^{-1}(H, C)) = (\widetilde{g} \circ \widetilde{f})^{-1}(H, C)$ is soft pre*-generalized closed in X. Hence $\widetilde{g} \circ \widetilde{f}$ is soft pre*-generalized irresolute. □

Theorem 14. *If $\widetilde{f} : (X, \widetilde{\tau}, A) \to (Y, \widetilde{\sigma}, B)$ is strongly soft* pre**-generalized continuous and $\widetilde{g} : (Y, \widetilde{\sigma}, B) \to (Z, \widetilde{\mu}, C)$ is soft* pre**-generalized irresolute, then $\widetilde{g} \circ \widetilde{f} : (X, \widetilde{\tau}, A) \longrightarrow (Z, \widetilde{\mu}, C)$ is strongly soft* pre**-generalized continuous.*

Proof. Let (H, C) be soft pre*-generalized closed set in Z. Since $\widetilde{g}$ is soft pre*-generalized irresolute, $\widetilde{g}^{-1}(H, C)$ is soft pre*-generalized closed in Y. Also, since $\widetilde{f}$ is strongly soft pre*-generalized continuous, $\widetilde{f}^{-1}(\widetilde{g}^{-1}(H, C)) = (\widetilde{g} \circ \widetilde{f})^{-1}(H, C)$ is soft closed in X. Hence $\widetilde{g} \circ \widetilde{f}$ is strongly soft pre*-generalized continuous. □

Theorem 15. *If* $\widetilde{f} : (X, \widetilde{\tau}, A) \to (Y, \widetilde{\sigma}, B)$ *is contra soft* pre*-*generalized continuous and* $\widetilde{g} : (Y, \widetilde{\sigma}, B) \to (Z, \widetilde{\mu}, C)$ *is strongly soft* pre*-*generalized continuous, then* $\widetilde{g} \circ \widetilde{f} : (X, \widetilde{\tau}, A) \to (Z, \widetilde{\mu}, C)$ *is contra soft* pre*-*generalized irresolute.*

Proof. Let (H, C) be soft pre*-generalized open in Z. Since $\widetilde{g}$ is strongly soft pre*-generalized continuous, $\widetilde{g}^{-1}(H, C)$ is soft open in Y. Since $\widetilde{f}$ is contra soft pre*-generalized continuous, $\widetilde{f}^{-1}(\widetilde{g}^{-1}(H, C))$ is soft pre*-generalized closed in X. Thus $\widetilde{f}^{-1}(\widetilde{g}^{-1}(H, C)) = (\widetilde{g} \circ \widetilde{f})^{-1}(H, C)$ is soft pre*-generalized closed in X. Hence $\widetilde{g} \circ \widetilde{f}$ is contra soft pre*-generalized irresolute. □

Theorem 16. *If* $\widetilde{f} : (X, \widetilde{\tau}, A) \to (Y, \widetilde{\sigma}, B)$ *is strongly soft* pre*-*generalized continuous, then* $\widetilde{f}$ *is soft* pre*-*generalized irresolute.*

Proof. Let (G, B) be soft pre*-generalized closed in Y. Since $\widetilde{f}$ is strongly soft pre*-generalized continuous, $\widetilde{f}^{-1}(G, B)$ is soft closed in X. By Remark 1, $\widetilde{f}^{-1}(G, B)$ is soft pre*-generalized closed in X. Hence $\widetilde{f}$ is soft pre*-generalized irresolute. □

Theorem 17. *If* $\widetilde{f} : (X, \widetilde{\tau}, A) \to (Y, \widetilde{\sigma}, B)$ *and* $\widetilde{g} : (Y, \widetilde{\sigma}, B) \to (Z, \widetilde{\mu}, C)$ *are strongly soft* pre*-*generalized continuous, then their composition* $\widetilde{g} \circ \widetilde{f}$ *is also strongly soft* pre*-*generalized continuous.*

Proof. Let (H, C) be soft pre*-generalized closed in Z. Since $\widetilde{g}$ is strongly soft pre*-generalized continuous, $\widetilde{g}^{-1}(H, C)$ is soft closed in Y. By remark, $\widetilde{g}^{-1}(H, C)$ is soft pre*-generalized closed in Y. Since $\widetilde{f}$ is strongly soft pre*-generalized continuous, $\widetilde{f}^{-1}(\widetilde{g}^{-1}(H, C))$ is soft closed in X. This implies $\widetilde{f}^{-1}(\widetilde{g}^{-1}(H, C)) = (\widetilde{g} \circ \widetilde{f})^{-1}(H, C)$ is soft closed in X. Hence $\widetilde{g} \circ \widetilde{f}$ is strongly soft pre*-generalized continuous. □

6 Perfectly soft pre*-generalized continuous functions

Definition 9. A soft function $\widetilde{f} : (X, \widetilde{\tau}, A) \to (Y, \widetilde{\sigma}, B)$ is said to be *perfectly soft* **pre***-*generalized continuous* if the inverse image of every soft pre*-generalized closed set in $(Y, \widetilde{\sigma}, B)$ is both soft open and soft closed in $(X, \widetilde{\tau}, A)$.

Example 4. Let $X = \{a, b\}$, $Y = \{x, y\}$, $A = \{e_1, e_2\}$ and $B = \{k_1, k_2\}$. Define $u : X \to Y$ and $p : A \to B$ as $u(a) = x$, $u(b) = y$, $p(e_1) = k_1$ and $p(e_2) = k_2$. Consider the soft topologies $\widetilde{\tau} = \{F_1, F_2, F_3, F_4, F_5, F_6, F_7, F_8, F_9, F_{10}, F_{11}, F_{12}, F_{13}, F_{14}, F_{15}, F_{16}\}$, $\widetilde{\tau}^{\widetilde{c}} = \{F_1, F_2, F_3, F_4, F_5, F_6, F_7, F_8, F_9, F_{10}, F_{11}, F_{12}, F_{13}, F_{14}, F_{15}, F_{16}\}$ and $\widetilde{\sigma} = \{G_1, G_3, G_{16}\}$, $\widetilde{\sigma}^{\widetilde{c}} = \{G_1, G_{14}, G_{16}\}$. Here, $S_t p^* gc(Y, \widetilde{\sigma}, B) = \{G_1, G_2, G_5, G_6, G_9, G_{10}, G_{13}, G_{14}, G_{16}\}$. Let $\widetilde{f} : (X, \widetilde{\tau}, A) \to (Y, \widetilde{\sigma}, B)$ be a soft mapping. Then $\widetilde{f}^{-1}(G_2) = F_2$, $\widetilde{f}^{-1}(G_5) = F_5$, $\widetilde{f}^{-1}(G_6) = F_6$, $\widetilde{f}^{-1}(G_9) = F_9$, $\widetilde{f}^{-1}(G_{10}) = F_{10}$, $\widetilde{f}^{-1}(G_{13}) = F_{13}$ and $\widetilde{f}^{-1}(G_{14}) = F_{14}$. Hence $\widetilde{f}$ is perfectly soft pre*-generalized continuous.

Theorem 18. *A function $\widetilde{f} : (X, \widetilde{\tau}, A) \to (Y, \widetilde{\sigma}, B)$ is perfectly soft* pre*-generalized continuous if and only if the inverse image of every soft* pre*-generalized open set in Y is both soft open and soft closed in X.*

Proof. Let (G, B) be soft pre*-generalized open in Y. Then $(G, B)^{\widetilde{c}}$ is soft pre*-generalized closed in Y. Since $\widetilde{f}$ is perfectly soft pre*-generalized continuous, $\widetilde{f}^{-1}((G, B)^{\widetilde{c}})$ is soft clopen in X. That is $(\widetilde{f}^{-1}(G, B))^{\widetilde{c}}$ is soft clopen in X. Hence $\widetilde{f}^{-1}(G, B)$ is soft clopen in X.

Conversely, suppose the inverse image of every soft pre*-generalized open set in Y is soft clopen in X. Let (G, B) be soft pre*-generalized closed in Y. Then $(G, B)^{\widetilde{c}}$ is soft pre*-generalized open in Y. By assumption, $\widetilde{f}^{-1}((G, B)^{\widetilde{c}})$ is soft clopen in X. That is $(\widetilde{f}^{-1}(G, B))^{\widetilde{c}}$ is soft clopen in X. Hence $\widetilde{f}^{-1}(G, B)$ is soft clopen in X. Hence $\widetilde{f}$ is perfectly soft pre*-generalized continuous. □

Theorem 19. *If a function $\widetilde{f} : (X, \widetilde{\tau}, A) \to (Y, \widetilde{\sigma}, B)$ is perfectly soft* pre*-generalized continuous, then $\widetilde{f}$ is strongly soft* pre*-generalized continuous.*

Proof. Let (G, B) be soft pre*-generalized open in Y. Since $\widetilde{f}$ is perfectly soft pre*-generalized continuous, $\widetilde{f}^{-1}(G, B)$ is both soft closed and soft open in X. Hence $\widetilde{f}$ is strongly soft pre*-generalized continuous. □

Theorem 20. *If $\widetilde{f} : (X, \widetilde{\tau}, A) \to (Y, \widetilde{\sigma}, B)$ is perfectly soft* pre*-generalized continuous and $\widetilde{g} : (Y, \widetilde{\sigma}, B) \to (Z, \widetilde{\mu}, C)$ is strongly soft* pre*-generalized continuous, then their composition $\widetilde{g} \circ \widetilde{f}$ is perfectly soft* pre*-generalized continuous.*

Proof. Let (H, C) be soft pre*-generalized open in Z. Since $\widetilde{g}$ is strongly soft pre*-generalized continuous, $\widetilde{g}^{-1}(H, C)$ is soft open in Y. By remark, $\widetilde{g}^{-1}(H, C)$ is soft pre*-generalized open in Y. Since $\widetilde{f}$ is perfectly soft pre*-generalized continuous, $\widetilde{f}^{-1}(\widetilde{g}^{-1}(H, C)) = (\widetilde{g} \circ \widetilde{f})^{-1}(H, C)$ is soft clopen in X. Hence $\widetilde{g} \circ \widetilde{f}$ is perfectly soft pre*-generalized continuous. □

Theorem 21. *If $\widetilde{f} : (X, \widetilde{\tau}, A) \to (Y, \widetilde{\sigma}, B)$ is perfectly soft* pre*-generalized continuous and $\widetilde{g} : (Y, \widetilde{\sigma}, B) \to (Z, \widetilde{\mu}, C)$ is contra soft* pre*-generalized irresolute, then $\widetilde{g} \circ \widetilde{f} : (X, \widetilde{\tau}, A) \longrightarrow (Z, \widetilde{\mu}, C)$ is perfectly soft* pre*-generalized continuous.*

Proof. Let (H, C) be soft pre*-generalized open set in Z. Since $\widetilde{g}$ is contra soft pre*-generalized irresolute, $\widetilde{g}^{-1}(H, C)$ is soft pre*-generalized closed in Y. Also, since $\widetilde{f}$ is perfectly soft pre*-continuous, $\widetilde{f}^{-1}(\widetilde{g}^{-1}(H, C)) = (\widetilde{g} \circ \widetilde{f})^{-1}(H, C)$ is soft clopen in X. Hence $\widetilde{g} \circ \widetilde{f}$ is perfectly soft pre*-generalized continuous. □

Theorem 22. *Let $\widetilde{f} : (X, \widetilde{\tau}, A) \to (Y, \widetilde{\sigma}, B)$ be a strongly soft* pre*-generalized continuous function and $(X, \widetilde{\tau}, A)$ be a soft locally indiscrete space. Then $\widetilde{f}$ is perfectly soft* pre*-generalized continuous.*

Proof. Let (G, B) be a soft pre*-generalized set in Y. Since $\widetilde{f}$ is strongly soft pre*-generalized continuous, $\widetilde{f}^{-1}(G, B)$ is soft closed in X. Also, since X is a soft locally indiscrete space, $\widetilde{f}^{-1}(G,B)$ is soft open in X. Therefore, $\widetilde{f}^{-1}(G,B)$ is both soft open and soft closed in X. Hence $\widetilde{f}$ is perfectly soft pre*-generalized continuous. □

Theorem 23. *If* $\widetilde{f} : (X,\widetilde{\tau},A) \to (Y,\widetilde{\sigma},B)$ *is soft continuous and* $\widetilde{g} : (Y,\widetilde{\sigma},B) \to (Z,\widetilde{\mu},C)$ *is perfectly soft* pre**-generalized continuous, then their composition* $\widetilde{g} \circ \widetilde{f}$ *is strongly soft* pre**-generalized continuous.*

Proof. Let (H,C) be soft pre*-generalized open in Z. Since $\widetilde{g}$ is perfectly soft pre*-generalized continuous, $\widetilde{g}^{-1}(H,C)$ is both soft open and soft closed in Y. Since $\widetilde{f}$ is soft continuous, $\widetilde{f}^{-1}(\widetilde{g}^{-1}(H,C))$ is both soft open and soft closed in X. Thus $\widetilde{f}^{-1}(\widetilde{g}^{-1}(H,C)) = (\widetilde{g} \circ \widetilde{f})^{-1}(H,C)$ is both soft open and soft closed in X. Hence $\widetilde{g} \circ \widetilde{f}$ is perfectly soft pre*-generalized continuous. □

Theorem 24. *If* $\widetilde{f} : (X,\widetilde{\tau},A) \to (Y,\widetilde{\sigma},B)$ *and* $\widetilde{g} : (Y,\widetilde{\sigma},B) \to (Z,\widetilde{\mu},C)$ *are two perfectly soft* pre**-generalized continuous functions, then their composition* $\widetilde{g} \circ \widetilde{f}$ *is also perfectly soft* pre**-generalized continuous.*

Proof. Let (H,C) be soft pre*-generalized open in Z. Since $\widetilde{g}$ is perfectly soft pre*-generalized continuous, $\widetilde{g}^{-1}(H,C)$ is both soft open and soft closed. By Remark 1, $\widetilde{g}^{-1}(H,C)$ is soft pre*-generalized open in X. Since $\widetilde{f}$ is perfectly soft pre*-generalized continuous, $\widetilde{f}^{-1}(\widetilde{g}^{-1}(H,C)) = (\widetilde{g} \circ \widetilde{f})^{-1}(H,C)$ is soft clopen in X. Hence $\widetilde{g} \circ \widetilde{f}$ is perfectly soft pre*-generalized continuous. □

7 Totally soft pre*-generalized continuous functions

Definition 10. A soft function $\widetilde{f} : (X,\widetilde{\tau},A) \to (Y,\widetilde{\sigma},B)$ is said to be *totally soft* **pre***-*generalized continuous* if the inverse image of every soft closed set (G,B) in $(Y,\widetilde{\sigma},B)$ is both soft pre*-generalized open and soft pre*-generalized closed in $(X,\widetilde{\tau},A)$.

Example 5. Let $X = \{a,b\}$, $Y = \{x,y\}$, $A = \{e_1,e_2\}$ and $B = \{k_1,k_2\}$. Define $u : X \to Y$ and $p : A \to B$ as $u(a) = x$, $u(b) = y$, $p(e_1) = k_1$ and $p(e_2) = k_2$. Consider the soft topologies $\widetilde{\tau} = \{F_1,F_6,F_{14},F_{16}\}$, $\widetilde{\tau}^{\widetilde{c}} = \{F_1,F_3,F_{11},F_{16}\}$ and $\widetilde{\sigma} = \{G_1,G_5,G_7,G_{13},G_{15},G_{16}\}$, $\widetilde{\sigma}^{\widetilde{c}} = \{G_1,G_2,G_4,G_{10},G_{12},G_{16}\}$. Here, $S_tp^*gc(X,\widetilde{\tau},A) = \{F_1,F_2,F_3,F_4,F_5,F_7, F_9,F_{10},F_{11}, F_{12},F_{13}, F_{15},F_{16}\}$, $S_tp^*go(X,\widetilde{\tau},A) = \{F_1,F_2,F_4,F_5,F_6,F_7,F_8,F_{10},F_{12},F_{13},F_{14},F_{15},F_{16}\}$ Let $\widetilde{f} : (X,\widetilde{\tau},A) \to (Y,\widetilde{\sigma},B)$ be a soft mapping. Then $\widetilde{f}^{-1}(G_2) = F_2$, $\widetilde{f}^{-1}(G_4) = F_4$, $\widetilde{f}^{-1}(G_{10}) = F_{10}$ and $\widetilde{f}^{-1}(G_{12}) = F_{12}$. Hence $\widetilde{f}$ is totally soft pre*-generalized continuous.

Theorem 25. *A soft function* $\widetilde{f} : (X,\widetilde{\tau},A) \to (Y,\widetilde{\sigma},B)$ *is totally soft* pre**-generalized continuous if and only if the inverse image of every soft open set in* Y *is both soft* pre**-generalized open and soft* pre**-generalized closed in* X.

Proof. Let (G,B) be soft open in Y. Then $(G,B)^{\widetilde{c}}$ is soft closed in Y. Since $\widetilde{f}$ is totally soft pre*-generalized continuous, $\widetilde{f}^{-1}((G,B)^{\widetilde{c}})$ is both soft pre*-generalized open and soft pre*-generalized closed in X. That is $(\widetilde{f}^{-1}(G,B))^{\widetilde{c}}$ is both soft pre*-generalized open and soft pre*-generalized closed in X. Thus $\widetilde{f}^{-1}(G,B)$ is both soft pre*-generalized open and soft pre*-generalized closed in X.

Conversely, suppose the inverse image of every soft open set in Y is both soft pre*-generalized open and soft pre*-generalized closed in X. Let (G,B) be soft closed in Y. Then, $(G,B)^{\widetilde{c}}$ is soft open in Y. By assumption, $\widetilde{f}^{-1}((G,B)^{\widetilde{c}})$ is both soft pre*-generalized open and soft pre*-generalized closed in X. That is, $(\widetilde{f}^{-1}(G,B))^{\widetilde{c}}$ is both soft pre*-generalized open and soft pre*-generalized closed in X. Thus $\widetilde{f}^{-1}(G,B)$ is both soft pre*-generalized open and soft pre*-generalized closed in X. Hence $\widetilde{f}$ is totally soft pre*-generalized continuous. □

Theorem 26. *If $\widetilde{f} : (X,\widetilde{\tau},A) \to (Y,\widetilde{\sigma},B)$ is totally soft* pre**-generalized continuous, then $\widetilde{f}$ is soft* pre**-generalized continuous.*

Proof. Let (G,B) be soft open in Y. Since $\widetilde{f}$ is totally soft pre*-generalized continuous, $\widetilde{f}^{-1}(G,B)$ is both soft pre*-generalized closed and soft pre*-generalized open in X. Hence $\widetilde{f}$ is soft pre*-generalized continuous. □

Theorem 27. *If $\widetilde{f} : (X,\widetilde{\tau},A) \to (Y,\widetilde{\sigma},B)$ is perfectly soft* pre**-generalized continuous, then $\widetilde{f}$ is totally soft* pre**-generalized continuous.*

Proof. Let (G,B) be soft open in Y. By remark, (G,B) is soft pre*-generalized open in Y. Since $\widetilde{f}$ is perfectly soft pre*-generalized continuous, $\widetilde{f}^{-1}(G,B)$ is both soft closed and soft open in X. By Remark 1, $\widetilde{f}^{-1}(G,B)$ is both soft pre*-generalized closed and soft pre*-generalized open. Hence $\widetilde{f}$ is totally soft pre*-generalized continuous. □

Theorem 28. *If $\widetilde{f} : (X,\widetilde{\tau},A) \to (Y,\widetilde{\sigma},B)$ is totally soft* pre**-generalized continuous and $\widetilde{g} : (Y,\widetilde{\sigma},B) \to (Z,\widetilde{\mu},C)$ is soft continuous, then $\widetilde{g}\circ\widetilde{f} : (X,\widetilde{\tau},A) \longrightarrow (Z,\widetilde{\mu},C)$ is totally soft* pre**-generalized continuous.*

Proof. Let (H,C) be soft closed set in Z. Since $\widetilde{g}$ is soft continuous, $\widetilde{g}^{-1}(H,C)$ is soft closed in Y. Also, since $\widetilde{f}$ is totally soft pre*-generalized continuous, $\widetilde{f}^{-1}(\widetilde{g}^{-1}(H,C)) = (\widetilde{g}\circ\widetilde{f})^{-1}(H,C)$ is both soft pre*-generalized open and soft pre*-generalized closed in X. Hence $\widetilde{g}\circ\widetilde{f}$ is totally soft pre*-generalized continuous. □

8 Slightly soft pre*-generalized continuous functions

Definition 11. A soft function $\widetilde{f} : (X,\widetilde{\tau},A) \to (Y,\widetilde{\sigma},B)$ is said to be *slightly soft* **pre****-generalized continuous* if the inverse image of every soft clopen set (G,B) in $(Y,\widetilde{\sigma},B)$ is soft pre*-generalized closed in $(X,\widetilde{\tau},A)$.

Example 6. Let $X = \{a,b\}$, $Y = \{x,y\}$, $A = \{e_1,e_2\}$ and $B = \{k_1,k_2\}$. Define $u : X \to Y$ and $p : A \to B$ as $u(a) = y$, $u(b) = x$, $p(e_1) = k_1$ and $p(e_2) = k_2$. Consider the soft topologies $\widetilde{\tau} = \{F_1, F_4, F_8, F_{12}, F_{16}\}$, $\widetilde{\tau}^{\widetilde{c}} = \{F_1, F_5, F_9, F_{13}, F_{16}\}$ and $\widetilde{\sigma} = \{G_1, G_7, G_{10}, G_{16}\}$, $\widetilde{\sigma}^{\widetilde{c}} = \{G_1, G_7, G_{10}, G_{16}\}$. Here, $S_t p^* gc(X, \widetilde{\tau}, A) = \{F_1, F_2, F_3, F_5, F_6, F_7, F_9, F_{10}, F_{11}, F_{13}, F_{14}, F_{15}, F_{16}\}$ Let $\widetilde{f} : (X, \widetilde{\tau}, A) \to (Y, \widetilde{\sigma}, B)$ be a soft mapping. Then $\widetilde{f}^{-1}(G_7) = F_{10}$ and $\widetilde{f}^{-1}(G_{10}) = F_7$. Hence $\widetilde{f}$ is slightly soft pre*-generalized continuous.

Theorem 29. *A function $\widetilde{f} : (X, \widetilde{\tau}, A) \to (Y, \widetilde{\sigma}, B)$ is slightly soft* pre**-generalized continuous if and only if the inverse image of every soft clopen in $(Y, \widetilde{\sigma}, B)$ is soft* pre**-generalized open in X.*

Proof. Let (G, B) be soft clopen in Y. Then, $(G, B)^{\widetilde{c}}$ is soft clopen in Y. Since $\widetilde{f}$ is slightly soft pre*-generalized continuous, $\widetilde{f}^{-1}((G, B)^{\widetilde{c}})$ is soft pre*-generalized closed in X. That is, $(\widetilde{f}^{-1}(G, B))^{\widetilde{c}}$ is soft pre*-generalized closed in X. Thus, $\widetilde{f}^{-1}(G, B)$ is soft pre*-generalized open in X.

Conversely, suppose the inverse image of every soft clopen set in Y is soft pre*-generalized open in X. Let (G, B) be soft clopen in Y. Then, $(G, B)^{\widetilde{c}}$ is soft clopen in Y. By assumption, $\widetilde{f}^{-1}((G, B)^{\widetilde{c}})$ is soft pre*-generalized open in X. That is, $(\widetilde{f}^{-1}(G, B))^{\widetilde{c}}$ is soft pre*-generalized open in X. Thus $\widetilde{f}^{-1}(G, B)$ is soft pre*-generalized closed in X. Hence $\widetilde{f}$ is slightly soft pre*-generalized continuous. □

Theorem 30. *If $\widetilde{f} : (X, \widetilde{\tau}, A) \to (Y, \widetilde{\sigma}, B)$ is soft* pre**-generalized continuous, then $\widetilde{f}$ is slightly soft* pre**-generalized continuous.*

Proof. Let (G, B) be a soft clopen set in Y. Since $\widetilde{f}$ is soft pre*-generalized continuous, $\widetilde{f}^{-1}(G, B)$ is both soft pre*-generalized open and soft pre*-generalized closed in X. Hence $\widetilde{f}$ is slightly soft pre*-generalized continuous. □

Theorem 31. *If $\widetilde{f} : (X, \widetilde{\tau}, A) \to (Y, \widetilde{\sigma}, B)$ is slightly soft* pre**-generalized continuous and $(Y, \widetilde{\sigma}, B)$ is soft locally indiscrete then $\widetilde{f}$ is soft* pre**-generalized continuous.*

Proof. Let (G, B) be soft open set in Y. Since Y is soft locally indiscrete, (G, B) is soft closed. Since $\widetilde{f}$ is slightly soft pre*-generalized continuous, $\widetilde{f}^{-1}(G, B)$ is soft pre*-generalized closed in X. Hence $\widetilde{f}$ is soft pre*-generalized continuous. □

Theorem 32. *If $\widetilde{f} : (X, \widetilde{\tau}, A) \to (Y, \widetilde{\sigma}, B)$ is soft* pre**-generalized irresolute and $\widetilde{g} : (Y, \widetilde{\sigma}, B) \to (Z, \widetilde{\mu}, C)$ is slightly soft* pre**-generalized continuous, then $\widetilde{g} \circ \widetilde{f} : (X, \widetilde{\tau}, A) \longrightarrow (Z, \widetilde{\mu}, C)$ is slightly soft* pre**-generalized continuous.*

Proof. Let (H, C) be soft clopen set in Z. Since $\widetilde{g}$ is slightly soft pre*-generalized continuous, $\widetilde{g}^{-1}(H, C)$ is soft pre*-generalized closed in Y. Also, since $\widetilde{f}$ is soft pre*-generalized irresolute, $\widetilde{f}^{-1}(\widetilde{g}^{-1}(H, C)) = (\widetilde{g} \circ \widetilde{f})^{-1}(H, C)$ is soft pre*-generalized closed in X. Hence $\widetilde{g} \circ \widetilde{f}$ is slightly soft pre*-generalized continuous. □

Theorem 33. *If $\widetilde{f} : (X, \widetilde{\tau}, A) \to (Y, \widetilde{\sigma}, B)$ is slightly soft* pre**-generalized closed and $\widetilde{g} : (Y, \widetilde{\sigma}, B) \to (Z, \widetilde{\mu}, C)$ is perfectly soft* pre**-generalized continuous, then $\widetilde{g} \circ \widetilde{f} : (X, \widetilde{\tau}, A) \longrightarrow (Z, \widetilde{\mu}, C)$ is soft* pre**-generalized irresolute.*

Proof. Let (H,C) be soft pre*-generalized closed set in Z. Since $\widetilde{g}$ is perfectly soft pre*-generalized continuous, $\widetilde{g}^{-1}(H,C)$ is soft clopen in Y. Also, since $\widetilde{f}$ is slightly soft pre*-generalized continuous, $\widetilde{f}^{-1}(\widetilde{g}^{-1}(H,C)) = (\widetilde{g} \circ \widetilde{f})^{-1}(H,C)$ is soft pre*-generalized closed in X. Hence $\widetilde{g} \circ \widetilde{f}$ is soft pre*-generalized irresolute. □

Bibliography

[1] C. G. Aras, A. Sonmez and H. Cakalli, On soft mappings, 2013, arXiv:1305.4545.

[2] I. Arockiarani and A. Selvi, On soft contra πg-continuous functions in soft topological spaces, *Int. J. Math. Trends Technol.* **19**(1) (2015).

[3] I. Arockiarani and A. Selvi, On soft slightly πg-continuous functions, *J. Progressive Res. Math.* **3**(2) (2015).

[4] A. Kharal and B. Ahmad, Mappings on soft classes, *New Math. Nat. Comput.* **7** (2010), 471–481.

[5] D. Molostsov, Soft set theory – first results, *Comput. Math. Appl.* **37** (1999), 19–31.

[6] C. Reena and M. Karthika, A New class of generalized closed sets in soft topological spaces, in: *International Virtual Conference on Current Scenario in Modern Mathematics (ICCSMM-2022). Proceedings*, 2022, ISBN: 978-81-948552-9-3.

[7] C. Reena and M. Karthika, Soft pre*-generalized continuous functions in soft topological spaces, *Ratio Math.* **44** (2022), 145–154.

Ekta Jain, Dhanpal Singh, and S. D. Ram

Modeling of piston problem for generalized Chaplygin gas

Abstract: The present study focuses to examine how a plane piston behaves in a generalized Chaplygin gas when it moves at a constant speed. A methodical perturbation approach is utilized to derive an analytical solution for the flow field, extending up to the first order of ϵ. The solutions for the fundamental state are derived for weak shock waves. Additionally, an approximate solution for the case of strong shock waves is presented and discussed. The obtained solution is significantly impacted by the presence of generalized Chaplygin gas, adding an intriguing dimension from both a physical and scientific perspective.

Keywords: Euler equations, generalized Chaplygin gas, piston problem, perturbation method

MSC 2010: 35L60, 35L65, 74H10, 74J40

1 Introduction

In the early 1990s, it was known that the universe could have sufficient energy density to halt its expansion and collapse, or that it could have an energy density that would cause it to continue expanding forever, but gravity would slow the expansion over time. Then, in 1998, the observations of extremely distant supernovae revealed that the universe was not only expanding, but also accelerating. Physicists are still unsure of the proper explanation, but they proposed a hypothetical form of energy which is known as dark energy to describe accelerated expansion. According to estimates, roughly 68 % of the universe's total energy is made up of dark energy and about 27 % of matter is dark matter. The nature of the universe's dark sector is still a mystery. To unify the dark sector of the universe, it is considered as a single component that acts as both dark energy and

Acknowledgement: The author Ekta Jain acknowledges the financial support from the Department of Mathematics, University of Delhi, Delhi-110007, India, under the UGC-Non-Net Fellowship scheme.

Ekta Jain, Department of Mathematics, University of Delhi, admin office, Faculty of Mathematical Sciences, Guru Tegh Bahadur Road, Delhi-110007, Delhi, India, e-mail: ejain@maths.du.ac.in
Dhanpal Singh, Department of Mathematics, Keshav Mahavidyalaya, University of Delhi, admin office, H-4-5 Zone, Pitampura, Near Sainik Vihar, Delhi-110034, Delhi, India, e-mail: drdhanpalsingh@keshav.du.ac.in
S. D. Ram, Department of Mathematics, Shivaji College, University of Delhi, admin office, Raja Garden, Ring Road, Delhi-110027, Delhi, India, e-mail: sdram.apm@shivaji.du.ac.in

https://doi.org/10.1515/9783111313634-031

dark matter. One way to achieve the unification of dark energy and dark matter is to use a hypothetical substance called Chaplygin gas. In the initial stages, the one-dimensional system of Chaplygin gas equations was proposed by Tsien [17], Karman [10] and Chaplygin [2] as appropriate approximation to determine the lifting force on a wing of an airplane in aerodynamics.

The Chaplygin gas is an ideal fluid that initially behaves as a pressure less fluid and later as a cosmological constant. Pure Chaplygin gas with an exotic equation of state is characterized by a negative pressure and given as

$$p = -\frac{B}{\rho}, \tag{31.1}$$

where p is the pressure, ρ is the density, and B is a positive parameter.

The pure Chaplygin gas has been extended to the generalized Chaplygin gas with the following equation of state

$$p = -\frac{B}{\rho^{\alpha}}, \quad 0 < \alpha \leq 1. \tag{31.2}$$

The one-dimensional Euler equations in isentropic gas dynamics with Chaplygin gas has been widely utilized in various fields due to their immense physical significance. Brenier [1] obtained concentrated solutions to the Riemann problem for an isentropic Chaplygin gas dynamics system. The delta shock waves that are solutions to the two-dimensional Reimann problem for the Chaplygin gas dynamic system have been examined by Guo et al. [7]. Gupta et al. [8] used the differential constraints method to arrive at the closed-form solution of the generalized Riemann problem for the one-dimensional Euler's equation for Chaplygin gas. In this paper, we discuss the one-dimensional piston problem of the isentropic Euler equations with generalized Chaplygin gas. The piston problem for the initial-boundary value problem of the Euler equations has received a great deal of attention in the study of the mathematical theory of compressible fluid dynamics. In general, rarefaction waves form when the piston is pulled back; in contrast, compressed waves enter the gas when the piston is shoved forward. It is explained in [5, 14, 20, 6, 11, 18] by using a large tube having a piston opened at one end and closed at the other end, and considering the gas inside the tube is stationary with a uniform density and pressure, whatever turn of the piston will end the gas to move in response. Chester [4] used a relatively low Mach number compressible fluid circulation model to theoretically examine the gas dynamic disturbances caused in a closed tube filled with gases, which is brought on by an oscillating piston. When analyzing the problem of propagation of weak and strong shock waves brought on by the movement of a piston in a field of weak gravity, Wen-rui [19] employed the perturbation method. Singh et al. [16] used perturbation scheme to examine the flow pattern caused by the motion of a plane piston moving at a constant speed in a non-ideal gas with a weak gravitational field. Using the same technique, Chaudhary et al. [3] studied how dust particles affected weak and strong shock waves with respect to the shock front and the piston position. In recent years, numerous authors have focused on discussing the problem of piston motion in Chaplygin gas. Qu et al. [13] carried out a study on the piston problem's high Mach number limit for the entire set of Euler equations for isentropic flow in case of polytropic gas. For Chaplygin–Euler equations, Qu and Yuan [12] investigated the integral weak solution of euler equations for a problem of piston's motion in Chaplygin gas. In their work, a shock wave solution is demonstrated to exist while the piston is traveling ahead at subsonic pace as well as a Radon measure solution is constructed if the piston is going ahead during its sonic or supersonic speed. In 2023, Singh et al. [15] adopted an analytical technique to ascertain a precise solution for one-dimensional Euler equations in adiabatic gas dynamics with perfect fluid under the influence of weak gravity. Huang et al. [9] resolved the piston problem for the modified Chaplygin gas. They rigorously demonstrated the universal existence and uniqueness of a shock wave that separates constant states ahead of the piston once it drives forward into the gas. They confirmed that the piston is in front of only rarefaction wave of first family and that concentration never occurs as the piston gets pulled away from the gas.

In the present article, implementing the perturbation method, it is determined how generalized Chaplygin gas affects weak and strong shocks within the piston position and the shock front. A Cartesian coordinate is used to discuss the one-dimensional dynamical process driven by a piston for Chaplygin gas. The Rankine–Hugoniot conditions are derived up to first order. We presented particular examples to demonstrate clearly that for some typical strengths of the weak shock waves by the influence of Chaplygin gas, which changes the shock wave's strength highly. Further, the analysis is carried out as well in case of the hypersonic shock wave.

2 Basic piston problem

The governing equations for an unsteady one-dimensional motion of the fluid can be expressed in the following form

$$\rho_t + v\rho_x + \rho v_x = 0, \tag{31.3}$$

$$v_t + vv_x + \frac{1}{\rho}p_x = 0, \tag{31.4}$$

$$p_t + vp_x + a^2\rho v_x = 0, \tag{31.5}$$

where ρ, p, and v respectively, stand for the density, pressure, and velocity of specific fluid flows at place x and time t. The set of equations (31.3)–(31.5) is supplemented with the equation of state (31.2) for generalized Chaplygin gas. Here, a represents the local speed of sound, which is defined by $a = \sqrt{p'(\rho)}$, we have

$$a = \sqrt{\alpha B}\rho^{-\frac{\alpha+1}{2}}. \tag{31.6}$$

In accordance with the concept of similarity and analysis of dimensions [14], the dimension of velocity is capable of being written as follows

$$\eta = \frac{x}{t}. \tag{31.7}$$

We introduce a set of non-dimensional parameters in order to explore the basic flow properties.

$$\widetilde{x} = \frac{x}{x_\dagger}, \quad \widetilde{t} = \frac{t}{x_\dagger/v_\dagger}, \quad \widetilde{v} = \frac{v}{v_\dagger}, \quad \widetilde{a} = \frac{a}{v_\dagger}, \tag{31.8}$$

where the variables $x_\dagger$, $t_\dagger$, $a_\dagger$ and $v_\dagger$ stands for the space, time, sonic velocity and the typical velocity, respectively.

The system of equations (31.3)–(31.5) for the isentropic flow can be expressed in form of non-dimensional parameters specified by (31.8). Thus, we have

$$\frac{\partial a}{\partial t} + v\frac{\partial a}{\partial x} - \frac{(\alpha+1)a}{2}\frac{\partial v}{\partial x} = 0, \tag{31.9}$$

$$\frac{\partial v}{\partial t} + v\frac{\partial v}{\partial x} - \frac{2a}{\alpha+1}\frac{\partial a}{\partial x} = 0. \tag{31.10}$$

To develop the solution, let us expand the variables for flow in form of small parameter ϵ. Thus, we have

$$v = v^{(0)} + \epsilon v^{(1)} + \epsilon^2 v^{(2)} + \epsilon^3 v^{(3)} \dots, \tag{31.11}$$

$$\rho = \rho^{(0)} + \epsilon \rho^{(1)} + \epsilon^2 \rho^{(2)} + \epsilon^3 \rho^{(3)} \dots, \tag{31.12}$$

$$a = a^{(0)} + \epsilon a^{(1)} + \epsilon^2 a^{(2)} + \epsilon^3 a^{(3)} \dots. \tag{31.13}$$

Utilizing the equations (31.11)–(31.13) in equations (31.9) and (31.10) and gathering the terms of zero order, the following basic equation of zeroth order can be determined as

$$\frac{\partial a^{(0)}}{\partial t} + v^{(0)}\frac{\partial a^{(0)}}{\partial x} - \frac{\alpha+1}{2}a^{(0)}\frac{\partial v^{(0)}}{\partial x} = 0, \tag{31.14}$$

$$\frac{\partial v^{(0)}}{\partial t} + v^{(0)}\frac{\partial v^{(0)}}{\partial x} - \frac{2}{\alpha+1}a^{(0)}\frac{\partial a^{(0)}}{\partial x} = 0. \tag{31.15}$$

The solution of the basic equations (31.14) and (31.15) may be given by the Riemann flow. So as a result, we assume [19, 16, 3]

$$v^{(0)} = v^{(0)}(\eta), \ a^{(0)} = a^{(0)}(\eta), \tag{31.16}$$

and the equations (31.14) and (31.15) can be modified in respect of similarity variable η to produce the following equations

$$(v^{(0)} - \eta)\frac{\partial a^{(0)}}{\partial \eta} - \frac{\alpha+1}{2}a^{(0)}\frac{\partial v^{(0)}}{\partial \eta} = 0, \tag{31.17}$$

$$(v^{(0)} - \eta)\frac{\partial v^{(0)}}{\partial \eta} - \frac{2}{\alpha+1}a^{(0)}\frac{\partial a^{(0)}}{\partial \eta} = 0. \tag{31.18}$$

The solution of above system of equations (31.17) and (31.18) express Riemann flow in Chaplygin gas. Riemann invariant for the above system are given by

$$v^{(0)} \pm \frac{2}{\alpha+1}a^{(0)} = \text{const.} \tag{31.19}$$

A specific solution of the given problem is established taking into account the piston's constant velocity and uniform consideration of the flow variables in front of the shock, i. e.,

$$a^{(0)} = \text{const.}, \quad v^{(0)} = \text{const.} \tag{31.20}$$

Similarly, after gathering the first-order terms, we have

$$\frac{\partial a^{(1)}}{\partial t} + v^{(0)}\frac{\partial a^{(1)}}{\partial x} - \frac{\alpha+1}{2}a^{(0)}\frac{\partial v^{(1)}}{\partial x} = 0, \tag{31.21}$$

$$\frac{\partial v^{(1)}}{\partial t} + v^{(0)}\frac{\partial v^{(1)}}{\partial x} - \frac{2}{\alpha+1}a^{(0)}\frac{\partial a^{(1)}}{\partial x} = 0, \tag{31.22}$$

and so on for all higher-order terms. The solution for first-order relations must be the functions of η and t in accordance with the concept of dimensional analysis. So, we discuss the solutions with the following form

$$v^{(1)}(\eta, t) = t^{-m}f(\eta), \quad a^{(1)}(\eta, t) = t^{-r}g(\eta). \tag{31.23}$$

Using equation (31.23) in equations (31.21) and (31.22), we have

$$m = r = 1.$$

When we adopt $v_{\dagger}$ as v_p, which is a notation for a plane piston velocity, and also assume it to be a constant, then the fundamental velocity becomes

$$v^{(0)} = 1,$$

so, at the piston, the boundary condition is

$$f(1) = 0. \tag{31.24}$$

If we assume the non-dimensional quantity $\zeta = \frac{x/t}{V^{\dagger}}$, then using the relations (31.23), the perturbation equations (31.21)–(31.22) may be reduced in terms of ζ as

$$£_1(f,g) \equiv (1-\zeta)\frac{\partial g(\zeta)}{\partial \zeta} - \frac{\alpha+1}{2}a^{(0)}\frac{\partial f(\zeta)}{\partial \zeta} - g(\zeta) = 0, \tag{31.25}$$

$$£_2(f,g) \equiv (1-\zeta)\frac{\partial f(\zeta)}{\partial \zeta} - \frac{2}{\alpha+1}a^{(0)}\frac{\partial g(\zeta)}{\partial \zeta} - f(\zeta) = 0. \tag{31.26}$$

3 Rankine–Hugoniot (RH) conditions

The Rankine–Hugoniot conditions for the shock wave are provided through the principle of conservation of energy, momentum, and mass. In case of Chaplygin gas, we can express the RH relations as follows:

$$\rho = \left(\frac{\alpha-1}{\alpha+1}\right)\left(1 - \frac{2}{\alpha+1}\frac{a_a^2}{U^2}\right)^{-1}\rho_a, \tag{31.27}$$

$$v = \frac{2}{(\alpha-1)}\frac{1}{U}(a_a^2 - U^2), \tag{31.28}$$

$$p = p_a + \frac{2\rho_a}{\alpha-1}(a_a^2 - U^2), \tag{31.29}$$

where U symbolizes the shock speed and the subscript "a" indicates the quantity assessed ahead of the shock front. The zero-order RH relations can be stated by expanding the variables p and U in terms of ϵ, analogous to those in equations (31.11)–(31.13) as

$$\rho^{(0)} = \left(\frac{\alpha-1}{\alpha+1}\right)\left(1 - \frac{2}{\alpha+1}\frac{a_a^2}{U^{(0)^2}}\right)^{-1}\rho_a, \tag{31.30}$$

$$v^{(0)} = \frac{2}{(\alpha-1)}\frac{1}{U^{(0)}}(a_a^2 - U^{(0)^2}), \tag{31.31}$$

$$p^{(0)} = p_a + \frac{2\rho_a}{\alpha-1}(a_a^2 - U^{(0)^2}). \tag{31.32}$$

Also, the first-order jump relations are

$$\rho_s^{(1)} = \frac{-\frac{4(\alpha-1)}{(\alpha+1)^2}\left(\frac{a_a}{U^{(0)}}\right)^2\rho_a}{\left(1-\left(\frac{a_a}{U^{(0)}}\right)^2\frac{2}{\alpha+1}\right)^2}\frac{U^{(1)}}{U^{(0)}}, \tag{31.33}$$

$$v_s^{(1)} = \frac{2}{1-\alpha}\frac{U^{(1)}}{U^{(0)^2}}(U^{(0)^2} + a_a^2), \tag{31.34}$$

$$p_s^{(1)} = \frac{4}{1-\alpha}U^{(0)}U^{(1)}\rho_a, \tag{31.35}$$

where value assessed at the point of shock $\zeta = \zeta_s$ is indicated by the subscript "s".

With the help of (31.11)–(31.13), we can expand the sonic velocity. Thus, the sound speed for zeroth order is obtained as

$$a^{(0)} = \sqrt{\alpha B \rho^{(0)^{-(\alpha+1)}}}, \tag{31.36}$$

and the first-order speed of sound can be written as

$$a^{(1)} = \left(\frac{p^{(1)}}{p^{(0)}} - \frac{\rho^{(1)}}{\rho^{(0)}} \right) \frac{a^{(0)}}{2}, \tag{31.37}$$

by substituting equations (31.33) and (31.35) into the equation (31.37), the local speed of sound at the shock is determined as

$$\begin{aligned} a_s^{(1)} &= \frac{a^{(0)}}{2} \left[-\frac{4\rho_a U^{(0)^2}}{(\alpha-1)p_a + 2(a_a^2 - U^{(0)^2})\rho_a} + \frac{4a_a^2}{(\alpha+1)U^{(0)^2} - 2a_a^2} \right] \\ &\quad \times \frac{(U^{(0)^2} - a_a^2)}{(U^{(0)^2} + a_a^2)} \frac{v_s^{(1)}}{v^{(0)}}. \end{aligned} \tag{31.38}$$

Using (31.36), the relationship (31.38) gives the expression, which connects the two functions $f(\zeta_s)$ and $g(\zeta_s)$ in the following way

$$f(\zeta_s) = \beta g(\zeta_s), \tag{31.39}$$

where the coefficient β is given by the following relation

$$\begin{aligned} \frac{1}{\beta} &= \frac{2a^{(0)}}{v^{(0)}} \left[-\frac{\rho_a U^{(0)^2}}{(\alpha-1)p_a + 2\rho_a(a_a^2 - U^{(0)^2})} + \frac{a_a^2}{(\alpha+1)U^{(0)^2} - 2a_a^2} \right] \\ &\quad \times \frac{(U^{(0)^2} - a_a^2)}{(U^{(0)^2} + a_a^2)}. \end{aligned} \tag{31.40}$$

Using the boundary conditions (31.24), (31.39) and the equation (31.38), perturbation state equations (31.25)–(31.26) are solved for the region, where ζ is given by $1 \le \zeta \le U^{(0)}/v_p$.

The relations (31.30)–(31.32) are used to determine the solutions of basic state, which are displayed in Tables 31.1–31.3 for the following values of $a_a/U^{(0)} = 0.25, 0.50, 0.625$.

Table 31.1: Solution for $a_a/U^{(0)} = 0.25$.

	$\alpha = 0.75$	$\alpha = 0.5$	$\alpha = 0.25$
$U^{(0)}/v_p$	0.1333300	0.2666000	0.4000000
$a^{(0)}/v_p$	−0.8017342	−0.5953523	−0.3674234
a_a/v_p	0.0333333	0.0666667	0.1000000
β	−0.5534480	−0.4882950	−0.3387323

Table 31.2: Solution for $a_a/U^{(0)} = 0.50$.

	$\alpha = 0.75$	$\alpha = 0.5$	$\alpha = 0.25$
$U^{(0)}/v_p$	0.1666667	0.3333333	0.5000000
$a^{(0)}/v_p$	−0.7682950	−0.5270460	−0.2500000
a_a/v_p	0.0833333	0.1666667	0.2500000
β	−0.0361551	−0.2529820	−0.3000000

Table 31.3: Solution for $a_a/U^{(0)} = 0.625$.

	$\alpha = 0.75$	$\alpha = 0.5$	$\alpha = 0.25$
$U^{(0)}/v_p$	0.2051280	0.4102560	0.6153850
$a^{(0)}/v_p$	−0.7297000	−0.4476180	−0.0608130
a_a/v_p	0.1282050	0.2564100	0.3846150
β	0.1475770	−0.1974500	−1.1047000

4 Certain examples for weak shock waves

Perturbation state equations (31.25) and (31.26) with the boundary conditions (31.24) and (31.39) are linear. The two elementary solutions listed below may be sufficient to solve the plane piston problem.

$$£_1(f_1, g_1) = 0, \quad £_2(f_1,, g_1) = -\frac{1}{\zeta^2}, \quad f_1(1) = 0, \quad g_1(1) = 0, \tag{31.41}$$

and

$$£_1(f_2,, g_2) = 0, \quad £_2(f_2, g_2) = 0, \quad f_2(1) = 0, \quad g_2(1) = 1. \tag{31.42}$$

Since the differential equations (31.25), (31.26), and expression (31.39) are linear, the solution to the problem of piston motion in generalized Chaplygin gas can be obtained as

$$f(\zeta) = f_1(\zeta) + Cf_2(\zeta), \quad g(\zeta) = g_1(\zeta) + Cg_2(\zeta), \tag{31.43}$$

where C is an arbitrary constant that can be found using equation (31.39) and represented as

$$C = \frac{f_1(\zeta_s) - \beta g_1(\zeta_s)}{\beta g_2(\zeta_s) - f_2(\zeta_s)}.$$

It is easy to check that above solution satisfies perturbation state equations (31.25) and (31.26) with the boundary conditions (31.24) and (31.39). Both the elementary solutions have two boundary conditions at the piston.

5 Approximation of strong shocks

The flow region gets restricted for strong shock wave and given by $\zeta - 1 \le \zeta_s - 1 \ll 1$. Then both equations (31.25) and (31.26) may be simplified as

$$-\frac{\alpha+1}{2}a^{(0)}f'(\zeta) - g(\zeta) = 0, \tag{31.44}$$

$$-\frac{2}{\alpha+1}a^{(0)}g'(\zeta) - f(\zeta) = 0. \tag{31.45}$$

On differentiating equation (31.44) and using equation (31.45) to cancel the function $g'(\zeta)$, we find the equation for the function $f(\zeta)$ that can be written as

$$\frac{\partial^2 f}{\partial \zeta^2} + \frac{1}{\zeta}\frac{\partial f}{\partial \zeta} - \frac{1}{a^{(0)^2}\zeta^2}f = 0. \tag{31.46}$$

Equation (31.46) is an Euler equation, which is uniform. So, the general solution of equation (31.46) can be obtained as

$$f(\zeta) = c_1\zeta^{\varphi} + c_2\zeta^{-\varphi}. \tag{31.47}$$

To find the distribution of $g(\zeta)$, we use above equation (31.47) into the equation (31.44). We have

$$g(\zeta) = \frac{\alpha+1}{2\zeta}(-c_1\zeta^{\varphi} + c_2\zeta^{-\varphi}), \tag{31.48}$$

where $\varphi = 1/a^{(0)}$ and c_1, c_2 are arbitrary constants. These constants can be determined by applying the boundary conditions. When we explore the piston problem, where the boundary conditions are set as (31.24) and (31.39) specifically defined at the piston, we find that the functions $f(\zeta) \approx 0$ and $g(\zeta) \approx 0$ at the shock $\zeta = \zeta_s$. Now, we discuss the solution of piston problem for the boundary conditions $f(\zeta) = 0$ and $g(\zeta) = 1$ at $\zeta = 1$. By applying these boundary conditions, c_1 and c_2 are determined as

$$c_1 = -\frac{1}{\alpha+1}, \tag{31.49}$$

$$c_2 = \frac{1}{\alpha+1}. \tag{31.50}$$

With the help of obtained results for the equation (31.23), we obtain the solutions of first order and represent as

$$v^{(1)}(x,t) = \frac{1}{x}\left[c_1\left(\frac{x}{t}\right)^{1+\varphi} + c_2\left(\frac{x}{t}\right)^{1-\varphi}\right], \tag{31.51}$$

$$a^{(1)}(x,t) = \frac{\alpha+1}{2x}\left[-c_1\left(\frac{x}{t}\right)^{\varphi} + c_2\left(\frac{x}{t}\right)^{-\varphi}\right]. \tag{31.52}$$

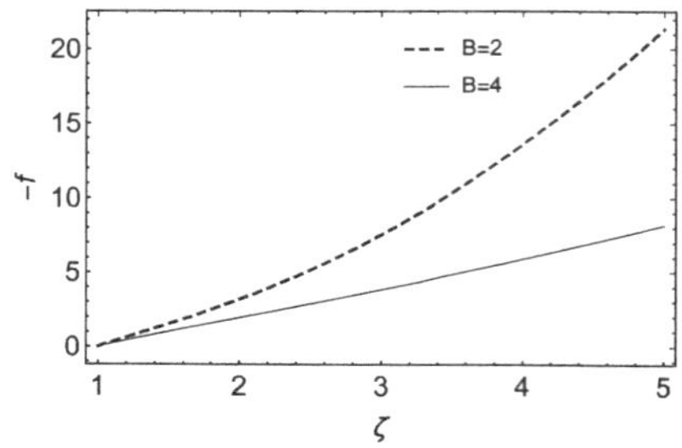

(a) Distribution of the function $f(\zeta)$

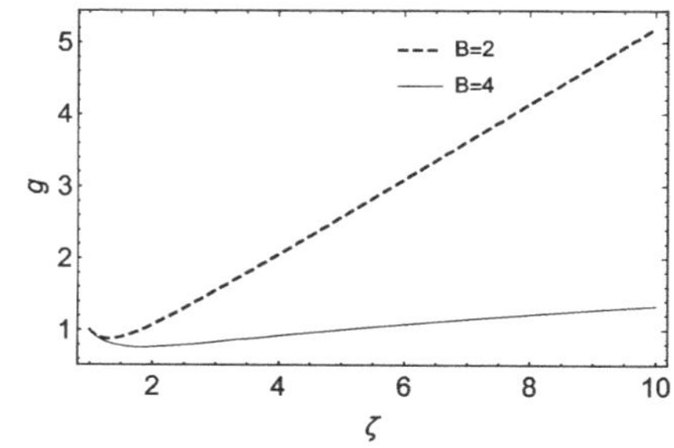

(b) Distribution of the function $g(\zeta)$

Figure 31.1: For $\alpha = 0.2$ with different values of B.

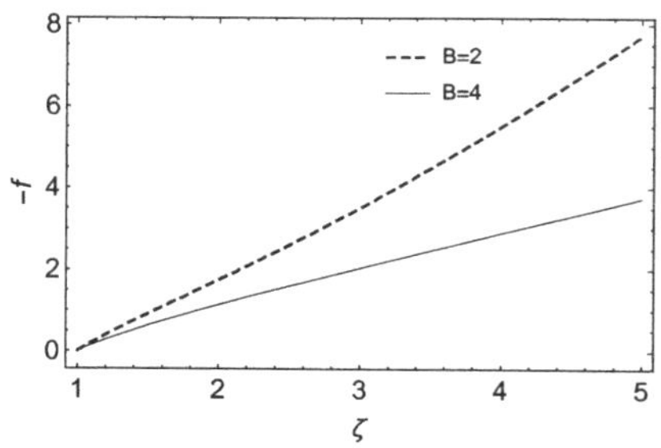

(a) Distribution of the function $f(\zeta)$

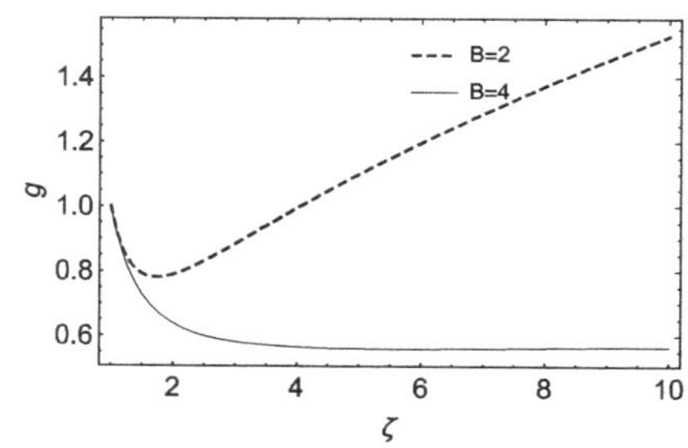

(b) Distribution of the function $g(\zeta)$

Figure 31.2: For $\alpha = 0.4$ with different values of B.

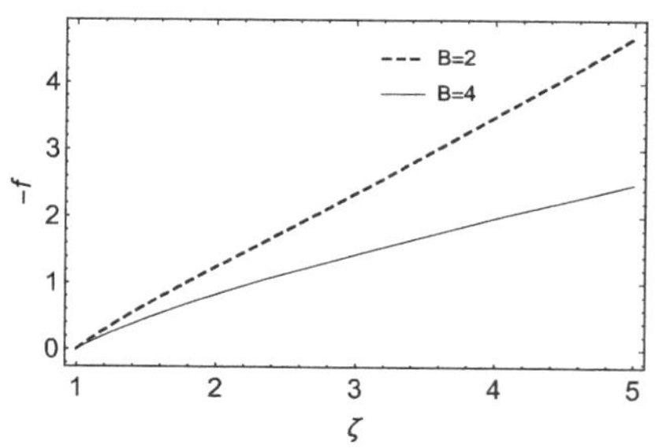

(a) Distribution of the function $f(\zeta)$

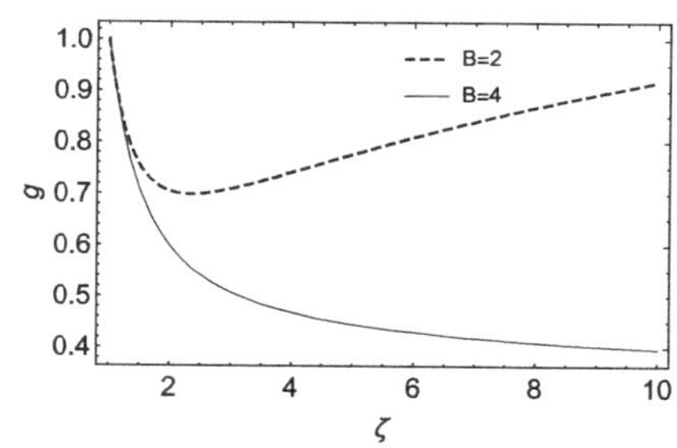

(b) Distribution of the function $g(\zeta)$

Figure 31.3: For $\alpha = 0.6$ with different values of B.

The functions $f(\zeta)$ and $g(\zeta)$ for the values of c_1 and c_2 given by the equations (31.49) and (31.50) are graphically shown in Figures 31.1–31.3 for different values of B and α. The plots in Figures 31.1a–31.3a demonstrate that the function $f(\zeta)$ exhibits a monotonically decreasing behavior with respect to ζ. This pattern of the flow indicate that kinetic energy of the gas between the piston and the shock wave is exhausted. However, from

Figures 31.1b–31.3b, It is observed that the graph of $g(\zeta)$ exhibits the characteristics of an upward parabola in close proximity to the piston, and subsequently, it undergoes an increase as it moves farther away. Hence, internal energy of the gas undergoes an initial deceleration near the piston, followed by a subsequent acceleration as the distance from the piston increases. The acceleration trend of internal energy intensifies with a decrease in both the values of B and α. Internal energy is contributing to the strengthening of the shock wave due to the presence of generalized Chaplygn gas.

6 Summary

In the current study, one-dimensional piston problem for generalized Chaplygin gas is discussed. The basic state solution is computed for three different value of $\frac{a_a}{U^{(0)}}$ as shown in Tables 31.1–31.3. In the first part of the article, we have grabbed certain examples of elementary solutions for weak shock wave, which is greatly impacted by generalized Chaplygin gas. Further, we also explored the case of strong shock wave in restricted region $\zeta - 1 \leq \zeta_s - 1 \ll 1$. In that region, the solution of strong shock wave problem represented by equations (31.47) and (31.48). The distribution of functions $f(\zeta)$ and $g(\zeta)$ sketched graphically as shown in the Figures 31.1–31.3 for distinct values of B and α. The figures lead us to the conclusion that kinetic energy of the gas depletes near the shock. The internal energy of the gas initially decreases from zero to a minimum positive value and then increases as the distance from the piston grows. When the values of B and α are reduced, the internal energy begins to increase earlier in the process, and does so rapidly. In this scenario, it is noteworthy that the internal energy does not exhaust throughout the entire process. The preceding discussion highlights that the selected model is profoundly influenced by the presence of generalized Chaplygin gas, a presence of dark energy that is scientifically justified.

Bibliography

[1] Y. Brenier, Solutions with concentration to the riemann problem for the one-dimensional chaplygin gas equations, *J. Math. Fluid Mech.* **7** (2005), S326–S331.

[2] S. Chaplygin, On gas jets, *Trans. by M. Slud, Brown University* **21** (1944), 1–127.

[3] J. Chaudhary, S. Ram and L. Singh, The plane piston problem with weak gravitational field in a dusty gas, *J. King Saud Univ., Sci.* **31**(4) (2019), 1027–1033.

[4] W. Chester, Resonant oscillations in closed tubes, *J. Fluid Mech.* **18**(1) (1964), 44–64.

[5] R. Courant and K. Friedrichs, *Supersonic Flow and Shock. Waves*, Interscience Publishers, Inc., New York, 1948.

[6] C. Dafermos, *Hyperbolic Conservation Laws in Continuum Physics*, Springer Book Archive-Mathematics, 2000.

[7] L. Guo, W. Sheng and T. Zhang, The two-dimensional riemann problem for isentropicchaplygin gas dynamic system, *Commun. Pure Appl. Anal.* **9**(2) (2009), 431–458.

[8] P. Gupta, R. Chaturvedi and L. Singh, The generalized riemann problem for the chaplygin gas equation, *Eur. J. Mech. B, Fluids* **82** (2020), 61–65.

[9] M. Huang, Y. Wang and Z. Shao, Piston problem for the isentropic euler equations for a modified chaplygin gas, *Phys. Fluids* **35**(1) (2023).

[10] T. Karman, Compressibility effects in aerodynamics, *J. Aeronaut. Sci.* **8**(9) (1941), 337–356.

[11] T. Li, Global classical solutions for quasilinear hyperbolic systems, 1994.

[12] A. Qu and H. Yuan, Measure solutions of one-dimensional piston problem for compressible euler equations of chaplygin gas, *J. Math. Anal. Appl.* **481**(1) (2020), 123486.

[13] A. Qu, H. Yuan and Q. Zhao, High mach number limit of one-dimensional piston problem for non-isentropic compressible euler equations: Polytropic gas, *J. Math. Phys.* **61**(1) (2020).

[14] L. Sedov, *Similarity and Dimensional Methods in Mechanics*, 1959.

[15] D. Singh, E. Jain and S. Ram, The behavior of weak shock waves under the influence of weak gravitational field, *Indian J. Phys.* **98**(3) (2023), 797–802.

[16] L. Singh, S. Ram and D. Singh, Flow pattern induced by the plane piston moving in a non-ideal gas with weak gravitational field, *Ain Shams Eng. J.* **2**(2) (2011), 125–131.

[17] H. Tsien, Two-dimensional subsonic flow of compressible fluids, *J. Aeronaut. Sci.* **6**(10) (1939), 399–407.

[18] M. Wang and D. Kassoy, Dynamic compression and weak shock formation in an inert gas due to fast piston acceleration, *J. Fluid Mech.* **220** (1990), 267–292.

[19] H. Wen-rui, The plane piston problem in a weak gravitational field, *Appl. Math. Mech.* **6**(1) (1985), 75–85.

[20] G. Whitham, *Linear and Nonlinear Waves*, John Wiley & Sons, 1974.

J. K. Kushwaha and R. Vishwakarma

A new estimation of the degree of approximation of functions belonging to Lipschitz class by Borel–Euler summability method of Fourier series

Abstract: Product summability methods are widely used by several researcher to find degree of approximation of functions belonging to various classes. Recently, the authors of [14] have used Borel–Euler summability method of Fourier series. But they did not find the degree of approximation of functions by Borel–Euler product summability method. In this chapter, we have determined the degree of approximation of functions belonging to Lipschitz class by Borel–Euler (BE^1) product summability method.

Keywords: Estimation of function, degree of approximation, Lipschitz class, Fourier series, Borel–Euler summability method, Borel mean, Euler mean

MSC 2010: 42B05, 42B08

1 Introduction and Definitions

Approximation theory plays an important role in the field of Mathematical Analysis. The estimation of degree of approximation of functions by using summability means have been done by several researchers such as ([1–13]) within last few decades. Now, we are going to estimate the degree of approximation of functions belonging to Lipschitz class by Borel-Euler summability method of Fourier series .

Let $f(x)$ be a 2π periodic and Lebesgue integrable function. Then its Fourier series is given by ([15])

$$S_n(f;x) = \frac{1}{2}a_0 + \sum_{k=0}^{n}(a_k \cos kx + b_k \sin kx) = \sum_{k=0}^{n} A_k(f;x). \tag{32.1}$$

An infinite series $\sum_{n=0}^{\infty} u_n$ with partial sum s_n is said to be $(E,1)$ summable if

$$E_n^1 = \frac{1}{2^n}\sum_{k=0}^{n}\binom{n}{k}s_n \to s \quad \text{as} \quad n \to \infty.$$

J. K. Kushwaha, R. Vishwakarma, Department of Mathematics and Statistics, DDU Gorakhpur University, Gorakhpur, U. P., India, e-mails: k.jitendrakumar@yahool.com, vishwakarmaradha89@gmail.com

https://doi.org/10.1515/9783111313634-032

The Borel B-summability is defined as

$$B_p = e^{-p} \sum_{m=0}^{\infty} \frac{p^m}{m!} S_n.$$

A function $f \in \mathrm{Lip}\,\alpha$. If, $f(x+t) - f(x) = o(|t|^{\alpha})$ for $0 < \alpha \le 1$.

The degree of approximation of a function f by a trigonometric polynomial t_n of order n under the norm $\|.\|_{\infty}$ is defined by Zygmund [17] with

$$\|t_n - f\|_{\infty} = \sup\{|t_n(x) - f(x)| : x \in R\}$$

and the best approximation $E_n(f)$ of a function $f \in L_p$ is defined by the equality

$$E_n(f) = \min_{t_n} \|t_n - f\|_p.$$

We use following notation throughout the chapter

$$\phi(t) = f(x+1) + f(x-t) - 2f(x).$$

2 Lemma

Lemma 1. For $0 < t < \frac{\pi}{p+1}$ and $\sin t/2 \ge \frac{t}{\pi}$

$$\begin{aligned}
|K(p,t)| &= \left|\frac{1}{2\pi} \frac{e^{-p\sin^2 t/2}}{\sin t/2} \sin\left(\frac{t + p\sin t}{2}\right)\right| \\
&\le \left|\frac{e^{-p\frac{t^2}{\pi^2}}}{2\pi(t/\pi)} \sin\left(\frac{t + p\sin t}{2}\right)\right| \\
&\le \left|\frac{e^{-p\frac{t^2}{\pi^2}}}{2t} \sin\left(\frac{t + pt}{2}\right)\right| \\
&\le \left|\frac{e^{-p\frac{t^2}{\pi^2}}}{2t} \sin\frac{(p+1)t}{2}\right| \\
&= \frac{1}{2t}\left[1 - \frac{pt^2}{\pi^2} + \frac{(\frac{pt^2}{\pi^2})^2}{2!} + \cdots\right]\left[(1+p)\frac{t}{2} - \frac{((1+p)\frac{t}{2})^3}{3!} + \cdots\right] \\
&= \frac{1}{2t}\left[(1+p)\frac{t}{2} + \cdots\right] \\
&= O(p+1).
\end{aligned}$$

Lemma 2. For $\frac{\pi}{p+1} < t < \pi$

$$\begin{aligned}|K(p,t)| &= \left|\frac{1}{2\pi}\frac{e^{-p\sin^2 t/2}}{\sin t/2}\sin\frac{(t+p\sin t)}{2}\right| \\ &\leq \left|\frac{1}{2\pi}\frac{1}{(t/\pi)}\sin\left(\frac{t+p\sin t}{2}\right)\right| \\ &= \frac{1}{2t}\left|\sin\left(\frac{t+p\sin t}{2}\right)\right| \\ &= O\left(\frac{1}{t}\right).\end{aligned}$$

3 Theorem

If f is a 2π periodic function belonging to Lip α class, then its degree of approximation by Borel–Euler summability method of Fourier series is given by

$$\|t_E^B - f\| = O\left(\frac{1}{(p+1)^\alpha}\right), \quad 0 < \alpha < 1.$$

Proof of the theorem. Let S_n be the nth partial sum of the Fourier series (32.1), following [16], we have

$$S_n - S = \frac{1}{2\pi}\int_0^\pi \frac{\phi(t)\sin(n+\frac{1}{2})t}{\sin t/2}dt.$$

The $(E,1)$ transform of S_n, we get

$$E_n^1 = \frac{1}{2\pi}2^{-n}\int_0^\pi \frac{\phi(t)}{\sin(t/2)}\sum_{\nu=0}^{n}\binom{n}{\nu}\sin\left(\nu+\frac{1}{2}\right)tdt. \tag{32.2}$$

Now, let

$$S = \sum_{\nu=0}^{n}\binom{n}{\nu}\sin\left(\nu+\frac{1}{2}\right)t,$$

$$S = \sin\left(\frac{t}{2}\right) + n\sin\left(\frac{3t}{2}\right) + \frac{n(n-1)}{2}\sin\left(\frac{5t}{2}\right) + \cdots + \sin\left(n+\frac{1}{2}\right)t$$

and writing cosine series as-

$$C = \cos\left(\frac{t}{2}\right) + n\cos\left(\frac{3t}{2}\right) + \frac{n(n-1)}{2}\cos\left(\frac{5t}{2}\right) + \cdots + \cos\left(n+\frac{1}{2}\right)t,$$

$$
\begin{aligned}
C + iS &= e^{it/2} + ne^{3it/2} + \frac{n(n-1)}{2}e^{5it/2} + \cdots + e^{i(n+\frac{1}{2})t} \\
&= e^{it/2}\left[1 + ne^{it} + \frac{n(n-1)}{2}e^{2it} + \cdots + e^{nit}\right] \\
&= e^{it/2}(1 + e^{it})^n \\
&= e^{it/2}[1 + \cos t + i\sin t]^n \\
&= e^{it/2}[2\cos^2 t/2 + 2i\sin t/2\cos t/2]^n \\
&= e^{it/2}[2\cos t/2(\cos t/2 + i\sin t/2)]^n \\
&= e^{it/2}2^n \cos^n t/2e^{int/2}, \\
C + is &= 2^n \cos^n t/2e^{\frac{i(n+1)}{2}t}, \\
C &= 2^n \cos^n t/2\cos\frac{(n+1)}{2}t, && (32.3) \\
S &= 2^n \cos^n t/2\sin\frac{(n+1)}{2}t && (32.4)
\end{aligned}
$$

so, equation (32.2) can be written as

$$
E_n^1 = \frac{1}{2\pi}\int_0^{\pi} \phi(t)\frac{\cos^n t/2\sin(\frac{n+1}{2})t}{\sin t/2}. \qquad (32.5)
$$

Now taking Borel transforms of E_n^1, we get

$$
\begin{aligned}
B_p.E_n^1 &= \frac{1}{e^p}\sum_{m=0}^{\infty}\frac{p^m}{m!}\left\{\frac{1}{2\pi}\int_0^{\pi}\frac{\phi(t)\cos^m t/2\sin(\frac{m+1}{2})t}{\sin t/2}dt\right\} \\
&= \frac{1}{2\pi}e^{-p}\int_0^{\pi}\frac{\phi(t)}{\sin t/2}\left\{\sum_{m=0}^{\infty}\frac{p^m\cos^m t/2\sin(\frac{m+1}{2})t}{m!}\right\}dt && (32.6)
\end{aligned}
$$

Now we consider

$$
S' = \sin\left(\frac{t}{2}\right) + \frac{p\cos\frac{t}{2}\sin(\frac{2t}{2})}{1!} + \frac{p^2\cos t/2\sin(\frac{3t}{2})}{2!} + \cdots
$$

also

$$
C' = \cos\left(\frac{t}{2}\right) + \frac{p\cos(t/2)\cos(\frac{2t}{2})}{1!} + \frac{p^2\cos(t/2)\cos(\frac{3t}{2})}{2!} + \cdots
$$

so,

$$
\begin{aligned}
C' + iS' &= e^{it/2} + \frac{p}{1!}\cos t/2e^{2it/2} + \frac{p^2}{2!}\cos^2 t/2e^{\frac{3it}{2}} + \cdots \\
&= e^{it/2}\left[1 + p\cos t/2e^{it/2} + \frac{p^2}{2!}\cos^2 t/2e^{\frac{2it}{2}} + \cdots\right]
\end{aligned}
$$

$$
\begin{aligned}
&= e^{it/2}\exp(p\cos t/2e^{it/2})\\
&= e^{it/2}\exp(p\cos t/2(\cos t/2 + i\sin t/2))\\
&= e^{it/2}\exp\{p\cos^2 t/2 + ip\sin t/2\cos t/2\}\\
&= e^{p\cos^2 t/2}.e^{\frac{i(t+p\sin(t))}{2}}
\end{aligned}
$$

$$\text{or} \quad C' = e^{p\cos^2 t/2}\cos\left(\frac{t+p\sin t}{2}\right) \tag{32.7}$$

$$\text{and} \quad S' = e^{p\cos^2 t/2}\sin\left(\frac{t+p\sin t}{2}\right) \tag{32.8}$$

so, equation (32.6) becomes,

$$
\begin{aligned}
B_pE_n^1 &= \frac{1}{2\pi}e^{-p}\int_0^{\pi}\frac{\phi(t)}{\sin t/2}e^{p\cos^2 t/2}\sin\left(\frac{t+p\sin t}{2}\right)dt\\
&= \frac{1}{2\pi}\int_0^{\pi}\frac{\phi(t)}{\sin t/2}\exp\left\{-p\left(\frac{1-\cos t}{2}\right)\right\}\sin\left(\frac{t+p\sin t}{2}\right)dt,
\end{aligned}
$$

$$B_pE_n^1 = \frac{1}{2\pi}\int_0^{\pi}\phi(t)K(p,t)dt. \tag{32.9}$$

The infinite series $\sum_{n=0}^{\infty} u_n$ is said to be $B_pE_n^1$ summable to S if

$$B_pE_n^1 \to s \quad \text{as} \quad p \to \infty.$$

Now, we write

$$
\begin{aligned}
|t_E^B - f| &= \int_0^{\pi}\phi(t)K(p,t)dt\\
&= \int_0^{\pi/(p+1)}\phi(t)K(p,t)dt + \int_{\pi/(p+1)}^{\pi}\phi(t)K(p,t)dt\\
&= I_1 + I_2,
\end{aligned}
$$

then

$$
\begin{aligned}
|I_1| &= \left|\int_0^{\pi/(p+1)}\phi(t)K(p,t)dt\right|\\
&= \left|\int_0^{\pi/(p+1)}t^{\alpha}O(p+1)\right|
\end{aligned}
$$

$$
\begin{aligned}
&= O(p+1)\left[\frac{t^{\alpha+1}}{\alpha+1}\right]_0^{\frac{\pi}{(p+1)}} \\
&= O(p+1)\left[\frac{\pi^{\alpha+1}}{(\alpha+1)(p+1)^{\alpha+1}}\right] \\
&= O\left(\frac{1}{(p+1)^{\alpha}}\right)
\end{aligned}
$$

$$
\begin{aligned}
|I_2| &= \left|\int_{\pi/(p+1)}^{\pi} \phi(t)K(p,t)dt\right| \\
&= \left|\int_{\pi/(p+1)}^{\pi} t^{\alpha}.\frac{1}{t}dt\right| \\
&= \left|\int_{\pi/(p+1)}^{\pi} t^{\alpha-1}dt\right| \\
&= \left[\frac{t^{\alpha}}{\alpha}\right]_{\pi/(p+1)}^{\pi} \\
&= O\left(\frac{1}{(p+1)^{\alpha}}\right) \quad \text{as} \quad p \longrightarrow \infty
\end{aligned}
$$

so

$$
\begin{aligned}
|t_E^B - f| &= O\left(\frac{1}{p^{\alpha}}\right) + O\left(\frac{1}{p^{\alpha}}\right) \\
&= O\left(\frac{1}{(p+1)^{\alpha}}\right) \quad \text{as} \quad p \longrightarrow \infty.
\end{aligned} \tag{32.10}
$$

Therefore, the degree of approximation of functions belonging to Lipschitz class by Borel–Euler summability method of Fourier series is given by

$$
\|t_E^B - f\|_{\infty} = O\left(\frac{1}{(p+1)^{\alpha}}\right).
$$

□

4 Conclusion

In this chapter, we have introduced the product summability of Fourier Series using Borel–Euler product summability means. The present theorem is the estimation of degree of approximation of functions belonging to Lipschitz class. These results may motivate other researchers to carry out the research in the field of summability theory.

Bibliography

[1] P. Chandra, Trigonometric approximation of functions in L_p norm, *J. Math. Anal. Appl.* **275** (2002), 13–26.

[2] G. H. Hardy, *Divergent Series*, 1st, Oxford University press, 1949.

[3] S. Izumi, Study of Cesaro summability of Fourier series, *Tohoku Math. J.* **51** (1949), 144–149.

[4] T. Laying, A Tauberian theorem for the general Euler–Borel summability method, *Can. J. Math.* **44**(5) (1992).

[5] H. Lebesgue, Recherches Sur la convergence des series de Fourier, *Math. Ann.* **61** (1905), 251–280.

[6] V. N. Mishra, On the degree of approximation of signals (functions) belonging to the weighted $W(L_p, \xi(t)), (p \geq 1)$-class by almost matrix summability method of its conjugate Fourier series, *Int. J. Appl. Math. Mech.* **5**(7) (2009), 16–27.

[7] V. N. Mishra, On the degree of approximation of signals (functions) belonging to generalized weighted $W(L_p, \xi(t)), (p \geq 1)$- class by product summability method, *J. Int. Acad. Phys. Sci.* **14**(4) (2010), 413–423.

[8] V. N. Mishra, H. H. Khan, K. Khatri and L. N. Mishra, On approximation of conjugate of signals (functions) belonging to the generalized Weighted $W(L_r, \xi(t)), (r \geq 1)$-class by product summability means of conjugate series of Fourier series, *Int. J. Math. Anal.* **6**(35) (2012), 1703–1715.

[9] V. N. Mishra, K. Khtri and L. N. Mishra, Product summability transform of conjugate series of Fourier series, *Int. J. Math. Math. Sci.* **2012** (2012), Article ID 298923, 13 pages.

[10] H. K. Nigam, A study on approximation of conjugate of functions belonging to Lipschitz class and generalized Lipschitz class by product summability means of conjugate series of Fourier series, *Thai J. Math.* **10**(2) (2012), 275–287.

[11] H. K. Nigam and K. Sarma, Approximation of conjugate functions belonging to Lip(α, γ) class and to weighted $W(L_r; \xi(t))$ class by product means of conjugate Fourier series, *Eur. J. Pure Appl. Math.* **3** (2011), 276–286.

[12] B. E. Rhaodes, On the degree of approximation of functions belonging to Lipschitz class by Hausdroff means of its Fourier series, *Tamkang J. Math.* **34**(3) (2003), 245–247.

[13] K. Saxena and M. Prabhakar, A study of double Euler summability method of Fourier series and it's conjugate series, *Int. J. Sci. Innov. Math. Res.* **4**(1) (2016), 46–52.

[14] K. Saxena and M. Prabhakar, A new criterion for Borel–Euler summability method of Fourier series, *Asian Res. J. Math.* **8**(2) (2018), 1–2; Article no. ARJOM. 33839, ISSN: 2456-477X.

[15] E. C. Titchmarch, *The Theory of Functions*, Oxford Uni Press, 1952.

[16] A. Zygmund, *Trigonometrical Series I-II*, Warsaw, 1935.

[17] A. Zygmund, *Trigonometric Series*, 2nd Rev. edn, Vol. 1, Cambridge Univ. Press, Cambridge, New York, 1959.

Pankaj Kumar, Raminder Pal Singh, and Dipesh

Modeling the effect of discrete delay in SIR epidemic model with logistic growth

Abstract: In this work, a discrete delayed SIR epidemic model with logistic growth is analyzed. A threshold value is obtained, which depicts the dynamics and outcomes of the disease. A basic reproductive number R_0 is defined with this a disease-free and non-zero endemic equilibrium is discussed. A disease-free equilibrium is shown to be asymptotically stable for a given value of the delay parameter τ. The disease disappears when the basic fundamental reproductive number $R_0 < 1$. For a certain value of the delay parameter τ, it is shown that the equilibrium E^* is asymptotically stable and devoid of diseases. When $R_0 > 1$, the disease persists, The stability switches are seen for $\tau \neq 0$ and Hopf bifurcation occurred when τ going beyond some critical point value. The delayed system for DFE (disease-free equilibrium) at $R_0 = 1$, has also been shown to be linearly neutrally stable. We also analyze the sensitivity of the state variables with respect to the fundamental reproductive number R_0. We conclude with some MATLAB-based numerical simulations to demonstrate the analytical findings.

Keywords: Basic reproduction number, SIR model, stability, bifurcation, incubation period, epidemic model

1 Introduction

To trace the mathematical modeling of epidemics back through history, we need to start with Daniel Bernoulli [3–5, 12]. By separating the population into three groups: susceptible (S), infected (I), and recovered (R), Kermack and Mckendric [15] propose a conventional SIR epidemic model. This SIR model is helpful for analyzing illnesses nowadays. A large body of work has been devoted to the study of nonlinear incidence epidemic models, and the SIR disease transmission model has a similar amount of background research [7, 44]. Numerous models illustrate the dynamics of disease using simple differential equations with no time lag. To depict the genuine dynamical behavior of models that rely on system history, time delays are suitable. In reality, delays in epidemic models make them more realistic by describing disease latency or immunity. The delay

Pankaj Kumar, Department of Mathematics, Lovely Professional University, Room No. 302, Block No. 55, Phagwara, Punjab 144-411, India, e-mail: pankaj.kumar1@lpu.co.in
Raminder Pal Singh, Department of Mathematics, Lovely Professional University, Room No. 508, Block No. 38, Phagwara, Punjab 144-411, India, e-mail: raminder.19058@gmail.com
Dipesh, Department of Mathematics, SR University, Room No. 3112, Block No. 3, Warangal, Telangana 506-371, India, e-mail: dipeshdalal2411@gmail.com

https://doi.org/10.1515/9783111313634-033

affects population dynamics. In many real-world processes, especially biological ones, state variable dynamics depend on past values. Delay may affect infectious dynamics. Many sicknesses spread slowly due to immunity period delays [20, 37, 38], infection period delays [27, 41], and incubation period delays [1, 21, 24]. Stability, attractivity, persistence, periodic oscillation, bifurcation, and chaos are studied extensively in population models with time delay. Hopf bifurcation periodic solutions are of special importance [6, 14, 30, 31, 35, 40]. Many researchers have outlined the delay differential equation in plant dynamics using competiting mathematical model [8, 10, 32], considered the delay as a transient immunity interval, and found the endemic equilibrium to be globally stable. This indicated that the delay did not have any impact on the system, which had a nonlinear incidence rate. [9] proved that a nonmonotonic incidence rate delayed SIRS pandemic model is robust on a global scale, this showed that delay could not alter nonmonotonic system dynamics, and also use the Lyapunov function technique for the global stability analysis [23]. Using a Lyapunov function, [11] investigated the global stability of the SIR epidemic model with an incubation period and nonlinear incidence rate. Delays affect system dynamics, causing Hopf bifurcation. In [1, 17, 22, 35, 40, 42, 43, 45], a generic nonlinear differential equation problem with a delay was addressed by discussing its Hopf bifurcation analysis. The directional analysis for the solution of the delayed differential equation using the normal form theory and manifold reduction proposed by Hassard et al. [45].

The basic reproductive number R_0, or sometimes called reproduction number, is one of the key parameters for calculating the stability of SIR model. Mathematical analysis is employed to establish a threshold value R_0, which indicates when the disease is nearing an endemic level and when the solutions begin to vary. The values of R_0 and incubation period duration totally control the global dynamics of the system. Disease always dies out and the DFE is globally asymptotically stable if $R_0 < 1$, but if $R_0 > 1$ the disease continues. In [2, 18, 19, 25, 26, 28, 29, 31, 36, 39, 43] stability, bifurcation, and sensitivity analysis have been performed with special reference to the reproduction numbers. By using the incubation time delay as a bifurcation parameter, this research investigates the endemic equilibrium's local stability. It identifies the circumstances under which the system must be both absolutely and conditionally stable. In numerical simulations, it has been proven that a system with a time delay has a rich, complicated dynamic and that as the delay rises, a series of periodic solutions develops, showing complex periodic patterns.

This chapter is organized as follows: Section 3 offers the delayed SIR epidemic model and reproduction number. Sections 3–6 offer findings on nonzero endemic equilibrium, positivity, boundedness of solutions, and system permanence. Section 8 includes equilibria of system (33.2) and the crucial value for endemic equilibrium. Section 7 addresses the global stability of the disease-free equilibrium and the lifetime of the endemic equilibrium. Section 9 presents the direction and stability of Hopf bifurcation to the system of equation (33.2) by applying manifold reduction and normal form theory. Sections 11 and 13 of this research report include numerical analysis and easy explanation.

2 Novelty of the study

In most of the earlier studies, SIR model is applied to assess the spread of an infectious disease and predict the number of infected, removed and recovered populations and deaths in the communities. The simplest forms of these models are ordinary differential equations (ODEs). However, the identification of infected individuals is not immediate, but takes an incubation time period to mature the disease, which is described by delay parameter. In this manuscript, not only identification is assumed to be delayed, but there is assumed to be delay in susceptible population as well. The overall impact of this delay parameter is studied here with the help of repeating cycles of bifurcation phenomenon.

3 Basic properties and model formulation

The following are the presumptions upon which the delayed model is based. The three-dimensional model tracks the progression of the susceptible people $S(t)$, the infected individuals $I(t)$, and the recovered individuals $R(t)$ at a given time t. Thus, the overall population at time t remains the same, with carrying capacity K. The susceptible individuals become infected when they come in contact with infected individuals at the rate of α_2. We hypothesized that a susceptible person comes into touch with an infected one at time $t-\tau$ and becomes infectious at time τ later. Under these conditions, the SIR epidemic model with a bilinear incidence rate is expressed as:

$$\begin{aligned}\frac{dS}{dt} &= \alpha_1 S(t)\left(1-\frac{S(t)}{K}\right) - \alpha_2 S(t-\tau)I(t-\tau),\\ \frac{dI}{dt} &= \alpha_2 S(t-\tau)I(t-\tau) - (\alpha_3+\alpha_4+\alpha_5)I(t),\\ \frac{dR}{dt} &= \alpha_3 I(t) - \alpha_5 R(t),\end{aligned} \tag{33.1}$$

where $S(0)>0, I(0)>0, R(0)>0$ for all t, $S(t-\tau)=k_1, I(t-\tau)=k_2, k_1, k_2$ are constants for $t\in[0,\tau]$.

The parameters used in the model (33.1) are described as below:

α_1	Intrinsic birth rate	(day^{-1})
α_2	Rate of transmission	(day^{-1})
α_3	Recovery rate	(day^{-1})
α_4	Disease mortality rate	(day^{-1})
α_5	Natural death rate	(day^{-1})
K	Carrying capacity of population	(day^{-1})
τ	Time lag	(day^{-1}).

It is expected in model (33.1) that those affected who are given effective treatment would recover, joining the recovery class after they develop temporary immunity to the disease. Note that since this model is used to track human populations, it is supposed that none of the parameters and variables are negative constants. For simplicity, the system of equations (33.1) is not dimensionalized:

$$S' = \frac{S}{K}, \quad I' = \frac{I}{K}, \quad R' = \frac{R}{K}, \quad t' = \alpha_2 K t,$$
$$(\alpha_1)' = \frac{\alpha_1}{\alpha_2 K}, \quad (\alpha_3)' = \frac{\alpha_3}{\alpha_2 K}, \quad (\alpha_4)' = \frac{\alpha_4}{\alpha_2 K}, \quad (\alpha_5)' = \frac{\alpha_5}{\alpha_2 K}.$$

Dropping the sign "′", the system of quations (33.1) becomes as:

$$\begin{aligned}\frac{dS}{dt} &= \alpha_1 S(1-S) - S(t-\tau)I(t-\tau),\\ \frac{dI}{dt} &= S(t-\tau)I(t-\tau) - (\alpha_3+\alpha_4+\alpha_5)I,\\ \frac{dR}{dt} &= \alpha_3 I - \alpha_5 R.\end{aligned} \tag{33.2}$$

With initial conditions $\xi = (\xi_1, \xi_2, \xi_3)$ of system (33.2) defined in the Banach space of continuous mapping:

$$C_+ = \{\xi \in C([-\tau,0], \mathbb{R}^3_+) : \xi_1(\theta) = S(\theta), \xi_2(\theta) = I(\theta), \xi_3(\theta) = R(\theta), \theta \in [-\tau, 0]\}$$

where $\xi_i(0) > 0$, $i = 1, 2, 3$, the feasible area of the system (33.2) is defined as:

$$\Lambda = \{(S, I, R) \in \mathbb{R}^3 : S(0) > 0, I(0) > 0, R(0) > 0\}.$$

It is easy to show that the feasible region Λ is positively not changeable with respect to system (33.1). System (33.2) has DFE $E_0(1, 0, 0)$. The basic reproductive number is defined as:

$$R_0 = \frac{1}{r}, \quad \text{where } r = (\alpha_3 + \alpha_4 + \alpha_5). \tag{33.3}$$

4 Non-zero endemic equilibrium

It is assumed that at equilibrium point $S(t-\tau) \approx S$, $I(t-\tau) \approx I$. The non-zero endemic equilibrium $E^*(S^*, I^*, R^*)$ is calculated as:

$$\frac{dI^*}{dt} = 0 \Rightarrow S^* = (\alpha_3 + \alpha_4 + \alpha_5) = r,$$

$$\frac{dS^*}{dt} = 0 \Rightarrow I^* = \alpha_1(1-r) = \alpha_1 r(R_0 - 1),$$
$$\frac{dR^*}{dt} = 0 \Rightarrow R^* = \frac{\alpha_3 I^*}{\alpha_5} = \frac{\alpha_3\alpha_1(1-r)}{\alpha_5} = \frac{\alpha_3\alpha_1 r(R_0-1)}{\alpha_5}.$$

Thus, for $R_0 > 1$, we have a unique non-zero endemic equilibrium:

$$E^*(S^*, I^*, R^*) = E^*\left(r, \alpha_1(1-r), \frac{\alpha_3\alpha_1(1-r)}{\alpha_5}\right) = E^*\left(r, \alpha_1 r(R_0-1), \frac{\alpha_3\alpha_1 r(R_0-1)}{\alpha_5}\right).$$

5 Positivity of solutions

The term "positivity" refers to the stability of a system. For positive solutions, one must show that the system of equations (33.2) provides all of the system's solutions, where initial conditions are $S(0) > 0, I(0) > 0, R(0) > 0$ for all t, and $S(t-\tau) = k_1, I(t-\tau) = k_2$, where k_1, k_2 are constants for $t \in [0, \tau]$. The model solutions $(S(t), I(t), R(t))$ remain positive for all time $t > 0$.

Now, we have $\frac{dR}{dt} = \alpha_3 I(t) - \alpha_5 R(t)$, i. e., $\frac{dR}{dt} \geq -\alpha_5 R(t)$, i. e., $\frac{dR}{R(t)} \geq -\alpha_5 dt$, i. e., $R(t) = c_0 e^{-\alpha_5 t}$. Here, c_0 is the integration constant. As a result, $R(t) > 0$ for all time t. So, S and I have a similar argument.

6 Boundedness of solutions

Lemma ([16]). *All feasible solutions of the system* (33.2) *are bounded and enter the region*

$$\Lambda = \left\{(S, I, R) \in \mathbb{R}^3 : 0 \leq S + I + R \leq \frac{\alpha_1 + 1}{\mu_m}\right\}$$

as $t \to \infty$, *where* $\mu_m = \min\{1, \alpha_4, \alpha_5\}$ *for all positive initial values* $S(0) > 0$, $I(0) > 0$, $R(0) > 0$ *for all* t *and* $S(t-\tau) = k_1$, $I(t-\tau) = k_2$, k_1, k_2 *are constant for* $t \in [0, \tau]$.

With $R_0 = 1/r$, where $r = (\alpha_3 + \alpha_4 + \alpha_5)$, for the system (33.2) there always exists the equilibrium $E_0(1, 0, 0)$. If $R_0 > 1$, we have a unique non-zero endemic equilibrium:

$$E^*(S^*, I^*, R^*) = E^*\left(r, \alpha_1(1-r), \frac{\alpha_3\alpha_1(1-r)}{\alpha_5}\right) = E^*\left(r, \alpha_1 r(R_0-1), \frac{\alpha_3\alpha_1 r(R_0-1)}{\alpha_5}\right),$$

where $r = (\alpha_3 + \alpha_4 + \alpha_5)$.

7 Permanence of the system

When $E^*(S^*, I^*, R^*)$ is in equilibrium, The set of equations (33.2) simplifies to;

$$\begin{aligned}\frac{dS^*}{dt} &= \alpha_1 S^*(1 - S^*) - S^*(t-\tau)I^*(t-\tau),\\ \frac{dI^*}{dt} &= S^*(t-\tau)I^*(t-\tau) - (\alpha_3 + \alpha_4 + \alpha_5)I^*(t),\\ \frac{dR^*}{dt} &= \alpha_3 I^*(t) - \alpha_5 R^*(t).\end{aligned} \tag{33.4}$$

All possible solutions to systems (33.2) and (33.4) have been shown to be constrained by Y. Takeuchi in 2000. The transcendent equation about the equilibrium E^* is given by:

$$\begin{vmatrix} \lambda - m_1 & -m_2 & -m_3 \\ -m_4 & \lambda - m_5 & -m_6 \\ -m_7 & -m_8 & \lambda - m_9 \end{vmatrix} = 0, \tag{33.5}$$

where $m_1 = \alpha_1 - 2\alpha_1 S^* - I^* e^{-\lambda\tau}$, $m_2 = I^* e^{-\lambda\tau}$, $m_3 = 0$, $m_4 = -e^{-\lambda\tau} S^*$, $m_5 = S^* e^{-\lambda\tau} - r$, $m_6 = \alpha_3$, $m_7 = 0$, $m_8 = 0$, $m_9 = -\alpha_5$. The equation (33.5) after expanding becomes:

$$(\lambda + \alpha_5)(\lambda^2 - (m_1 + m_5)\lambda + m_1 m_5 + S^* I^* e^{-2\lambda\tau}) = 0. \tag{33.6}$$

After further expanding, it becomes:

$$\lambda^3 + m_1\lambda^2 + m_2\lambda + m_3 + (m_4\lambda^2 + m_5\lambda + m_6)e^{-\lambda\tau} = 0, \tag{33.7}$$

where $m_1 = (r - \alpha_1 + 2\alpha_1 S^* + \alpha_5)$, $m_2 = (2r\alpha_1 S^* - r\alpha_1 + \alpha_5(r - \alpha_1 + 2\alpha_1 S^*))$, $m_3 = \alpha_5(2r\alpha_1 S^* - r\alpha_1)$, $m_4 = (I^* - S^*)$, $m_5 = (\alpha_1 S^* - 2\alpha_1 S^{*2} + rI^*)$, All $m_1, m_2, m_3, m_4, m_5 > 0$.

8 Local and global stability of disease-free equilibrium

Theorem 8.1. *The disease-free equilibrium (DFE) $E_0(1,0,0)$ of the system* (33.2) *is,*

(a) *Absolutely stable if $R_0 < 1$,*

(b) *linearly neutrally stable if $R_0 = 1$, and*

(c) *unstable if $R_0 > 1$.*

Proof. When $\tau \neq 0$, the transcendental equation at the DFE, $E_0(1,0,0)$ of system (33.2) will be of the form as;

$$(\lambda + \alpha_1)(\lambda + \alpha_5)(\lambda + r - e^{-\lambda\tau}) = 0 \tag{33.8}$$

$$\Rightarrow (\lambda + \alpha_1) = 0,\ (\lambda + \alpha_5) = 0,\ (\lambda + r - e^{-\lambda\tau}) = 0$$
$$\Rightarrow \lambda = -\alpha_1,\ \lambda = -\alpha_5,\ \lambda = -r + e^{-\lambda\tau} \tag{33.9}$$

The system (33.2) has two negative real roots given by $\lambda_1 = -\alpha_1, \lambda_2 = -\alpha_5$, and the third root is coming from the equation $\lambda = -r + e^{-\lambda\tau}$.

(a) When $R_0 < 1$, from the third equation of (33.9)

$$\lambda = -r + e^{-\lambda\tau} \Rightarrow R_e(\lambda) = e^{-R_e(\lambda)}\cos(pq\tau\lambda) - r$$

let

$$R_e(\lambda) \geq 0 \quad \Rightarrow$$
$$R_e(\lambda) \leq 1 - r \quad \Rightarrow R_e(\lambda) \leq r(R_0 - 1) \tag{33.10}$$
$$\text{As } R_0 < 1,\ \text{we have } R_e(\lambda) < 0. \tag{33.11}$$

Hence, the DFE of the system of equations (33.2) is locally asymptotically stable.

(b) If $R_0 = 1$, from (33.9) we have $R_e(\lambda) = 0$. This shows that the root has a negative real part except $\lambda = 0$, which gives the fact that E_0 is a linear neutrally stable state when $R_0 = 1$.

(c) If $R_0 > 1$, let $\lambda + r - e^{-\lambda\tau} = f(\lambda)$, because $f(0) = r - 1 = -r(R_0 - 1) < 0, f(\infty) > 0$, there exist a positive value of λ. This clearly shows that the DFE of the system of equations (33.2) is asymptotically unstable.

In addition to the above when $\tau = 0$, and $R_0 < 1$, the third root of the characteristic equation of system (33.2) will be $\lambda = -r + 1 = r(R_0 - 1) < 0$, hence the system becomes locally asymptotically stable. Hence the theorem. □

Theorem 8.2. *If the basic reproductive number $R_0 < 1$ then the DFE $E_0(1, 0, 0)$ is globally stable for any value of time delay parameter τ.*

Proof. For $t > 1$, let l_i be the translation of roots of the system if equations (33.2) with initial conditions $l_1(\theta) = S(\theta), l_2(\theta) = I(\theta), l_3(\theta) = R(\theta), \theta \in [-\tau, 0]$. We can define a Lyapunov function as:

$$L(t) = I(t) + \int_{(t-\tau)}^{t} S(\theta)I(\theta)d\theta. \tag{33.12}$$

Further, $L(t) > 0$ along with the solution of the system of equations (33.2), also $L(t) = 0$, iff both $I(t)$ and $R(t)$ are zero. The derivative of (33.12) is as;

$$\frac{dL(t)}{dt} = \frac{dI(t)}{dt} + (S(t)I(t) - S(t-\tau)I(t-\tau))$$
$$= S(t-\tau)I(t-\tau) - rI(t) + (S(t)I(t) - S(t-\tau)I(t-\tau)) = S(t)I(t) - rI(t).$$

As $S(t) \le 1$, so (33.12) implies $\frac{dL(t)}{dt} = S(t)I(t) - rI(t) \le rI(t)(R_0 - 1) \le 0$, as $R_0 < 1$. So, we have $\frac{dL(t)}{dt} \le 0$, as $R_0 < 1$, and $\frac{dL(t)}{dt} = 0$, if $I(t) = 0, R(t) = 0$.

The maximum invariance set $\{(S(t), I(t), R(t)) : \frac{dL(t)}{dt} = 0\}$, is a singleton set $\{E_0(1,0,0)\}$, when $R_0 < 1$. Therefore, by [33], the DFE $E_0(1,0,0)$ is globally asymptotically stable. Hence, the theorem. □

9 Bifurcation analysis

In this part of the paper, we will find the conditions for Hopf bifurcations to exist using τ as a parameter for the result. In this part, we assume that the basic reproduction number $R_0 > 1$, so the endemic equilibrium $E^*(S^*, I^*, R^*) = E^*(r, \alpha_1 r(R_0 - 1), (\alpha_3\alpha_1 r(R_0 - 1))/\alpha_5)$ exists. The transcendental equation of (33.2) is given by (33.7).

Case 1. When $\tau = 0$, the characteristic equation (33.6) takes the form:

$$(\lambda + \alpha_5)[\lambda^2 - (m_1 + m_5)\lambda + m_1 m_5 + S^* I^*] = 0, \tag{33.13}$$

where $m_1 = \alpha_1 - 2\alpha_1 S^* - I^*$ and $m_5 = S^* - r$.

Now $m_1 + m_5 = \alpha_1 - 2\alpha_1 S^* - I^* + S^* - r = \alpha_1 - 2\alpha_1 r - \alpha_1(1-r) + r - r = -\alpha_1 r$.

Therefore, $-(m_1 + m_5) = \alpha_1 r > 0$ and $m_1 m_5 + S^* I^* = (\alpha_1 - 2\alpha_1 S^* - I^*)(S^* - r) + S^* I^* = r\alpha_1(1-r) = \alpha_1 r^2(R_0 - 1) > 0$.

Therefore, all the roots of equation (33.13) are all negative. By Routh–Hurwitz's criteria, $E^*(S^*, I^*, R^*)$ is a locally asymptotically stable equilibrium.

Now we will find the positive roots of equation (33.7) for $\tau > 0$. Let $i\omega$ be the roots of the transcendental equation (33.7). So, we have:

$$(i\omega)^3 + m_1(i\omega)^2 + m_2 i\omega + m_3 + [m_4(i\omega)^2 + m_5 i\omega + m_6]e^{-\lambda i\omega} = 0,$$
$$-i(\omega)^3 - m_1(\omega)^2 + m_2 i\omega + m_3 + [-m_4(\omega)^2 + m_5 i\omega + m_6](\cos(\omega\tau) - i\sin(\omega\tau)) = 0.$$

We have

$$(m_6 - m_4(\omega)^2)\cos(\omega\tau) + m_5\omega\sin(\omega\tau) = (m_1(\omega)^2 - m_3), \tag{33.14}$$
$$m_5\omega\cos(\omega\tau) - (m_6 - m_4(\omega)^2)\sin(\omega\tau) = ((\omega)^3 - m_2\omega). \tag{33.15}$$

By solving (33.14) and (33.15), we get:

$$\omega^6 + (m_1^2 - m_4^2 - 2m_2)\omega^4 + (m_2^2 - 2m_1 m_3 + 2m_4 m_6 - m_5^2)\omega^2 + (m_3^2 - m_6^2) = 0$$
$$\Rightarrow x^3 + \beta_1 x^2 + \beta_2 x + \beta_3 = 0, \tag{33.16}$$

where $\beta_1 = m_1^2 - m_4^2 - 2m_2$, $\beta_2 = m_2^2 - 2m_1 m_3 + 2m_4 m_6 - m_5^2$, $\beta_3 = m_3^2 - m_6^2$, and $x = \omega^2$.

Now $\beta_3 = m_3^2 - m_6^2 = (\alpha_5(2r\alpha_1 S^* - r\alpha_1))^2 - (\alpha_5(\alpha_1 S^* - 2\alpha_1 S^2 + rI^*))^2$.

$\beta_3 = -\alpha_5^2\alpha_1^2 r^2(r-1)(5r-3)$, as $R_0 > 1 \Rightarrow r < 1$, when $r < 3/5$, we have $\beta_3 < 0$. Therefore, there exists only one pair of purely complex roots of the equation (33.16) say $x = \pm i\omega_0$. At the same time, by substituting $\omega = \omega_0$ in (33.14)–(33.15) and solving it for the delay parameter τ, we have:

$$\tau_p = \frac{1}{\omega_0}\left\{\arcsin\left(\frac{m_5\omega_0(m_1(\omega_0)^2 - m_3) - ((\omega_0)^3 - m_2\omega_0)(m_6 - m_4(\omega_0)^2)}{(m_5\omega_0)^2 + (m_6 - m_4(\omega_0)^2)^2}\right) + 2l\pi\right\}, \tag{33.17}$$

where $l = 0, 1, 2, 3, \ldots$.

Hence, by the lemma in [13], all the roots of the equation (33.7) have negative real parts for $\tau \in [0, \tau_0)$, so for the bifurcation analysis, we will show that there exists at least one eigenvalue that has a positive real part for $\tau > \tau_0$, and $(\frac{d\,\mathrm{Re}\,\lambda(\tau)}{d\tau})_{\tau=\tau_0} \neq 0$. By differentiating equation (33.7) with respect to τ and after rearranging the terms, we have:

$$\left(\frac{d\lambda}{d\tau}\right)^{-1} = \frac{(3\lambda^2 + 2m_1\lambda + m_2)e^{\lambda\tau}}{\lambda(m_4\lambda^2 + m_5\lambda + m_6)} + \frac{(2m_4\lambda + m_5)}{\lambda(m_4\lambda^2 + m_5\lambda + m_6)} - \frac{\tau}{\lambda}$$

$$\begin{aligned}
\mathrm{sign}\left(\frac{d\,\mathrm{Re}\,\lambda(\tau)}{d\tau}\bigg|_{\tau=\tau_0}\right) &= \mathrm{sign}\left(\mathrm{Re}\left[\left(\frac{d\lambda}{d\tau}\right)^{-1}\right]\bigg|_{\lambda=i\omega_0}\right)\\
&= \mathrm{sign}\left(\mathrm{Re}\left(\frac{(3\lambda^2 + 2m_1\lambda + m_2)e^{\lambda\tau}}{\lambda(m_4\lambda^2 + m_5\lambda + m_6)}\right) + \mathrm{Re}\left(\frac{2m_4\lambda + m_5}{\lambda(m_4\lambda^2 + m_5\lambda + m_6)}\right)\right)\\
&= \mathrm{sign}\left(\frac{3\omega_0^6 + (m_1^2 - m_4^2 - 2m_2)\omega_0^4 + (2m_2^2 - 2m_1m_3 + 2m_4m_6 - m_5^2)\omega_0^2}{(m_5\omega_0)^2 + (m_6 - m_4(\omega_0)^2)^2}\right)\\
&= \mathrm{sign}\left(\frac{3\omega_0^6 + \beta_1\omega_0^4 + \beta_2\omega_0^2}{(m_5\omega_0)^2 + (m_6 - m_4(\omega_0)^2)^2}\right) = \mathrm{sign}\left(\frac{\omega_0^2(3\omega_0^4 + \beta_1\omega_0^2 + \beta_2)}{(m_5\omega_0)^2 + (m_6 - m_4(\omega_0)^2)^2}\right).
\end{aligned}$$

Because $r > 0$, and when $r < 3/5$, $\beta_2 > 0$. Also, if $r < 1/2$ (i. e., if $R_0 > 2$), both $\beta_1 > 0$ and $\beta_2 > 0$. Therefore, $\frac{\omega_0^2(3\omega_0^4 + (\beta_1)\omega_0^2 + \beta_2)}{(m_5\omega_0)^2 + (m_6 - m_4\omega_0^2)^2} > 0$.

This implies that $(\frac{d\,\mathrm{Re}\,\lambda(\tau)}{d\tau})_{\tau=\tau_0} > 0$, which ensures that the conditions for transversality hold, and Hopf bifurcation occurs at $E^*(r, \alpha_1 r(R_0-1), (\alpha_3\alpha_1 r(R_0-1))/\alpha_5)$. Following from the above, we draw the following conclusions:

Theorem 9.1. *For the non-zero endemic positive equilibrium E^* given by*

$$E^*\left(r, \alpha_1 r(R_0 - 1), \frac{\alpha_3\alpha_1 r(R_0 - 1)}{\alpha_5}\right)$$

the following is what we have found.

If $1 < R_0 \leq 2$, then the equilibrium $E^*(r, \alpha_1 r(R_0 - 1), \frac{\alpha_3\alpha_1 r(R_0-1)}{\alpha_5})$ is asymptotically stable.

If $R_0 > 2$, then the equilibrium $E^*(r, a_1 r(R_0 - 1), \frac{a_3 a_1 r(R_0-1)}{a_5})$ is conditionally stable. Specifically, when we have a delay critical parameter value τ_0, the equilibrium $E^*(r, a_1 r(R_0 - 1), \frac{a_3 a_1 r(R_0-1)}{a_5})$ is asymptotically stable if $\tau \in [0, \tau_0)$. For $\tau > \tau_0$, the system undergoes a Hopf bifurcation at $E^*(r, a_1 r(R_0 - 1), \frac{a_3 a_1 r(R_0-1)}{a_5})$. Furthermore, at $\tau = \tau_p$, $p = 1, 2, 3, \ldots$.

10 Direction and stability analysis of Hopf bifurcation

In the last section, at the critical point values, a set of recurrent solutions obtained which bifurcates from the positive steady state E^*. It's also worth looking at the direction, stability, and period of these bifurcating periodic solutions. Using the normal form theory and manifold reduction proposed by K. R. Schneider, Hassard, B. D. [32], an explicit formula for finding the characteristics of the Hopf-bifurcation at the critical value τ_j will be derived in this section.

Let $v_1 = S - S^*$, $v_2 = I - I^*$, $v_3 = R - R^*$, and the delay τ can be normalized by time scaling $t \to t/\tau$. The system of equations (33.2) is transformed into

$$
\begin{aligned}
\frac{dv_1}{dt} &= (a_1 - 2a_1 S^*)v_1 - a_1 v_1^2 + (a_1 S^* - a_1 S^{*2}) - (I^* v_1 + S^* v_2 + v_1 v_2 + I^* S^*)(t-1),\\
\frac{dv_2}{dt} &= (I^* v_1 + S^* v_2 + v_1 v_2 + I^* S^*)(t-1) - r(v_2 + I^*),\\
\frac{dv_3}{dt} &= a_3(v_2 + I^*) - a_5(v_3 + R^*).
\end{aligned} \tag{33.18}
$$

Thus, in the phase $D = D([-1, 0], R_+^3)$ work can be done. Without losing generality, we denote the critical value τ_j by τ_0. Let $\tau = \tau_0 + \mu$, then the system of equations (33.18) has $\mu = 0$ as a Hopf bifurcation value. Rewrite this system as follows for ease of notation:

$$
v'(t) = L_\mu(v_t) + F(\mu, v_t), \tag{33.19}
$$

where $v(t) = (v_1(t), v_2(t), v_3(t))^T \in R^3$, $v_t(\theta) \in D$ is defined by $v_t(\theta) = v_t(t + \theta)$, and

$$
\begin{aligned}
L_\mu &: D \to R,\\
F &: R \times D \to R
\end{aligned}
$$

are given, respectively, by

$$L_\mu \emptyset = (\tau_0 + \mu)\begin{bmatrix} (\alpha_1 - 2\alpha_1 S^*) & 0 & 0 \\ 0 & -r & 0 \\ 0 & \alpha_3 & -\alpha_5 \end{bmatrix} \emptyset(0)$$
$$+ (\tau_0 + \mu)\begin{bmatrix} -I^* & -S^* & c \\ I^* & S^* & 0 \\ 0 & 0 & 0 \end{bmatrix} \emptyset(-1),$$
$$F(\mu, \emptyset) = (\tau_0 + \mu)\begin{bmatrix} F_1 \\ F_2 \\ F_3 \end{bmatrix},$$

where $F_1 = -\alpha_1 \emptyset_1^2(0) - \emptyset_1(-1)\emptyset_2(-1)$, $F_2 = -\emptyset_1(-1)\emptyset_2(-1)$, $F_3 = 0$, $\emptyset(\theta) = (\emptyset_1(\theta), \emptyset_2(\theta), \emptyset_3(\theta))^T \in D((-1,0), R_+^3)$.

By the representation theorem given by Riesz [34], there exists a function $\eta(\theta,\mu)$ of bounded variation for $\theta \in [-1,0]$, such that $L_\mu \emptyset = \int_{-1}^{0} d\eta(\theta,0)\emptyset(\theta)$ for $\emptyset \in D$. In fact, choose

$$L_\mu \emptyset = (\tau_0 + \mu)\begin{bmatrix} (\alpha_1 - 2\alpha_1 S^*) & 0 & 0 \\ 0 & -r & 0 \\ 0 & \alpha_3 & -\alpha_5 \end{bmatrix} \delta(\theta) + (\tau_0 + \mu)\begin{bmatrix} -I^* & -S^* & c \\ I^* & S^* & 0 \\ 0 & 0 & 0 \end{bmatrix} \delta(\theta + 1).$$

Here, δ represents the Dirac delta function. For $\emptyset \in D([-1,0], R_+^3)$, we define

$$A(\mu)\emptyset = \begin{cases} \frac{d\emptyset(\theta)}{d\theta}, & \theta \in [-1,0)] \\ \int_{-1}^{0} d\eta(\theta,0)\emptyset(\theta), & \theta = 0. \end{cases}$$

And

$$F(\mu)\emptyset = \begin{cases} 0, & \theta \in [-1,0) \\ F(\mu,\emptyset), & \theta = 0. \end{cases}$$

Then the system (33.19) is equivalent to

$$v'(t) = A(\mu)v_t + F(\mu)v_t. \tag{33.20}$$

For $\psi \in D^1([-1,0], R_+^3)$, define

$$A^*\psi(s) = \begin{cases} -\frac{d\psi(s)}{ds}, & s \in [-1,0) \\ \int_{-1}^{0} d\eta^T(-t,0)\psi(-t), & s = 0. \end{cases}$$

And the bilinear inner product

$$\langle \psi(s), \emptyset(\theta) \rangle = \overline{\psi(0)}\emptyset(0) - \int_{-1}^{0} \int_{\xi=\theta}^{\theta} \overline{\psi(\xi - \theta)} d\eta(\theta)\phi(\xi) d\xi. \tag{33.21}$$

Since $i\omega_0$ are eigenvalues of $A(0)$ and A^*, $A = A(0)$ are adjoint operators. Thus, they become the eigenvalues of A^*. Suppose that $q(\theta) = q(0)e^{i\omega_0\theta}$ is an eigenvector of $A(0)$ corresponding to the eigenvalue $i\omega_0$, then $A(0)q(0) = i\omega_0 q(0)$. If $\theta = 0$, $[i\omega_0 I - \int_{-1}^{0} e^{i\omega_0\theta} d\eta(\theta)]q(0) = 0$, then $q(0) = (1, u_1, u_2)^T$, where $u_1 = \frac{i\omega_0-(a_1-2a_1S^*)}{i\omega_0+r}$ and $u_2 = \frac{a_3(i\omega_0-(a_1-2a_1S^*))}{(i\omega_0+r)(i\omega_0+a_5)}$. In the same way it can be proved that $q^*(s) = D(1, u_1^*, u_2^*)e^{i\omega_0\tau_0 s}$ represents the eigenvalue of A^* corresponding to eigenvector $-i\omega_0$, where $u_1^* = \frac{-i\omega_0-(a_1-2a_1S^*)}{-i\omega_0+r}$ and $u_2^* = \frac{a_3(-i\omega_0-(a_1-2a_1S^*))}{(-i\omega_0+r)(i-\omega_0+a_5)}$. The value of D is required to assure that $\langle q^*(s), q(\theta)\rangle = 1$. From equation (33.21),

$$\begin{aligned}\langle q^*(s), q(\theta)\rangle &= \overline{D(1, \overline{u_1^*}, \overline{u_2^*})}(1, u_1, u_2)^T \\ &\quad - \int_{-1}^{0}\int_{\xi=\theta}^{\theta} \overline{D(1, \overline{u_1^*}, \overline{u_2^*})e^{-i\omega_0\tau_0(\xi-\theta)}} d\eta(\theta)(1, u, u)^T e^{i\omega_0\tau_0} d\xi \\ &= \overline{D}\left\{1 + u_1\overline{u_1^*} + u\overline{u_2^*} - \int_{-1}^{0} (1, \overline{u_1^*}, \overline{u_2^*})\theta e^{i\omega_0\tau_0\theta}(1, u, u)^T\right\} \\ &= \overline{D}\{1 + u_1\overline{u_1^*} + u\overline{u_2^*} - \tau_0\overline{u^*} W^*(u - a_1 u)e^{i\omega_0\tau_0}\}.\end{aligned}$$

Hence, choose $\overline{D} = \frac{1}{1+u_1\overline{u_1^*}+u\overline{u_2^*}+\tau_0\overline{u^*}W^*(u-a_1u)e^{i\omega_0\tau_0}}$ such that

$$\begin{aligned}\langle q^*(s), q(\theta)\rangle &= 1, \\ \langle q^*(s), \overline{q(\theta)}\rangle &= 0.\end{aligned}$$

By following [45] approach, we compute the coordinates describing the center manifold D_0 at $\mu = 0$ using the same notations manifold D_0 at $\mu = 0$. Let equation (33.19) have solution u_t with $\mu = 0$.

We define,

$$z(t) = \langle q^*(s), v_t(\theta)\rangle, W(t, \theta) = v_t(\theta) - 2\operatorname{Re}(z(t)q(\theta)). \tag{33.22}$$

On the center manifold C_0, $W(t, \theta) = W(z(t), \overline{z(t)}, \theta)$, where,

$$W(z, \overline{z}, \theta) = W_{20}(\theta)\frac{z^2}{2} + W_{11}(\theta)z\overline{z} + W_{02}(\theta)\frac{\overline{z}^2}{2} + \cdots.$$

In the direction of q^* and $\overline{q^*}$, the local coordinates for center manifold C_0 are z and $\overline{z}$. Note that u_t is real is the necessary condition for W is real. For solution $u_t \in D_0$ of equation (33.20), since $\mu = 0$,

$$z'(t) = i\omega_0\tau_0 z + \langle \overline{q^*}(\theta), F(0, W(z, \overline{z}, \theta) + 2\operatorname{Re}(z(t)q(\theta)))\rangle$$

$$= i\omega_0\tau_0 z + \overline{q^*}(0)F(0, W(z,\bar{z},0) + 2\,\mathrm{Re}(z(t)q(\theta)))$$
$$\equiv i\omega_0\tau_0 z + \overline{q^*}(0)F_0(z,\bar{z}).$$

Rewrite this equation as:

$$z'(t) = i\omega_0\tau_0 z(t) + g(z,\bar{z}) \tag{33.23}$$

Where,

$$g(z,\bar{z}) = \overline{q^*}(0)F_0(z,\bar{z}) = g_{20}(\theta)\frac{z^2}{2} + g_{11}(\theta)z\bar{z} + g_{02}(\theta)\frac{\bar{z}^2}{2} + g_{21}(\theta)\frac{z^2\bar{z}}{2} + \cdots. \tag{33.24}$$

As $v_t(\theta) = (v_{1t}, v_{2t}, v_{3t}) = W(t,\theta) + zq(\theta) + \overline{\bar{z}q(\theta)}$, $q(0) = (1, u_1, u_2)^T e^{i\omega_0\tau_0\theta}$, so

$$v_{1t}(0) = z + \bar{z} + W_{20}^{(1)}(0)\frac{z^2}{2} + W_{11}^{(1)}(0)z\bar{z} + W_{02}^{(1)}(0)\frac{\bar{z}^2}{2} + \cdots,$$
$$v_{2t}(0) = u_1 z + \overline{u_1 z} + W_{20}^{(2)}(0)\frac{z^2}{2} + W_{11}^{(2)}(0)z\bar{z} + W_{02}^{(2)}(0)\frac{\bar{z}^2}{2} + \cdots,$$
$$v_{3t}(0) = u_2 z + \overline{u_2 z} + W_{20}^{(3)}(0)\frac{z^2}{2} + W_{11}^{(3)}(0)z\bar{z} + W_{02}^{(3)}(0)\frac{\bar{z}^2}{2} + \cdots,$$
$$v_{1t}(-1) = ze^{-i\omega_0\tau_0} + \bar{z}e^{i\omega_0\tau_0} + W_{20}^{(1)}(-1)\frac{z^2}{2} + W_{11}^{(1)}(-1)z\bar{z} + W_{02}^{(1)}(-1)\frac{\bar{z}^2}{2} + \cdots,$$
$$v_{2t}(-1) = ue^{-i\omega_0\tau_0}z + \overline{u_1}e^{i\omega_0\tau_0}\bar{z} + W_{20}^{(2)}(-1)\frac{z^2}{2} + W_{11}^{(2)}(-1)z\bar{z} + W_{02}^{(2)}(-1)\frac{\bar{z}^2}{2} + \cdots.$$

Comparing coefficients with equation (33.24), we get:

$$g_{20} = \overline{D(1,u_1,u_2)f_{z^2}},$$
$$g_{02} = \overline{D(1,\overline{u_1},\overline{u_2}^*)f_{\bar{z}^2}},$$
$$g_{11} = \overline{D(1,\overline{u_1},\overline{u_2}^*)f_{z\bar{z}}},$$
$$g_{21} = \overline{D(1,\overline{u_1},\overline{u_2}^*)f_{z^2\bar{z}}}.$$

In order to find the value of g_{21}, the computation of $W_{20}(\theta)$ and $W_{11}(\theta)$ should be prioritized. From the equations (33.20) and (33.22), we have:

$$W' = [A - 2\,\mathrm{Re}(\overline{q^*})F_0] + H(z,\bar{z},\theta), \tag{33.25}$$

where

$$H(z,\bar{z},\theta) = H_{20}(\theta)\frac{z^2}{2} + H_{11}(\theta)z\bar{z} + H_{02}(\theta)\frac{\bar{z}^2}{2} + H_{21}(\theta)\frac{z^2\bar{z}}{2} + \cdots. \tag{33.26}$$

However, on D_0 near the origin $W' = W_z z' + W_{\bar{z}}\overline{z'}$. Simplifying the above series and equating the coefficients, we get:

$$[A - 2i\omega_0 1] W_{20}(\theta) = -H_{20}(\theta), \quad AW_{11}(\theta) = -H_{11}(\theta). \tag{33.27}$$

By equation (33.22), for $\theta \in [-1, 0)$,

$$H(z, \bar{z}, \theta) = -\overline{q^*} F_0 q - \overline{q^*} F_0 \bar{q} = -gq - \overline{gq}.$$

Comparing the coefficients with (33.23) for $\theta \in [-1, 0]$, we get:

$$\begin{aligned} H_{20}(\theta) &= -g_{20} q - \overline{g_{02} q} + F_{z^2}, \\ H_{11}(\theta) &= -g_{11} q - \overline{g_{11} q} + F_{z\bar{z}}. \end{aligned}$$

From equations (33.23), (33.26), and the definition of A, we obtain:

$$W_{20}(\theta) = 2i\omega_0 \tau_0 W_{20}(\theta) + g_{20} q + \overline{g_{02} q}.$$

Solving for $W_{20}(\theta)$:

$$\begin{aligned} W_{20}(\theta) = {} & \frac{ig_{20}}{\omega_0 \tau_0} q(0) e^{i\omega_0 \tau_0 \theta} + \frac{i\overline{g_{02}}}{3\omega_0 \tau_0} \overline{q(0)} e^{-i\omega_0 \tau_0 \theta} + E_1 e^{2i\omega_0 \theta} + E_2 e^{-2i\omega_0 \theta} \\ & + E_3 e^{2i\omega_0 \theta} + E_4 e^{-2i\omega_0 \theta}. \end{aligned}$$

Where,

$$\begin{aligned} E_1 &= \frac{i}{\omega_0 \tau_0} g_{20} + \frac{1}{3\omega_0 \tau_0} \overline{g_{02}} + \frac{2i}{\omega_0^2 \tau_0^2} F_{z^2}, \\ E_2 &= \frac{i}{\omega_0 \tau_0} g_{20} + \frac{1}{3\omega_0 \tau_0} \overline{g_{02}} - \frac{2i}{\omega_0^2 \tau_0^2} F_{z^2}, \\ E_3 &= \frac{i}{\omega_0 \tau_0} \overline{g_{02}} + \frac{1}{3\omega_0 \tau_0} g_{20} + \frac{2i}{\omega_0^2 \tau_0^2} F_{z^2}, \\ E_4 &= \frac{i}{\omega_0 \tau_0} \overline{g_{02}} + \frac{1}{3\omega_0 \tau_0} g_{20} - \frac{2i}{\omega_0^2 \tau_0^2} F_{z^2}. \end{aligned}$$

Similarly, we solve for $W_{11}(\theta)$:

$$W_{11}(\theta) = \frac{ig_{11}}{\omega_0} q(0) e^{i\omega_0 \tau_0 \theta} + \frac{i\overline{g_{11}}}{\omega_0} \overline{q(0)} e^{-i\omega_0 \tau_0 \theta} + E_5 e^{2i\omega_0 \theta} + E_6 e^{-2i\omega_0 \theta}.$$

Where,

$$\begin{aligned} E_5 &= \frac{i}{\omega_0} g_{11} + \frac{i}{\omega_0^2} F_{z\bar{z}}, \\ E_6 &= \frac{i}{\omega_0} \overline{g_{11}} - \frac{i}{\omega_0^2} F_{z\bar{z}}. \end{aligned}$$

Now, substitute $W_{20}(\theta)$ and $W_{11}(\theta)$ into the above system (33.27) and solve for g_{20}, g_{02}, g_{11}, and g_{21}:

$$g_{20} = \frac{-i\omega_0\tau_0}{\omega_0\tau_0 + 2i\omega_0}\left(F_{z^2} - \frac{1}{3}F_{z\bar{z}}\right),$$
$$g_{02} = \frac{-i\omega_0\tau_0}{\omega_0\tau_0 + 2i\omega_0}\left(F_{z^2} + \frac{1}{3}F_{z\bar{z}}\right),$$
$$g_{11} = \frac{-i\omega_0}{\omega_0 + 2i\omega_0\tau_0}F_{z\bar{z}},$$
$$g_{21} = \frac{-i\omega_0\tau_0}{\omega_0\tau_0 + 2i\omega_0}F_{z^2} - \frac{4}{3}\frac{\omega_0\tau_0}{\omega_0\tau_0 + 2i\omega_0}F_{z\bar{z}}.$$

Finally, from equation (33.24) and the definition of g, we have:

$$g = \overline{q^*}F_0 = g_{20}z^2 + g_{11}z\bar{z} + g_{02}\bar{z}^2 + g_{21}z^2\bar{z} + \cdots.$$

Where,

$$g_{20} = \overline{D(1, u_1, u_2)}f_{z^2},$$
$$g_{02} = \overline{D(1, \overline{u_1}, \overline{u_2}^*)}f_{\bar{z}^2},$$
$$g_{11} = \overline{D(1, \overline{u_1}, \overline{u_2}^*)}f_{z\bar{z}},$$
$$g_{21} = \overline{D(1, \overline{u_1}, \overline{u_2}^*)}f_{z^2\bar{z}}.$$
$$F_0 = F_{z^2}\frac{z^2}{2} + F_{z\bar{z}}z\bar{z} + F_{\bar{z}^2}\frac{\bar{z}^2}{2} + \cdots.$$

Hence combining the definition of A,

$$\int_{-1}^{0} d\eta(\theta)W_{20}(\theta) = 2i\omega_0\tau_0 W_{20}(0) + g_{20}q(0) + \bar{g}_{02}\bar{q}(0) - F_{z^2}$$

$$\Rightarrow \int_{-1}^{0} d\eta(\theta)W_{11}(\theta) = g_{11}q(0) - \bar{g}_{11}\bar{q}(0) - F_{z\bar{z}}.$$

$$\Rightarrow \left[i\omega_0\tau_0 I - \int_{-1}^{0} e^{i\omega_0\tau_0\theta}d\eta(\theta)\right]q(0) = 0$$

$$\Rightarrow \left[-i\omega_0\tau_0 I - \int_{-1}^{0} e^{-i\omega_0\tau_0\theta}d\eta(\theta)\right]\bar{q}(0) = 0,$$

$$\Rightarrow \left[2i\omega_0\tau_0 I - \int_{-1}^{0} e^{2i\omega_0\tau_0\theta}d\eta(\theta)\right]E_1 = F_{z^2}$$

$$\Rightarrow \int_{-1}^{0} d\eta(\theta)E_2 = F_{z\bar{z}}.$$

Hence,

$$\begin{bmatrix} 2i\omega_0 - (a_1 - 2a_1S^*) + I^*e^{-2i\omega_0\tau_0} & S^*e^{-2i\omega_0\tau_0} & -ce^{-2i\omega_0\tau_0} \\ -I^*e^{-2i\omega_0\tau_0} & (2i\omega_0 + r - S^*e^{-2i\omega_0\tau_0}) & 0 \\ 0 & -a_3 & (2i\omega_0 + a_5) \end{bmatrix} E_1$$

$$= -2\begin{bmatrix} u_1e^{-i\omega_0\tau_0\theta} \\ -u_1e^{-i\omega_0\tau_0\theta} \\ 0 \end{bmatrix}$$

$$\Rightarrow \begin{bmatrix} (a_1 - 2a_1S^*) + I^*e^{-2i\omega_0\tau_0} & S^*e^{-2i\omega_0\tau_0} & -ce^{-2i\omega_0\tau_0} \\ -I^*e^{-2i\omega_0\tau_0} & (r - S^*e^{-2i\omega_0\tau_0}) & 0 \\ 0 & -a_3 & (a_5) \end{bmatrix} E_2$$

$$= -2\begin{bmatrix} \mathrm{Re}\{u_1\}e^{i\omega_0\tau_0\theta} \\ -\mathrm{Re}\{u_1\}e^{i\omega_0\tau_0\theta} \\ 0 \end{bmatrix}.$$

The arguments can thus be used to express g_{21} and the parameters. Based on the above study, every g_{ij} can be found out by the parameters. As a result, the following quantities can be calculated:

$$C_1(0) = \frac{i}{2\omega_0\tau_0}\left(g_{11}g_{20} - 2|g_{11}|^2 - \frac{|g_{02}|^2}{3}\right) + \frac{g_{21}}{2}, \quad \mu_2 = -\frac{\mathrm{Re}\{C_1(0)\}}{\mathrm{Re}\{\lambda'(\tau_0)\}}, \quad \beta_2 = 2\,\mathrm{Re}\{C_0(0)\},$$
$$T_2 = -\frac{\mathrm{Im}\{C_1(0)\} + \mu_2\ \mathrm{Im}\{\lambda'(\tau_0)\}}{\omega_0\tau_0}. \tag{33.28}$$

Theorem 10.1. *The direction of the Hopf bifurcation is calculated by the value of μ_2: if $\mu_2 > 0$ ($\mu_2 < 0$), then the Hopf bifurcation is supercritical (subcritical), and the bifurcating periodic solutions exist for $\tau > \tau_0$ ($\tau < \tau_0$). The stability of bifurcating solutions is calculated by the value of β_2; the bifurcating recurring solutions are orbitally asymptotically stable (unstable) if $\beta_2 < 0$ ($\beta_2 > 0$). The value of T_2 defines the period of the bifurcating periodic solutions; the period increases (decreases) if $T_2 > 0$ ($T_2 < 0$).*

11 Numerical simulation and discussion

The purpose of this section is to provide some numerical simulations to demonstrate the theoretical findings gained in this work. Using MATLAB, we will display various bifurcation diagrams and investigate novel dynamical behaviors resulting from parameter alterations. For this endeavor, we consider hypothetical values of parameters fulfilling the requirements derived analytically in earlier sections.

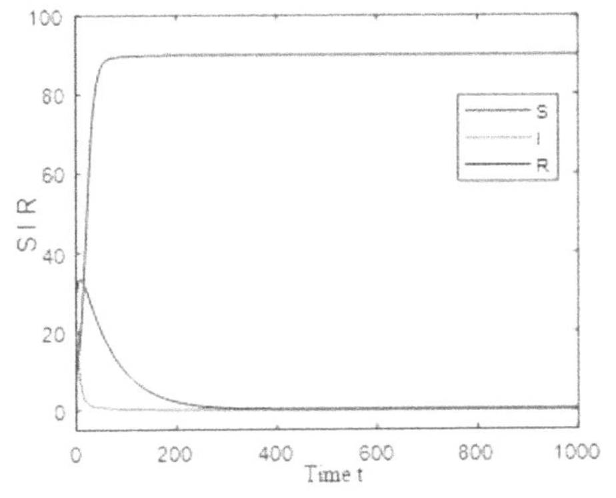

Figure 33.1: Figure a.

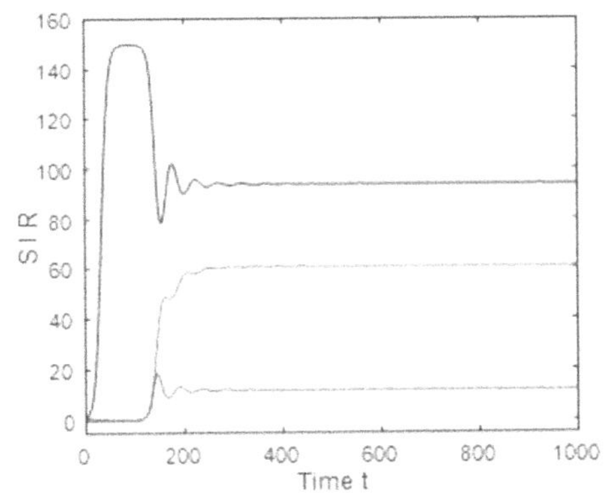

Figure 33.2: Figure b.

With the parameter set $K = 90$, $\alpha_1 = 0.145$, $\alpha_2 = 0.0045$, $\alpha_3 = 0.185$, $\alpha_4 = 0.25$, $\alpha_5 = 0.0347$, and $\tau = 1$, we get $R_0 = 0.5179 < 1$. The disease-free equilibrium (DFE) is globally stable, and the system's dynamics are plotted in Figure 33.1.

When $1 < R_0 = 1.597 \leq 2$, the system (33.1) exhibits asymptotic stability. The dynamics for $K = 150$, $\alpha_1 = 0.156$, $\alpha_2 = 0.051$, $\alpha_3 = 0.180$, $\alpha_4 = 0.2551$, and $\alpha_5 = 0.0346$ the dynamics of system are ploted in Figure 33.2.

For $R_0 > 2$, the dynamics of system (33.2) are simulated with $K = 20$, $\alpha_1 = 0.112$, $\alpha_2 = 0.0521$, $\alpha_3 = 0.0751$, $\alpha_4 = 0.0477$, and $\alpha_5 = 0.0374$. Direct computation yields $R_0 = 6.616 > 1$.

When $\tau = 0$, the endemic equilibrium E^* is locally asymptotically stable, and the equilibrium point is $E^*(3.1329, 1.7642, 3.5459)$ (see Figure 33.3–33.6). When $\tau = 0.85 < \tau_c = 1.0125$, the endemic equilibrium is also asymptotically stable (see Figure 33.7–33.8).

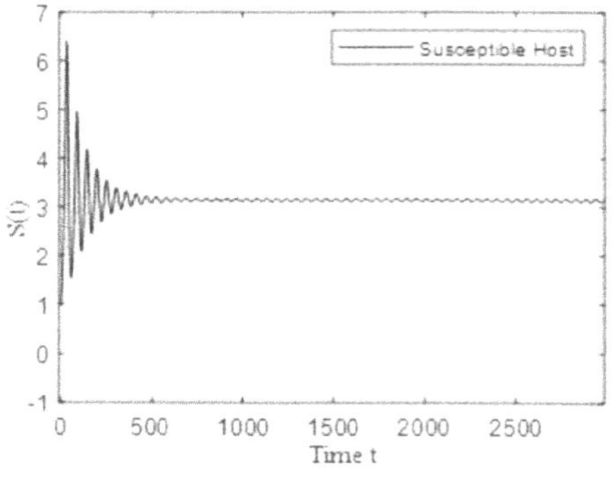

Figure 33.3: Figure c.

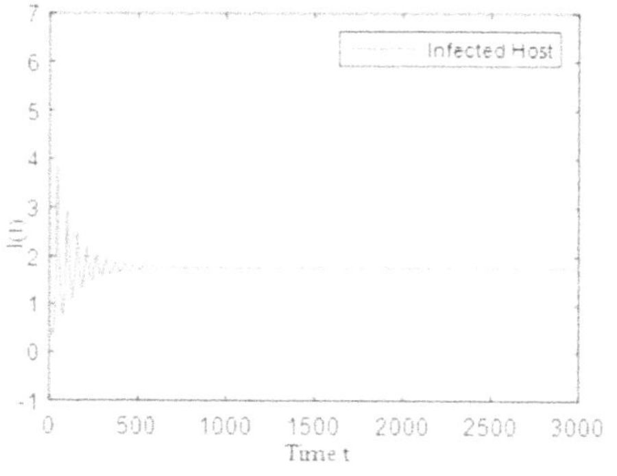

Figure 33.4: Figure d.

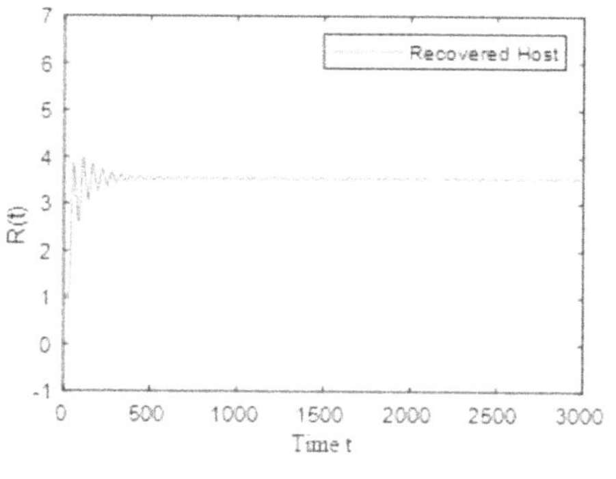

Figure 33.5: Figure e.

At $\tau = \tau_c = 1.0125$, the system exhibits Hopf bifurcation (see Figure 33.9–33.12). For $\tau = 1.27 > \tau_c = 1.0125$, the system becomes supercritical (see Figure 33.13–33.16). Figure 33.17 showing the Phase Space view of the dynamics of all the states of stability.

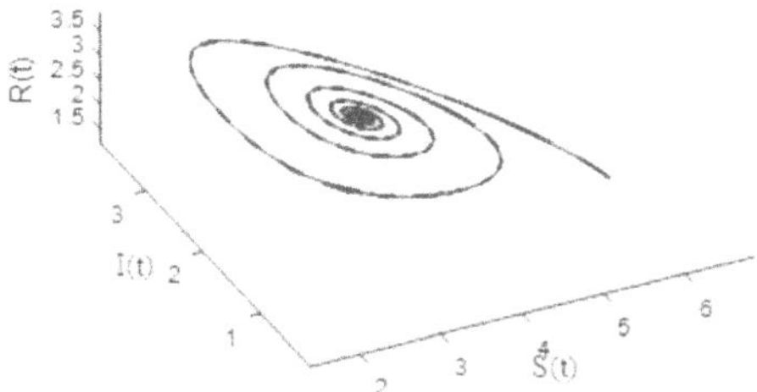

Figure 33.6: Figure f.

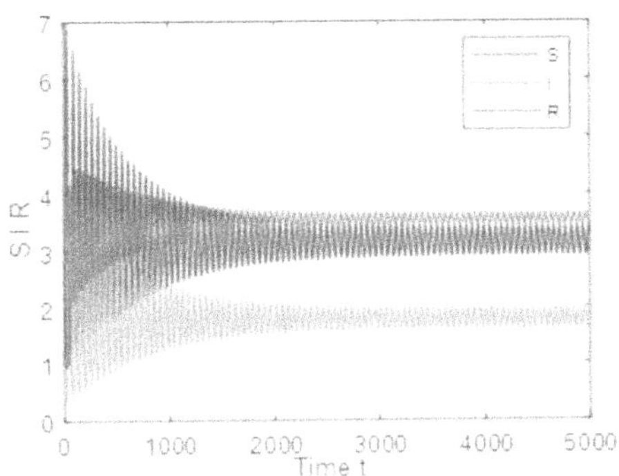

Figure 33.7: Figure g.

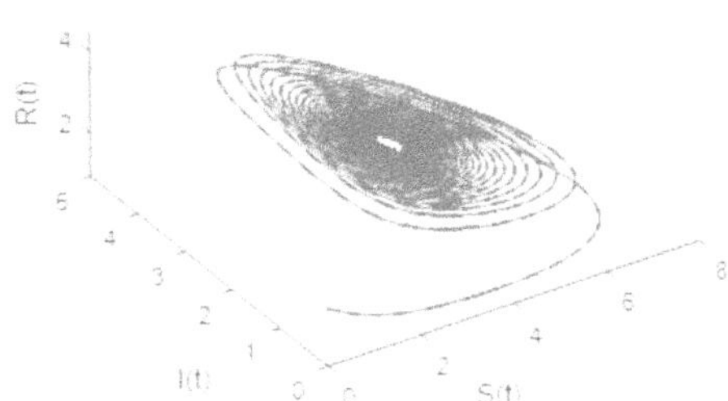

Figure 33.8: Figure h.

12 Significance of the work

It is a common fact that there is a lag between an individual's infection and contagious state. An individual is infected yet unable to spread the disease during the incubation phase of many infectious diseases. An improved portrayal of the dynamics of the disease is possible by include delays in the SIR model. Researchers and public health professionals may develop more realistic and complex simulations of disease propagation

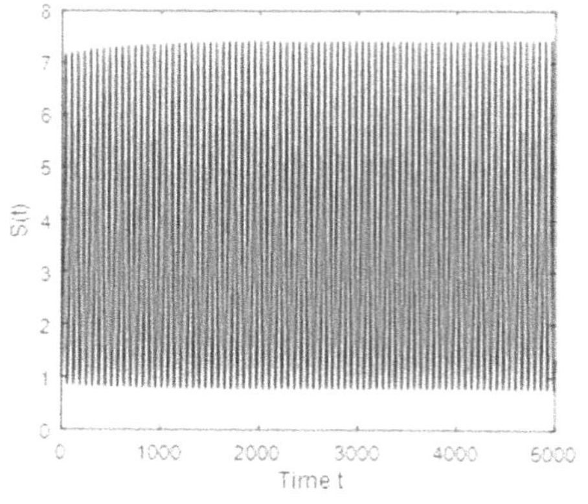

Figure 33.9: Figure i.

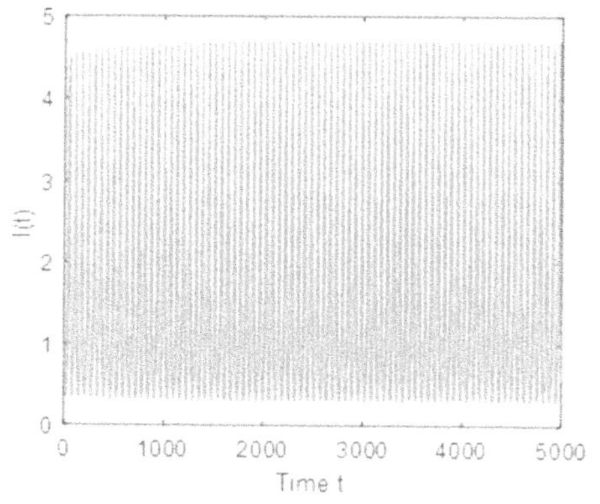

Figure 33.10: Figure j.

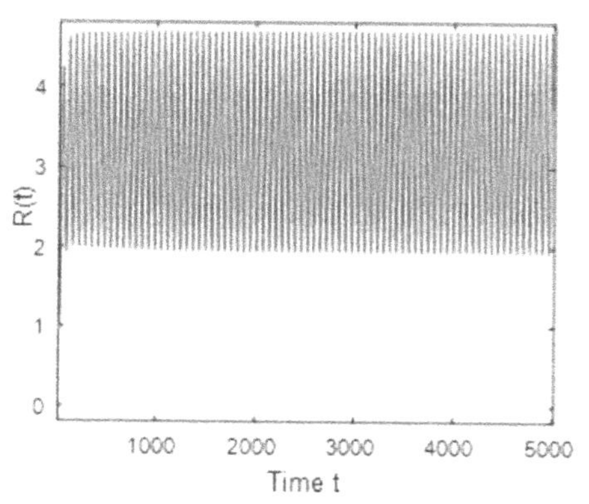

Figure 33.11: Figure k.

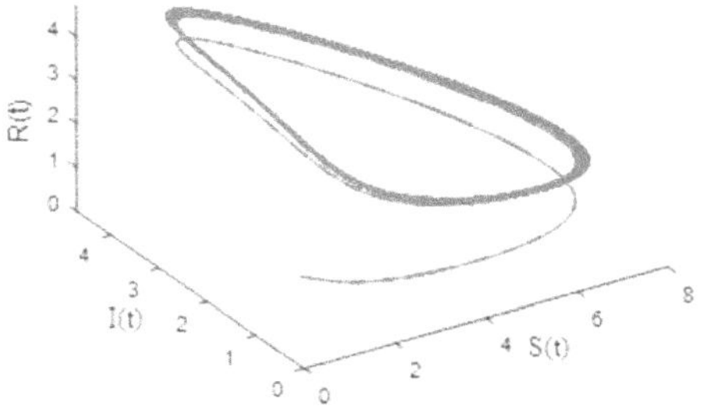

Figure 33.12: Figure l.

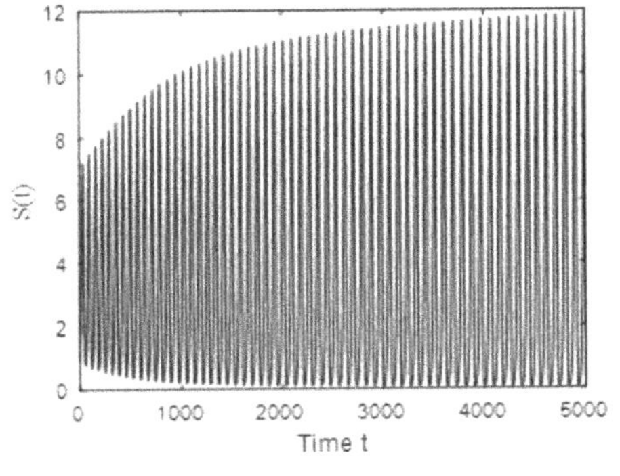

Figure 33.13: Figure m.

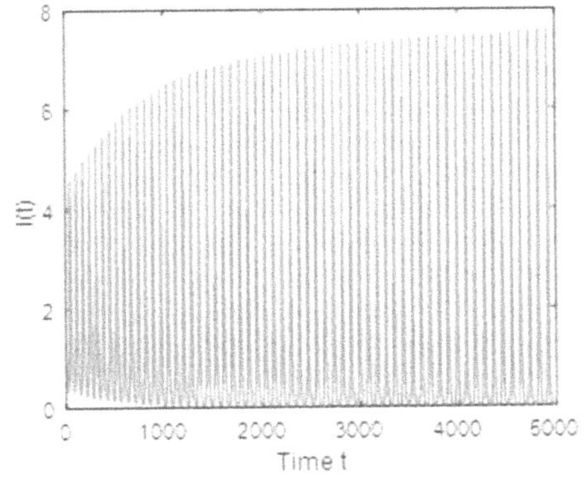

Figure 33.14: Figure n.

by incorporating delays into the SIR model. This can be useful for understanding the dynamics of an epidemic and helping decision makers make educated actions. Delay differential equation models or compartmental models with temporal delays are common terms used to describe models that take delays into consideration.

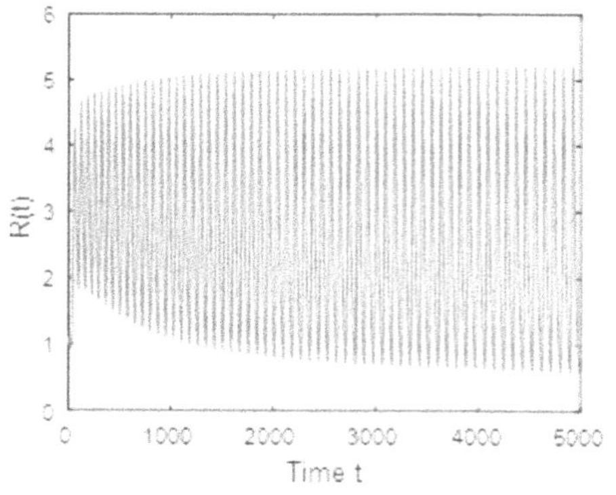

Figure 33.15: Figure o.

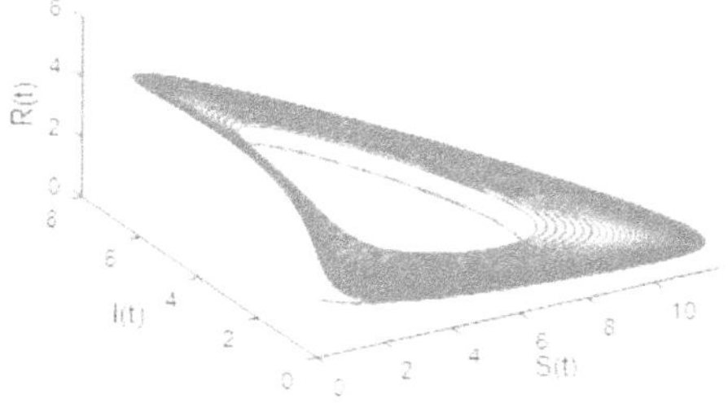

Figure 33.16: Figure p.

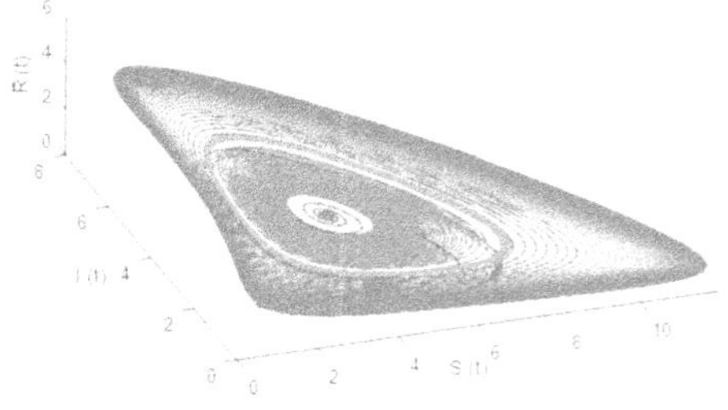

Figure 33.17: Figure q; Phase space view showing the dynamics of all the states of stability, i. e., global (green), asymptotic (purple), Hopf bifurcation (red) and super critical (blue).

13 Conclusion

This work investigates the global dynamic behaviors of an SIR model with an incubation period, where the susceptible population exhibits logistic expansion. The system (33.2) is characterized by a critical value R_0 that controls the disease behavior. The focus of this research is to understand the impact of the incubation period on the dynamic behavior of the system. The findings of this paper reveal that the proximity to an endemic value and the stability of solutions depend on two variables: the duration of the incubation period and the threshold value R_0. System (33.2) exhibits global asymptotic stability in the disease-free equilibrium E^* when $R_0 = 1$. However, when $R_0 > 1$, E^* loses stability, and the endemic equilibrium E^* becomes irreversible. The requisites for achieving local stability and Hopf bifurcation at E^* were derived by treating the delay time as a parameter. Specifically, when $1 < R_0 \leq 2$, E^* is absolutely stable; that is, E^* remains locally stable when τ is non-zero. On the other hand, when $R_0 > 2$, E^* is conditionally stable in the range range $\tau \in (0, \tau_0)$, as soon as the value of τ surpasses the critical value τ_0, i.e., $\tau > \tau_0, E^*$ becomes unstable and shows Hopf bifurcations. This paper shows that for short time delays in System (33.2), there is no effect on the positive invariance. Even the boundedness of solutions, the global stability of the disease-free equilibrium (DFE), or the permanence of the endemic equilibrium remains intact. However, for long delays, the time delay can destabilize the unique endemic equilibrium, leading to a variety of stable periodic solutions through Hopf bifurcation.

Bibliography

[1] A. Ashyani, H. Mohammadinejad and O. RabieiMotlagh, Hopf bifurcation analysis in a delayed system for cancer virotherapy, *Indag. Math.* **27**(1) (2016). https://doi.org/10.1016/j.indag.2015.10.015.

[2] F. Avram, R. Adenane, G. Bianchin and A. Halanay, Stability analysis of an eight parameter SIR-type model including loss of immunity, and disease and vaccination fatalities, *Mathematics* **10**(3) (2022). https://doi.org/10.3390/math10030402.

[3] N. Bacaër and N. Bacaër, Daniel Bernoulli, d'Alembert and the inoculation of smallpox (1760), in: *A Short History of Mathematical Population Dynamics*, 2011. https://doi.org/10.1007/978-0-85729-115-8_4.

[4] D. Bernoulli, Essai d'une nouvelle analyse de la mortalite causee par la petite verole et des avantages de l'inoculation pour la prevenir, in: *Histoire de l'Academie Royale des Sciences*, 1760, pp. 1–45.

[5] D. Bernoulli, Essai d'une nouvelle analyse de la mortalité causée par la petite vérole, et des avantages de l'inoculation pour la prévenir, in: *Die Werke von Daniel Bernoulli, Bd. 2 Analysis und Wahrscheinlichkeitsrechnung*, 1766.

[6] H. Cao, D. Yan and X. Xu, Hopf bifurcation for an SIR model with age structure, *Math. Model. Nat. Phenom.* **16** (2021). https://doi.org/10.1051/mmnp/2021003.

[7] K. L. Cooke, Stability analysis for a vector disease model, *Rocky Mt. J. Math.* **9**(1) (1979). https://doi.org/10.1216/RMJ-1979-9-1-31.

[8] Dipesh and P. Kumar, Modelling the stimulatory and inhibitory allelopathic effects on competing plant populations, *AIP Conf. Proc.* **2435**(1) (2022 Mar 18) AIP Publishing.

[9] Dipesh and P. Kumar, Investigating the impact of toxicity on plant growth dynamics through the zero of a fifth-degree exponential polynomial: A mathematical model using DDE, *Chaos Solitons Fractals* **171** (2023 Jun 1), 113457.

[10] Dipesh and P. Kumar, Delay differential equation model of forest biomass and competition between wood-based industries and synthetic-based industries, *Math. Methods Appl. Sci.* **46**(9) (2023 Jun), 10602–10616.

[11] Y. Enatsu, Y. Nakata and Y. Muroya, Lyapunov functional techniques for the global stability analysis of a delayed SIRS epidemic model, *Nonlinear Anal., Real World Appl.* **13**(5) (2012). https://doi.org/10.1016/j.nonrwa.2012.01.007.

[12] J.-P. Gabriel and P. de la Harpe, Daniel Bernoulli, pionnier des modèles mathématiques en médecine, in: *Histoire des mathématiques*, 2010.

[13] J. K. Hale, History Of Delay Equations, in: *Delay Differential Equations and Applications*, 2007. https://doi.org/10.1007/1-4020-3647-7_1.

[14] A. Kaddar and H. Talibi Alaoui, Local Hopf bifurcation and stability of limit cycle in a delayed Kaldor-Kalecki model, *Nonlinear Anal.: Model. Control* **14**(3) (2009). https://doi.org/10.15388/na.2009.14.3.14499.

[15] W. O. Kermack and A. G. McKendrick, A contribution to the mathematical theory of epidemics, *Proc. R. Soc. Lond. Ser. A, Contain. Pap. Math. Phys. Character* **115**(772) (1927). https://doi.org/10.1098/rspa.1927.0118.

[16] A. Kumar, K. Goel and Nilam, A deterministic time-delayed SIR epidemic model: mathematical modeling and analysis, *Theory Biosci.* **139**(1) (2020). https://doi.org/10.1007/s12064-019-00300-7.

[17] J. Liu, Hopf bifurcation analysis for an SIRS epidemic model with logistic growth and delays, *J. Appl. Math. Comput.* **50**(1–2) (2016). https://doi.org/10.1007/s12190-015-0884-8.

[18] L. Liu, A delayed SIR model with general nonlinear incidence rate, *Adv. Differ. Equ.* **2015**(1) (2015). https://doi.org/10.1186/s13662-015-0619-z.

[19] L. Liu and Y. Wang, Stability analysis for a delayed SIR model with a nonlinear incidence rate, *J. Nonlinear Sci. Appl.* **10**(11) (2017). https://doi.org/10.22436/jnsa.010.11.21.

[20] Q. Liu, Q. Chen and D. Jiang, The threshold of a stochastic delayed SIR epidemic model with temporary immunity, *Physica A: Stat. Mech. Appl.* **450** (2016). https://doi.org/10.1016/j.physa.2015.12.056.

[21] S. Liu, S. Ruan and X. Zhang, On avian influenza epidemic models with time delay, *Theory Biosci.* **134**(3–4) (2015). https://doi.org/10.1007/s12064-015-0212-8.

[22] C. C. McCluskey, Global stability for an SIR epidemic model with delay and nonlinear incidence, *Nonlinear Anal., Real World Appl.* **11**(4) (2010). https://doi.org/10.1016/j.nonrwa.2009.11.005.

[23] Y. Muroya, Y. Enatsu and Y. Nakata, Global stability of a delayed SIRS epidemic model with a non-monotonic incidence rate, *J. Math. Anal. Appl.* **377**(1) (2011). https://doi.org/10.1016/j.jmaa.2010.10.010.

[24] R. Naresh, A. Tripathi and D. Sharma, A nonlinear AIDS epidemic model with screening and time delay, *Appl. Math. Comput.* **217**(9) (2011). https://doi.org/10.1016/j.amc.2010.10.036.

[25] R. Naresh, A. Tripathi, J. M. Tchuenche and D. Sharma, Stability analysis of a time delayed SIR epidemic model with nonlinear incidence rate, *Comput. Math. Appl.* **58**(2) (2009). https://doi.org/10.1016/j.camwa.2009.03.110.

[26] K. R. Schneider, Hassard, B. D. / Kazarinoff, N. D. / Wan, Y.-H., Theory and Applications of Hopf Bifurcation. London Mathematical Society Lecture Note Series 41. Cambridge, Cambridge University Press 1981. 320 S., £ 15.00 A P/B. ISBN 0-521-23158-2, *Z. Angew. Math. Mech.* **62**(12) (1982). https://doi.org/10.1002/zamm.19820621221.

[27] R. P. Singh, Bifurcation and Stability Analysis of Delayed SIR Model, *J. Phys. Conf. Ser.* **2267**(1) (May 2022), 012011. https://doi.org/10.1088/1742-6596/2267/1/012011.

[28] A. Sirijampa, S. Chinviriyasit and W. Chinviriyasit, Hopf bifurcation analysis of a delayed SEIR epidemic model with infectious force in latent and infected period, *Adv. Differ. Equ.* **2018**(1) (2018). https://doi.org/10.1186/s13662-018-1805-6.

[29] Y. Song and Y. Peng, Stability and bifurcation analysis on a logistic model with discrete and distributed delays, *Appl. Math. Comput.* **181**(2) (2006). https://doi.org/10.1016/j.amc.2006.03.025.

[30] C. Sun, Z. Cao and Y. Lin, Analysis of stability and Hopf bifurcation for a viral infectious model with delay, *Chaos Solitons Fractals* **33**(1) (2007). https://doi.org/10.1016/j.chaos.2005.12.029.

[31] C. Sun, M. Han and Y. Lin, Analysis of stability and Hopf bifurcation for a delayed logistic equation, *Chaos Solitons Fractals* **31**(3) (2007). https://doi.org/10.1016/j.chaos.2005.10.019.

[32] Y. Takeuchi, W. Ma and E. Beretta, Global asymptotic properties of a delay SIR epidemic model with finite incubation times, *Nonlinear Anal.:Theory Methods Appl.* **42**(6) (2000). https://doi.org/10.1016/S0362-546X(99)00138-8.

[33] J. M. Tchuenche, A. Nwagwo and R. Levins, Global behaviour of an SIR epidemic model with time delay, *Math. Methods Appl. Sci.* **30**(6) (2007). https://doi.org/10.1002/mma.810.

[34] S. Tipsri and W. Chinviriyasit, The effect of time delay on the dynamics of an SEIR model with nonlinear incidence, *Chaos Solitons Fractals* **75** (2015). https://doi.org/10.1016/j.chaos.2015.02.017.

[35] J. Wu, X. S. Zhan, X. H. Zhang and H. L. Gao, Stability and Hopf bifurcation analysis on a numerical discretization of the distributed delay equation, *Chin. Phys. Lett.* **29**(5) (2012). https://doi.org/10.1088/0256-307X/29/5/050203.

[36] W. Xia, S. Kundu and S. Maitra, Dynamics of a delayed SEIQ epidemic model, *Adv. Differ. Equ.* **2018**(1) (2018). https://doi.org/10.1186/s13662-018-1791-8.

[37] C. Xu and X. Li, The threshold of a stochastic delayed SIRS epidemic model with temporary immunity and vaccination, *Chaos Solitons Fractals* **111** (2018). https://doi.org/10.1016/j.chaos.2017.12.027.

[38] R. Xu, Z. Ma and Z. Wang, Global stability of a delayed SIRS epidemic model with saturation incidence and temporary immunity, *Comput. Math. Appl.* **59**(9) (2010). https://doi.org/10.1016/j.camwa.2010.03.009.

[39] Y. Xue and T. Li, Stability and Hopf bifurcation for a delayed SIR epidemic model with logistic growth, *Abstr. Appl. Anal.* **2013** (2013). https://doi.org/10.1155/2013/916130.

[40] C. Yan and J. Jia, Hopf bifurcation of a delayed epidemic model with information variable and limited medical resources, *Abstr. Appl. Anal.* **2014** (2014). https://doi.org/10.1155/2014/109372.

[41] P. Yan and Z. Feng, Variability order of the latent and the infectious periods in a deterministic SEIR epidemic model and evaluation of control effectiveness, *Math. Biosci.* **224**(1) (2010). https://doi.org/10.1016/j.mbs.2009.12.007.

[42] J. Zhang, Z. Jin, J. Yan and G. Sun, Stability and Hopf bifurcation in a delayed competition system, *Nonlinear Anal.:Theory Methods Appl.* **70**(2) (2009). https://doi.org/10.1016/j.na.2008.01.002.

[43] T. Zhang, J. Liu and Z. Teng, Stability of Hopf bifurcation of a delayed SIRS epidemic model with stage structure, *Nonlinear Anal., Real World Appl.* **11**(1) (2010). https://doi.org/10.1016/j.nonrwa.2008.10.059.

[44] X. A. Zhang and L. Chen, Periodic solution of a class of epidemic models, *Comput. Math. Appl.* **38**(3) (1999). https://doi.org/10.1016/S0898-1221(99)00206-0.

[45] Y. Zhang, L. Li, J. Huang and Y. Liu, Stability and Hopf bifurcation analysis of a vector-borne disease model with two delays and reinfection, *Comput. Math. Methods Med.* **2021** (2021). https://doi.org/10.1155/2021/6648959.

Pintu Bhattacharya

Simple harmonic motion in view of quantum calculus

Abstract: In this paper, the simple harmonic motion (SHM) is analyzed extensively based on the basics of q-calculus. The equation of motion of a simple harmonic oscillator (SHO) is revisited in view of quantum calculus. In order to describe the SHM, q-differential equations are presented in accordance with the exact classical expressions. The solutions are obtained in terms of the elementary q-exponential function. It is also found that, in a limiting situation the solutions take the form as they are in general classical Newtonian mechanics. It is anticipated that this work may also be treated as a good pedagogical resource for further problems usually encountered in quantum calculus and mathematical physics.

Keywords: Projectile, quantum calculus, quantum derivative, quantum exponential, 2-D projectile

MSC 2010: 16P20, 16P60, 16P70

1 Introduction

The Leibniz notation $\frac{df(x)}{dx}$ is the ratio of two infinitesimals uses limits in calculating the derivative of a real function, whereas calculus without limits is known as quantum calculus [1–4]. A subsequent development in q-derivative and q-integral can also be useful to analyze physical systems in a different way as mentioned and described in Refs. [5–8]. In addition, q-calculus shows a different interesting way to find out exact solutions to several models in statistical mechanics, heat and wave equations [2, 3, 9]. Recently, the falling body problem, the very basic problem in Newtonian classical mechanics is also described in view of q-calculus [1]. This work is mainly motivated by the series of work as described in Refs. [1–4, 9]. In addition, this paper also expands the usefulness of the theoretical framework of quantum calculus in conventional classical physics.

Acknowledgement: The author acknowledges all the unknown reviewers for their careful reading of the manuscript. The author is also thankful for their valuable comments and suggestions to improve the quality of the manuscript. The manuscript is also prepared by the author only and there are no conflict of interests to disclose.

Pintu Bhattacharya, Department of Physics, L. N. College, Bhagwanpur, A constituent unit of B. R. A. Bihar University, Staff quarter, R. N. 5, L. N. College Street, Beside NH, Hajipur, Vaishali-844114, Bihar, India, e-mail: pbhattacharya89@gmail.com

https://doi.org/10.1515/9783111313634-034

2 Quantum calculus preliminaries

(I) Here $[m]_q = \frac{q^m-1}{q-1}$ for any integer m such that $\lim_{q\to 1}[m]_q = m$ (1st chapter in [8]). Thus, $[0]_q = 0$ and $[1]_q = 1$.

(II) If $f(t)$ is an arbitrary function, its q-differential is $d_q f(t) = f(qt) - f(t)$ and the q-derivative is $D_q f(t) = \frac{d_q f(t)}{d_q t} = \frac{f(qt)-f(t)}{(q-1)t}$. It is also to be noted that $\lim_{q\to 1} D_q f(t) = \frac{d_q f(t)}{d_q t} = \frac{df}{dt}$. Thus, $D_q t^m = \frac{(qt)^m - t^m}{(q-1)t} = \frac{q^m-1}{q-1} t^{m-1} = [m]_q t^{m-1}$.

(III) *q-factorial:* The quantum factorial or q-factorial $[m]_q$ is defined as (described in 3rd chapter of [8]) $[m]_q! = 1$, when $m = 0$ and $[m]_q! = [1]_q.[2]_q.[3]_q...[m-1]_q.[m]_q$.

(IV) *q-exponential:* $e_q(\pm i\omega t)$ is basically the small q-exponential function of t is represented as

$$e_q(\pm i\omega t) = \sum_{m=0}^{\infty} \frac{(\pm i\omega t)^m}{[m]_q!}.$$

(V) *q-sine and q-cosine function:* The q-analogous sine and cosine functions (as described in chapter 10 in Ref. [8]) can be expressed by considering their Euler expression in terms of exponential function as

$$\sin_q(\omega t) = \frac{e_q(i\omega t) - e_q(-i\omega t)}{2i} \quad \text{and} \quad \cos_q(\omega t) = \frac{e_q(i\omega t) + e_q(-i\omega t)}{2}.$$

And their derivative

$$D_q \sin_q(\omega t) = \omega \cos_q(\omega t) \quad \text{and} \quad D_q \cos_q(\omega t) = -\omega \sin_q(\omega t).$$

3 The problem

Any vibration under the action of a restoring force directly proportional and opposite to the displacement is simple harmonic motion (SHM) [6]. Suppose a particle of unit mass is undergoing SHM along the y-axis. If y is the displacement of the particle at any instant from the position of equilibrium, then from Newton's law the equation of motion is,

$$\frac{d^2y}{dt^2} = -\omega^2 y, \tag{34.1}$$

where ω is a real positive constant known as angular frequency having dimension $[T^{-1}]$ that represents the revolution per second. The classical solution of (34.1) is

$$y(t) = a\exp(-i\omega t) + b\exp(i\omega t) \tag{34.2}$$

or

$$y(t) = A\sin(\omega t) + B\cos(\omega t), \tag{34.3}$$

where a, b are constant of integration and $A = (a + b)$ and $B = i(a - b)$ are also constants to be determined from proper initial conditions. Considering the initial condition $y(t = 0) = y_0$ and $v(t = 0) = v_0$, (34.3) becomes,

$$y(t) = y_0\cos(\omega t) + \frac{v_0}{\omega}\sin(\omega t), \tag{34.4}$$

where y_0 and v_0 are initial displacement and velocity respectively.

4 Solution to the problem in quantum calculus

In view of q-calculus, (34.1) can be written as,

$$\frac{d_q^2 y}{d_q t^2} = -\omega^2 y. \tag{34.5}$$

In order to solve (34.5), we assume the solutions as

$$y = \sum_{m=0}^{+\infty} c_m t^m. \tag{34.6}$$

Therefore,

$$\frac{d_q y}{d_q t} = \sum_{m=0}^{+\infty} [m]_q c_m t^{m-1}. \tag{34.7}$$

Thus,

$$\frac{d_q^2 y}{d_q t^2} = \sum_{m=0}^{+\infty} [m]_q [m-1]_q c_m t^{m-2}. \tag{34.8}$$

Considering the definition of $[m]_q$, $[0]_q = 0$,

$$\frac{d_q^2 y}{d_q t^2} = \sum_{m=2}^{+\infty} [m]_q [m-1]_q c_m t^{m-2}, \tag{34.9}$$

$$\text{or,} \quad \frac{d_q^2 y}{d_q t^2} = \sum_{m=0}^{+\infty} [m+2]_q [m+1]_q c_{m+2} t^m. \tag{34.10}$$

Substituting, (34.10) in (34.5), we get

$$\sum_{m=0}^{+\infty}[m+2]_q[m+1]_q c_{m+2}t^m = -\omega^2\sum_{m=0}^{+\infty}c_m t^m. \tag{34.11}$$

This gives,

$$[2]_q[1]_q c_2 + \sum_{m=1}^{+\infty}[m+2]_q[m+1]_q c_{m+2}t^m = -\omega^2 c_0 - \omega^2\sum_{m=1}^{+\infty}c_m t^m. \tag{34.12}$$

Then, comparing the coefficient of t^0 and t^m,

$$c_2 = \frac{-\omega^2 c_0}{[2]_q[1]_q}, \tag{34.13}$$

and

$$c_{m+2} = \frac{-\omega^2 c_m}{[m+2]_q[m+1]_q}, \tag{34.14}$$

for $m \geq 0$.

Thus, from (34.14), we obtain,

$$c_3 = \frac{-\omega^2 c_1}{[3]_q[2]_q}, \tag{34.15}$$

$$c_4 = \frac{-\omega^2 c_2}{[4]_q[3]_q} = \frac{(-\omega^2)^2 c_0}{[4]_q[3]_q[2]_q[1]_q}, \tag{34.16}$$

$$c_5 = \frac{-\omega^2 c_3}{[5]_q[4]_q} = \frac{(-\omega^2)^2 c_1}{[5]_q[4]_q[3]_q[2]_q}. \tag{34.17}$$

And hence the solution in quantum calculus becomes,

$$y_q = c_0 + c_1 t + \frac{-\omega^2 c_0}{[2]_q[1]_q}t^2 + \frac{-\omega^2 c_1}{[3]_q[2]_q}t^3 + \frac{(-\omega^2)^2 c_0}{[4]_q[3]_q[2]_q[1]_q} + \frac{(-\omega^2)^2 c_1}{[5]_q[4]_q[3]_q[2]_q}. \tag{34.18}$$

As, $[1]_q = 1$, (34.18) can be rewritten as,

$$y_q = \left[c_0 + \frac{-\omega^2 c_0}{[2]_q!}t^2 + \frac{(-\omega^2)^2 c_0}{[4]_q!}t^4 + \cdots\right] + \left[c_1 t + \frac{-\omega^2 c_1}{[3]_q!}t^3 + \frac{(-\omega^2)^2 c_1}{[5]_q!}t^5 + \cdots\right]. \tag{34.19}$$

The term ωt is dimensionless. All the terms inside the second parenthesis must have dimension of displacement. Therefore, the dimension of c_1 must have the dimension of velocity. Hence to bring the solution in familiar form we need some modification. Thus, rearranging (34.19), the expression for y_q becomes,

$$y_q = \left[1 - \frac{(\omega t)^2}{[2]_q!} + \frac{(\omega t)^4}{[4]_q!} + \cdots\right](c_0) + \left[\omega t - \frac{(\omega t)^3}{[3]_q!} + \frac{(\omega t)^5}{[5]_q!} + \cdots\right]\left(\frac{c_1}{\omega}\right). \tag{34.20}$$

Writing (34.20) with some modification, y_q can be written as,

$$y_q = \left[1 + \frac{(i\omega t)^2}{[2]_q!} + \frac{(i\omega t)^4}{[4]_q!} + \cdots\right](c_0) + \left[i\omega t + \frac{(i\omega t)^3}{[3]_q!} - \frac{(i\omega t)^5}{[5]_q!} + \cdots\right]\left(\frac{c_1}{i\omega}\right). \tag{34.21}$$

Here, it is also to be noted that, $\frac{c_1}{\omega}$ has the dimension of length. So, $\frac{c_1}{\omega}$ can be treated as amplitude unambiguously. Considering, the expression (as described in Section 2) for $e_q^{i\omega t}$ and $e_q^{-i\omega t}$, (34.21) becomes,

$$y_q = \left[\frac{e_q^{i\omega t} + e_q^{-i\omega t}}{2}\right](c_0) + \left[\frac{e_q^{i\omega t} - e_q^{-i\omega t}}{2}\right]\left(\frac{c_1}{i\omega}\right), \tag{34.22}$$

$$\text{or,}\quad y_q = \frac{1}{2}\left(c_0 + \frac{c_1}{i\omega}\right)e_q^{i\omega t} + \frac{1}{2}\left(c_0 - \frac{c_1}{i\omega}\right)e_q^{-i\omega t}, \tag{34.23}$$

$$\text{or,}\quad y_q = c_0 \cos_q(\omega t) + \frac{c_1}{\omega}\sin_q(\omega t). \tag{34.24}$$

Writing $\frac{c_1}{\omega}$ as c_1', (34.24) can also be written as,

$$\text{or,}\quad y_q = c_0 \cos_q(\omega t) + c_1' \sin_q(\omega t). \tag{34.25}$$

As initially, at $t = 0$, the particle is at $y(t = 0) = y_0$ and $v(t = 0) = v_0$. Hence from (34.24) and considering the q-differential of

$$\text{or,}\quad y_q = \frac{v_0}{\omega}\sin_q(\omega t) + y_0 \cos_q(\omega t), \tag{34.26}$$

in (34.26), the expression for displacement (y_q) has the similarity with its conventional form, as obtained in (34.4).

5 Comparative analysis

In order to describe any classical problem in view of unconventional calculus, the dimensionality of all the kinematic parameters must be the essential point to be discussed. In this problem, it is found that, the dimensionality of all the parameters depend on the dimension of $e_q^{\pm i\omega t}$. As the dimension of ω is $[T^{-1}]$, ωt, $e_q^{\pm i\omega t}$ are dimensionless quantity. So, there is no issue regarding the dimensionality of any parameters. Here, it is important to be noted that (34.26) should exactly be reduced to the corresponding classical one as displayed in (34.4) if and only if $q \to 1$. In addition, (34.22) to (34.25) will also reduce to their exact corresponding classical forms if and only if $q \to 1$, and both $\sin_q(\omega t)$ and $\cos_q(\omega t)$ takes the corresponding conventional sine and cosine form when $q \to 1$.

6 Conclusion

The simplest and fundamental problem regarding the oscillation in Newtonian mechanics is studied in view of quantum calculus. Several more ways to find the equation of motion of a simple harmonic oscillator has been investigated, but the approach discussed in the present paper is completely new one. In theoretical viewpoint, we found that the general condition $q \to 1$ leads the q-solution to its corresponding classical one. Hence for the first time it is investigated for simple harmonic motion. As a generalization, the relation between q-exponential, q-sine, and q-cosine are verified in case of simple harmonic motion unambiguously. Moreover, the detailed dimensional analysis is presented to investigate the correctness of the dimensionality of all the deducted formulae. The present work can further be extended to deduce the equation of motion of many other dynamical systems in classical physics. Thus, the problem as described in the present paper in quantum calculus framework can also serve as a good pedagogical exercise to the introductory problem in classical mechanics.

Bibliography

[1] M. A. Abdullaziz, A. Ebaid, M. A. Wadha and G. Muhiuddin, The falling body problem in quantum calculu, *Front. Phys.* **8** (2020), Article 43.

[2] M. H. Annaby and Z. S. Mansour, *q-fractional Calculus and Equations*, Springer, 2012 and reference therein.

[3] N. Bettaibi and K. Mezlini, On the use of the q-Mellin transform to solve some q-heat and q-wave equations, *Int. J. Math. Arch.* **3**(2) (2012), 446–455.

[4] L. Chanchlani, S. Alha and J. Gupta, Generalization of Taylor's formula and differential transform method for composite fractional q-derivative, *Ramanujan J.* **48** (2019), 21–32.

[5] T. Ernst, A method for q-calculus, *J. Nonlin. Math. Phys. A Method* **10** (2003), 487–525.

[6] D. Halliday, R. Resnick and J. Walker, *Fundamentals of Physics*, Wiley, 1997.

[7] F. H. Jackson, On a q-definite integrals, *Q. J. Pure Appl. Math.* **41**(3) (1910), 193–203.

[8] V. G. Kac and P. Cheung, *Quantum Calculus*, Springer-Verlag, NY, 2002, https://link.springer.com/book/10.1007/978-1-4613-0071-7.

[9] Y. Tang and T. A. Zhang, Remark on the q-fractional order differential equations, *Appl. Math. Comput.* **350** (2019), 198–208.

Ishfaq Nazir

On Bernstein and Turán-type inequalities for polynomials

Abstract: The goal of this paper is to consider the class of polynomials $P(z) = \sum_{\nu=0}^{n} a_\nu z^\nu$ having all its zeros in the disk $|z| \leq k$, $k \geq 1$, and to obtain the results that improve as well as generalize some Bernstein and Turán-type inequalities by using certain coefficients of $P(z)$. Further, we extend the obtained results to the polar derivative of a polynomial. As a consequence, we also derive two results that improve and generalize Erdös–Lax type inequalities for the class of polynomials having no zeros in the disk $|z| < k$, $k \leq 1$.

Keywords: Polynomials, inequalities, zeros, polar derivative

MSC 2010: Primary 30A10, 30C10, Secondary 30D15

1 Introduction

The study of extremal problems of functions and the results where some approaches to obtaining polynomial inequalities for various norms and with various constraints on using different methods of the geometric function theory is a classical topic in analysis. The Erdös–Lax and Turán-type inequalities relating the norm of the derivative and the polynomial itself, as well as generalizing the classical polynomial inequalities, play a key role in the literature for proving the inverse theorems in approximation theory, and of course have their own intrinsic interest. These inequalities for constrained polynomials have been the subject of many research papers, which is witnessed by many recent articles (for example, see [4–6, 8, 9] and [12–14]). According to well-known inequality of Bernstein [2] on the derivative of a polynomial $P(z)$ of degree n, we have

$$\max_{|z|=1}|P'(z)| \leq n \max_{|z|=1}|P(z)|.$$

The result is best possible and equality holds for a polynomial having all its zeros at the origin.

If $P(z)$ is a polynomial of degree n having no zero in $|z| < 1$, Erdös conjectured and later Lax [10] proved that

$$\max_{|z|=1}|P'(z)| \leq \frac{n}{2} \max_{|z|=1}|P(z)|. \tag{35.1}$$

Acknowledgement: The authors thank the anonymous referee for helpful comments.

Ishfaq Nazir, Department of Mathematics, School of Chemical Engineering and Physical Sciences, Lovely Professional University, Block 38-302, Phagwara, Punjab, India-144411, e-mail: ishfaqnazir02@gmail.com

https://doi.org/10.1515/9783111313634-035

The inequality (35.1) is best possible and equality holds for $P(z) = a + bz^n$, where $|a| = |b|$. As an extension of (35.1), Malik [11] proved that if $P(z) \neq 0$ in $|z| < k$, $k \geq 1$, then

$$\max_{|z|=1}|P'(z)| \leq \frac{n}{1+k}\max_{|z|=1}|P(z)|. \tag{35.2}$$

For the class of polynomials not vanishing in $|z| < k$, $k \leq 1$, the precise estimate of maximum $|P'(z)|$ on $|z| = 1$ is not easily obtainable. For quite some time it was believed that if $P(z) \neq 0$ in $|z| < k$, $k \leq 1$, then the inequality analogous to (35.2) should be

$$\max_{|z|=1}|P'(z)| \leq \frac{n}{1+k^n}\max_{|z|=1}|P(z)| \tag{35.3}$$

until Professor Saff gave the example $P(z) = (z - \frac{1}{2})(z + \frac{1}{3})$ to counter this belief. In 1980, it was shown by Govil [3] that (35.3) holds with an additional hypothesis and proved the following result.

Theorem 1. *Let $P(z) = \sum_{\nu=0}^{n} a_\nu z^\nu$ be a polynomial of degree n having no zero in $|z| < k$, $k \leq 1$ and $Q(z) = z^n\overline{P(1/\overline{z})}$. If $|P'(z)|$ and $|Q'(z)|$ attain maximum at the same point on $|z| = 1$, then*

$$\max_{|z|=1}|P'(z)| \leq \frac{n}{1+k^n}\max_{|z|=1}|P(z)|. \tag{35.4}$$

The result is best possible and equality holds in (35.4) for $P(z) = z^n + k^n$.

P. Turán [16] showed that if a polynomial $P(z)$ of degree n has all its zeros in $|z| \leq 1$, then

$$\max_{|z|=1}|P'(z)| \geq \frac{n}{2}\max_{|z|=1}|P(z)|. \tag{35.5}$$

Aziz [1] considered the modulus of each zero of the underlying polynomial in the bound and generalized the inequality (35.5) to the class of polynomials having all their zeros in a closed disk of finite radius greater than or equal to unit length by proving that, if $P(z) = a_n \prod_{\nu=1}^{n}(z - z_\nu)$ is complex polynomial of degree n with $|z_\nu| \leq k$, $k \geq 1$, then

$$\max_{|z|=1}|P'(z)| \geq \frac{2}{1+k^n}\sum_{\nu=1}^{n}\frac{k}{k+|z_\nu|}\max_{|z|=1}|P(z)|. \tag{35.6}$$

Recently, Kumar [8] have proved generalization of (35.6) to the class of polynomials having all their zeros in the disc $|z| \leq k$, $k \geq 1$, by obtaining the bound, which involves the modulus of each zero of the underlying polynomial. More precisely, Kumar proved that if $P(z) = \sum_{\nu=0}^{n} a_\nu z^\nu = a_n \prod_{\nu=1}^{n}(z - z_\nu)$ is a polynomial of degree n having all its zeros in $|z| \leq k$, $k \geq 1$, then

$$\max_{|z|=1}|P'(z)| \geq \left(\frac{2}{1+k^n} + \frac{(|a_n|k^n - |a_0|)(k-1)}{(1+k^n)(|a_n|k^n + k|a_0|)}\right)\sum_{\nu=1}^{n}\frac{k}{k+|z_\nu|}\max_{|z|=1}|P(z)|. \tag{35.7}$$

Since $k \geq 1$, therefore $\frac{k}{k+|z_\nu|} \geq \frac{1}{2}$ for $1 \leq \nu \leq n$, the above inequality (35.7) gives, in particular, the following result.

Theorem 2. *If $P(z) = \sum_{\nu=0}^{n} a_\nu z^\nu = a_n \prod_{\nu=1}^{n}(z - z_\nu)$ is a polynomial of degree n that has all its zeros in the disk $|z| \leq k$, $k \geq 1$, then*

$$\max_{|z|=1}|P'(z)| \geq \frac{n(k+1)}{2(1+k^n)}\left(\frac{|a_n|k^n + |a_0|}{|a_n|k^n + k|a_0|}\right)\max_{|z|=1}|P(z)|. \tag{35.8}$$

Kumar [8] recently established the following result for the class of polynomials that have no zero in $|z| < k$, $k \leq 1$. In fact, he proved that:

Theorem 3. *Let $P(z) = \sum_{\nu=0}^{n} a_\nu z^\nu$ be a polynomial of degree n having no zero in $|z| < k$, $k \leq 1$ and $Q(z) = z^n\overline{P(1/\overline{z})}$. If $|P'(z)|$ and $|Q'(z)|$ attain maximum at the same point on $|z| = 1$, then*

$$\max_{|z|=1}|P'(z)| \leq \left[n - \frac{n(k+1)k^n}{2(1+k^n)}\left(\frac{|a_n|k^n + |a_0|}{|a_n|k^n + |a_0|k}\right)\right]\max_{|z|=1}|P(z)|. \tag{35.9}$$

Equality holds in (35.9) *for $P(z) = z^n + k^n$.*

Definition. Let $p(z)$ be a polynomial of degree n with complex coefficients and $\alpha \in \mathbb{C}$ be a complex number, then the polynomial

$$D_\alpha p(z) = np(z) + (\alpha - z)p'(z)$$

is called polar derivative of $p(z)$ with pole α. Note that $D_\alpha p(z)$ is a polynomial of degree $n-1$ and it is a generalization of the ordinary derivative in the sense that

$$\lim_{\alpha\to\infty}\frac{D_\alpha p(z)}{\alpha} = p'(z)$$

uniformly with respect to z for $|z| \leq R$, $R > 0$. For more information on polar derivatives of polynomials one can refer to monographs by Rahman and Schmeisser [15] or Milovanović et al. [12].

Bernstein-type inequalities for complex polynomials have been extended extensively from 'ordinary derivative' to 'polar derivative' of complex polynomials. For the latest publications in this direction one can refer to [4, 5, 8] and [9]. In this context, Kumar [8] have recently proved an extension of Theorem 2 to the polar derivative of a polynomial with the same restrictions by proving the following result.

Theorem 4. *If $P(z) = \sum_{\nu=0}^{n} a_\nu z^\nu$ is a polynomial of degree n that has all its zeros in the disk $|z| \leq k$, $k \geq 1$, then for any complex number α with $|\alpha| \geq k$,*

$$\max_{|z|=1}|D_\alpha P(z)| \geq \frac{n(|\alpha| - k)(1+k)}{2(1+k^n)}\left(\frac{|a_n|k^n + |a_0|}{|a_n|k^n + |a_0|k}\right)\max_{|z|=1}|P(z)|. \tag{35.10}$$

2 Main results

We begin by establishing the following result, which not only yields generalization of Theorem 2 and inequality (35.6), but also provides the sharpening of them.

Theorem 5. *If $P(z) = \sum_{\nu=0}^{n} a_\nu z^\nu$ be a polynomial of degree n that has all its zeros in the disk $|z| \leq k$, $k \geq 1$, then for $0 \leq t \leq 1$, we have*

$$\max_{|z|=1}|P'(z)| \geq \frac{n(k+1)}{2(1+k^n)}\left(\frac{|a_n|k^n + |a_0| + tm}{|a_n|k^n + (|a_0| + tm)k}\right)\left(\max_{|z|=1}|P(z)| + tm\right), \tag{35.11}$$

where $m = \min_{|z|=k}|P(z)|$. Equality holds in (35.11) for $P(z) = z^n + k^n$.

Remark 1. Since $P(z) = \sum_{\nu=0}^{n} a_\nu z^\nu$ has all its zeros in the disk $|z| \leq k$, $k \geq 1$ and if $z_1, z_2, \ldots, z_n$ are the zeros of $P(z)$, then

$$\left|\frac{a_0}{a_n}\right| = |z_1 z_2 \ldots z_n| = |z_1||z_2|\ldots|z_n| \leq k^n.$$

As we see in the proof of Theorem 5 (given in next section), we have for every β with $|\beta| \leq 1$, the polynomial $P(z) + \beta m$ has all its zeros in the disk $|z| \leq k$, $k \geq 1$, hence

$$\left|\frac{a_0 + \beta m}{a_n}\right| \leq k^n. \tag{35.12}$$

If in (35.12), we choose the argument of β suitably, we get

$$|a_0| + |\beta| m \leq k^n |a_n|. \tag{35.13}$$

If we take $|\beta| = t$ in (35.13), so that $0 \leq t \leq 1$, we get $|a_0| + tm \leq k^n|a_n|$.

Remark 2. Except for the case when $P(z)$ has all its zeros on $|z| = k$, $k \geq 1$, the bound obtained in Theorem 5 is always sharper than the bound obtained from inequality (35.6) and for this it needs to show that

$$\frac{k+1}{2(1+k^n)}\left(\frac{|a_n|k^n + |a_0| + tm}{|a_n|k^n + (|a_0| + tm)k}\right) \geq \frac{1}{1+k^n},$$

which is equivalent to

$$|a_0| + tm \leq k^n|a_n|,$$

which is true by Remark 1.

Remark 3. For $t = 0$, Theorem 5 reduces to Theorem 2.

Remark 4. It is clearly seen that Theorem 5 includes Turán's inequality (35.5) for the case when $t = 0$ and $k = 1$.

Theorem 5 generally provides much better information than Theorem 2 and inequality (35.6) regarding $\max_{|z|=1}|P'(z)|$, in case when $P(z)$ has all its zeros in $|z| < k$, $k \geq 1$. We illustrate this with the help of following example.

Example 1. Consider $P(z) = z^3 - z^2 + z - 1$, which is polynomial of degree 3. Clearly, $P(z)$ has all its zeros in $|z| \leq 1$. We take $k = 2$ and find that

$$\min_{|z|=2}|P(z)| = 5 \quad \text{and} \quad \max_{|z|=1}|P(z)| = 4.$$

By using inequality (35.6), for this polynomial, we obtain that

$$\max_{|z|=1}|P'(z)| \geq 1.33$$

and by Theorem 2, for this polynomial, we obtain

$$\max_{|z|=1}|P'(z)| \geq 1.8.$$

While Theorem 5 (with t=1) gives

$$\max_{|z|=1}|P'(z)| \geq 3.15,$$

which is much better than the bound given by inequality (35.6) and Theorem 2.

Using Theorem 5, we prove the following generalization of Theorem 3.

Theorem 6. *Let $P(z) = \sum_{v=0}^{n} a_v z^v$ be a polynomial of degree n that has no zero in $|z| < k$, $k \leq 1$ and $Q(z) = z^n \overline{P(1/\overline{z})}$. If $|P'(z)|$ and $|Q'(z)|$ attain maximum at the same point on $|z| = 1$, then for $0 \leq t \leq 1$, we have*

$$\begin{aligned}\max_{|z|=1}|P'(z)| \leq {}& \left[n - \frac{n(k+1)k^n}{2(1+k^n)}\left(\frac{|a_n|k^n + |a_0| + tmk^n}{|a_n|k^n + |a_0|k + tmk^n}\right)\right]\max_{|z|=1}|P(z)| \\ & - \frac{n(k+1)k^n}{2(1+k^n)}\left(\frac{|a_n|k^n + |a_0| + tmk^n}{|a_n|k^n + |a_0|k + tmk^n}\right)tm,\end{aligned} \tag{35.14}$$

where $m = \min_{|z|=1/k}|Q(z)|$. Equality holds in (35.14) for $P(z) = z^n + k^n$.

Remark 5. For $t = 0$, Theorem 6 reduces to Theorem 3.

Remark 6. As seen from Example 1, the polynomial $P(z) = z^3 - z^2 + z - 1$, has all its zeros on $|z| = 1$. Further,

$$Q(z) = z^n \overline{P\left(\frac{1}{\overline{z}}\right)} = -P(z).$$

So that $|P'(z)|$ and $|Q'(z)|$ attain maximum at the same point on $|z| = 1$. We take $k = \frac{1}{2}$, so that $P(z) \neq 0$ in $|z| < k = \frac{1}{2}$ and we find that

$$\max_{|z|=1}|P(z)| = 4 \quad \text{and} \quad \min_{|z|=\frac{1}{1/2}}|Q(z)| = \min_{|z|=2}|Q(z)| = 5$$

and we obtain the following estimates

$$\max_{|z|=1}|P'(z)| \leq 10.66 \quad \text{(By Theorem 1)},$$
$$\max_{|z|=1}|P'(z)| \leq 10.20 \quad \text{(By Theorem 3)},$$
$$\max_{|z|=1}|P'(z)| \leq 8.85 \quad \text{(By Theorem 6)}.$$

We next prove the following extension of Theorem 5 to the polar derivative of a polynomial with the same restrictions.

Theorem 7. *If $P(z) = \sum_{v=0}^{n} a_v z^v$ is a polynomial of degree n that has all its zeros in the disk $|z| \leq k$, $k \geq 1$, then for any complex number α with $|\alpha| \geq k$, and $0 \leq t \leq 1$, we have*

$$\max_{|z|=1}|D_\alpha P(z)| \geq \frac{n(|\alpha|-k)(1+k)}{2(1+k^n)}\left(\frac{|a_n|k^n + |a_0| + tm}{|a_n|k^n + k(|a_0| + tm)}\right) \times \left(\max_{|z|=1}|P(z)| + tm\right) - tmn, \tag{35.15}$$

where $m = \min_{|z|=k}|P(z)|$.

Remark 7. For $t = 0$, Theorem 7 reduces to Theorem 4.

Remark 8. If we divide (35.15) by $|\alpha|$ and take $|\alpha| \to \infty$, we get Theorem 5 and thus Theorem 7 contains Theorem 5.

Remark 9. If we divide (35.15) by $|\alpha|$ and take $|\alpha| \to \infty$ and $t = 0$, we get Theorem 2 and thus Theorem 7 also contains Theorem 2 (Kumar [8]).

In the same way, let us extend Theorem 6 to the polar derivative of a polynomial as follows.

Theorem 8. *Let $P(z) = \sum_{v=0}^{n} a_v z^v$ be a polynomial of degree n that has no zero in $|z| < k$, $k \leq 1$ and $Q(z) = z^n\overline{P(1/\overline{z})}$. If $|P'(z)|$ and $|Q'(z)|$ attain maximum at the same point on $|z| = 1$, then for any complex number $\alpha \in \mathbb{C}$ with $|\alpha| \geq 1$ and $0 \leq t \leq 1$, we have*

$$\max_{|z|=1}|D_\alpha P(z)| \leq \left[n|\alpha| - \frac{n(|\alpha|-1)(k+1)k^n}{2(1+k^n)}\left(\frac{|a_n|k^n + |a_0| + tmk^n}{|a_n|k^n + |a_0|k + tmk^n}\right)\right]\max_{|z|=1}|P(z)| - \frac{n(|\alpha|-1)(k+1)k^n}{2(1+k^n)}\left(\frac{|a_n|k^n + |a_0| + tmk^n}{|a_n|k^n + |a_0|k + tmk^n}\right)tm, \tag{35.16}$$

where $m = \min_{|z|=1/k}|Q(z)|$. Equality holds in (35.16) for $P(z) = z^n + k^n$.

Remark 10. If we divide (35.16) by $|\alpha|$ and take $|\alpha| \to \infty$, we get Theorem 6 and thus Theorem 8 contains Theorem 6.

For $t = 0$, we obtain following result proved by Kummar [8].

Corollary 1. *Let $P(z) = \sum_{v=0}^{n} a_v z^v$ be a polynomial of degree n that has no zero in $|z| < k$, $k \le 1$ and $Q(z) = z^n\overline{P(1/\overline{z})}$. If $|P'(z)|$ and $|Q'(z)|$ attain maximum at the same point on $|z| = 1$, then for any complex number $\alpha \in \mathbb{C}$ with $|\alpha| \ge 1$, we have*

$$\max_{|z|=1}|D_\alpha P(z)| \le \left[n|\alpha| - \frac{n(|\alpha|-1)(k+1)k^n}{2(1+k^n)}\left(\frac{|a_n|k^n + |a_0|}{|a_n|k^n + |a_0|k}\right)\right]\max_{|z|=1}|P(z)|, \tag{35.17}$$

where $m = \min_{|z|=k}|P(z)|$. Equality holds in (35.17) for $P(z) = z^n + k^n$.

For the proof our results, we need the following lemma due to Govil and Rahman [7].

Lemma 1. *If $P(z)$ is a polynomial of degree n, then on $|z| = 1$,*

$$|P'(z)| + |Q'(z)| \le n \max_{|z|=1}|P(z)|,$$

where $Q(z) = z^n\overline{P(1/\overline{z})}$.

3 Proofs of theorems

Proof of Theorem 5. If $P(z) = \sum_{v=0}^{n} a_v z^v$ has a zero on $|z| = k$, then $m = \min_{|z|=k}|P(z)| = 0$ and the result follows from Theorem 2 in this case. Henceforth, we suppose that $P(z)$ has all its zeros in $|z| < k$, $k \ge 1$.

Let $H(z) = P(kz)$ and $G(z) = z^n\overline{H(1/\overline{z})} = z^n\overline{P(k/\overline{z})}$, then all the zeros of $G(z)$ lie in $|z| > 1$ and $|H(z)| = |G(z)|$ for $|z| = 1$. This gives

$$\left|z^n\overline{P\left(\frac{k}{\overline{z}}\right)}\right| = |P(kz)| \ge m \quad \text{for } |z| = 1.$$

It follows by minimum modulus principle, that

$$\left|z^n\overline{P\left(\frac{k}{\overline{z}}\right)}\right| \ge m \quad \text{for } |z| \le 1.$$

Replacing z by $\frac{1}{\overline{z}}$, it implies that

$$|P(kz)| \ge m|z|^n \quad \text{for } |z| \ge 1$$

or

$$|P(z)| \ge m\left|\frac{z}{k}\right|^n \quad \text{for } |z| \ge k. \tag{35.18}$$

Now, consider the polynomial

$$F(z) = P(z) + \beta m,$$

where β is complex number with $|\beta| \le 1$, then all the zeros of $F(z)$ lie in $|z| \le k$. Because if for some $z = z_1$, with $|z_1| > k$, we have

$$F(z_1) = P(z_1) + \beta m = 0,$$

then

$$|P(z_1)| = |\beta m| \le m < m\left|\frac{z_1}{k}\right|^n,$$

which contradicts (35.18). Hence, for every complex number β with $|\beta| \le 1$, the polynomial

$$F(z) = P(z) + \beta m = (a_0 + \beta m) + \sum_{\nu=1}^{n} a_\nu z^\nu$$

has all its zeros in $|z| \le k$, where $k \ge 1$. Applying Theorem 2, to the polynomial $F(z)$, we get for every complex number β with $|\beta| \le 1$ and $|z| = 1$

$$\max_{|z|=1}|P'(z)| \ge \frac{n(k+1)}{2(1+k^n)}\left(\frac{|a_n|k^n + |a_0 + \beta m|}{|a_n|k^n + k|a_0 + \beta m|}\right)|P(z) + \beta m|. \tag{35.19}$$

For every $\beta \in \mathbb{C}$, we have

$$|a_0 + \beta m| \le |a_0| + |\beta| m.$$

Since the function $(\frac{x+|a_n|k^n}{xk+|a_n|k^n})$ is decreasing for $k \ge 1$, it follows that from (35.19) that for every β with $|\beta| \le 1$ and $|z| = 1$

$$\max_{|z|=1}|P'(z)| \ge \frac{n(k+1)}{2(1+k^n)}\left(\frac{|a_n|k^n + |a_0| + |\beta| m}{|a_n|k^n + k(|a_0| + |\beta| m)}\right)|P(z) + \beta m|. \tag{35.20}$$

Choosing argument of β on R. H. S of (35.20) such that

$$|P(z) + \beta m| = |P(z)| + |\beta| m,$$

we obtain from (35.20) that

$$\max_{|z|=1}|P'(z)| \ge \frac{n(k+1)}{2(1+k^n)}\left(\frac{|a_n|k^n + |a_0| + |\beta| m}{|a_n|k^n + k(|a_0| + |\beta| m)}\right)\left(\max_{|z|=1}|P(z)| + |\beta| m\right), \tag{35.21}$$

which on taking $|\beta| = t$, so that $0 \le t \le 1$ gives in particular (35.11).

This completes the proof of the Theorem 5. □

Proof of Theorem 6. Since $P(z)$ has no zero in $|z| < k$, $k \leq 1$, then $Q(z) = z^n\overline{P(\frac{1}{\bar{z}})}$ has all its zeros in $|z| \leq \frac{1}{k}$, $\frac{1}{k} \geq 1$. Hence by Theorem 5 and using the fact that $\max_{|z|=1}|Q(z)| = \max_{|z|=1}|P(z)|$, we have

$$\begin{aligned}\max_{|z|=1}|Q'(z)| &\geq \frac{n(1+1/k)}{2(1+1/k^n)}\left(\frac{|a_0|/k^n+|a_n|+tm}{|a_n|/k+|a_0|/k^n+tm/k}\right)\left(\max_{|z|=1}|P(z)|+tm\right)\\ &= \frac{n(1+k)k^n}{2(1+k^n)}\left(\frac{|a_n|k^n+|a_0|+tmk^n}{|a_n|k^n+k|a_0|+tmk^n}\right)\left(\max_{|z|=1}|P(z)|+tm\right),\end{aligned} \tag{35.22}$$

where $m = \min_{|z|=1/k}|Q(z)|$.

From Lemma 1, we have on $|z| = 1$

$$|P'(z)| + |Q'(z)| \leq n\max_{|z|=1}|P(z)|. \tag{35.23}$$

Since $|P'(z)|$ and $|Q'(z)|$ attains maximum at the same time point, we have

$$\max_{|z|=1}\{|P'(z)| + |Q'(z)|\} = \max_{|z|=1}|P'(z)| + \max_{|z|=1}|Q'(z)|. \tag{35.24}$$

Therefore from (35.22), (35.23) and (35.24), we have

$$\begin{aligned}\max_{|z|=1}|P'(z)| \leq{}& \left[n - \frac{n(k+1)k^n}{2(1+k^n)}\left(\frac{|a_n|k^n+|a_0|+tmk^n}{|a_n|k^n+|a_0|k+tmk^n}\right)\right]\max_{|z|=1}|P(z)|\\ &- \frac{n(k+1)k^n}{2(1+k^n)}\left(\frac{|a_n|k^n+|a_0|+tmk^n}{|a_n|k^n+|a_0|k+tmk^n}\right)tm.\end{aligned}$$

This completes the proof of Theorem 6. □

Proof of Theorem 7. If $P(z) = \sum_{\nu=0}^{n} a_\nu z^\nu$ has a zero on $|z| = k$, then $m = \max_{|z|=k}|P(z)| = 0$ and the result follows from Theorem 4 in this case. Henceforth, we suppose that $P(z)$ has all its zeros in $|z| < k$, $k \geq 1$.

Let $H(z) = P(kz)$ and $G(z) = z^n\overline{H(1/\bar{z})} = z^n\overline{P(k/\bar{z})}$, then all the zeros of $G(z)$ lie in $|z| > 1$ and $|H(z)| = |G(z)|$ for $|z| = 1$. This gives

$$\left|z^n\overline{P\left(\frac{k}{\bar{z}}\right)}\right| = |P(kz)| \geq m \quad \text{for } |z| = 1.$$

It follows by minimum modulus principle that

$$\left|z^n\overline{P\left(\frac{k}{\bar{z}}\right)}\right| \geq m \quad \text{for } |z| \leq 1.$$

Replacing z by $\frac{1}{\bar{z}}$, it implies that

$$|P(kz)| \geq m|z|^n \quad \text{for } |z| \geq 1$$

or

$$|P(z)| \geq m\left|\frac{z}{k}\right|^n \quad \text{for } |z| \geq k. \tag{35.25}$$

Now, consider the polynomial

$$F(z) = P(z) + \beta m,$$

where β is complex number with $|\beta| \leq 1$, then all the zeros of $F(z)$ lie in $|z| \leq k$. Because if for some $z = z_1$ with $|z_1| > k$, we have

$$F(z_1) = P(z_1) + \beta m = 0,$$

then

$$|P(z_1)| = |\beta m| \leq m < \left|\frac{z}{k}\right|^n,$$

which contradicts (35.25). Hence, for every complex number β with $|\beta| \leq 1$, the polynomial

$$F(z) = P(z) + \beta m = (a_0 + \beta m) + \sum_{\nu=1}^{n} a_\nu z^\nu$$

has all its zeros in $|z| \leq k$, where $k \geq 1$. Applying Theorem 4, to the polynomial $F(z)$, we get for every complex number β with $|\beta| \leq 1$ and $|z| = 1$

$$\max_{|z|=1}|D_\alpha(P(z) + \beta m)| \geq \frac{n(|\alpha| - k)(1+k)}{2(1+k^n)}\left(\frac{|a_n|k^n + |a_0 + \beta m|}{|a_n|k^n + |a_0 + \beta m|k}\right)|P(z) + \beta m|. \tag{35.26}$$

For every $\beta \in \mathbb{C}$, we have

$$|a_0 + \beta m| \leq |a_0| + |\beta|m.$$

Since the function $(\frac{x+|a_n|k^n}{xk+|a_n|k^n})$ is decreasing for $k \geq 1$, it follows that from (35.26) that for every β with $|\beta| \leq 1$ and $|z| = 1$

$$\max_{|z|=1}|D_\alpha P(z) + \beta mn| \geq \frac{n(|\alpha| - k)(1+k)}{2(1+k^n)}\left(\frac{|a_n|k^n + |a_0| + |\beta|m}{|a_n|k^n + k(|a_0| + |\beta|m)}\right)|P(z) + \beta m|. \tag{35.27}$$

Choosing argument of β on R. H. S of (35.27) such that

$$|P(z) + \beta m| = |P(z)| + |\beta|m$$

we obtain from (35.27) that

$$\max_{|z|=1}|D_\alpha P(z)| + |\beta|mn \geq \frac{n(|\alpha|-k)(1+k)}{2(1+k^n)}\left(\frac{|a_n|k^n + |a_0| + |\beta|m}{|a_n|k^n + k(|a_0| + |\beta|m)}\right)(|P(z)| + |\beta|m).$$

Hence

$$\max_{|z|=1}|D_\alpha P(z)| + |\beta|mn \geq \frac{n(|\alpha|-k)(1+k)}{2(1+k^n)}\left(\frac{|a_n|k^n + |a_0| + |\beta|m}{|a_n|k^n + k(|a_0| + |\beta|m)}\right)\left(\max_{|z|=1}|P(z)| + |\beta|m\right),$$

which on taking $|\beta| = t$, so that $0 \leq t \leq 1$ gives in particular (35.15).

This completes the proof of Theorem 7. □

Proof of Theorem 8. Note that for any complex number α with $|\alpha| \geq 1$, we have on $|z| = 1$

$$\begin{aligned}|D_\alpha P(z)| &= |nP(z) + (\alpha - z)P'(z)|\\ &= |nP(z) - zP'(z) + \alpha P'(z)|\\ &\leq |nP(z) - zP'(z)| + |\alpha||P'(z)|\\ &= |Q'(z)| + |\alpha||P'(z)|\\ &= n\max_{|z|=1}|P(z)| - |P'(z)| + |\alpha||P'(z)| \quad \text{(by Lemma 1)}\\ &= n\max_{|z|=1}|P(z)| + (|\alpha| - 1)|P'(z)|.\end{aligned}$$

Therefore using Theorem 6, we have

$$\begin{aligned}\max_{|z|=1}|D_\alpha P(z)| \leq n\max_{|z|=1}|P(z)| + (|\alpha|-1)\left[n - \frac{n(k+1)k^n}{2(1+k^n)}\left(\frac{|a_n|k^n + |a_0| + tmk^n}{|a_n|k^n + |a_0|k + tmk^n}\right)\right]\max_{|z|=1}|P(z)|\\ - \frac{n(k+1)(|\alpha|-1)k^n}{2(1+k^n)}\left(\frac{|a_n|k^n + |a_0| + tmk^n}{|a_n|k^n + |a_0|k + tmk^n}\right)tm.\end{aligned}$$

That is

$$\begin{aligned}\max_{|z|=1}|D_\alpha P(z)| \leq \left[n|\alpha| - \frac{n(|\alpha|-1)(k+1)k^n}{2(1+k^n)}\left(\frac{|a_n|k^n + |a_0| + tmk^n}{|a_n|k^n + |a_0|k + tmk^n}\right)\right]\max_{|z|=1}|P(z)|\\ - \frac{n(|\alpha|-1)(k+1)k^n}{2(1+k^n)}\left(\frac{|a_n|k^n + |a_0| + tmk^n}{|a_n|k^n + |a_0|k + tmk^n}\right)tm. \qquad (35.28)\end{aligned}$$

This completes the proof of Theorem 8. □

Bibliography

[1] A. Aziz, Inequalities for the derivative of a Polynomial, *Proc. Am. Math. Soc.* **89** (1983), 259–266.

[2] S. N. Bernstein, *Leçons sur les propriétés extrémales et la meilleure approximation des fonctions analytiques d'une variable réelle*, Gauthier-Villars, Paris, 1926.

[3] N. K. Govil, On a theorem of Bernstein, *Proc. Natl. Acad. Soc. India* **50** (1980), 50–52.

[4] N. K. Govil and P. Kumar, On Bernstein-type inequalities for the polar derivative of a polynomial, *Prog. Approx. Theory Appl. Complex Anal. Springer Optim. Appl.* **117** (2017), 41–74.
[5] N. K. Govil and P. Kumar, On L p inequalities involving polar derivative of a polynomial, *Acta Math. Hung.* **152** (2017), 130–139.
[6] N. K. Govil and P. Kumar, On sharpening of an inequality of Turán, *Appl. Anal. Discrete Math.* **13** (2019), 711–720.
[7] N. K. Govil and Q. I. Rahman, Functions of exponential type not vanishing in a half-plane and related polynomials, *Trans. Am. Math. Soc.* **137** (1969), 501–517.
[8] P. Kumar, Some inequalities concerning polynomials, *Complex Anal. Oper. Theory* **14** (2020), 1–11.
[9] P. Kumar, On Zygmund-type inequalities involving polar derivative of a lacunary-type polynomial, *Bull. Math. Soc. Sci. Math. Roum.* **61** (2019), 155–164.
[10] P. D. Lax, Proof of a conjecture of P. Erdös on the derivative of a polynomial, *Bull. Am. Math. Soc.* **50** (1944), 509–513.
[11] M. Malik, On the derivative of a polynomial, *J. Lond. Math. Soc.* **1** (1969), 57–60.
[12] G. V. Milovanovic, D. S. Mitrinovic and ThM. Rassias, *Topics in Polynomials: Extremal Properties, Inequalities, Zero*, World scientific Publishing Co., Singapore, 1994.
[13] A. Mir, A Turán-type inequality for polynomials, *Indian J. Pure Appl. Math.* (2021).
[14] A. Mir, A note on Turán-type inequalities for polynomials, *Bull. Math. Soc. Sci. Math. Roum.* **64**(112) (2021), 197–203.
[15] Q. I. Rahman and G. Schmeisser, *Analytic Theory of Polynomials*, 2002.
[16] P. Turán, Über die Ableitung von Polynomen, *Compos. Math.* **7** (1939), 89–95.

C. Jeslyn Sathya and Carmel Richard

A weak and strong form of $Ng * b$-continuous function in nano topological space

Abstract: The purpose of this paper is to study the relationship between nano $g * b$ continuous function with the existing functions. The concepts are discussed with theorems and counter examples.

Keywords: Nano $g * b$-closed sets, nano $g * b$-open sets, nano $g * b$-continuity

MSC 2010: 16P20, 16P60, 16P70

1 Introduction

Continuity of functions is one of the core concepts of topology. In general, a continuous function is one, for which small changes in the input result in small changes in the output. In 1970, Levine [2] introduced the concept of generalized closed sets in topological spaces. The notion of nano topology was introduced by Lellis Thivagar [9], which was defined in terms of approximations and boundary region of a subset of a universe using an equivalence relation on it. He also established the weak forms of nano open sets, namely nano α-open sets, nano pre-open sets, nano semi-open set [9]. Bhuvaneshwari et al. introduced and studied nano generalized closed sets and their properties in nano topological spaces [1]. Mary et al. introduced and studied the properties of nano gb closed sets in nano topological spaces [3]. Parimala et al. [4] introduced nano b open sets in nano topological spaces and some basic properties are analyzed. Sathishmohan et al. introduced the concept of nano pre and nano semi pre-neighborhoods in nano topological spaces and proved some of its main results [6, 5]. Vidhya et al. introduced a new class of generalized closed set, called N$g * b$-closed set, in topological spaces and study their basic properties and some relations with existing ones are investigated [7]. Lellis Thivagar and Carmel Richard [8] introduced the concept of nano continuity.

The aim of this paper is to continue the study of nano generalized closed sets in nano topological spaces. In particular, we define nano $g * b$-continuity using the notion of nano $g * b$-closed sets and establish its relationship with other various weak and strong forms of continuity in nano topological spaces.

C. Jeslyn Sathya, Department of Mathematics, Lady Doak College, 15/9,VNS Illam,Temcy Colony, Old Villangudi,Near RMR Mahal, Madurai-625018, Tamilnadu, India, e-mail: jazzjessy23@gmail.com
Carmel Richard, Department of Mathematics, Lady Doak College, Madurai-625002, Tamilnadu, India

https://doi.org/10.1515/9783111313634-036

2 Preliminaries

Definition 1 ([9]). Let U be a non-empty finite set of objects, called the universe, and R be an equivalence relation on U, named as indiscernibility relation. Then U is divided into disjoint equivalence classes. Elements belonging to the same equivalence class are said to be indiscernible with one another.

The pair (U, R) is said to be the approximation space. Let $X \subseteq U$. Then,

1. The lower approximation of X with respect to R is the set of all objects, which can be for certain classified as X with respect to R and is denoted by $L_R(X)$. $L_R(X) = \cup_{x \in U}\{R(x) : R(x) \subseteq X\}$, where $R(x)$ denotes the equivalence class determined by $x \in U$.
2. The upper approximation of X with respect to R is the set of all objects that can be possibly classified as X with respect to R and is denoted by $U_R(X)$. $U_R(X) = \cup_{x \in U}\{R(x) : R(x) \cap X \neq \phi\}$.
3. The boundary region of x with respect to R is the set of all objects that can be classified neither as X nor as not-X with respect to R and it is denoted by $B_R(X)$. $B_R(X) = U_R(X) - L_R(X)$.

Proposition 1 ([9]). *If (U, R) is an approximation space and $X, Y \subseteq U$, then*

1. $L_R(X) \subseteq X \subseteq U_R(X)$
2. $L_R(\phi) = U_R(\phi) = \phi$
3. $L_R(U) = U_R(U) = U$
4. $U_R(X \cup Y) = U_R(X) \cup U_R(Y)$
5. $U_R(X \cap Y) \subseteq U_R(X) \cap U_R(Y)$
6. $L_R(X \cup Y) \supseteq L_R(X) \cup L_R(Y)$
7. $L_R(X \cap Y) = L_R(X) \cap L_R(Y)$
8. $L_R(X) \subseteq L_R(Y)$ *and* $U_R(X) \subseteq U_R(Y)$ *whenever* $X \subseteq Y$
9. $U_R(X^C) = [L_R(X)]^C$ *and* $L_R(X^C) = [U_R(X)]^C$
10. $U_R(U_R(X)) = L_R(U_R(X)) = U_R(X)$
11. $L_R(L_R(X)) = U_R(L_R(X)) = L_R(X)$.

Definition 2 ([9]). Let U be the universe, R be an equivalence relation on U and $\tau_R = \{U, \phi, L_R(X), U_R(X), B_R(X)\}$ where $X \subseteq U$. Then τ_R satisfies the following axioms

1. U and $\phi \in \tau_R$.
2. The union of the elements of any sub-collection of τ_R is in τ_R.
3. The intersection of the elements of any finite sub collection of τ_R is in $\tau_R(X)$.

Then τ_R is a topology on U called the nano topology on U with respect to X. We call (U, τ_R) as nano topological space. The elements of $\tau_R(X)$ are called as nano open sets. The complement of the nano open sets is called nano closed sets.

Remark 1. If τ_R is the nano topology on U with respect to X, then the set $B = \{U, L_R(X), B_R(X)\}$ is the basis for τ_R.

Definition 3 ([9]). If $(U, \tau_R(X))$ is a nano topological space with respect to X, where $X \subseteq U$ and if $A \subseteq U$, then

1. The nano interior of A is defined as the union of all nano-open subsets of A and is denoted by $Nint(A)$. That is, $Nint(A)$ is the largest nano-open subset of A.
2. The nano closure of A is defined as the intersection of all nano-closed sets containing A and is denoted by $Ncl(A)$. That is, $Ncl(A)$ is the smallest nano-closed set containing A.

Definition 4 ([7]). Let $(U, \tau_R(X))$ be a nano topological space and $A \subseteq U$. Then A is said to be

1. N semi open if $A \subseteq Ncl(Nint(A))$
2. N pre open if $A \subseteq Nint(Ncl(A))$
3. N regular open if $A = Nint(Ncl(A))$
4. Nb open if $A \subseteq Ncl(Nint(A)) \cup Nint(Ncl(A))$.

Example 1. Let $U = \{a, b, c, d\}$ with $U/R = \{\{a, d\}, \{b\}, \{c\}\}$. Let $X = \{a, b\} \subseteq U$. Then $\tau_R(X) = \{U, \phi, \{b\}, \{a, b, d\}, \{a, d\}\}$. Then nano b-open $= \{U, \phi, \{a\}, \{b\}, \{d\}, \{b, d\}, \{b, c\}, \{a, d\}, \{a, b\}, \{a, b, c\}, \{a, c, d\}, \{b, c, d\}, \{a, b, d\}\}$.

Definition 5 ([7]). Let $(U, \tau_R(X))$ be a nano topological space and $A \subseteq U$. Then A is said to be

1. Ng closed if $Ncl(A) \subseteq G$, whenever $A \subseteq G$ and G is N open in U.
2. N gb closed if $Nbcl(A) \subseteq G$, whenever $A \subseteq G$ and G is N open in U.

Example 2. Let $U = \{a, b, c, d\}$ with $U/R = \{\{a, d\}, \{b\}, \{c\}\}$. Let $X = \{a, b\} \subseteq U$. Then $\tau_R(X) = \{U, \phi, \{b\}, \{a, b, d\}, \{a, d\}\}$.

Then nano g-open $= \{U, \phi, \{a\}, \{b\}, \{d\}, \{b, d\}, \{a, d\}, \{a, b\}, \{b, c, d\}, \{a, b, d\}\}$.

Then nano gb-open $= \{U, \phi, \{a\}, \{b\}, \{d\}, \{a, b\}, \{a, d\}, \{b, c\}, \{c, d\}, \{a, b, c\}, \{a, b, d\}, \{a, c, d\}, \{b, c, d\}\}$.

Definition 6 ([7]). A subset A of nano topological space $(U, \tau_R(X))$ is known as nano generalized star b-closed if $Nbcl(A) \subseteq G$, whenever $A \subseteq G$, where G is Ng-open.

Example 3. Let $U = \{a, b, c, d\}$ with $U/R = \{\{a, d\}, \{b\}, \{c\}\}$. Let $X = \{a, b\} \subseteq U$.

Then $\tau_R(X) = \{U, \phi, , \{b\}, \{a, b, d\}, \{a, d\}\}$.

Then nano $g * b$-closed $= \{U, \phi, \{a\}, \{b\}, \{c\}, \{d\}, \{a, c\}, \{a, d\}, \{b, c\}, \{c, d\}, \{a, b, c\}, \{a, c, d\}, \{b, c, d\}\}$

Definition 7 ([8, 5]). Let $(U, \tau_R(X))$ and $(V, \tau'_R(Y))$ be nano topological spaces. Then a mapping $f : (U, \tau_R(X)) \to (V, \tau'_R(Y))$ is said to be

(i) nano continuous if $f^{-1}(B)$ is nano open in U for every nano-open set B in V.

(ii) nano α-continuous if $f^{-1}(B)$ is nano α-open in U for every nano-open set B in V.
(iii) nano semi-continuous if $f^{-1}(B)$ is nano semi-open in U for every nano-open set B in V.
(iv) nano pre-continuous if $f^{-1}(B)$ is nano pre-open in U for every nano-open set B in V.

3 Nano $g * b$-continuous

Definition 8. Let $(U, \tau_R(X))$ and $(V, \tau'_R(Y))$ be nano topological spaces. Then a mapping $f : (U, \tau_R(X)) \to (V, \tau'_R(Y))$ is nano b continuous on U if the inverse image of every nano open set in V is nano b-open in U.

Definition 9. Let $(U, \tau_R(X))$ and $(V, \tau'_R(Y))$ be nano topological spaces. Then a mapping $f : (U, \tau_R(X)) \to (V, \tau'_R(Y))$ is nano g-continuous on U if the inverse image of every nano open set in V is nano g-open in U.

Definition 10. Let $(U, \tau_R(X))$ and $(V, \tau'_R(Y))$ be nano topological spaces. Then a mapping $f : (U, \tau_R(X)) \to (V, \tau'_R(Y))$ is nano gb-continuous on U if the inverse image of every nano-open set in V is nano gb-open in U.

Definition 11. Let $(U, \tau_R(X))$ and $(V, \tau'_R(Y))$ be nano topological spaces. Then a mapping $f : (U, \tau_R(X)) \to (V, \tau'_R(Y))$ is nano $g * b$ continuous on U if the inverse image of every nano-open set in V is nano $g * b$-open in U.

Theorem 1. *Every nano continuous function is nano $g * b$-continuous.*

Proof. Let $f : (U, \tau_R(X)) \to (V, \tau'_R(Y))$ be a nano continuous function. Let M be nano closed set in V. Since f is nano continuous, $f^{-1}(M)$ is nano closed in X. As every nano closed is nano $g * b$-closed, $f^{-1}(M)$ is nano $g * b$-closed. Hence, f is nano $g * b$-continuous. □

Remark 2. The converse of the above theorem is not true, as seen from the following example.

Example 4. Let $U = \{a, b, c, d\}$ with $U/R = \{\{a, d\}, \{b\}, \{c\}\}$. Let $X = \{a, b\} \subseteq U$. Then $\tau_R(X) = \{U, \phi, \{b\}, \{a, b, d\}, \{a, d\}\}$ and $\tau_R^c(X) = \{U, \phi, \{a, c, d\}, \{c\}, \{b, c\}\}$. Let $V = \{x, y, z, w\}$ with $V/R = \{\{x\}, \{y, z\}, \{w\}\}$ and $Y = \{x, z\}$. Then $\tau'_R(Y) = \{V, \phi, \{x\}, \{x, y, z\}, \{y, z\}\}$ and $\tau'^c_R(Y) = \{V, \phi, \{y, z, w\}, \{w\}, \{x, w\}\}$. Define $f : U \to V$ as $f(a) = z, f(b) = y, f(c) = w, f(d) = x$. Then nano $g * b$-closed $= \{U, \phi, \{a\}, \{b\}, \{c\}, \{d\}, \{a, c\}, \{a, d\}, \{b, c\}, \{c, d\}, \{a, b, c\}, \{a, c, d\}, \{b, c, d\}\}$. Then $f^{-1}(\{y, z, w\}) = \{b, a, c\}, f^{-1}(\{w\}) = \{c\}$ and $f^{-1}(\{x, w\}) = \{d, c\}$. Then, f is nano $g * b$-continuous, but f is not nano continuous, since $f^{-1}(x, w) = \{d, c\}$ is not nano-closed in U. Thus, a nano $g * b$-continuous function is not nano-continuous.

Theorem 2. *Every nano pre-continuous function is nano $g * b$-continuous.*

Proof. Let $f : (U, \tau_R(X)) \to (V, \tau'_R(Y))$ be a nano pre-continuous function. Let M be nano closed set in V. Since f is nano pre-continuous, $f^{-1}(M)$ is nano pre-closed in U. As every nano pre-closed set is nano $g * b$-closed, $f^{-1}(M)$ is nano $g * b$-closed. Hence, f is nano $g * b$-continuous. □

Remark 3. The converse of the above theorem is not true, which can be shown by the following example.

Example 5. Let $U = \{a, b, c, d\}$ with $U/R = \{\{a, d\}, \{b\}, \{c\}\}$. Let $X = \{a, b\} \subseteq U$. Then $\tau_R(X) = \{U, \phi, \{b\}, \{a, b, d\}, \{a, d\}\}$ and $\tau_R^c(X) = \{U, \phi, \{a, c, d\}, \{c\}, \{b, c\}\}$. Let $V = \{x, y, z, w\}$ with $V/R = \{\{x\}, \{y, z\}, \{w\}\}$ and $Y = \{x, z\}$. Then $\tau'_R(Y) = \{V, \phi, \{x\}, \{x, y, z\}, \{y, z\}\}$ and $\tau_R'^c(Y) = \{V, \phi, \{y, z, w\}, \{w\}, \{x, w\}\}$. Define $f : U \to V$ as $f(a) = z, f(b) = y, f(c) = w, f(d) = x$. Then nano pre-closed $= \{U, \phi\{a, c, d\}, \{b, c\}\}$ and nano $g * b$-closed $= \{U, \phi, \{a\}, \{b\}, \{c\}, \{d\}, \{a, c\}, \{a, d\}, \{b, c\}, \{c, d\}, \{a, b, c\}, \{a, c, d\}, \{b, c, d\}\}$. Then $f^{-1}(\{y, z, w\}) = \{b, a, d\}, f^{-1}(\{w\}) = \{c\}$ and $f^{-1}(\{x, w\}) = \{d, c\}$. Then, f is nano $g *$ b-continuous, but f is not nano pre continuous, since $f^{-1}(x, w) = \{d, c\}$ not nano pre-closed in U. Thus, a nano $g * b$-continuous function is not nano pre-continuous.

Theorem 3. *Every nano semi-continuous function is nano $g * b$-continuous.*

Proof. Let $f : (U, \tau_R(X)) \to (V, \tau'_R(Y))$ be a nano semi-continuous function. Let M be nano closed set in V. Since f is nano semi-continuous, $f^{-1}(M)$ is nano semi-closed in X. As every nano semi-closed is nano $g * b$-closed, $f^{-1}(M)$ is nano $g * b$-closed. Hence, f is nano $g * b$-continuous. □

Remark 4. The converse of the above theorem is not true, as seen from the following example.

Example 6. Let $U = \{a, b, c, d\}$ with $U/R = \{\{a, d\}, \{b\}, \{c\}\}$. Let $X = \{a, b\} \subseteq U$. Then $\tau_R(X) = \{U, \phi, \{b\}, \{a, b, d\}, \{a, d\}\}$ and $\tau_R^c(X) = \{U, \phi, \{a, c, d\}, \{c\}, \{b, c\}\}$. Let $V = \{x, y, z, w\}$ with $V/R = \{\{x\}, \{y, z\}, \{w\}\}$ and $Y = \{x, z\}$. Then $\tau'_R(Y) = \{V, \phi, \{x\}, \{x, y, z\}, \{y, z\}\}$ and $\tau_R'^c(Y) = \{V, \phi, \{y, z, w\}, \{w\}, \{x, w\}\}$. Define $f : U \to V$ as $f(a) = z, f(b) = y, f(c) = w$, $f(d) = x$. Then nano semi-closed $= \{U, \phi, \{a, d\}, \{b\}\}$ and nano $g * b$-closed $= \{U, \phi, \{a\}, \{b\}, \{c\}, \{d\}, \{a, c\}, \{a, d\}, \{b, c\}, \{c, d\}, \{a, b, c\}, \{a, c, d\}, \{b, c, d\}\}$. Then $f^{-1}(\{y, z, w\}) = \{b, a, d\}$, $f^{-1}(\{w\}) = \{c\}$ and $f^{-1}(\{x, w\}) = \{d, c\}$. Then, f is nano $g * b$-continuous, but f is not nano semi-continuous since $f^{-1}(x, w) = \{d, c\}$ is not nano semi-closed in U. Thus, a nano $g * b$-continuous function is not nano semi-continuous

Theorem 4. *Every nano α-continuous function is nano $g * b$-continuous.*

Proof. Let $f : (U, \tau_R(X)) \to (V, \tau'_R(Y))$ be a nano α-continuous function. Let M be nano closed set in V. Since f is nano α-continuous $f^{-1}(M)$ is nano α-closed in X. As every nano α-closed is nano $g * b$-closed, $f^{-1}(M)$ is nano $g * b$-closed. Hence, f is nano $g * b$-continuous. □

Remark 5. The converse of the above theorem is not true, as seen from the following example.

Example 7. Let $U = \{a,b,c,d\}$ with $U/R = \{\{a,d\},\{b\},\{c\}\}$. Let $X = \{a,b\} \subseteq U$. Then $\tau_R(X) = \{U,\phi,\{b\},\{a,b,d\},\{a,d\}\}$ and $\tau_R^c(X) = \{U,\phi,\{a,c,d\},\{c\},\{b,c\}\}$. Let $V = \{x,y,z,w\}$ with $V/R = \{\{x\},\{y,z\},\{w\}\}$ and $Y = \{x,z\}$. Then $\tau_R'(Y) = \{V,\phi,\{x\},\{x,y,z\},\{y,z\}\}$ and $\tau_R'^c(Y) = \{V,\phi,\{y,z,w\},\{w\},\{x,w\}\}$. Define $f : U \to V$ as $f(a) = z, f(b) = y, f(c) = w, f(d) = x$. Then nano α-closed $= \{U,\phi,\{b,c\},\{c\}\}$ and nano $g*b$-closed $= \{U,\phi,\{a\},\{b\},\{c\},\{d\},\{a,c\},\{a,d\},\{b,c\},\{c,d\},\{a,b,c\},\{a,c,d\},\{b,c,d\}\}$. Then $f^{-1}(\{y,z,w\}) = \{b,a,d\}$, $f^{-1}(\{w\}) = \{c\}$ and $f^{-1}(\{x,w\}) = \{d,c\}$. Then, f is nano $g*b$-continuous but f is not nano α-continuous since $f^{-1}(x,w) = \{d,c\}$ is not nano α-closed in U. Thus, a nano $g*$ b-continuous function is not nano α-continuous.

Theorem 5. *Every nano b-continuous function is nano $g*b$-continuous.*

Proof. Let $f : (U,\tau_R(X)) \to (V,\tau_R'(Y))$ be a nano b-continuous function. Let M be nano closed set in V. Since f is nano b-continuous $f^{-1}(M)$ is nano b-closed in X. As every nano b-closed is nano $g*b$-closed, $f^{-1}(M)$ is nano $g*b$-closed. Hence, f is nano $g*b$-continuous. □

Remark 6. The converse of the above theorem is not true, as seen from the following example.

Example 8. Let $U = \{a,b,c,d\}$ with $U/R = \{\{a,d\},\{b\},\{c\}\}$. Let $X = \{a,b\} \subseteq U$. Then $\tau_R(X) = \{U,\phi,\{b\},\{a,b,d\},\{a,d\}\}$ and $\tau_R^c(X) = \{U,\phi,\{a,c,d\},\{c\},\{b,c\}\}$. Let $V = \{x,y,z,w\}$ with $V/R = \{\{x\},\{y,z\},\{w\}\}$ and $Y = \{x,z\}$. Then $\tau_R'(Y) = \{V,\phi,\{x\},\{x,y,z\},\{y,z\}\}$ and $\tau_R'^c(Y) = \{V,\phi,\{y,z,w\},\{w\},\{x,w\}\}$. Define $f : U \to V$ as $f(a) = z, f(b) = y, f(c) = w, f(d) = x$. Then nano b-closed $= \{U,\phi,\{a\},\{b\},\{c\},\{d\},\{a,c\},\{a,d\},\{b,c\},\{c,d\},\{a,b,c\},\{a,c,d\},\{b,c,d\}\}$ and nano $g*b$-closed $= \{U,\phi,\{a\},\{b\},\{c\},\{d\},\{a,c\},\{a,d\},\{b,c\},\{c,d\},\{a,b,c\},\{a,c,d\},\{b,c,d\}\}$. Then $f^{-1}(\{y,z,w\}) = \{b,a,d\}$, $f^{-1}(\{w\}) = \{c\}$ and $f^{-1}(\{x,w\}) = \{d,c\}$. Then, f is nano $g*b$-continuous and f is not nano b-continuous, $f^{-1}(\{y,z,w\}) = \{b,a,d\}$ is not nano b-closed in U. Hence every nano $g*b$-continuous function is not nano b-continuous.

Theorem 6. *Every nano g-continuous function is nano $g*b$-continuous.*

Proof. Let $f : (U,\tau_R(X)) \to (V,\tau_R'(Y))$ be a nano g-continuous function. Let M be nano closed set in V. Since f is nano g-continuous $f^{-1}(M)$ is nano g-closed in X. As every nano g-closed is nano $g*b$-closed, $f^{-1}(M)$ is nano $g*b$-closed. Hence, f is nano $g*b$-continuous. □

Remark 7. The converse of the above theorem is not true, as seen from the following example.

Example 9. Let $U = \{a,b,c,d\}$ with $U/R = \{\{a,d\},\{b\},\{c\}\}$. Let $X = \{a,b\} \subseteq U$. Then $\tau_R(X) = \{U,\phi,\{b\},\{a,b,d\},\{a,d\}\}$ and $\tau_R^c(X) = \{U,\phi,\{a,c,d\},\{c\},\{b,c\}\}$. Let $V = \{x,y,z,w\}$

with $V/R = \{\{x\}, \{y,z\}, \{w\}\}$ and $Y = \{x,z\}$. Then $\tau'_R(Y) = \{V, \phi, \{x\}, \{x,y,z\}, \{y,z\}\}$ and $\tau'^{c}_R(Y) = \{V, \phi, \{y,z,w\}, \{w\}, \{x,w\}\}$. Define $f : U \to V$ as $f(a) = z, f(b) = y, f(c) = w$, $f(d) = x$. Then nano g-closed $= \{U, \phi, \{a\}, \{b,c\}, \{c\}, \{a,c\}, \{c,d\}, \{a,b,c\}, \{a,c,d\}, \{b,c,d\}\}$ and nano $g * b$-closed $= \{U, \phi, \{a\}, \{b\}, \{c\}, \{d\}, \{a,c\}, \{a,d\}, \{b,c\}, \{c,d\}, \{a,b,c\}, \{a,c,d\}, \{b,c,d\}\}$. Then $f^{-1}(\{y,z,w\}) = \{b,a,d\}, f^{-1}(\{w\}) = \{c\}$ and $f^{-1}(\{x,w\}) = \{d,c\}$. Then, f is nano $g * b$-continuous, but f is not nano g-continuous, since $f^{-1}(y,z,w) = \{a,b,d\}$ is not nano g-closed in U. Thus, a nano $g * b$-continuous function is not nano g-continuous.

Theorem 7. *Every nano gp-continuous function is nano $g * b$-continuous.*

Proof. Let $f : (U, \tau_R(X)) \to (V, \tau'_R(Y))$ be a nano gp-continuous function. Let M be nano closed set in V. Since f is nano *gp*-continuous $f^{-1}(M)$ is nano *gp*-closed in X. As every nano *gp*-closed is nano $g * b$-closed, $f^{-1}(M)$ is nano $g * b$-closed. Hence, f is nano $g *$ b-continuous. □

Remark 8. The converse of the above theorem is not true, as seen from the following example.

Example 10. Let $U = \{a,b,c,d\}$ with $U/R = \{\{a,d\}, \{b\}, \{c\}\}$. Let $X = \{a,b\} \subseteq U$. Then $\tau_R(X) = \{U, \phi, \{b\}, \{a,b,d\}, \{a,d\}\}$ and $\tau^c_R(X) = \{U, \phi, \{a,c,d\}, \{c\}, \{b,c\}\}$. Let $V = \{x,y,z,w\}$ with $V/R = \{\{x\}, \{y,z\}, \{w\}\}$ and $Y = \{x,z\}$. Then $\tau'_R(Y) = \{V, \phi, \{x\}, \{x,y,z\}, \{y,z\}\}$ and $\tau'^{c}_R(Y) = \{V, \phi, \{y,z,w\}, \{w\}, \{x,w\}\}$. Define $f : U \to V$ as $f(a) = z, f(b) = y, f(c) = w, f(d) = x$. Then nano *gp*-closed $= \{U, \phi, \{c\}, \{a,c\}, \{c,d\}, \{b,c\}, \{c,d\}, \{a,b,c\}, \{a,c,d\}, \{b,c,d\}\}$ and nano $g * b$-closed $= \{U, \phi, \{a\}, \{b\}, \{c\}, \{d\}, \{a,c\}, \{a,d\}, \{b,c\}, \{c,d\}, \{a,b,c\}, \{a,c,d\}, \{b,c,d\}\}$. Then $f^{-1}(\{y,z,w\}) = \{b,a,d\}, f^{-1}(\{w\}) = \{c\}$ and $f^{-1}(\{x,w\}) = \{d,c\}$. Then, f is nano $g * b$-continuous but f is not nano *gp*-continuous since $f^{-1}(y,z,w) = \{a,b,d\}$ is not nano *gp*-closed in U. Thus, a nano $g * b$-continuous function is not nano *gp*-continuous

Theorem 8. *Every nano gs-continuous function is nano $g * b$-continuous.*

Proof. Let $f : (U, \tau_R(X)) \to (V, \tau'_R(Y))$ be a nano *gs*-continuous function. Let M be nano *gs*-closed set in V. Since f is nano *gs*-continuous, $f^{-1}(M)$ is nano *gs*-closed in X. As every nano *gs*-closed is nano $g * b$-closed, $f^{-1}(M)$ is nano $g * b$-closed. Hence, f is nano $g *$ b-continuous. □

Remark 9. The converse of the above theorem is not true, as seen from the following example.

Example 11. Let $U = \{a,b,c,d\}$ with $U/R = \{\{a,d\}, \{b\}, \{c\}\}$. Let $X = \{a,b\} \subseteq U$. Then $\tau_R(X) = \{U, \phi, \{b\}, \{a,b,d\}, \{a,d\}\}$ and $\tau^c_R(X) = \{U, \phi, \{a,c,d\}, \{c\}, \{b,c\}\}$. Let $V = \{x,y,z,w\}$ with $V/R = \{\{x\}, \{y,z\}, \{w\}\}$ and $Y = \{x,z\}$. Then $\tau'_R(Y) = \{V, \phi, \{x\}, \{x,y,z\}, \{y,z\}\}$ and $\tau'^{c}_R(Y) = \{V, \phi, \{y,z,w\}, \{w\}, \{x,w\}\}$. Define $f : U \to V$ as $f(a) = z, f(b) = y, f(c) = w$, $f(d) = x$. Then nano *gs*-closed $= \{U, \phi, \{a\}, \{b\}, \{c\}, \{d\}, \{a,c\}, \{a,d\}, \{b,c\}\{c,d\}, \{b,d\}, \{a,b,c\}, \{a,c,d\}, \{b,c,d\}\}$ and nano $g * b$-closed $= \{U, \phi, \{a\}, \{b\}, \{c\}, \{d\}, \{a,c\}, \{a,d\}, \{b,c\}, \{c,d\}, \{a,b,c\}, \{a,c,d\}, \{b,c,d\}\}$. Then $f^{-1}(\{y,z,w\}) = \{b,a,d\}, f^{-1}(\{w\}) = \{c\}$ and

$f^{-1}(\{x, w\}) = \{d, c\}$. Then, f is nano $g * b$-continuous, but f is not nano gs-continuous, since $f^{-1}(y, z, w) = \{a, b, d\}$ is not nano gs-closed in U. Thus, a nano $g * b$-continuous function is not nano gs-continuous

Theorem 9. *Every nano ga-continuous function is nano $g * b$-continuous.*

Proof. Let $f : (U, \tau_R(X)) \to (V, \tau'_R(Y))$ be a nano ga-continuous function. Let M be nano closed set in V. Since f is nano ga-continuous $f^{-1}(M)$ is nano ga-closed in X. As every nano ga-closed is nano $g * b$-closed, $f^{-1}(M)$ is nano $g * b$-closed. Hence, f is nano $g * b$-continuous. □

Remark 10. The converse of the above theorem is not true, as seen from the following example.

Example 12. Let $U = \{a, b, c, d\}$ with $U/R = \{\{a, d\}, \{b\}, \{c\}\}$. Let $X = \{a, b\} \subseteq U$. Then $\tau_R(X) = \{U, \phi, \{b\}, \{a, b, d\}, \{a, d\}\}$ and $\tau_R^c(X) = \{U, \phi, \{a, c, d\}, \{c\}, \{b, c\}\}$. Let $V = \{x, y, z, w\}$ with $V/R = \{\{x\}, \{y, z\}, \{w\}\}$ and $Y = \{x, z\}$. Then $\tau'_R(Y) = \{V, \phi, \{x\}, \{x, y, z\}, \{y, z\}\}$ and $\tau_R'^c(Y) = \{V, \phi, \{y, z, w\}, \{w\}, \{x, w\}\}$. Define $f : U \to V$ as $f(a) = z, f(b) = y, f(c) = w, f(d) = x$. Then nano ga-closed $= \{U, \phi, \{c\}, \{d\}, \{a, c\}, \{a, d\}, \{b, c\}\{c, d\}, \{b, d\}, \{a, b, c\}, \{a, c, d\}, \{b, c, d\}\}$ and nano $g * b$-closed $= \{U, \phi, \{a\}, \{b\}, \{c\}, \{d\}, \{a, c\}, \{a, d\}, \{b, c\}, \{c, d\}, \{a, b, c\}, \{a, c, d\}, \{b, c, d\}\}$. Then $f^{-1}(\{y, z, w\}) = \{b, a, d\}, f^{-1}(\{w\}) = \{c\}$ and $f^{-1}(\{x, w\}) = \{d, c\}$. Then, f is nano $g * b$-continuous, but f is not nano ga-continuous, since $f^{-1}(y, z) = \{a, b, d\}$ is not nano ga-closed in U. Thus, a nano $g * b$-continuous function is not nano ga-continuous

Theorem 10. *Every nano gb-continuous function is nano $g * b$-continuous.*

Proof. Let $f : (U, \tau_R(X)) \to (V, \tau'_R(Y))$ be a nano $g * b$-continuous function. Let M be nano closed set in V. Since f is nano gb-continuous $f^{-1}(M)$ is nano gb-closed in X. As every nano gb-closed is nano $g * b$-closed, $f^{-1}(M)$ is nano $g * b$-closed. Hence, f is nano $g * b$-continuous. □

Remark 11. The converse of the above theorem is not true, as seen from the following example.

Example 13. Let $U = \{a, b, c, d\}$ with $U/R = \{\{a, d\}, \{b\}, \{c\}\}$. Let $X = \{a, b\} \subseteq U$. Then $\tau_R(X) = \{U, \phi, \{b\}, \{a, b, d\}, \{a, d\}\}$ and $\tau_R^c(X) = \{U, \phi, \{a, c, d\}, \{c\}, \{b, c\}\}$. Let $V = \{x, y, z, w\}$ with $V/R = \{\{x\}, \{y, z\}, \{w\}\}$ and $Y = \{x, z\}$. Then $\tau'_R(Y) = \{V, \phi, \{x\}, \{x, y, z\}, \{y, z\}\}$ and $\tau_R'^c(Y) = \{V, \phi, \{y, z, w\}, \{w\}, \{x, w\}\}$. Define $f : U \to V$ as $f(a) = z, f(b) = y, f(c) = w, f(d) = x$. Then nano gb-closed $= \{U, \phi\{a\}, \{b\}, \{c\}, \{d\}, \{a, b\}, \{a, d\}, \{b, c\}, \{c, d\}\{a, c, d\}, \{a, b, c\}, \{b, c, d\}\}$ and nano $g * b$-closed $= \{U, \phi, \{a\}, \{b\}, \{c\}, \{d\}, \{a, c\}, \{a, d\}, \{b, c\}, \{c, d\}, \{a, b, c\}, \{a, c, d\}, \{b, c, d\}\}$. Then $f^{-1}(\{y, z, w\}) = \{b, a, d\}, f^{-1}(\{w\}) = \{c\}$ and $f^{-1}(\{x, w\}) = \{d, c\}$. Then, f is nano $g*b$-continuous, but f is not nano gb-continuous, since $f^{-1}(y, z, w) = \{a, b, d\}$ is not nano gb-closed in U. Thus, a nano $g * b$-continuous function is not nano gb-continuous.

The above results are summarized in the following diagram.

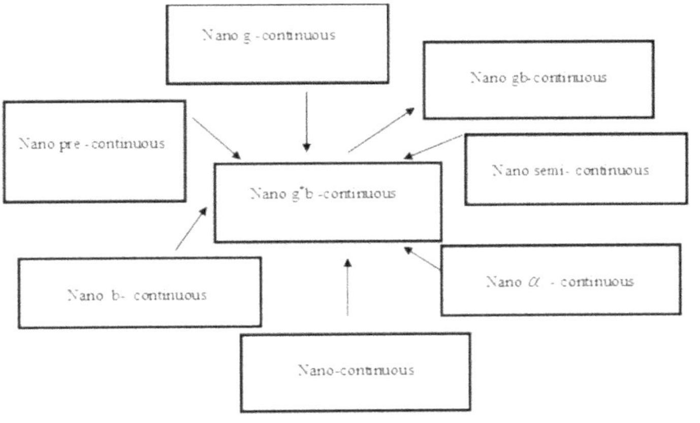

4 Conclusion

The study of nano $g * b$-continuous function is done using the concept of $g * b$-closed set. This can be extended to different types of continuity in nano topological spaces.

Bibliography

[1] K. Bhuvaneswari and K. M. Gnanapriya, Nano generalized closed sets in nano topological spaces, *Int. J. Sci. Res. Publ.* **4**(5) (2014), 1–3.

[2] N. Levine, Generalized closed sets in topology, *Rend. Circ. Mat. Palermo* **19** (1970), 89–96.

[3] A. Mary and I. Arockiarani, Properties of nano gb-closed maps, *IOSR J. Math.* **11**(2) (2015), 21–24.

[4] M. Parimala, C. Indirani and S. Jafari, On nano b-open sets in nano topological spaces, *Jordan J. Math. Stat.* **9**(3) (2016), 173–184.

[5] P. Sathishmohan, V. Rajendran, A. Devika and R. Vani, On nano semi continuity and nano pre continuity, *Int. J. Appl. Res.* **3**(2) (2017), 76–79.

[6] P. Sathishmohan, V. Rajendran, P. Dhanasekaran and C. V. Kumar, More on nano pre-neighbourhoods in nano topological-spaces, *J. Appl. Sci. Comput.* **5**(11) (2018), 899–907.

[7] P. Sathishmohan, V. Rajendran, C. V. Kumar and P. Dhanasekaran, On nano semi pre neighbourhoods in nano topological spaces, *Malaya J. Mat.* **6**(01) (2018), 294–298.

[8] M. Thivagar and C. Richard, On nano continuity, *Math. Theory Model.* **3**(7) (2013), 7–32.

[9] M. L. Thivagar and C. Richard, On nano forms of weakly open sets, *Int. J. Math. Stat. Invent.* **1**(1) (2013), 31–37.

R. Singh, Suditi, M. K. Patel, N. Bhardwaj, and A. K. Singh

Securing networks with graph theory: an approach to cyber security

Abstract: In light of increasing cyber threats, this study introduces algorithms by integrating graph theory and Python programming. This paper introduces two novel algorithms, developed through Python, that uses graph theory to use encryption mechanisms, providing a dual layer of security by integrating adjacency matrix-based encryption with traditional methods. Algorithm's working is shown through an example. Another algorithm for anomaly detection is proposed, which is supported by an example of having 20 nodes. The article presents the algorithm's development and implementation in securing network communications against cyber threats. This research works toward developing more resilient network security infrastructures by bridging the gap between theoretical graph concepts and their practical implementation in cybersecurity. The contributions lie not only in the algorithm's design, but also in its demonstration of graph theory's versatility in addressing cybersecurity challenges.

Keywords: Cyber security, graph theory, Python algorithm, adjacency matrix, encryption

MSC 2010: 94C15, 68P25

1 Introduction

With more developers than ever utilising it for the IT sector, Python has developed into one of the top programming languages in the world [8]. Yet, due to the robust packages it provides, Python is primarily employed in the cyber security industry. Conventional programming languages like C, C++ and others compel programmers to focus on supporting functions and data structures rather than algorithm design [12]. Python is an algorithm-oriented language that is desperately needed in the cyber security education field. Our research shows that the use of analyzed algorithms in Python, enables the creation of

R. Singh, Amity Institute of Applied Sciences, Amity University Uttar Pradesh, Lincoln - C, 801, Grand Omaxe, Sector-93 B, 201303 Noida, UP, India, e-mail: rsingh7@amity.edu
Suditi, Amity Institute of Applied Sciences, Amity University Uttar Pradesh, 201303 Noida, UP, India, e-mail: learnpython008@gmail.com
M. K. Patel, Department of Mathematics, NIT Nagaland, Dimapur, India
N. Bhardwaj, Department of Mathematics, SSBSR, Sharda University, 1059 Wisteria, Gaur Saundaryam, near GAur City Mall, 201318 Greater Noida West, Uttar Pradesh, India, e-mail: nehabhr1807@gmail.com
A. K. Singh, Department of Applied Sciences and Humanities, KNIT, KNI Kasba, 228118 Sultanpur, U.P., India, e-mail: akhilesh.singh@knit.ac.in

https://doi.org/10.1515/9783111313634-037

adaptable and useful techniques and algorithms that address the majority of current cyber security concerns. At the core of our framework is an algorithm, which investigates the coding using graphs. Because threats and technology are constantly evolving, if cyber security professionals do not adapt their skills accordingly, they will lose their relevance and be unable to provide the essential defenses that businesses are becoming increasingly dependent on [5]. One area of research that has gained attention is the application of linear algebra theory in networked control systems, specifically for detecting stochastic cyber-attacks, as discussed in [14]. In addition, mathematical modeling and simulation have been explored for cyber defense, as highlighted in [18]. The challenges and issues related to modern network security are discussed in [20].

Other research works have explored various topics related to cyber security, such as information security technology of big data systems [23], evaluating IT security investments [3], security features in databases [2] and encryption algorithms using graph theory [15]. Graph theory has also been explored for network security, including interference reduction in wireless networks [7], authentication schemes [1] and network security [19].

The motivation behind this article is: as cybersecurity threats become more sophisticated, exploiting these vulnerabilities, the limitations of traditional security measures become evident. Graph theory is on one branches of mathematics having profound ability to model complex network relationships and Python's versatility in implementing these theoretical models into practical security tools. Together, they offer a promising avenue for developing advanced cybersecurity mechanisms capable of countering the evolving threats in our interconnected world. This synergy of graph theory and Python programming enhances our defense capabilities and paves the way for innovative security strategies.

While effective to a certain extent, traditional encryption algorithms often fall short against complex intrusion tactics used by modern cybercriminals. The application of graph theory in cybersecurity, particularly in encryption and anomaly detection, has not been fully explored. This gap highlights the need for novel methodologies that use mathematical principles to enhance encryption techniques and anomaly detection mechanisms in network security.

As a future scope, we propose fuzzy graph theory that applies fuzzy sets to graph theory. Fuzzy graph theory allows for more flexible representation of uncertainty in graph-related problems, such as graph coloring or shortest path algorithms. Fuzzy sets have been used to generalize various concepts, and generalization in nearness spaces [10, 9, 21, 17]. To enhance the precision and efficacy of research on compound cyber attacks, [22] integrates a knowledge graph for the detection of attacks and develops a cybersecurity knowledge graph informed by knowing previously identified attacks. The CVSS is employed to quantify the exploitability and severity of each vulnerability [6]. A study on real data sets regarding serious cyber attacks happened before and during the pandemic due to Covid-19 is done in [13]. Furthermore, Li et al. [4] have recently explored the application of deep learning techniques for the identification of threats through the analysis of system logs. The concept of number theory used in public cipher can be referred from [11].

The article is structured as follows: In Section 2, we study the mathematical principles and methods of graph theory that are relevant to cyber security, specifically those involved in the process. It is assumed that the readers are familiar with Python programming, as this knowledge is required for the implementation of the proposed model discussed in this article. Section 3, focused around the presentation of our proposed model, which is implemented using python programming. We provide a detailed description of the workings of the algorithm and demonstrate its effectiveness. Finally, in Section 4, we discuss the motivation behind our research and offer concluding remarks on the potential applications of graph theory in cyber security.

2 Mathematical principles and methods in graph theory for cyber security

Graph theory, defined in [16], explores the properties and characteristics of graphs. Graphs are mathematical structures employed to illustrate the pairwise connections among objects. Essentially, a graph is composed of vertices (also referred to as nodes or points) linked by edges (also known as connections or lines) (as shown in Figure 37.1). The manner in which edges connect vertices distinguishes undirected graphs from directed graphs. In undirected graphs, edges connect vertices symmetrically, while in directed graphs, edges connect vertices asymmetrically.

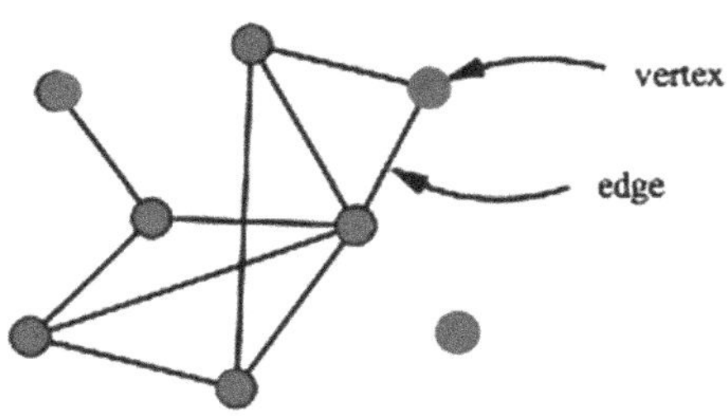

Figure 37.1: Graph showing vertex and edges [6].

A graph is an ordered pair consisting of (i) a set of vertices (also known as nodes or points) and (ii) a set of edges (also known as links or lines), which are unordered pairs of vertices (i. e., an edge is associated with two distinct vertices) [5].

Graph theory provides a mathematical framework for analyzing networks and relationships between objects, and it has many applications in cyber security. Here are some of the main principles and methods in graph theory that are important for cyber security:

2.1 Graph models: Graphs are used to model network topologies and relationships between nodes, such as computers, routers, and servers. These models can be used to analyze the vulnerability of a network to cyber-attacks.

2.2 Centrality measures: Centrality measures are used to identify the most important nodes in a network. In the context of cyber security, this can help identify critical nodes that are vulnerable to attacks or that can be used to compromise the network [11].

2.3 Degree centrality metric: The degree centrality metric defines the significance of a node within a graph based on its degree, which is directly proportional to the number of connections it has. In other words, nodes with higher degrees are considered more important in the graph. For instance, when analyzing large-scale social networks like Twitter, if we calculate the in-degree centrality for different nodes, those with high in-degree centrality (such as popular celebrities with a large number of followers) have a substantial influence potential, making them suitable candidates for public influence or commercial promotion. Consequently, celebrities and influential individuals are often financially compensated on platforms like Instagram and Twitter to endorse specific products or express certain viewpoints. This compensation reflects the acknowledgment by commercial entities that these individuals possess a high in-degree centrality, enabling them to swiftly influence or reach a significant number of individuals.

2.4 Closeness centrality metric: Closeness centrality is defined as a metric that quantifies a node's importance based on its closeness to all other nodes in the graph. It measures the sum of geodesic distances between a given node and all other nodes in the network. Geodesic distance refers to the shortest path between two nodes, considering the network structure as a graph. The computation of closeness centrality involves calculating the inverse of the sum of geodesic distances for a node. A higher closeness centrality value indicates that a node is closer, on average, to all other nodes in the graph, suggesting a greater importance or centrality within the network.

2.5 Betweenness centrality: Betweenness centrality is a metric that measures the significance of a node within a network by quantifying how frequently it occurs on the shortest paths between all pairs of nodes in a graph. Nodes with higher BC scores are considered more important due to their potential to bridge different parts of the network and control information flow. To illustrate the practical application of betweenness centrality, let us see a case study using a follower graph, depicted in Figure 37.2. By analyzing BC scores within this graph, we gain insights into the nodes that act as influential intermediaries, connecting different clusters and facilitating information flow.

Mathematically, betweenness centrality $B(j)$ of a node j in a graph is defined as below:

$$B(j) = \sum_{c,d} (g_{cjd} / g_{cd})$$

c, d is any pair of nodes in the graph, g_{cjd} is the number of shortest paths from node "c" to "d" passing through "j", g_{cd} is the number of shortest paths from node "c" to "d".

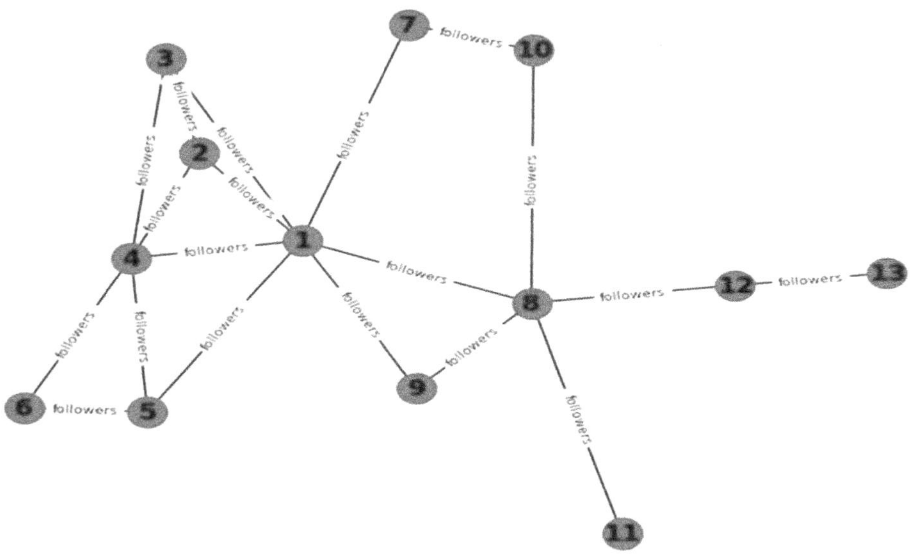

Figure 37.2: Graph depicting followers.

On analyzing we note that, BC's for different nodes from 1 to 13 are: 0.5909090909090909, 0.0, 0.0, 0.10606060606060606, 0.0606060606060606, 0.0, 0.045454545454545456, 0.5, 0.0, 0.030303030303030304, 0.0, 0.16666666666666669, 0.0.

Looking at node 1, we can observe that it has the highest BC among all other nodes in the graph. Based on this analysis, node 1, node 4, node 8, and node 12 seem to be the most important nodes in terms of betweenness centrality. These nodes act as key connectors or bottlenecks for information flow in the graph, and their removal or disruption could have a significant impact on the network's connectivity.

As previously mentioned, if we examine the graph of our followers (Figure 37.2), node 1 has a very high BC. If we were to remove it, the network would be severely disrupted, as nodes 2, 3, 4, 5, 6 would be unable to communicate with nodes 7, 8, 9, 10, 11, 12, 13 and vice versa, resulting in two isolated subgraphs. This comprehension highlights the significance of nodes with high BCs. Finding bridge nodes in graphs is a sample application of BC. Nodes with a high BC are on the shortest paths between a large number of pairings of nodes and are, therefore, crucial to communication in a graph because they connect a large number of nodes with one another. Eliminating these entities from the network would wreak havoc on the network's connectivity and communication.

A real-life use case of the above application is the transportation and logistics optimization. By representing transportation networks as graphs, companies can analyze various graph metrics to identify bottlenecks, optimize routes, and improve overall efficiency in supply chain management. For example, a shipping company can model their transportation network as a graph, with nodes representing locations (e. g., warehouses, ports, distribution centers) and edges representing routes or connections between them. They can then apply graph algorithms and metrics to identify critical nodes and edges in the network.

One key metric that can be used is the betweenness centrality, which identifies nodes that act as bridges or critical connectors within the network. By identifying these nodes, the company can focus on improving the efficiency of those particular locations or routes to alleviate congestion and delays.

2.6 Connectivity and resilience: Graph theory provides tools for analyzing the connectivity and resilience of a network. For example, it can be used to identify critical links or nodes that, if compromised, would result in the network becoming disconnected.

2.7 Random graph models: These models are used to study the properties of networks with varying degrees of randomness. In cyber security, this can be used to analyze the likelihood of certain types of attacks, such as random scanning or brute-force attacks.

2.8 Graph coloring: It is used to assign colors to nodes in a graph such that no two adjacent nodes have the same color. This principle is used in cyber security to assign security levels to nodes or to ensure that sensitive information is not transmitted over unsecured channels.

3 Model description

In this study, we demonstrated encryption and decryption algorithms in Python function. The application was used to transmit messages. Table 37.1 displays an example of the message.

3.1 Proposed Python algorithm

Step 1. Getting the word from the user.

Step 2. Take the ascii value in binary format for each character.

Step 3. Construct the graph $G(v, e)$, where G is the graph, v are the vertices, and e are the edges.

- i) Count the number of ones (1's bit) for each binary number, then save the results in an array of one dimension (m).

ii) Construct a symmetric adjacency matrix (C) whose size is equal to the number of ones-per-bit (1's) in array m.
iii) Save zeros for the matrix C's diagonal elements.
iv) For each binary number, count (l) the number of zeros (0's bit) that come after each one (1's bit).
v) If a binary stream contains m zeroes (0's bit) and then a one's (1's bit), then $m = m + 1$.
vi) If the final component of a binary number equals 1, place 1 in the final component of row 1 of C.

Step 4. Construct an array of one dimension, D, that extends the upper main diagonal matrix C.
Step 5. In adjacency matrix (C), row one, add the final member to D.
Step 6. The receiver side receives the adjacency matrix. (output).
Step 7. The aforementioned actions are continued till the end vertex of the Hamiltonian circuit tracing.

3.2 Working of the algorithm

Let us take an example: Suppose the word we are feeding is MATH.

The ascii value of the alphabet M is 77; its binary number is 1101101. The one-dimensional array formed is $[1, 2, 1, 2, 1]$. For A, the ascii value is 65, and its binary number is 1100001. The one d array formed is $[1, 5, 1]$. For T, the ascii value is 84, and its binary number is 1110100. The one d array formed is $[1, 1, 2, 3]$. For H, the ascii value is 72; its binary number is 1101000 The one d array formed is $[1, 2, 4]$. The values of the one-dimensional array are basically the values that we will put in the upper main diagonal of the adjacency matrix. In the case of undirected graphs, the matrix is symmetric about the diagonal because for every edge (i, j) there is also an edge (j, i). Table 37.1 demonstrates how the algorithm is supposed to work.

Table 37.1: Outlines of the process: Algorithm involving several steps to construct a graph based on the user input.

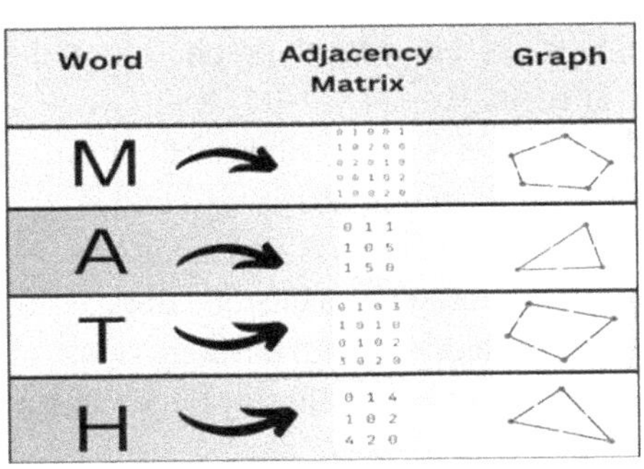

4 Graph theory in anomaly detection

Network intrusions refer to unauthorized or malicious activities that occur within a computer network, such as unauthorized access, data breaches, or other security breaches. Detecting and preventing these intrusions is crucial to ensure the security and integrity of the network and the systems connected to it. The code utilizes graph theory to represent the network as a graph, where nodes represent different elements of the network (e. g., devices, hosts, or IP addresses), and edges represent the connections or relationships between them. By analyzing the network graph, patterns, anomalies, or suspicious behavior can be identified, aiding in the detection of network intrusions. Anomaly detection techniques, such as the Isolation Forest algorithm used in the code, are employed to identify nodes in the network that exhibit anomalous behavior compared to normal network behavior. By detecting these anomalies, potential network intrusions can be identified and investigated further. The code further visualizes the network graph, highlighting the detected intrusions, which provides a visual representation of the network and helps in understanding the potential security threats and their impact on the overall network.

4.1 Proposed Python algorithm

Step 1. Define a function to create a network graph based on the specific requirements or dataset: Specify the number of nodes and edges, or load the graph from a file. Create the network graph object and add nodes and edges as required. Return the network graph.

Step 2. Define a function to preprocess the data or extract features from the network graph: Identify relevant attributes or properties of the nodes and edges. Transform the data into a suitable format for anomaly detection algorithms. Return the preprocessed data or features.

Step 3. Apply a graph-based anomaly detection algorithm: Choose an appropriate algorithm based on the characteristics of the network graph and the type of anomalies to be detected. Train the algorithm using the preprocessed data or features. Apply the trained model to detect anomalies in the network graph. Obtain the anomaly scores or labels for each node or edge.

Step 4. Identify and classify network intrusions: Set a threshold or criteria for determining anomalies. Classify the nodes or edges as normal or intrusive based on the anomaly scores or labels. Create a list or store the identified intrusions.

Step 5. Visualize the network graph and intrusions: Use graph visualization libraries or tools to plot the network graph. Highlight the intrusions with different colors or markers to distinguish them from normal nodes or edges. Display the visualization.

Step 6. Implement error handling and exception handling mechanisms: Add appropriate error handling codes to handle any potential errors or exceptions during data processing, training, or visualization. Include logging or error reporting mechanisms to capture and handle errors effectively.

Step 7. Implement additional security measures: Consider incorporating other security measures like access control, encryption, or network monitoring tools to enhance overall cyber security.

Step 8. Test and evaluate the performance: Assess the accuracy, precision, recall, and other relevant metrics of the intrusion detection system. Iterate and refine the algorithms, parameters, or preprocessing steps based on the evaluation results.

Step 9. Document and present the findings.

4.2 Working of the algorithm

For 20 nodes, an edge probability of 0.2, the result obtained is given in Figure 37.3.

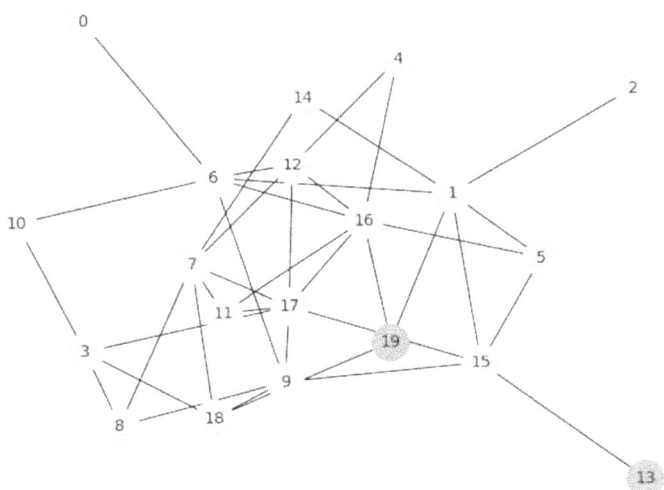

Figure 37.3: Network intrusion detected to be 13 and 19.

5 Concluding remarks

As Python facilitates the implementation of the algorithm, and this study introduces algorithms that integrate graph theory with Python programming to help network security, the implementation of the present work is easy and can be extended to work on large data sets also. When we send messages or emails to someone so the receiver only receives the data and messages and no tampering in between happens, in this investigation, we have demonstrated the mathematical discipline involved in the back-end operation of these phenomena. Through this work, we contribute to the body of knowledge in cybersecurity, providing an approach that uses the power of graph theory and Python programming to use network defenses.

Bibliography

[1] H. O. Abdullah and M. Eftekhari, Cryptanalysis and improvements on some graph-based authentication schemes, *J. Discrete Math. Sci. Cryptogr.* **16**(4–5) (2013), 297–306.

[2] J. Browning, Security Features in the Teradata Database, Teradata Corporation, 2015.

[3] H. Cavusoglu, B. Mishra and S. Raghunathan, A model for evaluating IT security investments, *Commun. ACM* **47**(7) (2004), 87–92.

[4] X. Chen, P. Yadav, R. Singh and S. M. N. Islam, ES structure based on soft J-subset, *Mathematics* **11**(4) (2023), 853. https://doi.org/10.3390/math11040853.

[5] F. Cremer, B. Sheehan, M. Fortmann, A. N. Kia, M. Mullins, F. Murphy and S. Materne, Cyber risk and cybersecurity: a systematic review of data availability, *Geneva Pap. Risk Insur., Issues Pract.* **47**(3) (2022), 698–736. https://doi.org/10.1057/s41288-022-00266-6.

[6] S. Facchinetti, S. A. Osmetti and C. Tarantola, Network models for cyber attacks evaluation, *Socio-Econ. Plan. Sci.* **87**(Part B) (2023), 101584. https://doi.org/10.1016/j.seps.2023.101584.

[7] M. A. Hassan and A. Chickade, A review of interference reduction in wireless networks using graph coloring methods, *Int. J. Appl. Graph Theory Wirel. Ad Hoc Netw. Sens. Netw. (GRAPH-HOC)* **3**(1) (2011), 58–67.

[8] Khaleel and Al-Shumam, A study of graph theory applications in IT security, *Iraqi J. Sci.* **61**(10) (2020), 2705–2714. https://doi.org/10.24996/ijs.2020.61.10.28.

[9] M. Khare and R. Singh, L-contiguities and their order structure, *Fuzzy Sets Syst.* **158** (2007), 399–408.

[10] M. Khare and R. Singh, Complete ψ-grills and (L, n)-merotopies, *Fuzzy Sets Syst.* **159** (2008), 620–628.

[11] A. Kosek, An exploration of mathematical applications in cryptography, Thesis, The Ohio State University, 2015.

[12] K. Kuk, P. Milić, P. Spalević and M. Gocić, Algorithm design in Python for cybersecurity, in: *ERK'2019*, Portorož, 2019, pp. 243–246.

[13] T. Li, Y. Jiang, C. Lin, M. S. Obaidat, Y. Shen and J. Ma, DeepAG: attack graph construction and threats prediction with bi-directional deep learning, *IEEE Trans. Dependable Secure Comput.* **20**(1) (2023), 740–757.

[14] Y. Li, H. Voos, M. Darouach and C. Hua, An application of linear algebra theory in networked control systems: stochastic cyber-attacks detection approach, *IMA J. Math. Control Inf.* **33**(4) (2016), 1081–1102. https://doi.org/10.1093/imamci/dnv026.

[15] W. Mahmoud and A. Etaiwi, Encryption algorithm using graph theory, *J. Sci. Res. Rep.* **3**(19) (2014), 2519–2527.

[16] J. C. Meza, S. Campbell and D. H. Bailey, Mathematical and statistical opportunities in cyber security, 2009, arXiv:0904.1616.

[17] Y. Qi, Z. Gu, A. Li, X. Zhang, M. Shafiq, Y. Mei and K. Lin, Cybersecurity knowledge graph enabled attack chain detection for cyber-physical systems, *Comput. Electr. Eng.* **108** (2023), 108660. https://doi.org/10.1016/j.compeleceng.2023.108660.

[18] D. K. Saini, Cyber defense: mathematical modeling and simulation, *Int. J. Appl. Phys. Math.* **2**(5) (2012), 312–315.

[19] S. Sen and S. Samanta, Network security using graph theory, *Int. J. Innov. Res. Technol.* **1**(4) (2014), 223–230.

[20] R. Singh, A. Jain and K. S. Saini, Network security: issues and challenges, *Int. J. Adv. Sci. Technol.* **29**(10s) (2020), 2403–2408, Retrieved from http://sersc.org/journals/index.php/IJAST/article/view/16872.

[21] R. Singh and A. K. Umrao, On finite order nearness in soft set theory, *WSEAS Trans. Math.* **18** (2019), 118–122.

[22] J. Zhang, W. Wang and E. Zio, Study on the application of graph theory algorithms and attack graphs in cybersecurity assessment, in: *2023 7th International Conference on System Reliability and Safety (ICSRS)*, Bologna, Italy, 2023, pp. 558–564. https://doi.org/10.1109/ICSRS59833.2023.10381005.

[23] H. Zhu, J. Wu, H. Shi, C. Wang and Y. Li, Discussion on information security technology of big data system, *J. Phys. Conf. Ser.* **1087**(1) (2018), 1–6.

De Gruyter Proceedings in Mathematics

Bruce M. Landman, Florian Luca, Melvyn B. Nathanson, Jaroslav Nešetřil, Aaron Robertson (Eds.)
Combinatorial Number Theory. Proceedings of the Integers Conference 2023, 2024
ISBN 978-3-11-139540-1, e-ISBN 978-3-11-139559-3

Oleg V. Kaptsov, Evgeniy I. Kaptsov (Eds.)
Mathematical Models and Integration Methods. Seminar Proceedings on Applications to Mechanics and Physics, 2024
ISBN 978-3-11-154625-4, e-ISBN 978-3-11-154666-7

Toke Knudsen, Jessica Carter (Eds.)
Mastering the History of Pure and Applied Mathematics. Essays in Honor of Jesper Lützen, 2024
ISBN 978-3-11-076990-6, e-ISBN 978-3-11-076996-8

Manoj Kumar Patel, Ratnesh Kumar Mishra, Shiv Datt Kumar (Eds.)
Advances in Pure and Applied Algebra. Proceedings of the CONIAPS XXVII International Conference 2021, 2023
ISBN 978-3-11-078572-2, e-ISBN 978-3-11-078580-7

Richard J. Nowakowski, Bruce M. Landman, Florian Luca, Melvyn B. Nathanson, Jaroslav Nešetřil, Aaron Robertson (Eds.)
Combinatorial Game Theory. A Special Collection in Honor of Elwyn Berlekamp, John H. Conway and Richard K. Guy, 2022
ISBN 978-3-11-075534-3, e-ISBN 978-3-11-075541-1

Bruce M. Landman, Florian Luca, Melvyn B. Nathanson, Jaroslav Nešetřil, Aaron Robertson (Eds.)
Number Theory and Combinatorics. A Collection in Honor of the Mathematics of Ronald Graham, 2022
ISBN 978-3-11-075343-1, e-ISBN 978-3-11-075421-6

Kağan Kurşungöz, Ayberk Zeytin (Eds.)
Number Theory. Proceedings of the Journées Arithmétiques, 2019, XXXI, held at Istanbul University, 2021
ISBN 978-3-11-076029-3, e-ISBN 978-3-11-076111-5

Aref Jeribi (Ed.)
Operator Theory. Proceedings of the International Conference on Operator Theory, Hammamet, Tunisia, April 30 – May 3, 2018
ISBN 978-3-11-059686-1, e-ISBN 978-3-11-059819-3

Printed in the USA
CPSIA information can be obtained
at www.ICGtesting.com
LVHW081131151024
793848LV00006B/94

9 783111 304373